HUTCHINSON

NATIONS
OF THE
WORLD

2001

2001

HUTCHINSON

NATIONS
OF THE
WORLD

Helicon

Helicon Publishing Ltd
Clarendon House
Shoe Lane
Oxford OX1 2DP
United Kingdom
e-mail: admin@helicon.co.uk
Web site: www.helicon.co.uk

First published 2001

ISBN 1-85986-382-5

British Library Cataloguing in Publication Data
A catalogue record for this book is available from the British Library.

Front cover photos (left to right):
Hong Kong © Corbis; Flags © Superstock; Vietnamese worker © Stone.
Typeset and designed by Robert Updegraff
Printed in Italy by Giunti Industrie Grafiche

Editorial Director
Hilary McGlynn

Managing Editor
Elena Softley

Project Editor
Ruth Collier

Content Development Manager
Claire Lishman

Technical Editor
Tracey Auden

Production Manager
John Normansell

Production Controller
Stacey Penny

Cartographic Production Manager
Caroline Beckley

Cartography
Ben Brown
Rachel Hopper
Adam Meara
Nikki Sargeant

CONTENTS

PREFACE

The *Hutchinson Nations of the World* is arranged alphabetically, with an entry for each of the 192 sovereign nations. Over 50 categories of information are provided for each country, grouped under the general headings of Government, Economy and Resources, Population and Society, Transport, Chronology, and Practical Information. Each country has a map showing its major towns and rivers, plus an accompanying locator map showing its location within a continent. The national flag of each country is also shown. At the beginning of the book is a political world map and six physical continent maps showing the geographical position of every country. The colour of individual country maps is linked to the colour of their continent as shown on the world map at the start.

Arrangement of entries

Entries are arranged alphabetically, as if there were no spaces between words. Country names beginning 'St' are ordered as if they were spelt 'Saint'.

Foreign names

Where a place name has a popular English language version, the English language version is used in the text, and shown in brackets on the map.

Abbreviations

GDP	–	gross domestic product
GNP	–	gross national product
PPP	–	purchasing power parity
km	–	kilometre
sq km	–	square kilometre
mi	–	mile
sq mi	–	square mile
m	–	metre
est	–	estimate

Country map legend   *(The continental map legend can be found on each continental map)*

Political type & boundaries

ONTARIO	state or province
———	international boundary
– – –	international boundary in water
–·–·–	undefined/disputed boundary or ceasefire/demarcation line
———	state or province boundary

Communications

———	motorway
———	main road
— ––	other road or track
✈	international airport

Topographic features

▲ Mount Ziel 1510	elevation above sea level
✕ Khyber Pass 1080	mountain pass

Hydrographic features

	river, canal
	seasonal river
Niagara Falls Kariba Dam	waterfall, dam
	lake, seasonal lake
	salt lake, seasonal salt lake

Cities, towns & capitals

▣ CHICAGO	over 3 million
▢ HAMBURG	1–3 million
○ Bulawayo	250 000–1 million
● Antofogasta	100 000–250 000
◉ Ajaccio	25 000–100 000
• Indian Springs	under 25 000
LONDON	country capital
Edinburgh	state or province capital

Cultural features

......	ancient wall

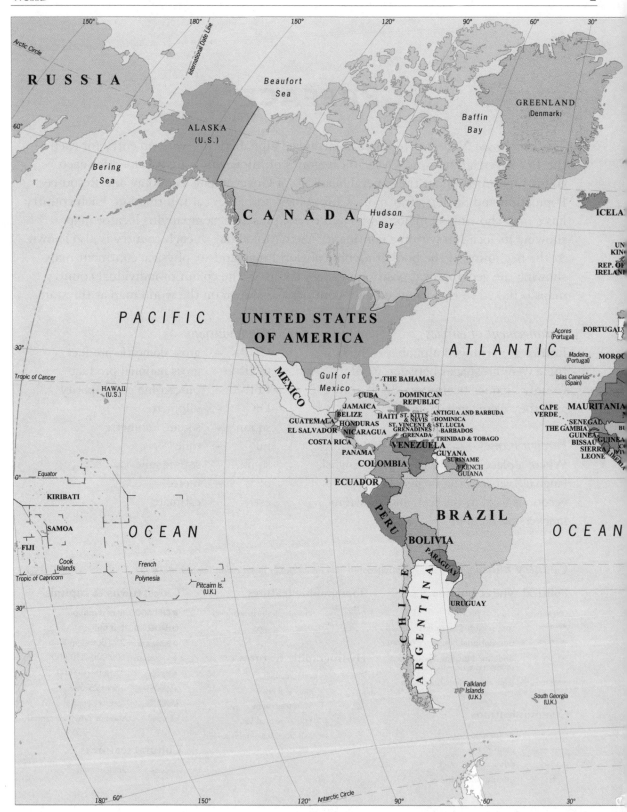

ARCTIC OCEAN

Svalbard (Norway)

Severnaya Zemlya

Zemlya Frantsa-Iosifa

Novaya Zemlya

Barents Sea

RUSSIA

Sea of Okhotsk

Bering Sea

ALASKA (U.S.)

Arctic Circle

NORWAY SWEDEN FINLAND

ESTONIA
LATVIA
RUS.-LITHUANIA
BELARUS
POLAND

KAZAKHSTAN

MONGOLIA

NORTH KOREA
SOUTH KOREA

JAPAN

PACIFIC

GERMANY
CZECH REP.
AUSTRIA SLOVAK.
SLOV. HUNGARY MOLDOVA
CRO. ROMANIA
BOSN. YUG. BULGARIA
ITALY MAC.
GREECE
TURKEY

UZBEKISTAN KYRGYZSTAN
TURKMENISTAN TAJIKISTAN

CHINA

OCEAN

MALTA
TUNISIA
CYPRUS LEBANON
ISRAEL
JORDAN

GEORGIA
ARMENIA AZER.

SYRIA
IRAQ

IRAN

AFGHANISTAN

PAKISTAN

NEPAL BHUTAN

BANGLA-DESH

MYANMAR (BURMA)

TAIWAN

Tropic of Cancer

LIBYA

EGYPT

KUWAIT
BAHRAIN
QATAR
U.A.E.

SAUDI ARABIA

Arabian Sea

INDIA

VIETNAM
LAOS
THAILAND
CAMBODIA

PHILIPPINES

Northern Mariana Islands (U.S.)

MARSHALL ISLANDS

NIGER
CHAD
SUDAN

ALGERIA

ERITREA YEMEN OMAN
DJIBOUTI

CENTRAL AFRICAN REPUBLIC

CAMEROON
EQUATORIAL GUINEA
GABON
SÃO TOMÉ & PRÍNCIPE
CONGO

ETHIOPIA

SOMALIA

MALDIVES

SRI LANKA

BRUNEI

MALAYSIA
SINGAPORE

PALAU

FEDERATED STATES OF MICRONESIA

Equator

NAURU

UGANDA
DEMOCRATIC REPUBLIC OF CONGO
RWANDA
BURUNDI
KENYA

TANZANIA

SEYCHELLES

INDIAN

INDONESIA

PAPUA NEW GUINEA

SOLOMON ISLANDS

TUVALU

ANGOLA
ZAMBIA
ZIMBABWE

NAMIBIA
BOTSWANA

COMOROS

MOZAMBIQUE

MADAGASCAR

Réunion (France)

MAURITIUS

OCEAN

AUSTRALIA

VANUATU

Nouvelle Calédonie (New-Caledonia) (France)

FIJI

Tropic of Capricorn

SOUTH AFRICA
SWAZILAND
LESOTHO

Îles Kerguélen (France)

NEW ZEALAND

SOUTHERN OCEAN

Country Abbreviations

ALB.	ALBANIA	NETH.	NETHERLANDS
AZER.	AZERBAIJAN	RUS.	RUSSIA
BEL.	BELGIUM	SLOV.	SLOVENIA
BOSN.	BOSNIA-HERZEGOVINA	SLOVAK.	SLOVAK REPUBLIC
CRO.	CROATIA	SWITZ.	SWITZERLAND
LUX.	LUXEMBOURG	U.A.E.	UNITED ARAB EMIRATES
MAC.	MACEDONIA	YUG.	YUGOSLAVIA

ANTARCTICA

Antarctic Circle

International Date Line

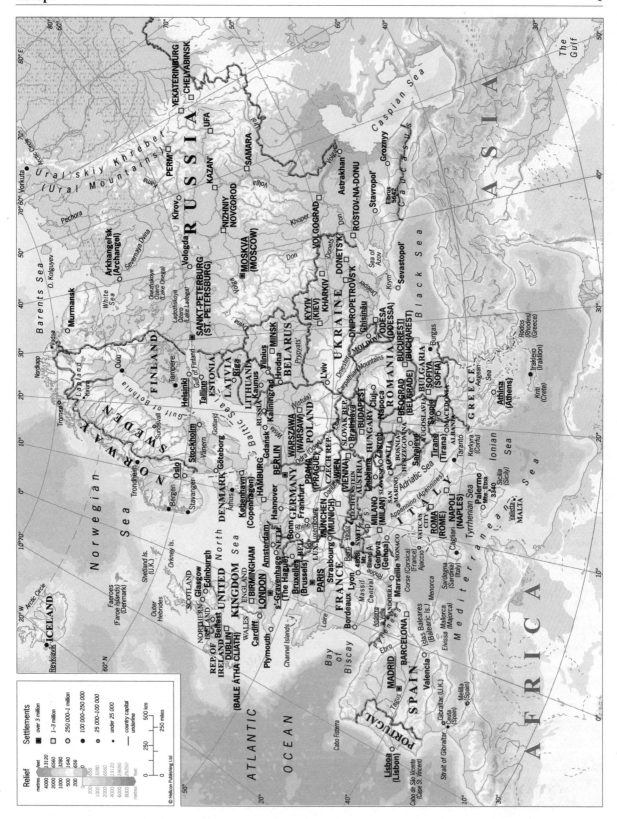

Settlements

■ over 3 million
□ 1–3 million
○ 250 000–1 million
● 100 000–250 000
○ 25 000–100 000
· under 25 000
— country capital
underline

Relief

metres	feet
4000	13120
2000	6560
1000	3280
500	1640
200	656
0	
200	656
2000	6560
4000	13120
6000	19690
8000	26250
metres	feet

© Helicon Publishing Ltd

0   250   500 km
0   250 miles

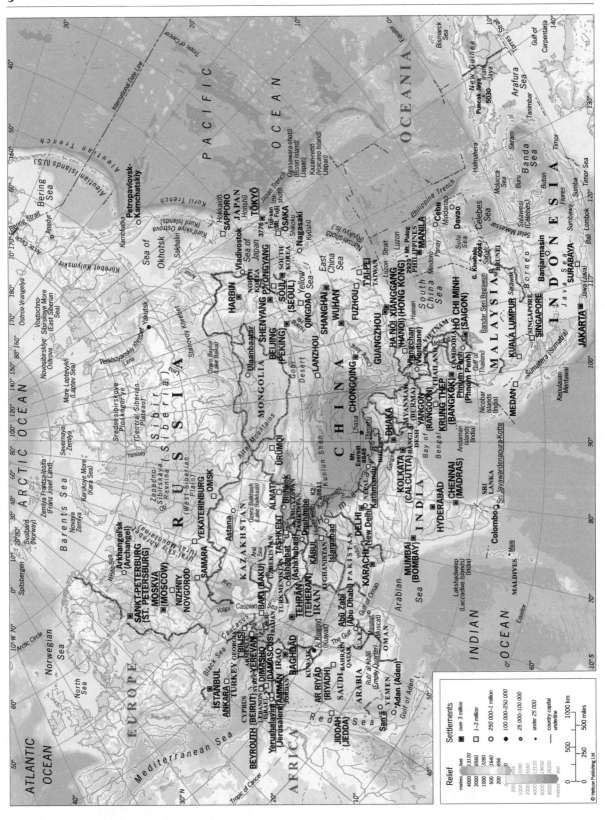

Relief
metres feet
4000 13120
2000 6560
1000 3280
500 1640
200 656
0 0

feet metres
656 200
3280 1000
6560 2000
13120 4000
19690 6000
26250 8000

Settlements
■ over 3 million
□ 1–3 million
◻ 250 000–1 million
● 100 000–250 000
◦ 25 000–100 000
• under 25 000
□ country capital
underline

1000 km
500
250
0

1000 miles
500
0

© Helicon Publishing Ltd

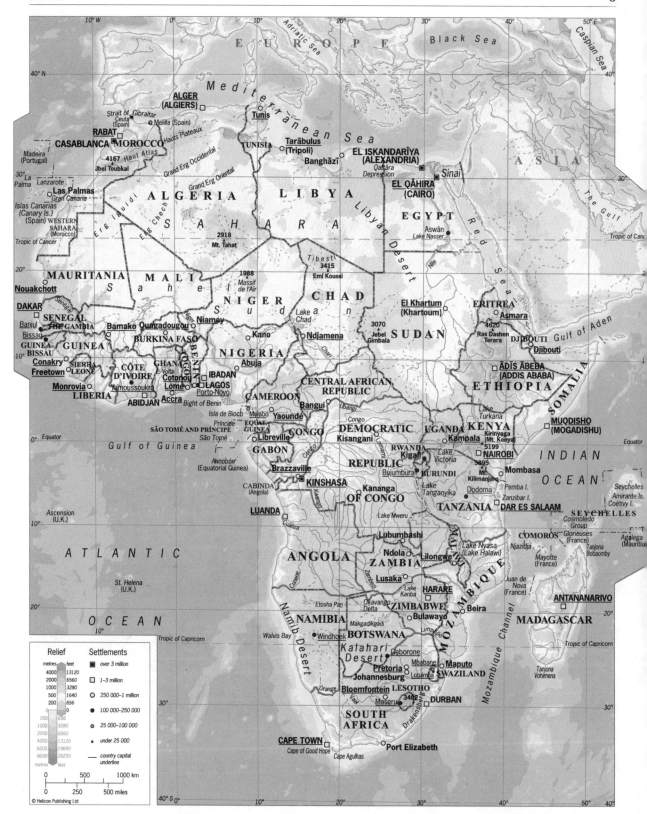

Relief

metres	feet
4000	13120
2000	6560
1000	3280
500	1640
200	656
0	0
200	656
1000	3280
2000	6560
4000	13120
6000	19690
8000	26250
metres	feet

Settlements

- ■ over 3 million
- □ 1–3 million
- ○ 250 000–1 million
- ● 100 000–250 000
- ◉ 25 000–100 000
- ∙ under 25 000
- — country capital underline

0   500   1000 km

0   250   500 miles

© Helicon Publishing Ltd

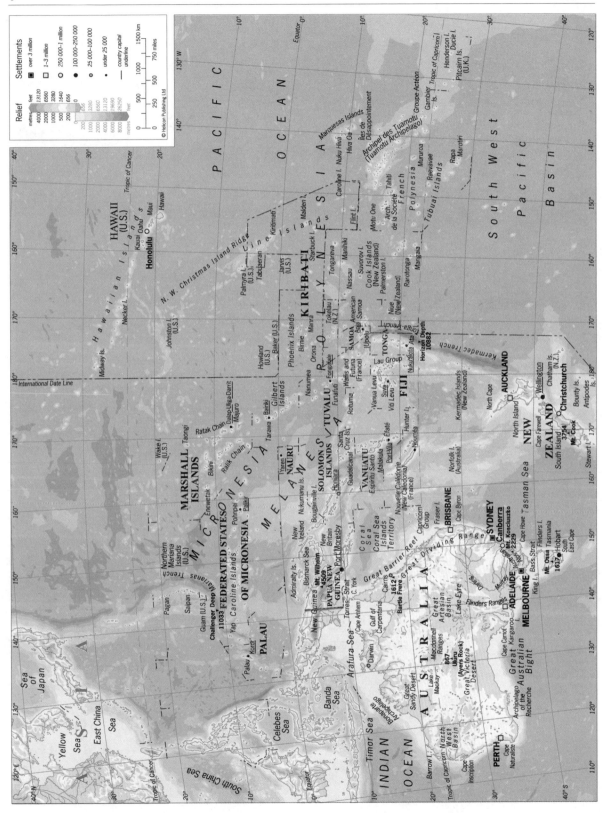

Settlements

■	over 3 million
□	1–3 million
○	250 000–1 million
●	100 000–250 000
○	25 000–100 000
•	under 25 000
—	country capital
_	underline

Relief

metres	feet
4000	13120
2000	6560
1000	3280
500	1640
200	656
0	0
	656
1000	3280
2000	6560
4000	13120
6000	19690
8000	26250
metres	feet

© Helicon Publishing Ltd

1500 km
1000
750 miles
500
250
500
250
0

PACIFIC OCEAN

Tropic of Cancer

Equator

130° W

140°

150°

160°

170°

180°

170°

160°

150°

140°

130°

120°

HAWAII (U.S.)
Kauai Oahu Maui
Honolulu ○ Hawaii

Midway Is.

Necker I.
Johnston I. (U.S.)

N. W. Christmas Island Ridge

Line Islands
Kiritimati
Palmyra I. (U.S.)
Tabuaeran
Jarvis (U.S.)
Malden I.

Marquesas Islands
Caroline I. Nuku Hiva
Hiva Oa
Îles de Désappointement
Groupe Actéon
Gambier Is.
Tropic of Capricorn
Henderson I.
Pitcairn Is. (U.K.)
Ducie I. (U.K.)

KIRIBATI
Starbuck I.
Flint I.

Motu One
Arch. Tahiti
de la Société
Manihiki

POLYNESIA
French Polynesia
Mururoa
Mangaia
Marotiri
Rapa
Tubuai Islands
Raevavae

South West Pacific Basin

Phoenix Islands
Birnie
Orona
Manra
Tokelau (N.Z.)
Wallis and Futuna (France)

Nassau
Suvorov I.
Cook Islands (New Zealand)
Palmerston I.

SAMOA
Upolu
American Samoa
Niue (New Zealand)
Rarotonga

Howland (U.S.)
Baker (U.S.)
Gilbert Islands
Nanumea
Tarawa
Bairiki
Banaba

TUVALU
Funafuti
Rotuma
TONGA
Lau Group
Ata
Horizon Depth 10882
Nuku'alofa

Kermadec Trench
Tonga Trench

NAURU
Yaren

SOLOMON
ISLANDS
Nukumanu Is.
Bougainville I.
Honiara
Guadalcanal
Santa Cruz Is.

FIJI
Vanua Levu
Suva
Viti Levu

VANUATU
Espíritu Santo
Malakula
Port-Vila
Efate
Hunter I.

Nouméa
Nouvelle Calédonie
New Caledonia (France)

Norfolk I. (Australia)
Kermadec Islands (New Zealand)

AUCKLAND
North Cape
North Island
NEW ZEALAND
Cape Farewell
Wellington
Chatham Is. (N.Z.)
Christchurch
South Island
3754
Mt. Cook
Stewart I.
Bounty Is.
Antipodes
Is.

MARSHALL
ISLANDS
Taongi
Ratak Chain
Bikini
Eniwetok
Ralik Chain

Wake I. (U.S.)

Northern Mariana Islands (U.S.)
Pagan
Saipan
Guam (U.S.)

Marianas Trench
Challenger Deep 11033

FEDERATED STATES
OF MICRONESIA
Yap Caroline Islands
Pohnpei Palikir

MICRONESIA

MELANESIA

Admiralty Is.
New Ireland
Bismarck Sea
New Britain
Mt. Wilhelm 4509
PAPUA NEW
GUINEA
Port Moresby
Torres Strait
C. York

PALAU
Palau Koror

A
Sea of Japan
East China Sea
Yellow Sea
South China Sea

Banda Sea
Celebes Sea
Bonaparte Archipelago
Timor Sea
Arafura Sea

INDIAN OCEAN

Barrow I.
Cape Naturaliste
Cape Inscription
PERTH

North West Basin
Great Sandy Desert
Uluru (Ayers Rock) 867
Great Victoria Desert
Archipelago of the Recherche

AUSTRALIA

Darwin
Gulf of Carpentaria
Cairns
Bartle Frere 1612
Coral Sea
Coral Sea Islands Territory
Great Barrier Reef
Fraser I.
Cape Byron

BRISBANE
SYDNEY
Canberra
Mt. Kosciuszko 2229

Lake Eyre
MacDonnell Ranges
Macumba
Macumba
Lake Mackay
Great Artesian Basin
Great Dividing Range
Flinders Ranges
ADELAIDE
MELBOURNE
Capricorn Group
Cape Carnot
Kangaroo I.
Great Australian Bight
Tropic of Capricorn

Murray
Murrumbidgee
Darling

Bass Strait
King I.
Flinders I.
Hobart
Tasmania
Mt. Ossa 1617
South East Cape
Tasman Sea
Cape Howe

International Date Line

Tropic of Capricorn

Equator

40° N

30°

20°

Tropic of Cancer

20°

10°

0°

10°

20°

30°

40° S

120° E

130°

140°

150°

160°

170°

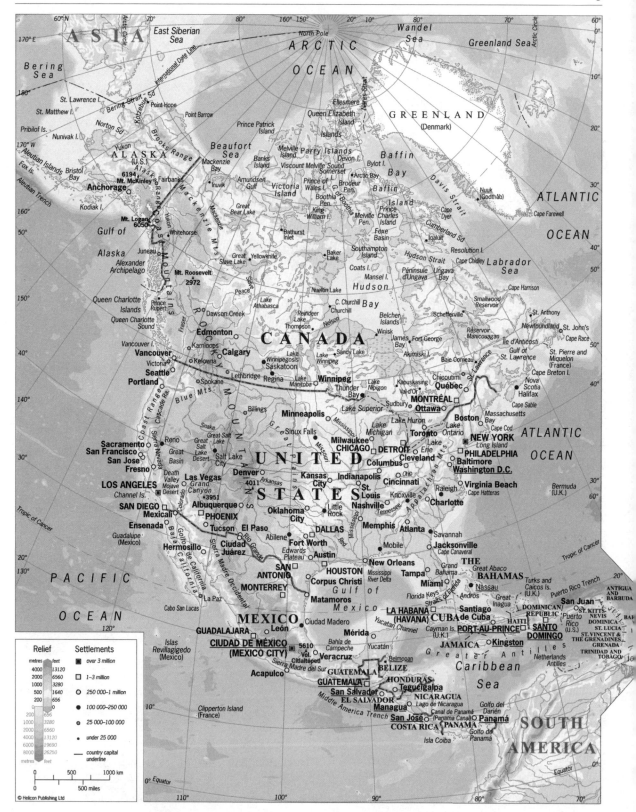

Relief

metres	feet
4000	13120
2000	6560
1000	3280
500	1640
200	656
0	0
200	656
1000	3280
2000	6560
4000	13120
6000	19690
8000	26250
metres	feet

Settlements

- ▣ over 3 million
- ☐ 1–3 million
- ○ 250 000–1 million
- ● 100 000–250 000
- ◦ 25 000–100 000
- · under 25 000
- — country capital
- underline

0 500 1000 km
0 500 miles

© Helicon Publishing Ltd

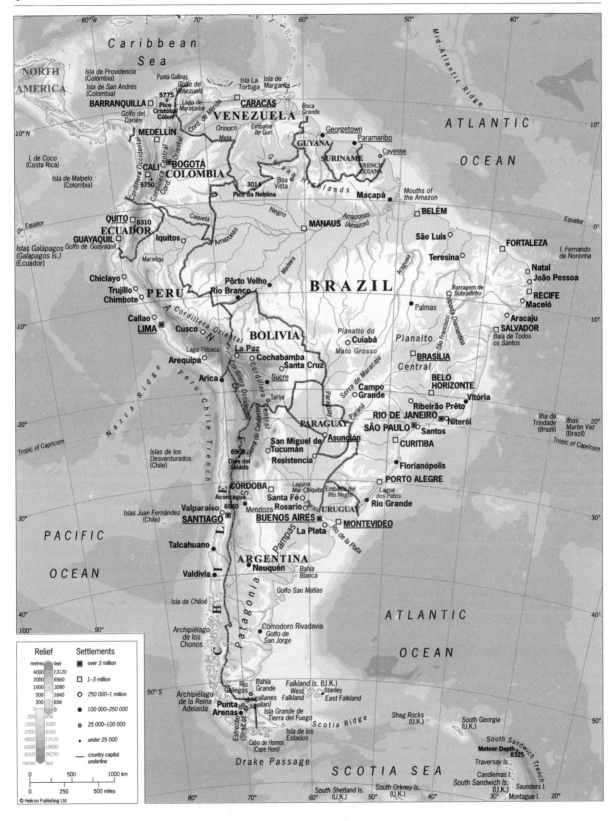

Relief

metres	feet
4000	13120
2000	6560
1000	3280
500	1640
200	656
0	0

200	656
1000	3280
2000	6560
4000	13120
6000	19690
8000	26250
metres	feet

Settlements

■ over 3 million
□ 1–3 million
○ 250 000–1 million
● 100 000–250 000
◉ 25 000–100 000
• under 25 000

— country capital
underline

0   500   1000 km

0   250   500 miles

© Helicon Publishing Ltd

AFGHANISTAN

Mountainous, landlocked country in south-central Asia, bounded north by Tajikistan, Turkmenistan, and Uzbekistan, west by Iran, and south and east by Pakistan, India, and China.

National name: *Dowlat-e Eslāmi-ye Afghānestān/Islamic State of Afghanistan*
Area: 652,225 sq km/251,825 sq mi
Capital: Kabul
Major towns/cities: Kandahar, Herat, Mazar-e-Sharif, Jalalabad, Konduz, Qal'eh-ye Now
Physical features: mountainous in centre and northeast (Hindu Kush mountain range, Khyber and Salang passes, Wakhan salient, and Panjshir Valley), plains in north and southwest, Amudar'ya (Oxus) River, Helmand River, Lake Saberi

Green denotes Islam. Black represents the dark past. Wheat symbolizes the country's communist heritage. The Islamic year 1371 (1992) marks the date of the foundation of the new regime. Effective date: 3 December 1992.

Government

Head of state and government: Muhammad Rabbani from 1996
Political system: Islamic nationalist
Political executive: unlimited presidency
Administrative divisions: 31 provinces
Political parties: Hezb-i-Islami, Islamic fundamentalist Mujahedin, anti-Western; Jamiat-i-Islami, Islamic fundamentalist Mujahedin; National Liberation Front, moderate Mujahedin
Armed forces: approximately 400,000 (1998)
Conscription: compulsory for four years, with break of three years after second year (since 1992 conscription has been difficult to enforce and desertion is common)
Death penalty: retains and uses the death penalty for ordinary crimes
Defence spend: (% GDP) 14.5 (1998)
Education spend: (% GNP) 2.0 (1992)
Health spend: (% GDP) 3.2 (1997 est)

Economy and resources

Currency: afgháni
GDP: (US$) 21 billion (1999 est)
Real GDP growth: (% change on previous year) N/A
GNP: (US$) N/A
GNP per capita (PPP): (US$) 800 (1999 est)
Consumer price inflation: 240% (1996 est)
Unemployment: 8% (1995 est)
Foreign debt: (US$) 5.5 billion (1997)
Major trading partners: Kyrgyzstan, Russia, Azerbaijan, Japan, Singapore, Germany, Pakistan, Iran, India, UK
Resources: natural gas, coal, iron ore, barytes, lapis lazuli, salt, talc, copper, chrome, gold, silver, asbestos, small petroleum reserves
Industries: food products, cotton textiles, cement, coalmining, chemical fertilizers, small vehicle assembly plants, processed hides and skins, carpetmaking, sugar manufacture, leather and plastic goods
Exports: fruit and nuts, carpets, wool, karakul skins, cotton, natural gas. Principal market: Kyrgyzstan 37.3% (1995)

Imports: basic manufactured goods and foodstuffs (notably wheat), petroleum products, textiles, fertilizers, vehicles and spare parts. Principal source: Japan 25.6% (1995)
Arable land: 12.1% (1995)
Agricultural products: wheat, barley, maize, rice; livestock rearing (sheep, goats, cattle, and camels); world's leading opium producer (1995)

Population and society

Population: 22,720,000 (2000 est)
Population growth rate: 2.9% (1995–2000)
Population density: (per sq km) 34 (1999 est)
Urban population: (% of total) 22 (2000 est)
Age distribution: (% of total population) 0–14 44%, 15–59 51%, 60+ 5% (2000 est)
Ethnic groups: Pathans, or Pushtuns, comprise the largest ethnic group, 38% of the population, followed by the Tajiks (concentrated in the north, 25%), the Uzbeks (6%), and the Hazaras (19%)
Language: Pashto, Dari (both official), Uzbek, Turkmen, Balochi, Pashai
Religion: Muslim (84% Sunni, 15% Shiite), other 1%
Education: (compulsory years) 6
Literacy rate: 52% (men); 22% (women) (2000 est)
Labour force: 68% agriculture, 16% industry, 16% services (1992 est)
Life expectancy: 45 (men); 46 (women) (1995–2000)
Child mortality rate: (under 5, per 1,000 live births) 257 (1999–2000)
Physicians: 1 per 7,001 people (1993 est)
TV sets: (per 1,000 people) 13 (1997)
Radios: (per 1,000 people) 132 (1997)

Transport

Airports: two international airports: Kabul (Khwaja Rawash) and Kandahar; 18 domestic airports; total passenger km: 276 million (1995)
Railways: none (a trans-Afghan railway was proposed in an Afghan-Pakistan-Turkmen agreement of 1994)
Roads: total road network: 21,000 km/13,050 mi, of which 13.3% paved (1996 est); passenger cars: 1.4 per 1,000 people (1996 est)

Chronology

6th century BC: Part of Persian Empire under Cyrus II and Darius I. **329 BC:** Conquered by Alexander the Great. **323 BC:** Fell to the Seleucids, who ruled from Babylon. **304 BC:** Ruled by Mauryan dynasty in south and independent Bactria in north. **135 BC:** Central Asian tribes established Kusana dynasty. **3rd–7th centuries AD:** Decline of Kusana dynasty. Emergence of Sassanids as ruling power with Hepthalites (central Asian nomads) and western Turks also fighting for control. **642–11th century:** First Muslim invasion followed by a succession of Muslim dynasties, including Mahmud of Ghazni in 998. **1219–14th century:** Mongol invasions led by Genghis Khan and Tamerlane. **16th–18th centuries:** Much of Afghanistan came under the rule of the Mogul Empire under Babur (Zahir) and Nadir Shah. **1747:** Afghanistan became an independent emirate under Dost Muhammad. **1838–42:** First Afghan War, the first in a series of three wars between Britain and Afghanistan, instigated by Britain to counter the threat to British India from expanding Russian influence in Afghanistan. **1878–80:** Second Afghan War. **1919:** Afghanistan recovered full independence following the Third Afghan War. **1953:** Lt-Gen Daud Khan became prime minister and introduced social and economic reform programme. **1963:** Daud Khan forced to resign and constitutional monarchy established. **1973:** Monarchy overthrown in coup by Daud Khan. **1978:** Daud Khan assassinated in coup. Start of Muslim guerrilla (Mujahedin) resistance. **1979:** The USSR invaded the country to prop up the pro-Soviet government. **1986:** Partial Soviet troop withdrawal. **1988:** New non-Marxist constitution adopted. **1989:** Withdrawal of Soviet troops; continued Mujahedin resistance to communist People's Democratic Party of Afghanistan (PDPA) regime and civil war intensified. **1991:** US and Soviet military aid withdrawn. Mujahedin began talks with the Russians and Kabul government. **1992:** Mujahedin leader Burhanuddin Rabbani was elected president. **1993–94:** There was fighting around Kabul. **1996:** The Talibaan controlled two-thirds of the country, including Kabul; the country was split between the Talibaan-controlled fundamentalist south and the more liberal north; strict Islamic law was imposed. **1997:** The Talibaan was recognized as the legitimate government of Afghanistan by Pakistan and Saudi Arabia. **1998:** Two earthquakes in the north killed over 8,000 people. The USA launched a missile attack on a suspected terrorist site in retaliation for bombings of US embassies in Nairobi and Dar es Salaam. Talibaan extended its control in the north, massacring 6,000 people at Mazar-e-Sharif. **1999:** Fighting resumed in northern Afghanistan after a four-month lull. Intending to punish the Talibaan regime for failing to expel suspected terrorist Osama bin Laden, the United Nations (UN) imposed sanctions on Afghanistan in November, which provoked mobs to attack UN offices in the capital, Kabul. **2000:** Fighting continued between the Talibaan and the opposing United Islamic Front for Salvation of Afghanistan (UIFSA), led by Ahmed Shah Masud, and the Talibaan made further gains in the north. Pakistan closed its border with Afghanistan in November to prevent a further influx of refugees fleeing war and famine. The UN withdrew its aid workers and imposed tighter sanctions as bin Laden had still not been surrendered.

Practical information

Visa requirements: UK: visa required. USA: visa required
Time difference: GMT+4.5
Chief tourist attractions: Bamian (with its high statue of Buddha and thousands of painted caves); the Blue Mosque of Mazar; the suspended lakes of Bandi Amir; the Grand Mosque and minarets of Herat; the towns of Kandahar, Gereshk, and Baekh (ancient Bactria); the high mountains of the Hindu Kush
Major holidays: 27 April, 1 May, 19 August; variable: Eid-ul-Adha, Arafa, Ashora, end of Ramadan, New Year (Hindu), Prophet's Birthday, first day of Ramadan

ALBANIA

Located in southeastern Europe, bounded north by Yugoslavia, east by the Former Yugoslav Republic of Macedonia, south by Greece, and west and southwest by the Adriatic Sea.

National name: *Republika e Shqipërisë/Republic of Albania*
Area: 28,748 sq km/11,099 sq mi
Capital: Tirana
Major towns/cities: Durrës, Shkodër, Elbasan, Vlorë, Korçë
Major ports: Durrës
Physical features: mainly mountainous, with rivers flowing east–west, and a narrow coastal plain

Red represents the blood shed during the fight for independence. According to legend Albanians are descended from the eagle, which is the national emblem. Effective date: 22 May 1993.

Government

Head of state: Rexhep Mejdani from 1997
Head of government: Ilir Meta from 1999
Political system: emergent democracy
Political executive: limited presidency
Administrative divisions: 36 districts
Political parties: Democratic Party of Albania (PDS; formerly the Democratic Party: DP), moderate, market-oriented; Socialist Party of Albania (PSS), ex-communist; Human Rights Union (HMU), Greek minority party; Agrarian Party (AP); Christian Democratic Party (CDP); Democratic Alliance Party (DAP), centrist; Democratic Party
Armed forces: 54,000 (1998)
Conscription: compulsory for 15 months
Death penalty: death penalty abolished 2000
Defence spend: (% GDP) 6.6 (1998)
Education spend: (% GDP) 3.1 (1996)
Health spend: (% GDP) 3.5 (1997)

Economy and resources

Currency: lek
GDP: (US$) 3.06 billion (1999 est)
Real GDP growth: (% change on previous year) 7.3 (1999)
GNP: (US$) 2.9 billion (1999)
GNP per capita (PPP): (US$) 2,892 (1999)
Consumer price inflation: 0.4% (1999)
Unemployment: 20% (1998 est)
Foreign debt: (US$) 900 million (1999)
Major trading partners: Italy, Greece, Germany, Bulgaria, Austria, Turkey, Macedonia
Resources: chromite (one of world's largest producers), copper, coal, nickel, petroleum, and natural gas
Industries: food processing, mining, textiles, oil products, cement, energy generation
Exports: textiles and footwear, mineral products, base metals, foodstuffs, beverages and tobacco, vegetable products. Principal market: Italy 66.9% (1999)
Imports: textiles and footwear, machinery, fuels and minerals, plant and animal raw materials, chemical products. Principal source: Italy 37.6% (1999)
Arable land: 21.1% (1996)
Agricultural products: wheat, sugar beet, maize, potatoes, barley, sorghum, cotton, tobacco

Population and society

Population: 3,113,000 (2000 est)
Population growth rate: –0.4% (1995–2000)
Population density: (per sq km) 108 (1999 est)
Urban population: (% of total) 42 (2000 est)
Age distribution: (% of total population) 0–14 29%, 15–59 62%, 60+ 9% (2000 est)
Ethnic groups: 95% of Albanian, non-Slavic, descent; 3% ethnic Greek (concentrated in south)
Language: Albanian (official), Greek
Religion: Muslim, Albanian Orthodox, Roman Catholic
Education: (compulsory years) 8
Literacy rate: 92% (men); 77% (women) (2000 est)
Labour force: 48% of population: 55% agriculture, 23% industry, 22% services (1990)
Life expectancy: 70 (men); 76 (women) (1995–2000)
Child mortality rate: (under 5, per 1,000 live births) 43 (1995–2000)
Physicians: 1 per 530 people (1993 est)
Hospital beds: 1 per 327 people (1995 est)
TV sets: (per 1,000 people) 161 (1997)
Radios: (per 1,000 people) 217 (1997)
Internet users: (per 10,000 people) 6.5 (1999)
Personal computer users: (per 100 people) 0.5 (1999)

Transport

Airports: international airport: Tirana (Rinas); no regular domestic air service; total passenger km: 4 million (1995)
Railways: total length: around 720 km/447 mi (1994); total passenger km: 197 million (1995)
Roads: total road network: 18,000 km/11,185 mi, of which 30% paved (1996); passenger cars: 20.5 per 1,000 people (1996)

Chronology

2000 BC: Albania was part of Illyria. **168 BC:** Illyria was conquered by the Romans. **AD 395:** Became part of Byzantine Empire. **6th–14th centuries:** Byzantine decline exploited by Serbs, Normans, Slavs, Bulgarians, and Venetians. **1381:** Ottoman invasion of Albania followed by years of resistance to Turkish rule. **1468:** Resistance led by national hero Skanderbeg (George Kastrioti) largely collapsed, and Albania passed to Ottoman Empire. **15th–16th centuries:** Thousands fled to southern Italy to escape Ottoman rule; over half of the rest of the population converted to Islam. **1878:** Foundation of Albanian League promoted emergence of nationalism. **1912:** Achieved independence from Turkey as a result of First Balkan War and end of Ottoman Empire in Europe. **1914–20:** Occupied by Italy. **1925:** Declared itself a republic. **1928–39:** Monarchy of King Zog. **1939:** Italian occupation led by Benito Mussolini. **1943–44:** Under German rule following Italian surrender. **1946:** Proclaimed Communist People's Republic of Albania, with Enver Hoxha as premier. **1949:** Developed close links with Joseph Stalin in USSR and entered Comecon (Council for Mutual Economic Assistance). **1961:** Broke with USSR in wake of Nikita Khrushchev's denunciation of Stalin, and withdrew from Comecon. In 1978 Albania also severed diplomatic links with China. **1987:** Normal diplomatic relations restored with Canada, Greece, and West Germany. **1990–91:** The one-party system was abandoned in the face of popular protest; the first opposition party was formed, and the first multiparty elections were held. **1992:** Former communist officials were charged with corruption and abuse of power. Totalitarian and communist parties were banned. **1993:** Conflict began between ethnic Greeks and Albanians, followed by a purge of ethnic Greeks from the civil service and army. **1997:** Antigovernment riots; police killed demonstrators in the southern port of Vlorë. Southern Albania fell under rebel control. The government signed a World Bank and IMF rescue package to salvage the economy. **1998:** A new constitution came into effect. **1999:** Ilir Meta, a socialist, became prime minister.

Practical information

Visa requirements: UK: visa not required. USA: visa not required
Time difference: GMT +1
Chief tourist attractions: main tourist centres include Tirana, Durrës, and Popgradec. The ancient towns of Apollonia and Butrint are important archaeological sites, and there are many other towns of historic interest
Major holidays: 1–2, 11 January, 8 March, 1 May, 28 November, 25 December; variable: end of Ramadan, Easter Monday, Good Friday, Eid-ul-Adha, Orthodox Easter

ALGERIA

Located in north Africa, bounded east by Tunisia and Libya, southeast by Niger, southwest by Mali and Mauritania, northwest by Morocco, and north by the Mediterranean Sea.

National name: *Al-Jumhuriyyat al-Jaza'iriyya ad-Dimuqratiyya ash-Sha'biyya/Democratic People's Republic of Algeria*
Area: 2,381,741 sq km/919,590 sq mi
Capital: Algiers (Arabic al-Jaza'ir)
Major towns/cities: Oran, Annaba, Blida, Sétif, Constantine
Major ports: Oran (Ouahran), Annaba (Bône)
Physical features: coastal plains backed by mountains in north, Sahara desert in south; Atlas mountains, Barbary Coast, Chott Melrhir depression, Hoggar mountains

Red may suggest bloodshed or liberty. White symbolizes purity. Green represents Islam. Effective date: 3 July 1962.

Government

Head of state: Abdel Aziz Bouteflika from 1999
Head of government: Ali Benflis from 2000
Political system: military
Political executive: military
Administrative divisions: 48 departments
Political parties: National Democratic Rally (RND), left of centre; National Liberation Front (FLN), nationalist, socialist; Socialist Forces Front (FSS), Berber-based, left of centre; Islamic Front for Salvation (FIS), Islamic fundamentalist (banned from 1992); Movement for a Peacetime Society (MSP), formerly Hamas, fundamentalist
Armed forces: 122,000 (1998)
Conscription: compulsory for 18 months
Death penalty: retained and used for ordinary crimes
Defence spend: (% GDP) 4.8 (1998)
Education spend: (% GNP) 5.2 (1996)
Health spend: (% GDP) 3.1 (1997)

Economy and resources

Currency: Algerian dinar
GDP: (US$) 47.9 billion (1999)
Real GDP growth: (% change on previous year) 3.3 (1999)
GNP: (US$) 46.5 billion (1999)
GNP per capita (PPP): (US$) 4,753 (1999)
Consumer price inflation: 2.4% (1999)
Unemployment: 30% (1998)
Foreign debt: (US$) 28.1 billion (1999 est)
Major trading partners: France, Italy, Germany, USA, the Netherlands, Brazil, Spain, Turkey
Resources: natural gas and petroleum, iron ore, phosphates, lead, zinc, mercury, silver, salt, antimony, copper
Industries: food processing, machinery and transport equipment, textiles, cement, tobacco
Exports: crude oil, gas, vegetables, tobacco, hides, dates. Principal market: Italy 17.8% (1999)
Imports: machinery and transportation equipment, food and basic manufactures. Principal source: France 29.8% (1999)

Arable land: 3.2% (1996)
Agricultural products: wheat, barley, potatoes, citrus fruits, olives, grapes; livestock rearing (sheep and cattle)

Population and society

Population: 31,471,000 (2000 est)
Population growth rate: 2.3% (1995–2000); 2% (2000–05)
Population density: (per sq km) 13 (1999 est)
Urban population: (% of total) 60 (2000 est)
Age distribution: (% of total population) 0–14 37%, 15–59 57%, 60+ 6% (2000 est)
Ethnic groups: 99% of Arab Berber origin, the remainder of European descent, mainly French
Language: Arabic (official), Berber, French
Religion: Sunni Muslim (state religion) 99%, Christian and Jewish 1%
Education: (compulsory years) 9
Literacy rate: 78% (men); 57% (women) (2000 est)
Labour force: 28% of population: 26% agriculture, 31% industry, 43% services (1990)
Life expectancy: 68 (men); 70 (women) (1995–2000)
Child mortality rate: (under 5, per 1,000 live births) 51 (1999)
Physicians: 1 per 1,062 people (1993 est)
Hospital beds: 1 per 390 people (1993 est)
TV sets: (per 1,000 people) 67 (1997)
Radios: (per 1,000 people) 241 (1997)
Internet users: (per 10,000 people) 6.5 (1999)
Personal computer users: (per 100 people) 0.6 (1999)

Transport

Airports: international airports: Algiers (Houari Boumédienne), Annaba (El Mellah), Oran (Es Senia), Constantine (Ain El Bey); ten domestic airports; total passenger km: 2,855 million (1995)
Railways: total length: 4,772 km/2,965 mi; total passenger km: 3,166 million (1997 est)
Roads: total road network: 104,000 km/64,626 mi, of which 68.9% paved (1996 est); passenger cars: 24.8 per 1,000 people (1996 est)

Chronology

9th century BC: Part of Carthaginian Empire. **146 BC:** Conquered by Romans, who called the area Numidia. **6th century:** Part of the Byzantine Empire. **late 7th century:** Conquered by Muslim Arabs, who spread Islam as the basis of a new Berberized Arab-Islamic civilization. **1516:** Ottoman Turks expelled recent Christian Spanish invaders. **1816:** Anglo-Dutch forces bombarded Algiers as a reprisal against the Barbary pirates' attacks on Mediterranean shipping. **1830–47:** French occupation of Algiers, followed by extension of control to the north, overcoming fierce resistance from Amir Abd al-Qadir, a champion of Arab Algerian nationalism, and from Morocco. **1850–70:** The mountainous inland region, inhabited by the Kabyles, was occupied by the French. **1871:** There was a major rebellion against French rule as French settlers began to take over the best agricultural land.

1900–09: The Sahara region was subdued by France, who kept it under military rule. **1940:** Following France's defeat by Nazi Germany, Algeria became allied to the pro-Nazi Vichy regime during World War II. **1945:** 8,000 died following the ruthless suppression of an abortive uprising against French rule. **1954–62:** Battle of Algiers: bitter war of independence fought between the National Liberation Front (FLN) and the French colonial army. **1958:** French inability to resolve the civil war in Algeria toppled the Fourth Republic and brought to power, in Paris, Gen Charles de Gaulle, who accepted the principle of national self-determination. **1962:** Independence from France was achieved and a republic declared. Many French settlers fled. **1963:** A one-party state was established. **1976:** New Islamic-socialist constitution approved. **1988:** Riots took place in protest at austerity policies; 170 people were killed. A reform programme was introduced. Diplomatic relations were restored with Morocco after a 12-year break. **1989:** Constitutional changes introduced limited political pluralism. **1992:** The military took control of the government and a state of emergency was declared. **1993:** The civil strife worsened, with assassinations of politicians and other public figures. **1994:** The fundamentalists' campaign of violence intensified. **1996:** The constitution was amended to increase the president's powers and counter religious fundamentalism. Arabic was declared the official public language. **1998:** The violence continued. **1999:** Abdel Aziz Bouteflika was elected president. **2000:** Ali Benflis was appointed prime minister. The violence continued, averaging 200 deaths a month.

Practical information

Visa requirements: UK: visa required. USA visa required
Time difference: GMT +/–0
Chief tourist attractions: include the Mediterranean coast, the Atlas Mountains, and the desert; the Hoggar massif and the Tassili N'Ajjer (Plateau of Chasms) – both important centres of Tuareg culture
Major holidays: 1 January, 1 May, 19 June, 5 July, 1 November; variable: Eid-ul-Adha, Ashora, end of Ramadan, New Year (Muslim), Prophet's Birthday

ANDORRA

Landlocked country in the east Pyrenees, bounded north by France and south by Spain.

National name: *Principat d'Andorra/Principality of Andorra*
Area: 468 sq km/181 sq mi
Capital: Andorra la Vella
Major towns/cities: Les Escaldes, Escaldes-Engordany (a suburb of the capital)
Physical features: mountainous, with narrow valleys; the eastern Pyrenees; Valira River

Blue and red acknowledge Andorra's links with France. Red and yellow represent the influence of Spain. Effective date: c. 1866.

Government

Heads of state: Joan Marti i Alanis (bishop of Urgel, Spain; from 1971) and Jacques Chirac (president of France; from 1995)
Head of government: Marc Forne Molne from 1994
Political system: emergent democracy
Political executive: parliamentary
Administrative divisions: seven parishes
Political parties: National Democratic Grouping (AND; formerly the Democratic Party of Andorra: PDA), moderate, centrist; National Democratic Initiative (IND), left of centre; New Democracy Party (ND), centrist; National Andorran Coalition (CNA), centrist; Liberal Union (UL), right of centre
Armed forces: no standing army
Death penalty: abolished in 1990 (last execution in 1943)
Health spend: (% GDP) 7.5 (1997 est)

Economy and resources

Currency: French franc and Spanish peseta
GDP: (US$) 1.2 billion (1996 est)
Real GDP growth: (% change on previous year) 3 (1995)
GNP: (US$) 1 billion (1995 est)
GNP per capita (PPP): (US$) 18,000 (1996 est)
Consumer price inflation: 0.8% (1998)
Unemployment: 0% (1997 est)
Major trading partners: Spain, France, USA
Resources: iron, lead, aluminium, hydroelectric power
Industries: cigar and cigarette manufacturing, textiles, leather goods, wood products, processed foodstuffs, furniture, tourism, banking and financial services
Exports: cigars and cigarettes, furniture, electricity. Principal market: Spain 58% (1998)
Imports: foodstuffs, electricity, mineral fuels. Principal source: Spain 48% (1998)
Arable land: 2.2% (1995)

Agricultural products: tobacco, potatoes, rye, barley, oats, vegetables; livestock rearing (mainly sheep) and timber production

Population and society

Population: 78,000 (2000 est)
Population growth rate: 3.9% (1995–2000)
Population density: (per sq km) 146 (1999 est)
Urban population: (% of total) 93 (2000 est)
Age distribution: (% of total population) 0–14 16%, 15–59 69%, 60+ 15% (2000 est)
Ethnic groups: 20% Andorrans, 44% Spanish, 11% Portuguese, 7% French, 18% other
Language: Catalan (official), Spanish, French
Religion: Roman Catholic (92%)
Education: (compulsory years) 10
Literacy rate: 99% (men); 99% (women) (1995 est)
Labour force: 1% agriculture, 21% industry, 78% services (1998)
Life expectancy: 70 (men); 73 (women) (1994 est)
Physicians: 1 per 439 people (1996)
Hospital beds: 1 per 454 people (1996)
TV sets: (per 1,000 people) 403 (1997)
Radios: (per 1,000 people) 238 (1997)
Internet users: (per 10,000 people) 665.7 (1999)

Transport

Airports: international airports: none; closest airport for Andorran traffic 20 km/12.5 mi from Andorra at Seo de Urgel, Spain
Railways: none; there is a connecting bus service to stations in France and Spain
Roads: total road network: 220 km/137 mi, of which 55% paved (1995); passenger cars: 108.6 per 1,000 people (1996 est)

Chronology

AD 803: Holy Roman Emperor Charlemagne liberated Andorra from Muslim control. **819:** Louis I 'the Pious', the son of Charlemagne, granted control over the area to the Spanish bishop of Urgel. **1278:** A treaty was signed making Spanish bishop and French count joint rulers of Andorra. Through marriage the king of France later inherited the count's right. **1806:** After a temporary suspension during the French Revolution, from 1789 the feudal arrangement of dual allegiance to the French and Spanish rulers was re-established by the French emperor Napoleon Bonaparte. **1976:** The first political organization, the Democratic Party of Andorra, was formed. **1981:** The first prime minister was appointed by the General Council. **1991:** Links with the European Community (EC) were formalized. **1993:** A new constitution legalized political parties and introduced the first direct elections. Andorra became a member of the United Nations (UN). **1994:** Andorra joined the Council of Europe. **1997:** The Liberal Union (UL) won an assembly majority in a general election.

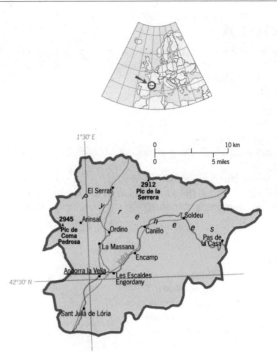

Practical information

Visa requirements: UK: visa not required. USA: visa not required
Time difference: GMT +1
Chief tourist attractions: attractive mountain scenery; winter-sports facilities available at five skiing centres; duty-free shopping facilities; the spa town of Les Escaldes
Major holidays: 1, 6, January, 19 March, 1 May, 24 June, 15 August, 8 September, 1, 4, November, 8, 25–26 December; variable: Ascension Thursday, Carnival, Corpus Christi, Good Friday, Easter Monday, Whit Monday

ANGOLA

Located in southwest Africa, bounded west by the Atlantic Ocean, north and northeast by Congo (formerly Zaire), east by Zambia, and south by Namibia. The Cabinda enclave, a district of Angola, is bounded west by the Atlantic Ocean, north by the Congo River, and east and south by Congo.

National name: *República de Angolo/Republic of Angola*
Area: 1,246,700 sq km/481,350 sq mi
Capital: Luanda (and chief port)
Major towns/cities: Lobito, Benguela, Huambo, Lubango, Malanje, Namibe, Kuito
Major ports: Huambo, Lubango, Malanje
Physical features: narrow coastal plain rises to vast interior plateau with rainforest in northwest; desert in south; Cuanza, Cuito, Cubango, and Cunene rivers

The yellow of the emblem is said to denote Angola's natural wealth. Red is said to stand for the blood spilt by the freedom fighters. Black represents Africa. Effective date: 11 November 1975.

Government

Head of state: José Eduardo dos Santos from 1979, who is also head of government from 1999
Political system: emergent democracy
Political executive: limited presidency
Administrative divisions: 18 provinces
Political parties: People's Movement for the Liberation of Angola – Workers' Party (MPLA–PT), Marxist-Leninist; National Union for the Total Independence of Angola (UNITA), conservative; National Front for the Liberation of Angola (FNLA), conservative
Armed forces: 114,000; plus a paramilitary force of approximately 15,000 (1998)
Conscription: military service is compulsory for two years
Death penalty: abolished in 1992
Defence spend: (% GDP) 11.7 (1998)
Education spend: (% GNP) N/A
Health spend: (% GDP) 3.6 (1997)

Economy and resources

Currency: kwanza
GDP: (US$) 5.9 billion (1999)
Real GDP growth: (% change on previous year) 2.7 (1999)
GNP: (US$) 2.7 billion (1999)
GNP per capita (PPP): (US$) 632 (1999)
Consumer price inflation: 248.2% (1999)
Unemployment: 21.5% (1995)
Foreign debt: (US$) 11.9 billion (1999)
Major trading partners: Portugal, USA, Germany, France, Japan, Brazil, the Netherlands
Resources: petroleum, diamonds, granite, iron ore, marble, salt, phosphates, manganese, copper

Industries: mining, petroleum refining, food processing, textiles, construction materials
Exports: petroleum and petroleum products, diamonds, gas. Principal market: USA 63% (1998)
Imports: foodstuffs, transport equipment, base metals, electrical equipment. Principal source: Portugal 20% (1998)
Arable land: 2.4% (1996)
Agricultural products: coffee, sugar cane, bananas, cassava, maize, sweet potatoes

Population and society

Population: 12,878,000 (2000 est)
Population growth rate: 3.2% (1995–2000); 3.1% (2000–05)
Population density: (per sq km) 10 (1999 est)
Urban population: (% of total) 34 (2000 est)
Age distribution: (% of total population) 0–14 47%, 15–59 48%, 60+ 5% (2000 est)
Ethnic groups: nine main ethnic groups (Bakonga, Quimbundo, Lunda-Quioco (or Tchokwe), Ovimnundo, Ganguela, Nhaneca-Huambe, Ambo, Herero and the Xindonga), and about 100 subgroups. 2% mestizo (mixed European and Native American), 1% European. A major exodus of Europeans in the 1970s left only around 30,000 Europeans, mainly Portuguese.
Language: Portuguese (official), Bantu, other native dialects
Religion: Roman Catholic 38%, Protestant 15%, animist 47%
Education: (compulsory years) 8
Literacy rate: 56% (men); 28% (women) (1998 est)
Labour force: 72.7% agriculture, 7.2% industry, 20.1% services (1997 est)
Life expectancy: 45 (men); 48 (women) (1995–2000)

Practical information

Visa requirements: UK/USA: visa required
Time difference: GMT +1
Chief tourist attractions: travel within Angola remains unsafe due to the presence of undisciplined armed troops, unexploded landmines, hostile actions against aircraft, and widespread banditry
Major holidays: 1 January, 4 February, 27 March, 14 April, 1 May, 1 August, 17 September, 11 November, 1, 10, 25 December

Child mortality rate: (under 5, per 1,000 live births)
208 (1995–2000)
Physicians: 1 per 23,725 people (1993 est)
Hospital beds: 1 per 774 people (1993 est)
TV sets: (per 1,000 people) 14 (1998)
Radios: (per 1,000 people) 54 (1997)
Internet users: (per 10,000 people) 8.0 (1999)
Personal computer users: (per 100 people) 0.1 (1999)

Transport

Airports: international airports: Luanda (4 de Fevereio);
domestic services to all major towns; total passenger km:
926 million (1997 est)
Railways: total length: 2,771 km/1,722 mi; total
passenger km: 246 million (1991)
Roads: total road network: 72,626 km/45,130 mi, of
which 25% paved (1996); passenger cars: 18 per 1,000
people (1996 est)

Chronology

14th century: The powerful Kongo kingdom controlled much of northern Angola. **early 16th century:** The
Kongo ruler King Afonso I adopted Christianity and sought relations with Portuguese traders. **1575 and 1617:**
Portugal secured control over the ports of Luanda and Benguela and began to penetrate inland, meeting resistance
from Queen Nzinga, the Ndonga ruler. **17th–18th centuries:** Inland, the
Lunda peoples established powerful kingdoms that stretched into southern
Congo. The Portuguese made Angola a key centre for the export of slaves;
over 1 million were shipped to Brazil 1580–1680. **1836:** The slave trade
was officially abolished. **1885–1915:** Military campaigns were waged by
Portugal to conquer the interior. **1951:** Angola became an overseas
territory of Portugal. **1956:** The People's Movement for the Liberation of
Angola (MPLA), a socialist guerrilla independence movement based in the
Congo, was formed. **1961:** 50,000 people were massacred in rebellions
on coffee plantations. Forced labour was abolished. There was
an armed struggle for independence. **1962:** The
National Front for the Liberation of Angola (FNLA), a
nationalist guerrilla movement, was formed. **1966:** The
National Union for the Total Independence of Angola
(UNITA) was formed in southeastern Angola as a
breakaway from the FNLA. **1975:** Independence from
Portugal was achieved. The MPLA (backed by Cuba)
proclaimed the People's Republic of Angola. The
FNLA and UNITA (backed by South Africa and the
USA) proclaimed the People's Democratic Republic
of Angola. **1976:** The MPLA gained control of most
of the country. South African troops withdrew, but
Cuban units remained as the civil war continued.
1980: UNITA guerrillas, aided by South Africa,
continued raids against the government and bases of
the Namibian South West Africa People's Organization
(SWAPO) in Angola. **1988:** A peace treaty providing
for the withdrawal of all foreign troops was signed with
South Africa and Cuba. **1989:** A ceasefire agreed with
UNITA broke down and guerrilla activity resumed. **1991:**
A peace agreement ended the civil war. An amnesty was
declared for all political prisoners, and there was a new
multiparty constitution. **1992:** A MPLA general election
victory was fiercely disputed by UNITA, and plunged the country
into renewed civil war. **1993:** The MPLA government was recognized by the USA. United
Nations (UN) sanctions were imposed against UNITA. **1994:** A peace treaty was signed by the government and
UNITA representatives. **1995:** UN peacekeepers were drafted in. **1996:** UNITA leader Jonas Savimbi rejected an
offer of the vice presidency. President dos Santos appointed Fernando Franca van Dunem as his new prime minister.
1997: After some delay a national unity government was eventually sworn in but was boycotted by Savimbi. **1998:**
UNITA was demilitarized and transformed into a political party, but after UNITA was accused of massacres, UNITA
ministers were suspended and the peace process threatened. **2000:** Fighting between government forces and UNITA
rebels continued in the south and east, with government troops making significant gains in the east.

ANTIGUA AND BARBUDA

Three islands in the eastern Caribbean (Antigua, Barbuda, and uninhabited Redonda).

Area: 440 sq km/169 sq mi (Antigua 280 sq km/108 sq mi, Barbuda 161 sq km/62 sq mi, plus Redonda 1 sq km/0.4 sq mi)
Capital: St John's (on Antigua) (and chief port)
Major towns/cities: Codrington (on Barbuda)
Physical features: low-lying tropical islands of limestone and coral with some higher volcanic outcrops; no rivers and low rainfall result in frequent droughts and deforestation. Antigua is the largest of the Leeward Islands; Redonda is an uninhabited island of volcanic rock rising to 305 m/1,000 ft

Black reflects the African origins of the islanders. Red stands for the vigour of the people. White represents hope. Effective date: 27 February 1967.

Government

Head of state: Queen Elizabeth II from 1981, represented by Governor General James B Carlisle from 1993
Head of government: Lester Bird from 1994
Political system: liberal democracy
Political executive: parliamentary
Administrative divisions: six parishes and two dependencies
Political parties: Antigua Labour Party (ALP), moderate left of centre; United Progressive Party (UPP), centrist; Barbuda People's Movement (BPM), left of centre
Armed forces: 200 (1998); US government leases two military bases on Antigua
Conscription: military service is voluntary
Death penalty: retained and used for ordinary crimes
Defence spend: (% GDP) 0.6% (1998)
Education spend: (% GNP) 3.7 (1988); N/A (1990–95)
Health spend: (% GDP) 6.4 (1997)

Economy and resources

Currency: East Caribbean dollar
GDP: (US$) 622 million (1999)
Real GDP growth: (% change on previous year) 2.8 (1999 est)
GNP: (US$) 570 million (1999 est)
GNP per capita (PPP): (US$) 8,959 (1999 est)
Consumer price inflation: 1.6% (1999 est)
Unemployment: 7% (1997)
Foreign debt: (US$) 357 million (1998)
Major trading partners: Barbados, USA, UK, Canada, Trinidad and Tobago, Guyana
Industries: oil refining, food and beverage products, paint, bedding, furniture, electrical components. Tourism is the main economic activity.
Exports: petroleum products, food, manufactures, machinery and transport equipment. Principal market: Barbados 15% (1999 est)

Imports: petroleum, food and live animals, machinery and transport equipment, manufactures, chemicals. Principal source: USA 27% (1998 est)
Arable land: 18.2% (1995)
Agricultural products: cucumbers, pumpkins, mangoes, coconuts, limes, melons, pineapples, cotton; fishing

Population and society

Population: 68,000 (2000 est)
Population growth rate: 0.5% (1995–2000)
Population density: (per sq km) 246 (1999 est)
Urban population: (% of total) 37 (2000 est)
Age distribution: (% of total population) 0–14 25%, 15–59 69%, 60+ 6% (1999)
Ethnic groups: population almost entirely of black African descent, British, Portuguese, Lebanese, Syrian
Language: English (official), local dialects
Religion: Christian (mostly Anglican)
Education: (compulsory years) 11
Literacy rate: 94% (men); 90% (women) (1998 est)
Labour force: 11% agriculture, 19.7% industry, 69.3% services (1991)
Life expectancy: 72 (men); 76 (women) (1998 est)
Child mortality rate: (under 5, per 1,000 live births) 22 (1995)
Physicians: 1 per 1,316 people (1993)
Hospital beds: 1 per 364 people (1990)
TV sets: (per 1,000 people) 462 (1997)
Radios: (per 1,000 people) 537 (1997)
Internet users: (per 10,000 people) 535.5 (1999)

Transport

Airports: international airports: St John's (V C Bird International); one airstrip on Barbuda; total passenger km: 252 million (1995)
Railways: none
Roads: total road network: 250 km/155 mi (1996 est); passenger cars: 195 per 1,000 people (1994)

Chronology

1493: Antigua, peopled by American Indian Caribs, was visited by Christopher Columbus. **1632:** Antigua was colonized by British settlers from St Kitts. **1667:** The Treaty of Breda ceded Antigua to Britain. **1674:** Christopher Codrington, a sugar planter from Barbados, established sugar plantations and acquired Barbuda island on lease from the British monarch in 1685; Africans were brought in as slaves. **1834:** Antigua's slaves were freed. **1860:** Barbuda was annexed. **1871–1956:** Antigua and Barbuda were administered as part of the Leeward Islands federation. **1958–62:** Antigua and Barbuda became part of the West Indies Federation. **1967:** Antigua and Barbuda became an associated state within the Commonwealth. **1969:** A separatist movement developed on Barbuda. **1981:** Independence from Britain was achieved. **1983:** Antigua and Barbuda assisted in the US invasion of Grenada. **1994:** General elections were won by the ALP, with Lester Bird becoming prime minister.

Practical information

Visa requirements: UK: visa not required. USA: visa not required
Time difference: GMT –4
Chief tourist attractions: over 300 beaches; the historic Nelson's Dockyard in English Harbour (a national park); cruise-ship facilities; Barbuda is less developed, but offers pink sandy beaches, beauty, and wildlife; animal attractions; international sailing regatta and carnival week
Major holidays: 1 January, 1 July, 1 November, 25–26 December; variable: Good Friday, Easter Monday, Whit Monday, Labour Day (May), CARICOM (July), Carnival (August)

ARGENTINA

Located in South America, bounded west and south by Chile, north by Bolivia, and east by Paraguay, Brazil, Uruguay, and the Atlantic Ocean.

National name: *República Argentina/Argentine Republic*
Area: 2,780,400 sq km/1,073,518 sq mi
Capital: Buenos Aires
Major towns/cities: Rosario, Córdoba, San Miguel de Tucumán, Mendoza, Santa Fé, La Plata
Major ports: La Plata and Bahía Blanca
Physical features: mountains in west, forest and savannah in north, pampas (treeless plains) in east-central area, Patagonian plateau in south; rivers Colorado, Salado, Paraná, Uruguay; Río de La Plata estuary; Andes mountains, with Aconcagua the highest peak in western hemisphere; Iguaçu Falls
Territories: disputed claim to the Falkland Islands (*Islas Malvinas*), and part of Antarctica

The 'Sun of May' was added in 1818. The blue bands are a shade known as 'celeste', said to be the colour of the sky which inspired Argentine revolutionary Manuel Belgrano before battle. Effective date: 16 August 1985.

Government

Head of state and government: Fernando de la Rua from 1999
Political system: liberal democracy
Political executive: limited presidency
Administrative divisions: 23 provinces and one federal district (Buenos Aires)
Political parties: Radical Civic Union Party (UCR), moderate centrist; Justicialist Party (PJ), right-wing Perónist; Movement for Dignity and Independence (Modin), right wing; Front for a Country in Solidarity (Frepaso), left of centre
Armed forces: 73,000 plus paramilitary gendarmerie of 31,200 (1998)
Conscription: abolished in 1995
Death penalty: abolished for ordinary crimes in 1984; laws provide for the death penalty for exceptional crimes only
Defence spend: (% GDP) 1.8 (1998)
Education spend: (% GNP) 3.5 (1996)
Health spend: (% GDP) 8.2 (1997)

Economy and resources

Currency: peso (= 10,000 australs, which it replaced in 1992)
GDP: (US$) 282.9 billion (1999)
Real GDP growth: (% change on previous year) –3.1 (1999)
GNP: (US$) 277.9 billion (1999)
GNP per capita (PPP): (US$) 11,324 (1999)
Consumer price inflation: –1.2 (1999)
Unemployment: 15.4% (1997)
Foreign debt: (US$) 148.6 billion (1999)
Major trading partners: Brazil, USA, the Netherlands, Germany, France, Chile, Italy
Resources: coal, crude oil, natural gas, iron ore, lead ore, zinc ore, tin, gold, silver, uranium ore, marble, borates, granite
Industries: petroleum and petroleum products, primary iron, crude steel, sulphuric acid, synthetic rubber, paper and paper products, crude oil, cement, cigarettes, motor vehicles
Exports: meat and meat products, prepared animal fodder, cereals, petroleum and petroleum products, soybeans, vegetable oils and fats. Principal market: Brazil

24.4% (1999)
Imports: machinery and transport equipment, chemicals and mineral products. Principal sources: Brazil 21.9% (1999)
Arable land: 9.1% (1996)
Agricultural products: wheat, maize, soybeans, sugar cane, rice, sorghum, potatoes, tobacco, sunflowers, cotton, vine fruits, citrus fruit; livestock production (chiefly cattle)

Population and society

Population: 37,032,000 (2000 est)
Population growth rate: 1.3% (1995–2000); 1.1% (2000–05)
Population density: (per sq km) 13 (1999 est)
Urban population: (% of total) 90 (2000 est)
Age distribution: (% of total population) 0–14 28%, 15–59 59%, 60+ 13% (2000 est)
Ethnic groups: 85% of European descent, mainly Spanish; 15% mestizo (offspring of Spanish–American and American Indian parents)
Language: Spanish (official) (95%), Italian (3%), English, German, French
Religion: predominantly Roman Catholic (state-supported), 2% protestant, 2% Jewish
Education: (compulsory years) 7; age limits 7–16
Literacy rate: 97% (men); 97% (women) (2000 est)
Labour force: 38% of population: 12% agriculture, 32% industry, 55% services (1996)
Life expectancy: 70 (men); 77 (women) (1995–2000)
Child mortality rate: (under 5, per 1,000 live births) 25 (1995–2000)
Physicians: 1 per 330 people (1993 est)
Hospital beds: 1 per 233 people (1995)
TV sets: (per 1,000 people) 289 (1997)
Radios: (per 1,000 people) 681 (1997)
Internet users: (per 10,000 people) 246.1 (1999)
Personal computer users: (per 100 people) 4.9 (1999)

Transport

Airports: international airports: Buenos Aires, Aeroparque Jorge Newbery, Córdoba, Corrientes, El Plumerillo, Ezeiza, Jujuy, Resistencia, Río Gallegos, Salta,

San Carlos de Bariloche; domestic services to all major towns; total passenger km: 13,957 million (1997 est) **Railways:** total length: 34,115 km/21,199 mi; total passenger km: 7,180 million (1995)

Roads: total road network: 218,276 km/135,637 mi, of which 29.1% paved (1996); passenger cars: 135 per 1,000 people (1995 est)

Chronology

1516: The Spanish navigator Juan Diaz de Solis discovered Río de La Plata. **1536:** Buenos Aires was founded, but was soon abandoned because of attacks by American Indians. **1580:** Buenos Aires was re-established as part of the Spanish province of Asunción. **1617:** Buenos Aires became a separate province within the Spanish viceroyalty of Lima. **1776:** The Spanish South American Empire was reorganized: Atlantic regions became viceroyalty of La Plata, with Buenos Aires as capital. **1810:** After the French conquest of Spain, Buenos Aires junta took over government of viceroyalty. **1816:** Independence was proclaimed, as the United Provinces of Río de La Plata, but Bolivia and Uruguay soon seceded; civil war followed between federalists and those who wanted a unitary state. **1835–52:** Dictatorship of Gen Juan Manuel Rosas. **1853:** Adoption of federal constitution based on US model; Buenos Aires refused to join confederation. **1861:** Buenos Aires was incorporated into the Argentine confederation by force. **1865–70:** Argentina took part in the War of Triple Alliance against Paraguay. **late 19th century:** Large-scale European immigration and economic development. **1880:** Buenos Aires became the national capital. **1880–1916:** The government was dominated by an oligarchy of conservative landowners. **1916:** The secret ballot was introduced and the Radical Party of Hipólito Irigoyen won elections, beginning a period of 14 years in government. **1930:** A military coup ushered in a series of conservative governments sustained by violence and fraud. **1946:** Col Juan Perón won presidential elections; he secured working-class support through welfare measures, trade unionism, and the popularity of his wife, Eva Perón (Evita). **1949:** A new constitution abolished federalism and increased powers of president. **1952:** Death of Evita. Support for Perón began to decline. **1955:** Perón was overthrown; the constitution of 1853 was restored. **1966–70:** Dictatorship of Gen Juan Carlos Ongania. **1973:** Perónist Party won free elections; Perón returned from exile in Spain to become president. **1974:** Perón died and was succeeded by his third wife, Isabel Perón. **1976:** A coup resulted in rule by a military junta. **1976–83:** The military regime conducted murderous campaign ('Dirty War') against left-wing elements. More than 8,000 people disappeared. **1982:** Argentina invaded the Falkland Islands but was defeated by the UK. **1983:** Return to civilian rule; an investigation into the 'Dirty War' was launched. **1989:** Annual inflation reached 12,000%. Carlos Menem won the presidential elections. **1990:** Full diplomatic relations with the UK were restored. **1991:** The government introduced the peso to replace the austral. **1995:** Carlos Menem was elected for a second term as president. **1999:** Falkland Islanders held their first talks with Argentina since 1982. Fernando de la Rua won the presidential elections. **2000:** There were protests against spending cuts that aimed to bring the economy into line with targets set by the International Monetary Fund (IMF).

Practical information

Visa requirements: UK: visa not required for tourist visits; visa required for business purposes. USA: visa not required for tourist visits; visa required for business purposes
Time difference: GMT –3
Chief tourist attractions: include the Andes

Mountains; lake district centred on Bariloche; Atlantic beaches; Patagonia; Mar del Plata beaches; Iguaçu Falls; the Pampas; Tierra del Fuego
Major holidays: 1 January, 1, 25 May, 10, 20 June, 9 July, 17 August, 12 October, 8, 25, 31 December; variable: Good Friday, Holy Thursday

ARMENIA

Located in western Asia, bounded east by Azerbaijan, north by Georgia, west by Turkey, and south by Iran.

National name: *Hayastani Hanrapetoutioun/Republic of Armenia*
Area: 29,800 sq km/11,505 sq mi
Capital: Yerevan
Major towns/cities: Gyumri (formerly Leninakan), Vanadzor (formerly Kirovakan), Hrazdan, Aboyvan
Physical features: mainly mountainous (including Mount Ararat), wooded

Red stands for bloodshed. Orange symbolizes agriculture. Blue represents the sky and hope. Effective date: 24 August 1990.

Government

Head of state: Robert Kocharian from 1998
Head of government: Andranik Markaryan from 2000
Political system: authoritarian nationalist
Political executive: unlimited presidency
Administrative divisions: 11 provinces, including the capital, Yerevan
Political parties: Armenian Pan-National Movement (APM), nationalist, left of centre; Armenian Revolutionary Federation (ARF), centrist (banned in 1994); Communist Party of Armenia (banned 1991–92); National Unity, opposition coalition; Armenian Christian Democratic Union (CDU), moderately right wing; Huchak Armenian Social Democratic Party
Armed forces: 53,400 (1998)
Conscription: compulsory for 18 months
Death penalty: retained and used for ordinary crimes
Defence spend: (% GDP) 8.4 (1998)
Education spend: (% GNP) 2 (1996)
Health spend: (% GDP) 7.9 (1997)

Economy and resources

Currency: dram (replaced Russian rouble in 1993)
GDP: (US$) 1.9 billion (1999)
Real GDP growth: (% change on previous year) 3.3 (1999)
GNP: (US$) 1.9 billion (1999)
GNP per capita (PPP): (US$) 2,210 (1999)
Consumer price inflation: 0.8% (1999)
Unemployment: 9.3% (1997)
Foreign debt: (US$) 800 million (1998)
Major trading partners: Belgium, Russia, Iran, USA
Resources: copper, zinc, molybdenum, iron, silver, marble, granite
Industries: food processing and beverages, fertilizers, synthetic rubber, machinery and metal products, textiles, garments
Exports: machinery and metalworking products, chemical and petroleum products, precious or semi-precious metals and stone, base metals, equipment Principal market: Belgium 36.1% (1999)

Imports: light industrial products, petroleum and derivatives, industrial raw materials, vegetable products. Principal source: Russia 17.3% (1999)
Arable land: 21.2% (1996)
Agricultural products: potatoes, vegetables, fruits, cotton, almonds, olives, figs, cereals; livestock rearing (sheep and cattle)

Population and society

Population: 3,520,000 (2000 est)
Population growth rate: –0.3% (1995–2000)
Population density: (per sq km) 118 (1999 est)
Urban population: (% of total) 70 (2000 est)
Age distribution: (% of total population) 0–14 25%, 15–59 62%, 60+ 13% (2000 est)
Ethnic groups: 93% of Armenian ethnic descent, 3% Azeri, 2% Russian, and 2% Kurdish
Language: Armenian (official)
Religion: Armenian Orthodox
Education: (compulsory years) 9
Literacy rate: 99% (men); 98% (women) (2000 est)
Labour force: 32.2% agriculture, 32.8% industry, 35% services (1993)
Life expectancy: 67 (men); 74 (women) (1995–2000)
Child mortality rate: (under 5, per 1,000 live births) 33 (1995–2000)
Physicians: 1 per 293.1 people (1996)
Hospital beds: 1 per 140.5 people (1996)
TV sets: (per 1,000 people) 218 (1997)
Radios: (per 1,000 people) 224 (1997)
Internet users: (per 10,000 people) 85.1 (1999)
Personal computer users: (per 100 people) 0.6 (1999)

Transport

Airports: international airports Yerevan (Zvarnots); domestic services to most major towns
Railways: total length: 830 km/516 mi; total passenger km: 166 million (1995)
Roads: total road network: 8,580 km/5,331 mi, of which 100% paved; passenger cars: 0.3 per 1,000 people (1996 est)

Chronology

6th century BC: Armenian peoples moved into the area, which was then part of the Persian Empire. *c.* **94–56 BC:** Under King Tigranes II 'the Great', Armenia reached the height of its power, becoming the strongest state in the eastern Roman empire. *c.* **AD 300:** Christianity became the state religion when the local ruler was converted by St Gregory the Illuminator. *c.* **AD 390:** Armenia was divided between Byzantine Armenia, which became part of the Byzantine Empire, and Persarmenia, under Persian control. **886–1045:** Became independent under the Bagratid monarchy. **13th century:** After being overrun by the Mongols, a substantially independent Little Armenia survived until 1375. **early 16th century:** Conquered by Muslim Ottoman Turks. **1813–28:** Russia took control of eastern Armenia. **late 19th century:** Revival in Armenian culture and national spirit, provoking Ottoman backlash in western Armenia and international concern at Armenian maltreatment: the 'Armenian Question'. **1894–96:** Armenians were massacred by Turkish soldiers in an attempt to suppress unrest. **1915:** Suspected of pro-Russian sympathies, two-thirds of Armenia's population of 2 million were deported to Syria and Palestine. Around 600,000 to 1 million died en route: the survivors contributed towards an Armenian diaspora in Europe and North America. **1916:** Armenia was conquered by tsarist Russia and became part of a brief 'Transcaucasian Alliance' with Georgia and Azerbaijan. **1918:** Armenia became an independent republic. **1920:** Occupied by Red Army of Soviet Union (USSR), but western Armenia remained part of Turkey and northwest Iran. **1936:** Became constituent republic of USSR; rapid industrial development. **late 1980s:** Armenian 'national reawakening', encouraged by *glasnost* (openness) initiative of Soviet leader Mikhail Gorbachev. **1988:** Around 20,000 people died in an earthquake. **1989:** Strife-torn Nagorno-Karabakh was placed under direct rule from Moscow; civil war erupted with Azerbaijan over Nagorno-Karabakh and Nakhichevan, an Azerbaijani-peopled enclave in Armenia. **1990:** Independence was declared, but ignored by Moscow and the international community. **1991:** After the collapse of the USSR, Armenia joined the new Commonwealth of Independent States. Nagorno-Karabakh declared its independence. **1992:** Armenia was recognized as an independent state by the USA and admitted into the United Nations (UN). **1993:** Armenian forces gained control of more than a fifth of Azerbaijan, including much of Nagorno-Karabakh. **1994:** A Nagorno-Karabakh ceasefire ended the conflict. **1997:** There was border fighting with Azerbaijan. **1999:** Prime Minister Vazgen Sarkisian was assassinated in October when gunmen burst into parliament and shot him and seven other officials. He was replaced by his brother, Amen Sarkisian. **2000:** President Robert Kocharian dismissed Amen Sarkisian as prime minister and replaced him with Andranik Markarya.

Practical information

Visa requirements: UK: visa required. USA: visa required
Time difference: GMT +4
Chief tourist attractions: Armenian mountains in the Lesser Caucasus; Lake Sevan
Major holidays: 1, 6 January, 28–31 March, 24, 28 May, 21 September, 7 December; variable: Good Friday, Easter Monday

AUSTRALIA

Occupies all of the Earth's smallest continent; situated south of Indonesia, between the Pacific and Indian oceans.

National name: *Commonwealth of Australia*
Area: 7,682,850 sq km/2,966,136 sq mi
Capital: Canberra
Major towns/cities: Adelaide, Alice Springs, Brisbane, Darwin, Melbourne, Perth, Sydney, Hobart, Newcastle, Wollongong
Physical features: Ayers Rock; Arnhem Land; Gulf of Carpentaria; Cape York Peninsula; Great Australian Bight; Great Sandy Desert; Gibson Desert; Great Victoria Desert; Simpson Desert; the Great Barrier Reef; Great Dividing Range and Australian Alps in the east (Mount Kosciusko, 2,229 m/7,136 ft, Australia's highest peak). The fertile southeast region is watered by the Darling, Lachlan, Murrumbridgee, and Murray rivers. Lake Eyre basin and Nullarbor Plain in the south.
Territories: Norfolk Island, Christmas Island, Cocos (Keeling) Islands, Ashmore and Cartier Islands, Coral Sea Islands, Heard Island and McDonald Islands, Australian Antarctic Territory

The Union Jack marks Australia's historical links with Britain. The Southern Cross helped guide early European navigators to the continent. Effective date: 15 April 1954.

Government

Head of state: Queen Elizabeth II from 1952, represented by Governor General Sir William Deane from 1996
Head of government: John Howard from 1996
Political system: liberal democracy
Political executive: parliamentary
Administrative divisions: six states and two territories
Political parties: Australian Labor Party, moderate left of centre; Liberal Party of Australia, moderate, liberal, free enterprise; National Party of Australia (formerly Country Party), centrist non-metropolitan; Australian Democratic Party (AD), liberal, moderately left wing; One Nation (ON), right-wing, racist, and anti-immigrant
Armed forces: 57,400 (1998)
Conscription: military service is voluntary
Death penalty: abolished in 1985
Defence spend: (% GDP) 1.9 (1998)
Education spend: (% GNP) 5.6 (1996)
Health spend: (% GDP) 8.4 (1997)

Economy and resources

Currency: Australian dollar
GDP: (US$) 394 billion (1999)
Real GDP growth: (% change on previous year) 4.6 (1999)
GNP: (US$) 380.8 billion (1999)
GNP per capita (PPP): (US$) 22,448 (1999)
Consumer price inflation: 1.5% (1999)
Unemployment: 8% (1998)
Major trading partners: USA, Japan, EU, New Zealand, South Korea, China, Taiwan, Singapore
Resources: coal, iron ore (world's third-largest producer), bauxite, copper, zinc (world's second-largest producer), nickel (world's fifth-largest producer), uranium, gold, diamonds
Industries: mining, metal products, textiles, wood and paper products, chemical products, electrical machinery, transport equipment, printing, publishing and recording media, tourism, electronic communications

Exports: major world producer of raw materials: iron ore, aluminium, coal, nickel, zinc, lead, gold, tin, tungsten, uranium, crude oil; wool, meat, cereals, fruit, sugar, wine. Principal market: Japan 19.3% (1999)
Imports: processed industrial supplies, transport equipment and parts, road vehicles, petroleum and petroleum products, medicinal and pharmaceutical products, organic chemicals, consumer goods. Principal source: EU 23.9% (1999)
Arable land: 6.5% (1996)
Agricultural products: wheat, barley, oats, rice, sugar cane, fruit, grapes; livestock (cattle and sheep) and dairy products

Population and society

Population: 18,886,000 (2000 est)
Population growth rate: 1% (1995–2000)
Population density: (per sq km) 2 (1999 est)
Urban population: (% of total) 85 (2000 est)
Age distribution: (% of total population) 0–14 21%, 15–59 62%, 60+ 17% (2000 est)
Ethnic groups: 92% of European descent; 7% Asian, 1% Aborigine and other
Language: English (official), Aboriginal languages
Religion: Anglican 26%, Roman Catholic 26%, other Christian 24%
Education: (compulsory years) 10 or 11 (states vary)
Literacy rate: 99% (men); 99% (women) (2000 est)
Labour force: 5.2% agriculture, 22.1% industry, 72.7% services (1997)
Life expectancy: 76 (men); 82 (women) (1995–2000)
Child mortality rate: (under 5, per 1,000 live births) 7 (1995–2000)
Physicians: 1 per 400 people (1996)
Hospital beds: 1 per 115 people (1996)
TV sets: (per 1,000 people) 638 (1997)
Radios: (per 1,000 people) 1,376 (1997)
Internet users: (per 10,000 people) 3,172.7 (1999)
Personal computer users: (per 100 people) 47.1 (1999)

Transport

Airports: international airports: Sydney (NSW), Melbourne (Victoria), Canberra, Brisbane, Cairns (Queensland), Perth (Western Australia), Adelaide (South Australia), Hobart (Tasmania), Townsville (Queensland), Darwin (Northern Territory); domestic services to all major resorts and cities; total passenger km: 77,915 million (1997 est) **Railways:** total length: 37,295 km/23,175 mi; passengers carried: 447 million (1996) **Roads:** total road network: 913,000 km/567,338 mi, of which 38.7% paved (1996 est); passenger cars: 488 per 1,000 people (1996 est)

Chronology

c. **40,000 BC:** Aboriginal immigration from southern India, Sri Lanka, and Southeast Asia. **AD 1606:** First recorded sightings of Australia by Europeans including discovery of Cape York by Dutch explorer Willem Jansz in *Duyfken*. **1770:** Capt James Cook claimed New South Wales for Britain. **1788:** Sydney founded as British penal colony. **late 18th–19th centuries:** Great age of exploration. **1804:** Castle Hill Rising by Irish convicts in New South Wales. **1813:** Crossing of Blue Mountains removed major barrier to exploration of interior. **1825:** Tasmania seceded from New South Wales. **1829:** Western Australia colonized. **1836:** South Australia colonized. **1840–68:** End of convict transportation. **1850:** British Act of Parliament permitted Australian colonies to draft their own constitutions and achieve virtual self-government. **1851–61:** Gold rushes contributed to exploration and economic growth. **1851:** Victoria seceded from New South Wales. **1855:** Victoria achieved self-government. **1856:** New South Wales, South Australia, and Tasmania achieved self-government. **1859:** Queensland was formed from New South Wales and achieved self-government. **1890:** Western Australia achieved self-government. **1891:** Depression gave rise to the Australian Labor Party. **1901:** The Commonwealth of Australia was created. **1919:** Australia was given mandates over Papua New Guinea and the Solomon Islands. **1927:** The seat of federal government moved to Canberra. **1931:** Statute of Westminster confirmed Australian independence. **1933:** Western Australia's vote to secede was overruled. **1948–75:** Influx of around 2 million new immigrants, chiefly from continental Europe. **1967:** A referendum gave Australian Aborigines full citizenship rights. **1970s:** Japan became Australia's chief trading partner. **1974:** 'White Australia' immigration restrictions were abolished. **1975:** Papua New Guinea became independent. **1978:** Northern Territory achieved self-government. **1986:** The Australia Act was passed by British Parliament, eliminating the last vestiges of British legal authority in Australia. **1988:** A Free Trade Agreement was signed with New Zealand. **1992:** The Citizenship Act removed the oath of allegiance to the British crown. **1998:** John Howard's Liberal–National coalition government was re-elected. **1999:** Australians voted to keep the British queen as head of state, rather than become a republic. **2000:** Torrential rains in much of eastern Australia during November caused widespread flooding in rural areas.

Practical information

Visa requirements: UK: visa required. USA: visa required
Time difference: GMT +8/10
Chief tourist attractions: swimming and surfing on the Pacific beaches; skin-diving along the Great Barrier Reef; sailing from Sydney and other harbours; winter sports in the Australian Alps; summer sports in the Blue Mountains; Alice Springs and Ayers Rock in desert interior; unique wildlife
Major holidays: 1 January, 25 April, 25–26 December (except South Australia); variable: Good Friday, Easter Monday, Holy Saturday; additional days vary between states

AUSTRIA

Landlocked country in central Europe, bounded east by Hungary, south by Slovenia and Italy, west by Switzerland and Liechtenstein, northwest by Germany, north by the Czech Republic, and northeast by the Slovak Republic.

National name: *Republik Österreich/Republic of Austria*
Area: 83,859 sq km/32,367 sq mi
Capital: Vienna
Major towns/cities: Graz, Linz, Salzburg, Innsbruck, Klagenfurt
Physical features: landlocked mountainous state, with Alps in west and south (Austrian Alps, including Grossglockner and Brenner and Semmering passes, Lechtaler and Allgauer Alps north of River Inn, Carnic Alps on Italian border) and low relief in east where most of the population is concentrated; River Danube

Red and white have been Austria's national colours for over 800 years. Effective date: 27 April 1984.

Government

Head of state: Thomas Klestil from 1992
Head of government: Wolfgang Schüssel from 2000
Political system: liberal democracy
Political executive: parliamentary
Administrative divisions: nine provinces
Political parties: Social Democratic Party of Austria (SPÖ), democratic socialist; Austrian People's Party (ÖVP), progressive centrist; Freedom (formerly Freedom Party of Austria: FPÖ), right wing; United Green Party of Austria (VGÖ), conservative ecological; Green Alternative Party (ALV), radical ecological; Liberal Forum, moderately left wing; Communist Party of Austria
Armed forces: 45,500 (1998)
Conscription: 6 months
Death penalty: abolished in 1968
Defence spend: (% GDP) 0.8 (1998)
Education spend: (% GNP) 5.7 (1996)
Health spend: (% GDP) 8.3 (1997)

Economy and resources

Currency: schilling
GDP: (US$) 208.9 billion (1999)
Real GDP growth: (% change on previous year) 2.2 (1999)
GNP: (US$) 210 billion (1999)
GNP per capita (PPP): (US$) 23,808 (1999)
Consumer price inflation: 0.6% (1999)
Unemployment: 4.4% (1998)
Major trading partners: EU, Switzerland, USA, Japan, Eastern Europe
Resources: lignite, iron, kaolin, gypsum, talcum, magnesite, lead, zinc, forests
Industries: raw and rolled steel, machinery, cellulose, paper, cardboard, cement, fertilizers, viscose staple yarn, sawn timber, flat glass, salt, sugar, milk, margarine
Exports: dairy products, food products, wood and paper products, machinery and transport equipment, metal and metal products, chemical products. Principal market for exports: Germany 34.8% (1999)
Imports: petroleum and petroleum products, food and live animals, chemicals and related products, textiles, clothing. Principal source: Germany 41.7% (1999)
Arable land: 17.2% (1996)
Agricultural products: wheat, barley, rye, oats, potatoes, maize, sugar beet; dairy products

Population and society

Population: 8,211,000 (2000 est)
Population growth rate: 0.5% (1995–2000); 0.2% (2000–05)
Population density: (per sq km) 98 (1999 est)
Urban population: (% of total) 65 (2000 est)
Age distribution: (% of total population) 0–14 17%, 15–59 63%, 60+ 20% (2000 est)
Ethnic groups: 99% German, 0.3% Croatian, 0.2% Slovene
Language: German (official)
Religion: Roman Catholic 78%, Protestant 5%
Education: (compulsory years) 9
Literacy rate: 99% (men); 99% (women) (2000 est)
Labour force: 6.8% agriculture, 30.3% industry, 62.9% services (1997)
Life expectancy: 74 (men); 80 (women) (1995–2000)
Child mortality rate: (under 5, per 1,000 live births) 7 (1995–2000)
Physicians: 1 per 253 people (1996)
Hospital beds: 1 per 109 people (1996)
TV sets: (per 1,000 people) 516 (1998)
Radios: (per 1,000 people) 753 (1997)
Internet users: (per 10,000 people) 1,039.5 (1999)
Personal computer users: (per 100 people) 25.7 (1999)

Transport

Airports: international airports: Vienna (Wien-Schwechat), Graz (Thalerhof), Innsbruck (Kranebitten), Klagenfurt (Wörthersee), Linz (Hörsching), Salzburg (Maxglam); domestic services between the above; total passenger km: 10,047 million (1997 est)
Railways: total length: 6,185 km/3,843 mi; total passenger km: 8,647 million (1997)
Roads: total road network: 200,000 km/124,280 mi, of which 100% paved (1997); passenger cars: 468.4 per 1,000 people (1997)

Chronology

14 BC: Country south of River Danube conquered by Romans. **5th century AD:** The region was occupied by Vandals, Huns, Goths, Lombards, and Avars. **791:** Charlemagne conquered the Avars and established East Mark, the nucleus of the future Austrian Empire. **976:** Holy Roman Emperor Otto II granted East Mark to House of Babenburg, which ruled until 1246. **1282:** Holy Roman Emperor Rudolf of Habsburg seized Austria and invested his son as its duke; for over 500 years most rulers of Austria were elected Holy Roman Emperor. **1453:** Austria became an archduchy. **1519–56:** Emperor Charles V was both archduke of Austria and king of Spain; the Habsburgs were dominant in Europe. **1526:** Bohemia came under Habsburg rule. **1529:** Vienna was besieged by the Ottoman Turks. **1618–48:** Thirty Years' War: Habsburgs weakened by failure to secure control over Germany. **1683:** Polish-Austrian force led by Jan Sobieski defeated the Turks at Vienna. **1699:** Treaty of Karlowitz: Austrians expelled the Turks from Hungary, which came under Habsburg rule.

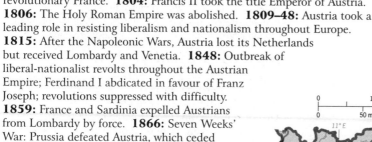

1713: By the Treaty of Utrecht, Austria obtained the Spanish Netherlands (Belgium) and political control over most of Italy. **1740–48:** War of Austrian Succession: Prussia (supported by France and Spain) attacked Austria (supported by Holland and England) on the pretext of disputing rights of Maria Theresa; Austria lost Silesia to Prussia. **1772:** Austria joined in partition of Poland, annexing Galicia. **1780–90:** 'Enlightened despotism': Joseph II tried to impose radical reforms. **1792:** Austria went to war with revolutionary France. **1804:** Francis II took the title Emperor of Austria.

1806: The Holy Roman Empire was abolished. **1809–48:** Austria took a leading role in resisting liberalism and nationalism throughout Europe.

1815: After the Napoleonic Wars, Austria lost its Netherlands but received Lombardy and Venetia. **1848:** Outbreak of liberal-nationalist revolts throughout the Austrian Empire; Ferdinand I abdicated in favour of Franz Joseph; revolutions suppressed with difficulty. **1859:** France and Sardinia expelled Austrians from Lombardy by force. **1866:** Seven Weeks' War: Prussia defeated Austria, which ceded Venetia to Italy. **1867:** Austria conceded equality to Hungary within the dual monarchy of Austria-Hungary. **1878:** Treaty of Berlin: Austria-Hungary occupied Bosnia-Herzegovina; annexed in 1908. **1914:** Archduke Franz Ferdinand, the heir to the throne, was assassinated by a Serbian nationalist; Austria-Hungary invaded Serbia, precipitating World War I. **1916:** Death of Franz Joseph; succeeded by Karl I. **1918:** Austria-Hungary collapsed in military defeat; empire dissolved; republic proclaimed. **1919:** Treaty of St Germain reduced Austria to its present boundaries and prohibited union with Germany. **1934:** Political instability culminated in brief civil war; right-wingers defeated socialists. **1938:** The *Anschluss*: Nazi Germany incorporated Austria into the Third Reich. **1945:** Following World War II, the victorious Allies divided Austria into four zones of occupation (US, British, French, and Soviet); the Second Republic was established under Karl Renner. **1955:** Austrian State Treaty ended occupation; Austria regained independence on condition of neutrality. **1960–70s:** Austria experienced rapid industrialization and prosperity. **1986:** Kurt Waldheim was elected president, despite allegations of war crimes during World War II. This led to some diplomatic isolation until Waldheim's replacement by Thomas Klestil in 1992. **1995:** Austria became a full member of the European Union (EU). **1998:** NATO membership was ruled out. **2000:** A new coalition government was elected, made up of the conservative People's Party, led by Wolfgang Schüssel, and the far-right Freedom Party, led by Jörg Haider. This was met with protests from across Europe and the imposition of diplomatic sanctions. Haider resigned in May, but his party remained part of the coalition. In September, sanctions were lifted after a favourable report on the country's human rights record.

Practical information

Visa requirements: UK: visa not required. USA: visa not required
Time difference: GMT +1
Chief tourist attractions: mountain scenery, enjoyed by visitors in both summer and winter; arts festivals at cultural centres of Vienna and Salzburg
Major holidays: 1, 6 January, 1 May, 15 August, 26 October, 1 November, 8, 24–26 December; variable: Ascension Thursday, Corpus Christi, Easter Monday, Whit Monday

AZERBAIJAN

Located in western Asia, bounded south by Iran, east by the Caspian Sea, west by Armenia and Georgia, and north by Russia.

National name: *Azärbaycan Respublikasi/Republic of Azerbaijan*
Area: 86,600 sq km/33,436 sq mi
Capital: Baku
Major towns/cities: Gäncä, Sumqayit, Nakhichevan, Xankändi, Mingacevir
Physical features: Caspian Sea with rich oil reserves; the country ranges from semidesert to the Caucasus Mountains

The emblem recalls the flag of Turkey, a long-standing ally. The points of the star represent the eight Turkic tribes of Azerbaijan. Effective date: 5 February 1991.

Government

Head of state: Geidar Aliyev from 1993
Head of government: Artur Rasizade from 1996
Political system: authoritarian nationalist
Political executive: unlimited presidency
Administrative divisions: 59 regions, 11 cities, and one autonomous republic (Nakhchyuan)
Political parties: Popular Front of Azerbaijan (FPA), democratic nationalist; New Azerbaijan, ex-communist; Communist Party of Azerbaijan (banned 1991–93); Muslim Democratic Party (Musavat), Islamic, pro-Turkic unity
Armed forces: 72,200 (1998)
Conscription: military service is for 17 months
Death penalty: abolished in 1998
Defence spend: (% GDP) 4.6 (1998)
Education spend: (% GNP) 3.0 (1997)
Health spend: (% GDP) 2.9 (1997)

Economy and resources

Currency: manat (replaced Russian rouble in 1993)
GDP: (US$) 4.5 billion (1999)
Real GDP growth: (% change on previous year) 7.2 (1999)
GNP: (US$) 4.4 billion (1999)
GNP per capita (PPP): (US$) 2,322 (1999)
Consumer price inflation: –8.6% (1999)
Unemployment: 1.1% (1997)
Foreign debt: (US$) 1 billion (1999)
Major trading partners: Italy, Russia, Turkey, Georgia, USA, Iran, Germany
Resources: petroleum, natural gas, iron ore, aluminium, copper, barytes, cobalt, precious metals, limestone, salt
Industries: petroleum extraction and refining, chemicals, petrochemicals, construction, machinery, food processing, textiles, timber
Exports: refined petroleum products, machinery, food products, textiles. Principal market: Italy 33.7% (1999)
Imports: industrial raw materials, processed food, machinery. Principal source: Russia 21.9% (1999)

Arable land: 18.5% (1996)
Agricultural products: grain, grapes and other fruit, vegetables, cotton, silk, tobacco; livestock rearing (cattle, sheep, and goats); fisheries (about 10 tonnes of caviar are produced annually); silkworm breeding

Population and society

Population: 7,734,000 (2000 est)
Population growth rate: 0.5% (1995–2000)
Population density: (per sq km) 89 (1999 est)
Urban population: (% of total) 57 (2000 est)
Age distribution: (% of total population) 0–14 29%, 15–59 63%, 60+ 8% (2000 est)
Ethnic groups: 83% of Azeri descent, 6% Russian, 6% Armenian
Language: Azeri (official), Russian
Religion: Shiite Muslim 68%, Sunni Muslim 27%, Russian Orthodox 3%, Armenian Orthodox 2%
Education: (compulsory years) 11
Literacy rate: 96% (men); 96% (women) (1997 est)
Labour force: 33.7% agriculture, 24.3% industry, 42% services (1991)
Life expectancy: 66 (men); 74 (women) (1995–2000)
Child mortality rate: (under 5, per 1,000 live births) 50 (1995–2000)
Physicians: 1 per 264 people (1998)
Hospital beds: 1 per 97 people (1998)
TV sets: (per 1,000 people) 254 (1998)
Radios: (per 1,000 people) 23 (1997)
Internet users: (per 10,000 people) 10.4 (1999)

Transport

Airports: international airports: Baku; total passenger km: 1,650 million (1995)
Railways: total length: 2,122 km/1,319 mi; total passenger km: 791 million (1995)
Roads: total road network: 45,870 km/28,504 mi, of which 93.8% paved (1997); passenger cars: 35.5 per 1,000 people (1997)

Chronology

4th century BC: Established as an independent state for the first time by Atrophates, a vassal of Alexander III of Macedon. **7th century AD:** Spread of Islam. **11th century:** Immigration by Oghuz Seljuk peoples, from the steppes to the northeast. **13th–14th centuries:** Incorporated within Mongol Empire; the Mongol ruler Tamerlane had his capital at Samarkand. **16th century:** Baku besieged and incorporated within Ottoman Empire, before falling under Persian dominance. **1805:** Khanates (chieftaincies), including Karabakh and Shirvan, which had won independence from Persia, gradually became Russian protectorates, being confirmed by the Treaty of Gulistan, which concluded the 1804–13 First Russo-Iranian War. **1828:** Under the Treaty of Turkmenchai, which concluded the Second Russo-Iranian War begun in 1826, Persia was granted control over southern and Russia over northern Azerbaijan. **late 19th century:** The petroleum industry developed, resulting in a large influx of Slav immigrants to Baku. **1917–18:** Member of anti-Bolshevik Transcaucasian Federation. **1918:** Became an independent republic. **1920:** Occupied by Red Army and subsequently forcibly secularized. **1922–36:** Became part of the Transcaucasian Federal Republic with Georgia and Armenia. **early 1930s:** Peasant uprisings against agricultural collectivization and Stalinist purges of the local Communist Party. **1936:** Became a constituent republic of the USSR. **late 1980s:** Growth in nationalist sentiment, taking advantage of the *glasnost* initiative of the reformist Soviet leader Mikhail Gorbachev. **1988:** Riots followed the request of Nagorno-Karabakh, an Armenian-peopled enclave within Azerbaijan, for transfer to Armenia. **1989:** Nagorno-Karabakh was placed under direct rule from Moscow; civil war broke out with Armenia over Nagorno-Karabakh. **1990:** Soviet troops were dispatched to Baku to restore order amid calls for secession from the USSR. **1991:** Independence was declared after the collapse of an anti-Gorbachev coup in Moscow, which had been supported by the Azeri communist leadership. Azerbaijan joined the new Commonwealth of Independent States (CIS); Nagorno-Karabakh declared independence. **1992:** Azerbaijan was admitted into the United Nations (UN). **1993:** Nagorno-Karabakh was overtaken by Armenian forces. **1995:** An attempted coup was foiled. A market-centred economic reform programme was introduced. **1997:** There was border fighting with Armenia. The extraction of oil from oilfields in the Caspian Sea began, operated by a consortium of 11 international oil companies. **1998:** A new pro-government grouping, Democratic Azerbaijan, was formed. Heidar Aliyev was re-elected president in a disputed poll. A Nagorno-Karabakh peace plan was rejected. **2000:** Heidar Aliyev was re-elected, although foreign observers denounced the election as deeply flawed.

Practical information

Visa requirements: UK: visa required. USA: visa required
Time difference: GMT +4

Chief tourist attractions: resorts on Caspian Sea and on Apsheron peninsula, near Baku
Major holidays: 1 January, 8 March, 28 May, 9, 18 October, 17 November, 31 December

BAHAMAS

A group of about 700 islands and about 2,400 uninhabited islets in the Caribbean, 80 km/50 mi from the southeast coast of Florida.

National name: *Commonwealth of the Bahamas*
Area: 13,880 sq km/5,383 sq mi
Capital: Nassau (on New Providence island)
Major towns/cities: Freeport City (on Grand Bahama)
Physical features: comprises 700 tropical coral islands and about
1,000 cays; the Exumas are a narrow spine of 365 islands; only 30 of the

Blue represents the Caribbean Sea. Yellow stands for the golden beaches. Effective date: 10 July 1973.

desert islands are inhabited; Blue Holes of Andros, the world's longest and deepest submarine caves
Principal islands: Andros, Grand Bahama, Abaco, Eleuthera, New Providence, Berry Islands, Bimini Islands, Great Inagua, Acklins Island, Exuma Islands, Mayguana, Crooked Island, Long Island, Cat Islands, Rum Cay, Watling (San Salvador) Island, Inagua Islands

Government

Head of state: Queen Elizabeth II from 1973, represented by Governor General Orville Turnquest from 1995
Head of government: Hubert Ingraham from 1992
Political system: liberal democracy
Political executive: parliamentary
Administrative divisions: 21 districts
Political parties: Progressive Liberal Party (PLP), centrist; Free National Movement (FNM), left of centre
Armed forces: 900 (1998) and 2,300 paramilitary forces
Conscription: military service is voluntary
Death penalty: retained and used for ordinary crimes
Defence spend: (% GDP) 0.6 (1998)
Education spend: (% GNP) 3.9 (1993–94)
Health spend: (% GDP) 5.9 (1997)

Economy and resources

Currency: Bahamian dollar
GDP: (US$) 4.56 billion (1999)
Real GDP growth: (% change on previous year) 6 (1999)
GNP: (US$) 4.1 billion (1999 est)
GNP per capita (PPP): (US$) 13,955 (1999 est)
Consumer price inflation: 1.3% (1999)
Unemployment: 9.8% (1997 est)
Foreign debt: (US$) 349 million (1998)
Major trading partners: USA, Italy, Switzerland, UK, Japan, Denmark
Resources: aragonite (extracted from seabed), chalk, salt
Industries: pharmaceutical chemicals, salt, rum, beer, cement, shipping, financial services, tourism
Exports: foodstuffs (fish), oil products and transhipments, chemicals, rum, salt. Principal market: USA 22.3% (1998)
Imports: machinery and transport equipment, basic manufactures, petroleum and products, chemicals. Principal source: USA 27.3% (1998)
Arable land: 1% (1995)

Agricultural products: sugar cane, cucumbers, tomatoes, pineapples, papayas, mangoes, avocados, limes and other citrus fruit; commercial fishing (conches and crustaceans)

Population and society

Population: 307,000 (2000 est)
Population growth rate: 1.8% (1995–2000)
Population density: (per sq km) 22 (1999 est)
Urban population: (% of total) 89 (2000 est)
Age distribution: (% of total population) 0–14 31%, 15–59 61%, 60+ 8% (2000 est)
Ethnic groups: about 85% of the population is of African origin, remainder mainly British, American, and Canadian
Language: English (official), Creole
Religion: Christian 94% (Baptist 32%, Roman Catholic 19%, Anglican 20%, other Protestant 23%)
Education: (compulsory years) 10
Literacy rate: 95% (men); 96% (women) (2000 est)
Labour force: 50% of population: 6.5% agriculture, 12.1% industry, 81.4% services (1993)
Life expectancy: 71 (men); 77 (women) (1995–2000)
Child mortality rate: (under 5, per 1,000 live births) 18 (1995–2000)
Physicians: 1 per 800 people (1994 est)
Hospital beds: 1 per 267 people (1993 est)
TV sets: (per 1,000 people) 226 (1997)
Radios: (per 1,000 people) 726 (1997)
Internet users: (per 10,000 people) 497.8 (1999)

Transport

Airports: international airports: Nassau, Freeport City, Moss Town; four domestic airports serve internal chartered flights; total passenger km: 220 million (1995)
Railways: none
Roads: total road network: 2,500 km/1,553 mi, of which 57.4% paved (1996 est); passenger cars: 161 per 1,000 people (1996 est)

Chronology

8th–9th centuries AD: Arawak Indians driven northwards to the islands by the Caribs. **1492:** Visited by Christopher Columbus; Arawaks deported to provide cheap labour for the gold and silver mines of Cuba and Hispaniola (Haiti). **1629:** King Charles I of England granted the islands to Robert Heath. **1666:** The colonization of New Providence island began. **1783:** Recovered after brief Spanish occupation and became a British colony, being settled during the American War of Independence by American loyalists, who brought with them black slaves. **1838:** Slaves were emancipated. **from 1950s:** Major development of the tourist trade. **1964:** Became internally self-governing. **1967:** First national assembly elections. **1973:** Full independence was achieved within the British Commonwealth. **1992:** A centre-left Free National Movement (FNM) led by Hubert Ingraham won an absolute majority in elections.

Practical information

Visa requirements: UK: visa not required. USA: visa not required
Time difference: GMT –5
Chief tourist attractions: mild climate and beautiful beaches

Major holidays: 1 January, 10 July, 25–26 December; variable: Good Friday, Easter Monday, Whit Monday, Labour Day (June), Emancipation (August), Discovery (October)

BAHRAIN

A group of islands in the Persian Gulf, between Saudi Arabia and Iran.

National name: *Dawlat al-Bahrayn/State of Bahrain*
Area: 688 sq km/266 sq mi
Capital: Al Manamah (on Bahrain island)
Major towns/cities: Al Muharraq, Jidd Hafs, Isa Town, Rifa'a, Sitra
Major ports: Mina Sulman
Physical features: archipelago of 35 islands in Arabian Gulf, composed largely of sand-covered limestone; generally poor and infertile soil; flat and hot; causeway linking Bahrain to mainland Saudi Arabia

White was added to the flag to identify Bahrain as a friendly state. Red was the traditional colour of the Kharijite Sect. Effective date: 19 August 1972.

Government

Head of state: Sheikh Hamad bin Isa al-Khalifa from 1999
Head of government: Sheikh Khalifa bin Salman al-Khalifa from 1970
Political system: absolutist
Political executive: absolute
Administrative divisions: 12 municipalities
Political parties: not permitted
Armed forces: 11,000 (1998)
Conscription: military service is voluntary
Death penalty: retained and used for ordinary crimes
Defence spend: (% GDP) 6.7 (1998)
Education spend: (% GNP) 4.4 (1997)
Health spend: (% GDP) 4.4 (1997)

Economy and resources

Currency: Bahraini dinar
GDP: (US$) 6.8 billion (1999 est)
Real GDP growth: (% change on previous year) 2.5 (1999)
GNP: (US$) 5.1 billion (1999 est)
GNP per capita (PPP): (US$) 11,527 (1999 est)
Consumer price inflation: –1.3% (1999)
Unemployment: 15% (1997 est)
Foreign debt: (US$) 2.7 billion (1999)
Major trading partners: India, Saudi Arabia, Japan, South Korea, Australia, USA, Singapore, UK
Resources: petroleum and natural gas
Industries: petroleum refining, aluminium smelting, petrochemicals, shipbuilding and repairs, electronics assembly, banking
Exports: petroleum and petroleum products, aluminium, chemicals, textiles. Principal market: India 17% (1998)
Imports: crude petroleum, machinery and transport equipment, chemicals, basic manufactures. Principal source: Saudi Arabia 57.4% (1998)
Arable land: 1.4% (1995)
Agricultural products: dates, tomatoes, melons, vegetables; poultry products and fishing

Population and society

Population: 617,000 (2000 est)
Population growth rate: 2% (1990–95); 1.5% (1995–2025)
Population density: (per sq km) 882 (1999 est)
Urban population: (% of total) 92 (2000 est)
Age distribution: (% of total population) 0–14 30%, 15–59 65%, 60+ 5% (2000 est)
Ethnic groups: 63% Bahraini, 13% Asian, 10% other Arab, 8% Iranian
Language: Arabic (official), Farsi, English, Urdu
Religion: 85% Muslim (Shiite 60%, Sunni 40%), Christian; Islam is the state religion
Education: (compulsory years) 12
Literacy rate: 91% (men); 83% (women) (2000 est)
Labour force: 45% of population: 2% agriculture, 30% industry, 68% services (1990)
Life expectancy: 71 (men); 75 (women) (1995–2000)
Child mortality rate: (under 5, per 1,000 live births) 22 (1995–2000)
Physicians: 1 per 775 people (1991)
Hospital beds: 1 per 368 people (1991 est)
TV sets: (per 1,000 people) 462 (1997)
Radios: (per 1,000 people) 580 (1997)
Internet users: (per 10,000 people) 526.2 (1999)
Personal computer users: (per 100 people) 10.5 (1999)

Transport

Airports: international airports: Al Muharraq (Bahrain); total passenger km: 2,766 million (1995)
Railways: none
Roads: total road network: 3,103 km/1,928 mi, of which 76.5% paved (1997); passenger cars: 241 per 1,000 people (1997)

Chronology

4th century AD: Became part of Persian (Iranian) Sassanian Empire. **7th century:** Adopted Islam. **8th century:** Came under Arab Abbasid control. **1521:** Seized by Portugal and held for eight decades, despite local unrest. **1602:** Fell under the control of a Persian Shiite dynasty. **1783:** Persian rule was overthrown and Bahrain became a sheikdom under the Sunni Muslim al-Khalifa dynasty, which originated from the same tribal federation, the Anaza, as the al-Saud family who now rule Saudi Arabia. **1816–20:** Friendship and peace treaties were signed with Britain, which sought to end piracy in the Gulf. **1861:** Became British protectorate; government shared between the ruling sheikh (Arab leader) and a British adviser. **1923:** British influence increased when Sheikh Isa al-Khalifa was deposed and Charles Belgrave was appointed as the dominating 'adviser' to the new ruler. **1928:** Sovereignty was claimed by Persia (Iran). **1930s:** Oil was discovered, providing the backbone for the country's wealth. **1953–56:** Council for National Unity was formed by Arab nationalists, but was suppressed after large demonstrations against British participation in the Suez War. **1968:** Britain announced its intention to withdraw its forces. Bahrain formed, with Qatar and the Trucial States of the United Arab Emirates, the Federation of Arab Emirates. **1970:** Iran accepted a United Nations (UN) report showing that Bahrain's inhabitants preferred independence to Iranian control. **1971:** Qatar and the Trucial States withdrew from the federation; Bahrain became an independent state under Sheikh Isa bin Sulman al-Khalifa, who assumed the title of emir. **1973:** A new constitution was adopted. **1975:** The national assembly was dissolved and political activists driven underground. The emir and his family assumed virtually absolute power. **early 1980s:** Tensions between the Sunni and Shiite Muslim communities were heightened by the Iranian Shiite Revolution of 1979. **1986:** A causeway opened linking the island with Saudi Arabia. **1991:** Bahrain joined a UN coalition that ousted Iraq from its occupation of Kuwait, and signed a defence cooperation agreement with the USA. **1995:** Prodemocracy demonstrations were violently suppressed, with 11 deaths. **1999:** Sheikh Hamad became Emir and head of state. **2000:** A new Shura (consultative council) includes women and non-Muslims for the first time.

Practical information

Visa requirements: UK: visa not required. USA: visa required
Time difference: GMT +3
Chief tourist attractions: Bahrain is the site of the ancient trading civilization of Dilmun, and there are several sites of archaeological importance
Major holidays: 1 January, 16 December; variable: Eid-ul-Adha, Ashora, end of Ramadan, New Year (Muslim), Prophet's Birthday

BANGLADESH

FORMERLY EAST BENGAL (UNTIL 1955), EAST PAKISTAN (1955–71)

Located in southern Asia, bounded north, west, and east by India, southeast by Myanmar, and south by the Bay of Bengal.

National name: *Gana Prajatantri Bangladesh/People's Republic of Bangladesh*
Area: 144,000 sq km/55,598 sq mi
Capital: Dhaka
Major towns/cities: Rajshahi, Khulna, Chittagong, Sylhet, Rangpur, Narayanganj
Major ports: Chittagong, Khulna
Physical features: flat delta of rivers Ganges (Padma) and Brahmaputra (Jamuna), the largest estuarine delta in the world; annual rainfall of 2,540 mm/100 in; some 75% of the land is less than 3 m/10 ft above sea level; hilly in extreme southeast and northeast

The red disc, set towards the hoist, recalls the fight for independence. Green represents Islam, fertility, and the country's youth. Effective date: 25 January 1972.

Government

Head of state: Shahabuddin Ahmed from 1996
Head of government: Sheikh Hasina Wazed from 1996
Political system: emergent democracy
Political executive: parliamentary
Administrative divisions: 64 districts within four divisions
Political parties: Bangladesh Nationalist Party (BNP), Islamic, right of centre; Awami League (AL), secular, moderate socialist; Jatiya Dal (National Party), Islamic nationalist
Armed forces: 121,000 (1998)
Conscription: military service is voluntary
Death penalty: retained and used for ordinary crimes
Defence spend: (% GDP) 1.9 (1998)
Education spend: (% GNP) 2.9 (1996)
Health spend: (% GDP) 4.9 (1997)

Economy and resources

Currency: taka
GDP: (US$) 45.8 billion (1999)
Real GDP growth: (% change on previous year) 5.2 (1999 est)
GNP: (US$) 47 billion (1999)
GNP per capita (PPP): (US$) 1,475 (1999)
Consumer price inflation: 6.3% (1999)
Unemployment: 2.5% (1996)
Foreign debt: (US$) 17.1 billion (1999 est)
Major trading partners: USA, India, Germany, China, Japan, Singapore, UK, Italy, France, Hong Kong
Resources: natural gas, coal, limestone, china clay, glass sand
Industries: textiles, food processing, industrial chemicals, petroleum refineries, cement
Exports: raw jute and jute goods, tea, clothing, leather and leather products, shrimps and frogs' legs. Principal market: USA 32.3% (1999)
Imports: wheat, crude petroleum and petroleum products, pharmaceuticals, cement, raw cotton, machinery and transport equipment. Principal source: India 13.5% (1999)

Arable land: 65.3% (1996)
Agricultural products: rice, jute, wheat, tobacco, tea; fishing and fish products

Population and society

Population: 129,155,000 (2000 est)
Population growth rate: 1.7% (1995–2000)
Population density: (per sq km) 881 (1999 est)
Urban population: (% of total) 25 (2000 est)
Age distribution: (% of total population) 0–14 35%, 15–59 60%, 60+ 5% (2000 est)
Ethnic groups: 98% of Bengali descent, quarter of a million Bihari, and around 1 million belonging to 'tribal' communities
Language: Bengali (official), English
Religion: Muslim 88%, Hindu 11%; Islam is the state religion
Education: (compulsory years) 5
Literacy rate: 52% (men); 30% (women) (2000 est)
Labour force: 59.1% agriculture, 9.5% industry, 31.4% services (1997)
Life expectancy: 58 (men); 58 (women) (1995–2000)
Child mortality rate: (under 5, per 1,000 live births) 111 (1995–2000)
Physicians: 1 per 12,884 people (1993 est)
Hospital beds: 1 per 5,479 people (1993 est)
TV sets: (per 1,000 people) 7 (1997)
Radios: (per 1,000 people) 50 (1997)
Internet users: (per 10,000 people) 2.4 (1999)
Personal computer users: (per 100 people) 0.1 (1999)

Transport

Airports: international airports: Dhaka (Zia), Chittagong, Sylhet; seven domestic airports; total passenger km: 2,129 million (1997 est)
Railways: total length: 2,706 km/1,682 mi; total passenger km: 3,754 million (1997)
Roads: total road network: 204,022 km/126,678 mi, of which 12.3% paved (1996); passenger cars: 1.0 per 1,000 people (1997)

Chronology

c. **1000 BC:** Arrival of Bang tribe in lower Ganges valley, establishing the kingdom of Banga (Bengal). **8th–12th centuries AD:** Bengal was ruled successively by the Buddhist Pala and Hindu Senha dynasties. **1199:** Bengal was invaded and briefly ruled by the Muslim Khiljis from Central Asia. **1576:** Bengal was conquered by the Muslim Mogul emperor Akbar. **1651:** The British East India Company established a commercial factory in Bengal. **1757:** Bengal came under de facto British rule after Robert Clive defeated the nawab (ruler) of Bengal at Battle of Plassey. **1905–12:** Bengal was briefly partitioned by the British Raj into a Muslim-dominated east and Hindu-dominated west. **1906:** The Muslim League (ML) was founded in Dhaka. **1947:** Bengal was formed into an eastern province of Pakistan on the partition of British India, with the ML administration in power. **1954:** The opposition United Front, dominated by the Awami League (AL) and campaigning for East Bengal's autonomy, trounced the ML in elections. **1955:** East Bengal was renamed East Pakistan. **1966:** Sheikh Mujibur Rahman of AL announced a Six-Point Programme of autonomy for East Pakistan. **1970:** 500,000 people were killed in a cyclone. The pro-autonomy AL secured an electoral victory in East Pakistan. **1971:** Bangladesh ('land of the Bangla speakers') emerged as an independent nation after a bloody civil war with Indian military intervention on the side of East Pakistan; 10 million refugees fled to India. **1974:** Hundreds of thousands died in a famine; a state of emergency was declared. **1975:** Martial law was imposed. **1978–79:** Elections were held and civilian rule restored. **1982:** Martial law was reimposed after a military coup. **1986:** Elections were held but disputed. Martial law ended. **1987:** A state of emergency was declared in response to demonstrations and violent strikes. **1988:** Assembly elections were boycotted by the main opposition parties. The state of emergency was lifted. Islam was made the state religion. Monsoon floods left 30 million people homeless and thousands dead. **1991:** A cyclone killed around 139,000 people and left up to 10 million homeless. Parliamentary government was restored. **1996:** Power was handed to a neutral caretaker government. A general election was won by the AL, led by Sheikh Hasina Wazed, and Shahabuddin Ahmed was appointed president. The BNP boycotted parliament. An agreement was made with India on the sharing of River Ganges water. **1998:** The BNP ended its boycott of parliament. Two-thirds of Bangladesh was devastated by floods; 1,300 people were killed. Opposition-supported general strikes sought the removal of Sheikh Hasina's government. **2000:** Ex-president Hussain Mohammad Ershad was fined US$1 million and sentenced to five years' imprisonment for corruption by the Dhaka high court.

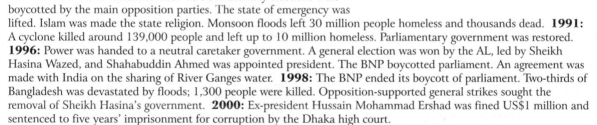

Practical information

Visa requirements: UK: visa required. USA: visa not required for a tourist visit of up to 15 days
Time difference: GMT +6
Chief tourist attractions: cities of Dhaka and Chittagong; Cox's Bazaar on the Bay of Bengal, the world's longest beach (120 km/74.5 mi); Tekhaf, at the southernmost point of Bangladesh
Major holidays: 21 February, 26 March, 1 May, 1 July, 7 November, 16, 25, 31 December; variable: Eid-ul-Adha, end of Ramadan, New Year (Bengali), New Year (Muslim), Prophet's Birthday, Jumat-ul-Wida (May), Shab-e-Barat (April), Buddah Purnima (April/May), Shab-I-Qadr (May), Durga-Puza (October)

BARBADOS

Island in the Caribbean, one of the Lesser Antilles, located about 483 km/300 mi north of Venezuela.

Area: 430 sq km/166 sq mi
Capital: Bridgetown
Major towns/cities: Speightstown, Holetown, Oistins
Physical features: most easterly island of the West Indies; surrounded by coral reefs; subject to hurricanes June–November; highest point Mount Hillaby 340 m/1,115 ft

Blue represents the sea and the sky. The points of the trident represent the three principles of democracy: government of, for, and by the people. Effective date: 30 November 1966.

Government

Head of state: Queen Elizabeth II from 1966, represented by Governor General Sir Clifford Straughn Husbands from 1996
Head of government: Owen Arthur from 1994
Political system: liberal democracy
Political executive: parliamentary
Administrative divisions: 11 parishes
Political parties: Barbados Labour Party (BLP), moderate left of centre; Democratic Labour Party (DLP), moderate left of centre; National Democratic Party (NDP), centrist
Armed forces: 600 (1998)
Conscription: military service is voluntary
Death penalty: retained and used for ordinary crimes
Defence spend: (% GDP) 0.5 (1998)
Education spend: (% GNP) 7.2 (1996)
Health spend: (% GDP) 7.3 (1997)

Economy and resources

Currency: Barbados dollar
GDP: (US$) 2.4 billion (1999 est)
Real GDP growth: (% change on previous year) 3.2 (1999 est)
GNP: (US$) 2.1 billion (1998)
GNP per capita (PPP): (US$) 12,260 (1998)
Consumer price inflation: 3% (1999 est)
Unemployment: 13.2% (1998 est)
Foreign debt: (US$) 582 million (1999 est)
Major trading partners: UK, USA, Trinidad and Tobago, Japan, Canada, Jamaica
Resources: petroleum and natural gas
Industries: sugar refining, food processing, industrial chemicals, beverages, tobacco, household appliances, electrical components, plastic products, electronic parts, tourism
Exports: sugar, molasses, syrup-rum, chemicals, electrical components. Principal market: UK 14.8% (1998)
Imports: machinery, foodstuffs, motor cars, construction materials, basic manufactures. Principal source: USA 30.7% (1998)
Arable land: 37.2% (1995)

Agricultural products: sugar cane, cotton, sweet potatoes, yams, carrots, and other vegetables; fishing (740 fishing vessels employed in 1994)

Population and society

Population: 270,000 (2000 est)
Population growth rate: 0.5% (1995–2000)
Population density: (per sq km) 625 (1999 est)
Urban population: (% of total) 50 (2000 est)
Age distribution: (% of total population) 0–14 21%, 15–59 66%, 60+ 13% (2000 est)
Ethnic groups: about 80% of African descent, about 16% mixed ethnicity, and 4% of European origin (mostly British)
Language: English (official), Bajan (a Barbadian English dialect)
Religion: 40% Anglican, 8% Pentecostal, 6% Methodist, 4% Roman Catholic
Education: (compulsory years) 12
Literacy rate: 98% (men); 97% (women) (1997)
Labour force: 5.3% agriculture, 19.2% industry, 75.5% services (1997)
Life expectancy: 74 (men); 79 (women) (1995–2000)
Child mortality rate: (under 5, per 1,000 live births) 14 (1995–2000)
Physicians: 1 per 1,100 people (1993)
Hospital beds: 1 per 137 people (1993)
TV sets: (per 1,000 people) 293 (1997)
Radios: (per 1,000 people) 915 (1997)
Internet users: (per 10,000 people) 222.8 (1999)
Personal computer users: (per 100 people) 7.8 (1999)

Transport

Airports: international airports: Bridgetown (Grantley Adams)
Railways: none
Roads: total road network: 1,650 km/1,029 mi, of which 95.9% paved (1996 est); passenger cars: 136 per 1,000 people (1996 est)

Chronology

1536: Visited by Portuguese explorer Pedro a Campos and the name Los Barbados ('The Bearded Ones') given in reference to its 'bearded' fig trees. Indigenous Arawak people were virtually wiped out, via epidemics, after contact with Europeans. **1627:** British colony established; developed as a sugar-plantation economy, initially on basis of black slaves brought in from West Africa. **1639:** The island's first parliament, the House of Assembly, was established. **1834:** The island's slaves were freed. **1937:** There was an outbreak of riots, followed by establishment of the Barbados Labour Party (BLP) by Grantley Adams, and moves towards a more independent political system. **1951:** Universal adult suffrage was introduced. The BLP won a general election. **1954:** Ministerial government was established, with BLP leader Adams as the first prime minister. **1955:** A group broke away from the BLP and formed the Democratic Labour Party (DLP). **1961:** Independence was achieved from Britain. **1966:** Barbados achieved full independence within the Commonwealth. **1967:** Barbados became a member of the United Nations (UN). **1972:** Diplomatic relations with Cuba were established. **1983:** Barbados supported the US invasion of Grenada. **1999:** The BLP gained a landslide victory in general elections, securing 26 of the 28 House of Assembly seats.

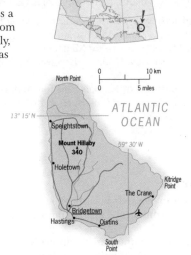

Practical information

Visa requirements: UK: visa not required (some visitors will require a business visa). USA: visa not required (some visitors will require a business visa)
Time difference: GMT –4
Chief tourist attractions: good climate; coral beaches; swimming; outdoor sports
Major holidays: 1 January, 30 November, 25–26 December; variable: Good Friday, Easter Monday, Whit Monday, Kadooment (August), May Holiday, United Nations (October)

BELARUS OR BYELORUSSIA OR BELORUSSIA

Located in east-central Europe, bounded south by Ukraine, east by
Russia, west by Poland, and north by Latvia and Lithuania.

National name: *Respublika Belarus/Republic of Belarus*
Area: 207,600 sq km/80,154 sq mi
Capital: Minsk (Belorussian Mensk)
Major towns/cities: Gomel, Vitsyebsk, Mahilyow, Babruysk, Hrodna,
Brest
Physical features: more than 25% forested; rivers Dvina, Dnieper and
its tributaries, including the Pripet and Beresina; the Pripet Marshes in
the east; mild and damp climate

The distinctive pattern in the hoist is designed to look
like woven cloth. Effective date: 7 June 1995.

Government

Head of state: Alexandr Lukashenko from 1994
Head of government: Uladzimir Yarmoshyn from
2000
Political system: authoritarian nationalist
Political executive: unlimited presidency
Administrative divisions: six regions (oblasts)
Political parties: Belarus Communist Party (BCP,
banned 1991–92); Belarus Patriotic Movement (BPM),
populist; Belorussian Popular Front (BPF;
Adradzhenne), moderate nationalist; Christian
Democratic Union of Belarus, centrist; Socialist Party
of Belarus, left of centre
Armed forces: 83,000 (1998)
Conscription: compulsory for 18 months
Death penalty: retained and used for ordinary crimes
Defence spend: (% GDP) 3.2 (1998)
Education spend: (% GNP) 2.9 (1997)
Health spend: (% GDP) 5.9 (1997)

Economy and resources

Currency: Belarus rouble, or zaichik
GDP: (US$) 25.7 billion (1999)
Real GDP growth: (% change on previous year) 3.4
(1999)
GNP: (US$) 26.8 billion (1999)
GNP per capita (PPP): (US$) 6,518 (1999)
Consumer price inflation: 294% (1999)
Unemployment: 2.3% (1998 est)
Foreign debt: (US$) 1.05 billion (1999 est)
Major trading partners: Russia, Ukraine,
Kazakhstan, Germany, Poland, Lithuania
Resources: petroleum, natural gas, peat, salt, coal,
lignite
Industries: machine building, metalworking,
electronics, chemicals, construction materials, food
processing, textiles
Exports: machinery, chemicals and petrochemicals,
iron and steel, light industrial goods. Principal market:
Russia 65.5% (1998)
Imports: petroleum, natural gas, chemicals, machinery,

processed foods. Principal source: Russia 53.8% (1998)
Arable land: 30% (1996)
Agricultural products: potatoes, grain, sugar beet;
livestock rearing (cattle and pigs) and dairy products.
Livestock sector accounts for approximately 60% of
agricultural output

Population and society

Population: 10,236,000 (2000 est)
Population growth rate: –0.3% (1995–2000)
Population density: (per sq km) 50 (1999 est)
Urban population: (% of total) 71 (2000 est)
Age distribution: (% of total population) 0–14 19%,
15–59 62%, 60+ 19% (2000 est)
Ethnic groups: 78% of Belorussian ('eastern Slav')
descent, 13% ethnic Russian, 4% Polish, 3% Ukranian,
1% Jewish
Language: Belorussian (official), Russian, Polish
Religion: 80% Eastern Orthodox; Baptist, Roman
Catholic Muslim, and Jewish minorities
Education: (compulsory years) 11
Literacy rate: 99% (men); 99% (women) (2000 est)
Labour force: 17.4% agriculture, 34.7% industry,
47.9% services (1997)
Life expectancy: 62 (men); 74 (women) (1995–2000)
Child mortality rate: (under 5, per 1,000 live births)
28 (1995–2000)
Physicians: 1 per 243 people (1995)
Hospital beds: 1 per 85 people (1995)
TV sets: (per 1,000 people) 315 (1997)
Radios: (per 1,000 people) 296 (1997)
Internet users: (per 10,000 people) 9.7 (1999)

Transport

Airports: international airports: Minsk; total
passenger km: 910 million (1997)
Railways: total length: 5,488 km/3,410 mi; total
passenger km: 12,909 million (1994)
Roads: total road network: 53,407 km/33, 237 mi, of
which 98.2% paved; passenger cars: 111 per 1,000
people (1997)

Chronology

5th–8th centuries: Settled by East Slavic tribes, ancestors of present-day Belorussians. **11th century:** Minsk was founded. **12th century:** Part of Kievan Russia, to the south, with independent Belarus state developing around Polotsk, on River Dvina. **14th century:** Incorporated within Slavonic Grand Duchy of Lithuania, to the west. **1569:** Union with Poland. **late 18th century:** Came under control of tsarist Russia as Belarussia ('White Russia'), following three partitions of Poland in 1772, 1793, and 1795. **1812:** Minsk was destroyed by French emperor Napoleon Bonaparte during his campaign against Russia. **1839:** The Belorussian Catholic Church was abolished. **1914–18:** Belarus was the site of fierce fighting between Germany and Russia during World War I. **1918–19:** Belarus was briefly independent from Russia. **1919–20:** Wars between Poland and Soviet Russia over control of Belarus. **1921:** West Belarus was ruled by Poland; East Belarus became a Soviet republic. **1930s:** Agriculture was collectivized despite peasant resistance; over 100,000 people, chiefly writers and intellectuals, shot in mass executions ordered by the Soviet dictator Joseph Stalin. **1939:** West Belarus was occupied by Soviet troops. **1941–44:** The Nazi occupation resulted in the death of 1.3 million people, including many Jews; Minsk was destroyed. **1945:** Belarus became a founding member of the United Nations (UN); much of West Belarus was incorporated into Soviet republic. **1950s–60s:** Large-scale immigration of ethnic Russians and 'Russification'. **1986:** Fallout from the nearby Chernobyl nuclear reactor in Ukraine rendered 20% of agricultural land unusable. **1989:** The Belorussian Popular Front was established as national identity was revived under the *glasnost* initiative of Soviet leader Mikhail Gorbachev. **1990:** Belorussian was established as the state language and republican sovereignty declared. **1991:** Independence was recognized by the USA; the Commonwealth of Independent States (CIS) was formed in Minsk. **1996:** An agreement on economic union was signed with Russia. Syargey Ling became prime minister. **1997:** There were prodemocracy demonstrations. **1998:** The Belarus rouble was devalued. A new left-wing and centrist political coalition was created. Food rationing was imposed as the economy deteriorated. Belarus signed a common policy with Russia on economic, foreign, and military matters. **2000:** President Lukashenka dismissed Prime Minister Syargey Ling and appointed Uladzimir Yarmoshyn. Lukashenka was re-elected in October, although foreign observers described the election as below international standards for fairness, and opposition leaders led popular protests.

Practical information

Visa requirements: UK: visa required. USA: visa required
Time difference: GMT +2

Chief tourist attractions: forests; lakes; wildlife
Major holidays: 1, 7 January, 8 March, 1, 9 May, 3, 27 July, 2 November, 25 December; variable: Good Friday, Easter Monday

BELGIUM

Located in Western Europe, bounded to the north by the Netherlands, to the northwest by the North Sea, to the south and west by France, and to the east by Luxembourg and Germany.

National name: *Royaume de Belgique* (French), *Koninkrijk België* (Flemish)/*Kingdom of Belgium*
Area: 30,510 sq km/11,779 sq mi
Capital: Brussels
Major towns/cities: Antwerp, Ghent, Liège, Charleroi, Bruges, Mons, Namur, Leuven
Major ports: Antwerp, Ostend, Zeebrugge
Physical features: fertile coastal plain in northwest, central rolling hills rise eastwards, hills and forest in southeast; Ardennes Forest; rivers Schelde and Meuse

Modelled on the French tricolour, the vertical stripes represent liberty and revolution. The almost square proportions of the flag are unusual. Effective date: 23 January 1831.

Government

Head of state: King Albert II from 1993
Head of government: Guy Verhofstadt from 1999
Political system: liberal democracy
Political executive: parliamentary
Administrative divisions: ten provinces within three regions (including the capital, Brussels)
Political parties: Flemish Christian Social Party (CVP), left of centre; French Social Christian Party (PSC), left of centre; Flemish Socialist Party (SP), left of centre; French Socialist Party (PS), left of centre; Flemish Liberal Party (PVV), moderate centrist; French Liberal Reform Party (PRL), moderate centrist; Flemish People's Party (VU), federalist; Flemish Vlaams Blok, right wing; Flemish Green Party (Agalev); French Green Party (Ecolo), ecological
Armed forces: 43,700 (1998)
Conscription: abolished in 1995
Death penalty: abolished in 1996
Defence spend: (% GDP) 1.5 (1998)
Education spend: (% GNP) 3.2 (1996)
Health spend: (% GDP) 7.6 (1997)

Economy and resources

Currency: Belgian franc
GDP: (US$) 248.4 billion (1999)
Real GDP growth: (% change on previous year) 2.5 (1999)
GNP: (US$) 250.6 billion (1999)
GNP per capita (PPP): (US$) 24,200 (1999)
Consumer price inflation: 1.1% (1999)
Unemployment: 8.8% (1998)
Major trading partners: Germany, the Netherlands, France, UK, Belgium, Luxembourg, USA
Resources: coal, coke, natural gas, iron
Industries: wrought and finished steel, cast iron, sugar refining, glassware, chemicals and related products, beer, textiles, rubber and plastic products
Exports: food, livestock and livestock products, gem diamonds, iron and steel manufacturers, machinery and transport equipment, chemicals and related products.

Principal market: Germany 17.8% (1999)
Imports: food and live animals, machinery and transport equipment, precious metals and stones, mineral fuels and lubricants, chemicals and related products. Principal source: Germany 17.7% (1999)
Arable land: 22% (1996)
Agricultural products: wheat, barley, potatoes, beet (sugar and fodder), fruit, tobacco; livestock (pigs and cattle) and dairy products

Population and society

Population: 10,161,000 (2000 est)
Population growth rate: 0.14% (1995–2000); 0.1% (2000–05)
Population density: (per sq km) 333 (1999 est)
Urban population: (% of total) 97 (2000 est)
Age distribution: (% of total population) 0–14 17%, 15–59 61%, 60+ 22% (2000 est)
Ethnic groups: mainly Flemings in north, Walloons in south
Language: Flemish (a Dutch dialect, known as *Vlaams*; official) (spoken by 56%, mainly in Flanders, in the north), French (especially the dialect Walloon; official) (spoken by 32%, mainly in Wallonia, in the south), German (0.6%; mainly near the eastern border)
Religion: Roman Catholic 75%, various Protestant denominations
Education: (compulsory years) 12
Literacy rate: 99% (men); 99% (women) (2000 est)
Labour force: 2.3% agriculture, 26% industry, 71.7% services (1997)
Life expectancy: 74 (men); 81 (women) (1995–2000)
Child mortality rate: (under 5, per 1,000 live births) 8 (1995–2000)
Physicians: 1 per 268 people (1996)
Hospital beds: 1 per 132 people (1996)
TV sets: (per 1,000 people) 510 (1997)
Radios: (per 1,000 people) 793 (1997)
Internet users: (per 10,000 people) 1,379.0 (1999)
Personal computer users: (per 100 people) 31.5 (1999)

Transport

Airports: international airports: Brussels (Zaventem), Antwerp (Deurne), Ostend, Liège, Charleroi; total passenger km: 11,277 million (1997 est)

Railways: total length: 3,368 km/2,093 mi; total passenger km: 6,984 million (1997)
Roads: total road network: 145,774 km/91,827 mi, of which 79.7% paved (1996); passenger cars: 433 per 1,000 people (1997)

Chronology

57 BC: Romans conquered the Belgae (the indigenous Celtic people), and formed the province of Belgica. **3rd–4th centuries AD:** The region was overrun by Franks and Saxons. **8th–9th centuries:** Part of Frankish Empire; peace and order fostered growth of Ghent, Bruges, and Brussels. **843:** Division of Holy Roman Empire; became part of Lotharingia, but frequent repartitioning followed. **10th–11th centuries:** Several feudal states emerged: Flanders, Hainaut, Namur, Brabant, Limburg, and Luxembourg, all nominally subject to French king or Holy Roman Emperor, but in practice independent. **12th century:** The economy began to flourish. **15th century:** One by one, the states came under rule of the dukes of Burgundy. **1477:** Passed into Habsburg dominions through the marriage of Mary of Burgundy to Maximilian, archduke of Austria. **1555:** Division of Habsburg dominions; Low Countries allotted to Spain. **1648:** Independence of Dutch Republic recognized; south retained by Spain. **1713:** Treaty of Utrecht transferred Spanish Netherlands to Austrian rule. **1792–97:** Austrian Netherlands invaded by revolutionary France and finally annexed. **1815:** The Congress of Vienna reunited north and south Netherlands as one kingdom under the House of Orange. **1830:** The largely French-speaking people in south rebelled against union with Holland and declared Belgian independence. **1831:** Leopold of Saxe-Coburg-Gotha became the first king of Belgium. **1839:** The Treaty of London recognized the independence of Belgium and guaranteed its neutrality. **1914–18:** Belgium was invaded and occupied by Germany. Belgian forces under King Albert I fought in conjunction with the Allies. **1919:** Belgium acquired the Eupen-Malmédy region from Germany. **1940:** Second invasion by Germany; King Leopold III ordered the Belgian army to capitulate. **1944–45:** Belgium was liberated. **1948:** Belgium formed the Benelux customs union with Luxembourg and the Netherlands. **1949:** Belgium was a founding member of the North Atlantic Treaty Organization (NATO). Brussels became its headquarters in 1967. **1958:** Belgium was a founding member of the European Economic Community (EEC), which made Brussels its headquarters. **1971:** The constitution was amended to safeguard cultural rights of Flemish- (Flanders in north) and French-speaking communities (Walloons in southeast). **1974:** Separate regional councils and ministerial committees were established for Flemings and Walloons. **1980:** There was violence over language divisions; regional assemblies for Flanders and Wallonia and a three-member executive for Brussels were created. **1999:** In the general election, Guy Verhofstadt became liberal prime minister of a coalition government together with socialists and Greens. **2000:** Local elections were marked by the rise of the far-right party Vlaams Blok, which campaigned against immigration.

Practical information

Visa requirements: UK: visa not required. USA: visa not required
Time difference: GMT +1
Chief tourist attractions: towns of historic and cultural interest include Bruges, Ghent, Antwerp, Liège, Namur, Tournai, and Durbuy; seaside towns; forested Ardennes region
Major holidays: 1 January, 30 November, 25–26 December; variable: Ascension Thursday, Easter Monday, Whit Monday, May, August, and November holidays

BELIZE FORMERLY BRITISH HONDURAS (UNTIL 1973)

Located in Central America, bounded north by Mexico, west and south by Guatemala, and east by the Caribbean Sea.

Area: 22,963 sq km/8,866 sq mi
Capital: Belmopan
Major towns/cities: Belize, Dangriga, Orange Walk, Corozal, San Ignacio
Major ports: Belize, Dangriga, Punta Gorda
Physical features: tropical swampy coastal plain, Maya Mountains in south; over 90% forested

Blue is the colour of the People's United Party. *Sub umbra floreo*, 'I flourish in the shade', is the national motto. Effective date: late 1980s.

Government

Head of state: Queen Elizabeth II from 1981, represented by Governor General Dr Norbert Colville Young from 1993
Head of government: Said Musa from 1998
Political system: liberal democracy
Political executive: parliamentary
Administrative divisions: six districts
Political parties: People's United Party (PUP), left of centre; United Democratic Party (UDP), moderate conservative; National Alliance for Belizean Rights (NABR), dissolved
Armed forces: 1,100; plus 700 militia reserves (1998)
Conscription: military service is voluntary
Death penalty: retained and used for ordinary crimes
Defence spend: (% GDP) 2.6 (1998)
Education spend: (% GNP) 5.0 (1996)
Health spend: (% GDP) 4.7 (1997)

Economy and resources

Currency: Belize dollar
GDP: (US$) 687 million (1999)
Real GDP growth: (% change on previous year) 6.4 (1999)
GNP: (US$) 673 million (1999)
GNP per capita (PPP): (US$) 4,492 (1999)
Consumer price inflation: −1.2% (1999)
Unemployment: 12.7% (1998 est)
Foreign debt: (US$) 337.6 million (1998)
Major trading partners: USA, UK, Mexico, Canada, EU, Caricom
Industries: clothing, agricultural products (particularly sugar cane for sugar and rum), timber, tobacco
Exports: sugar, clothes, citrus products, forestry and fish products, bananas. Principal market: USA 42% (1998)
Imports: foodstuffs, machinery and transport equipment, mineral fuels, chemicals, basic manufactures. Principal source: USA 58% (1998)
Arable land: 2.6% (1995)
Agricultural products: sugar cane, citrus fruits, bananas, maize, red kidney beans, rice; livestock rearing (cattle, pigs, and poultry); fishing; timber reserves

Population and society

Population: 241,000 (2000 est)
Population growth rate: 2.4% (1995–2000); 2.3% (2000–05)
Population density: (per sq km) 10 (1999 est)
Urban population: (% of total) 54 (2000 est)
Age distribution: (% of total population) 0–14 40%, 15–59 53%, 60+ 7% (2000 est)
Ethnic groups: 44% mestizos, 30% Creoles, Maya 11%, Garifuna 7%, East Indians, Mennonites, Canadians and Europeans, including Spanish and British
Language: English (official), Spanish (widely spoken), Creole dialects
Religion: Roman Catholic 62%, Protestant 30%
Education: (compulsory years) 10
Literacy rate: 93% (men); 93% (women) (2000 est)
Labour force: 31% of population: 34% agriculture, 19% industry, 48% services (1990)
Life expectancy: 73 (men); 76 (women) (1995–2000)
Child mortality rate: (under 5, per 1,000 live births) 37 (1995–2000)
Physicians: 1 per 1,507 people (1996)
Hospital beds: 1 per 386 people (1996)
TV sets: (per 1,000 people) 178 (1997)
Radios: (per 1,000 people) 578 (1997)
Internet users: (per 10,000 people) 510.0 (1999)
Personal computer users: (per 100 people) 10.6 (1999)

Transport

Airports: international airports: Belize (Philip S W Goldson); domestic air services provide connections to major towns and offshore islands
Railways: none
Roads: total road network: 2,747 km/1,707 mi, of which 16% paved (1997); passenger cars: 42.2 per 1,000 people (1997)

Chronology

325–925 AD: Part of American Indian Maya civilization. **1600s:** Colonized by British buccaneers and log-cutters. **1862:** Formally declared a British colony, known as British Honduras. **1893:** Mexico renounced its longstanding claim to the territory. **1954:** Constitution adopted, providing for limited internal self-government. **1964:** Self-government was achieved. Universal adult suffrage and a two-chamber legislature were introduced. **1970:** The capital was moved from Belize City to the new town of Belmopan. **1973:** Name changed to Belize. **1975:** British troops sent to defend the long-disputed frontier with Guatemala. **1980:** The United Nations (UN) called for full independence. **1981:** Full independence was achieved. **1991:** Diplomatic relations were re-established with Guatemala, which finally recognized Belize's sovereignty. **1993:** The UK announced its intention to withdraw troops following the resolution of the border dispute with Guatemala. **1998:** The PUP won a sweeping victory in assembly elections, with Said Musa as prime minister.

Practical information

Visa requirements: UK: visa not required. USA: visa not required
Time difference: GMT –6
Chief tourist attractions: beaches and barrier reef; hunting and fishing; Mayan remains; nine major wildlife reserves (including the only reserves for jaguar and red-footed booby)
Major holidays: 1 January, 9 March, 1, 24 May, 10, 24 September, 12 October, 19 November, 25–26 December; variable: Good Friday, Easter Monday, Holy Saturday

BENIN FORMERLY DAHOMEY (1899–1975)

Located in west Africa, bounded east by Nigeria, north by Niger and Burkina Faso, west by Togo, and south by the Gulf of Guinea.

National name: *République du Bénin/Republic of Benin*
Area: 112,622 sq km/43,483 sq mi
Capital: Porto-Novo (official), Cotonou (de facto)
Major towns/cities: Abomey, Natitingou, Parakou, Kandi, Ouidah, Djougou, Bohicou, Cotonou
Major ports: Cotonou
Physical features: flat to undulating terrain; hot and humid in south; semiarid in north; coastal lagoons with fishing villages on stilts; Niger River in northeast

The pan-African colours express African unity. Effective date: 1 August 1990.

Government

Head of state: Mathieu Kerekou from 1996
Head of government: vacant from 1998
Political system: emergent democracy
Political executive: limited presidency
Administrative divisions: twelve departments
Political parties: Union for the Triumph of Democratic Renewal (UTDR); National Party for Democracy and Development (PNDD); Party for Democratic Renewal (PRD); Social Democratic Party (PSD); National Union for Solidarity and Progress (UNSP); National Democratic Rally (RND). The general orientation of most parties is left of centre
Armed forces: 4,800 (1998)
Conscription: by selective conscription for 18 months
Death penalty: retained and used for ordinary crimes
Defence spend: (% GDP) 1.4 (1998)
Education spend: (% GNP) 3.2 (1996)
Health spend: (% GDP) 3 (1997)

Economy and resources

Currency: franc CFA
GDP: (US$) 2.4 billion (1999)
Real GDP growth: (% change on previous year) 5 (1999)
GNP: (US$) 2.3 billion (1999)
GNP per capita (PPP): (US$) 886 (1999)
Consumer price inflation: 0.3% (1999)
Foreign debt: (US$) 1.7 billion (1998)
Major trading partners: France, Brazil, Libya, China, Indonesia, UK, Italy, Côte d'Ivoire
Resources: petroleum, limestone, marble
Industries: palm-oil processing, brewing, cement, cotton ginning, sugar refining, textiles
Exports: cotton, crude petroleum, palm oil and other palm products. Principal market: Brazil 13.9% (1999)
Imports: foodstuffs (particularly cereals), miscellaneous manufactured articles (notably cotton yarn and fabrics), fuels, machinery and transport equipment, beverages, tobacco. Principle source: France 38.2% (1999)
Arable land: 12.2% (1996)

Agricultural products: cotton, maize, yarns, cassava, sorghum, millet; fishing

Population and society

Population: 6,097,000 (2000 est)
Population growth rate: 2.7% (1995–2000)
Population density: (per sq km) 53 (1999 est)
Urban population: (% of total) 42 (2000 est)
Age distribution: (% of total population) 0–14 46%, 15–59 50%, 60+ 4% (2000 est)
Ethnic groups: 99% indigenous African, distributed among 42 ethnic groups, the largest being the Fon, Adja, Yoruba, and Braiba; small European (mainly French) community
Language: French (official), Fon (47%), Yoruba (9%) (both in the south), six major tribal languages in the north
Religion: animist 70%, Muslim 15%, Christian 15%
Education: (compulsory years) 6
Literacy rate: 57% (men); 25% (women) (2000 est)
Labour force: 46% of population: 64% agriculture, 8% industry, 28% services (1990)
Life expectancy: 52 (men); 55 (women) (1995–2000)
Child mortality rate: (under 5, per 1,000 live births) 133 (1995–2000)
Physicians: 1 per 16,000 people (1994 est)
Hospital beds: 1 per 4,182 people (1993)
TV sets: (per 1,000 people) 91 (1997)
Radios: (per 1,000 people) 108 (1997)

Transport

Airports: international airports: Cotonou (Cootonou-Cadjehoun); four domestic airports; total passenger km: 223 million (1995)
Railways: total length: 578 km/359 mi (1995); total passenger km: 116 million (1995)
Roads: total road network: 6,787 km/4,217 mi, of which 20% paved (1996); passenger cars: 6.8 per 1,000 people (1996)

Chronology

12th–13th centuries: The area was settled by a Ewe-speaking people called the Aja, who mixed with local peoples and gradually formed the Fon ethnic group. **16th century:** The Aja kingdom, called Great Ardha, was at its peak. **early 17th century:** The Kingdom of Dahomey was established in the south by Fon peoples, who defeated the neighbouring Dan; following contact with European traders, the kingdom became an intermediary in the slave trade. **1800–50:** King Dezo of Dahomey raised regiments of female soldiers to attack the Yoruba ('land of the big cities') kingdom of eastern Benin and southwest Nigeria in order to obtain slaves. **1857:** A French base was established at Grand-Popo. **1892–94:** War broke out between the French and Dahomey, after which the victorious French established a protectorate. **1899:** Incorporated in federation of French West Africa as Dahomey. **1914:** During World War I French troops from Dahomey participated in conquest of German-ruled Togoland to the west. **1940–44:** During World War II, along with the rest of French West Africa, the country supported the 'Free French' anti-Nazi resistance cause. **1960:** Independence achieved from France. **1960–77:** Acute political instability, with frequent switches from civilian to military rule, and regional ethnic disputes. **1975:** The name of the country was changed from Dahomey to Benin. **1989:** The army was deployed against antigovernment strikers and protesters, inspired by Eastern European revolutions; Marxist-Leninism was dropped as the official ideology and a market-centred economic reform programme adopted. **1990:** A referendum backed the establishment of multiparty politics. **1991:** In multiparty elections, the leader of the new Benin Renaissance Party (PRB), Nicéphore Soglo, became president and formed a ten-party coalition government. **1996:** Major Mathieu Kerekou became president. **1998:** Prime Minister Adrien Houngbedji resigned; no immediate successor was appointed.

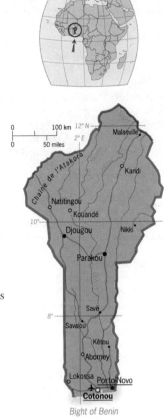

Bight of Benin

Practical information

Visa requirements: UK: visa required. USA: visa required
Time difference: GMT +1
Chief tourist attractions: national parks; game reserves

Major holidays: 1, 16 January, 1 April, 1 May, 26 October, 30 November, 25, 31 December; variable: Eid-ul-Adha, end of Ramadan, Good Friday, Easter Monday, Whit Monday

BHUTAN

Mountainous, landlocked country in the eastern Himalayas (southeast Asia), bounded north and west by Tibet (China) and to the south and east by India.

National name: *Druk-yul/Kingdom of Bhutan*
Area: 47,500 sq km/18,147 sq mi
Capital: Thimphu
Major towns/cities: Paro, Punakha, Mongar, Phuntsholing, Wangdiphodrang, Tashigang
Physical features: occupies southern slopes of the Himalayas; Gangkar Punsum (7,529 m/24,700 ft) is one of the world's highest unclimbed peaks; cut by valleys formed by tributaries of the Brahmaputra; thick forests in south

The wingless dragon holds jewels in its claws to represent prosperity. Saffron yellow symbolizes the power of the monarchy. Orange-red stands for Buddhism. Effective date: c. 1971.

Government

Head of state: Jigme Singye Wangchuk from 1972
Head of government: Lyonpo Jigme Thimley from 1998
Political system: absolutist
Political executive: absolute
Administrative divisions: 20 districts
Political parties: none officially; illegal Bhutan People's Party (BPP) and Bhutan National Democratic Party (BNDP), both ethnic Nepali
Armed forces: 6,000 (1998)
Conscription: military service is voluntary
Death penalty: retains the death penalty for ordinary crimes but can be considered abolitionist in practice (date of last known execution was 1964)
Defence spend: (% GDP) 4.5 (1998)
Education spend: (% GNP) 4.4 (1998)
Health spend: (% GDP) 7 (1997)

Economy and resources

Currency: ngultrum, although the Indian rupee is also accepted
GDP: (US$) 410 million (1999 est)
Real GDP growth: (% change on previous year) 7 (1999 est)
GNP: (US$) 399 million (1999)
GNP per capita (PPP): (US$) 1,496 (1999 est)
Consumer price inflation: 9% (1998)
Foreign debt: (US$) 120 million (1998)
Major trading partners: India, Bangladesh, Singapore, Europe
Resources: limestone, gypsum, coal, slate, dolomite, lead, talc, copper
Industries: food processing, cement, calcium carbide, textiles, tourism, cardamon, gypsum, timber, handicrafts, cement, fruit, electricity, precious stones, spices
Exports: cardamon, cement, timber, fruit, electricity, precious stones, spices. Principal market: India 94% (1998)
Imports: aircraft, mineral fuels, machinery and transport equipment, rice. Principal source: India 77% (1998)

Arable land: 2.8% (1995)
Agricultural products: potatoes, rice, apples, oranges, cardamoms; timber production

Population and society

Population: 2,124,000 (2000 est)
Population growth rate: 2.8% (1995–2000); 2.3% (2000–05)
Population density: (per sq km) 44 (1999 est)
Urban population: (% of total) 7 (2000 est)
Age distribution: (% of total population) 0–14 43%, 15–59 50%, 60+ 7% (2000 est)
Ethnic groups: 60% Bhotia, living principally in north and east (Tibetan descent), 32%; a substantial Nepali minority (about 30%) lives in the south – they are prohibited from moving into the Bhotia-dominated north; 10% indigenous or migrant tribes
Language: Dzongkha (a Tibetan dialect; official), Tibetan, Sharchop, Bumthap, Nepali, English
Religion: 70% Mahayana Buddhist (state religion), 25% Hindu
Education: not compulsory
Literacy rate: 61% (men); 34% (women) (2000 est)
Labour force: 94% agriculture, 1% industry, 5% services (1997 est)
Life expectancy: 60 (men); 62 (women) (1995–2000)
Child mortality rate: (under 5, per 1,000 live births) 96 (1995–2000)
Physicians: 1 per 6,000 people (1996)
Hospital beds: 1 per 688 people (1996)
TV sets: (per 1,000 people) 5.5 (1997)
Radios: (per 1,000 people) 18 (1997)
Internet users: (per 10,000 people) 7.6 (1999)
Personal computer users: (per 100 people) 0.5 (1999)

Transport

Airports: international airports: Paro; total passenger km: 5 million (1995)
Railways: none
Roads: total road network: 3,285 km/2,041 mi (1996)

Chronology

to 8th century: Effectively under Indian control. **16th century:**
Came under Tibetan rule. **1616–51:** Unified by Ngawang Namgyal,
leader of the Drukpa Kagyu (Thunder Dragon) Tibetan Buddhist
branch. **1720:** Came under Chinese rule. **1774:** Treaty signed with
East India Company. **1865:** Trade treaty with Britain signed after
invasion. **1907:** Ugyen Wangchuk, governor of Tongsa, became
Bhutan's first hereditary monarch. **1910:** Anglo-Bhutanese Treaty
signed, placing foreign relations under the 'guidance' of the British
government in India. **1949:** Indo-Bhutan Treaty of Friendship
signed, giving India continued influence over Bhutan's foreign
relations, but returning territory annexed in 1865. **1953:** The
national assembly (Tshogdu) was established. **1958:** Slavery was
abolished. **1959:** 4,000 Tibetan refugees were given asylum
after Chinese annexation of Tibet. **1968:** The first cabinet was
established. **1973:** Bhutan joined the nonaligned
movement. **1979:** Tibetan refugees were told to take
Bhutanese citizenship or leave; most stayed. **1983:**
Bhutan became a founding member of the South Asian
Regional Association for Cooperation. **1988:** The
Buddhist Dzongkha king imposed a 'code of conduct'
suppressing the customs of the large Hindu-Nepali
community in the south. **1990:** Hundreds of people were
allegedly killed during prodemocracy demonstrations. **1998:**
Political powers were ceded from the monarchy to the National
Assembly. Lyonpo Jigme Thimley became prime minister.

Practical information

Visa requirements: UK: visa required. USA: visa
required
Time difference: GMT +6
Chief tourist attractions: Bhutan is open only to

'controlled' tourism – many monasteries, mountains,
and other holy places remain inaccessible to tourists;
wildlife includes snow leopards and musk deer
Major holidays: 2 May, 2 June, 21 July, 11–13
November, 17 December

BOLIVIA

Landlocked country in central Andes mountains in South America, bounded north and east by Brazil, southeast by Paraguay, south by Argentina, and west by Chile and Peru.

National name: *República de Bolivia/Republic of Bolivia*
Area: 1,098,581 sq km/424,162 sq mi
Capital: La Paz (seat of government), Sucre (legal capital and seat of the judiciary)
Major towns/cities: Santa Cruz, Cochabamba, Oruro, El Alto, Potosí, Tarija
Physical features: high plateau (Altiplano) between mountain ridges (cordilleras); forest and lowlands (llano) in east; Andes; lakes Titicaca (the world's highest navigable lake, 3,800 m/12,500 ft) and Poopó

Red stands for Bolivia's animals and the valour of the liberating army. Green symbolizes fertility. Yellow represents Bolivia's mineral deposits. Effective date: c. 1966.

Government

Head of state and government: Hugo Banzer Suarez from 1997
Political system: liberal democracy
Political executive: limited presidency
Administrative divisions: nine departments
Political parties: National Revolutionary Movement (MNR), right of centre; Movement of the Revolutionary Left (MIR), left of centre; Nationalist Democratic Action Party (ADN), right wing; Solidarity and Civic Union (UCS), populist, free market; Patriotic Conscience Party, populist
Armed forces: 33,500 (1998)
Conscription: selective conscription for 12 months at the age of 18
Death penalty: abolished for ordinary crimes in 1997; laws provide for the death penalty for exceptional crimes only (last execution in 1974)
Defence spend: (% GDP) 1.8 (1998)
Education spend: (% GNP) 5.6 (1996)
Health spend: (% GDP) 5.8 (1997)

Economy and resources

Currency: boliviano
GDP: (US$) 8.5 billion (1999)
Real GDP growth: (% change on previous year) 0.6 (1999)
GNP: (US$) 8.2 billion (1999)
GNP per capita (PPP): (US$) 2,193 (1999)
Consumer price inflation: 2.2% (1999)
Unemployment: 4.2% (1996)
Foreign debt: (US$) 6.3 billion (1999 est)
Major trading partners: USA, Brazil, Argentina, Peru, Colombia, Chile, Ecuador, Sweden, Uruguay
Resources: petroleum, natural gas, tin (world's fifth-largest producer), zinc, silver, gold, lead, antimony, tungsten, copper
Industries: mining, food products, petroleum refining, tobacco, textiles
Exports: metallic minerals, natural gas, jewellery, soybeans, wood. Principal market: USA 20.9% (1999). Illegal trade in coca and its derivatives (mainly cocaine) was worth approximately $600 million in 1990 – almost equal to annual earnings from official exports.
Imports: industrial materials, machinery and transport equipment, consumer goods. Principal source: Brazil 25.4% (1999)
Arable land: 1.8% (1996)
Agricultural products: coffee, coca, soybeans, sugar cane, rice, chestnuts, maize, potatoes; livestock products (beef and hides); forest resources

Population and society

Population: 8,329,000 (2000 est)
Population growth rate: 2.3% (1995–2000); 2.2% (2000–05)
Population density: (per sq km) 7 (1999 est)
Urban population: (% of total) 63 (2000 est)
Age distribution: (% of total population) 0–14 40%, 15–59 53%, 60+ 7% (2000 est)
Ethnic groups: 30% Quechua Indians, 25% Aymara Indians, 25–30% mixed, 5–15% of European descent
Language: Spanish (official) (4%), Aymara, Quechua
Religion: Roman Catholic 90% (state-recognized)
Education: (compulsory years) 8
Literacy rate: 92% (men); 79% (women) (2000 est)
Labour force: 40% of population: 47% agriculture, 19% industry, 34% services (1993)
Life expectancy: 60 (men); 63 (women) (1995–2000)
Child mortality rate: (under 5, per 1,000 live births) 88 (1995–2000)
Physicians: 1 per 2,348 people (1993 est)
Hospital beds: 1 per 709 people (1993 est)
TV sets: (per 1,000 people) 115 (1997)
Radios: (per 1,000 people) 675 (1997)
Internet users: (per 10,000 people) 43.0 (1999)
Personal computer users: (per 100 people) 1.2 (1999)

Transport

Airports: international airports: La Paz (El Alto), Santa Cruz (Viru-Viru); 28 domestic airports; total passenger km: 1,891 million (1997 est)
Railways: total length: 3,652 km/2,269 mi; total passenger km: 240 million (1995)
Roads: total road network: 49,400 km/30,697 mi, of which 5.5% paved (1996 est); passenger cars: 32.1 per 1,000 people (1996)

Chronology

c. **AD 600:** Development of sophisticated civilization at Tiahuanaco, south of Lake Titicaca. *c.* **1200:** Tiahuanaco culture was succeeded by smaller Aymara-speaking kingdoms. **16th century:** Became incorporated within westerly Quechua-speaking Inca civilization, centred in Peru. **1538:** Conquered by Spanish and, known as 'Upper Peru', became part of the Viceroyalty of Peru, whose capital was at Lima (Peru); Charcas (now Sucre) became the local capital. **1545:** Silver was discovered at Potosí in the southwest, which developed into chief silver-mining town and most important city in South America in the 17th and 18th centuries. **1776:** Transferred to the Viceroyalty of La Plata, with its capital in Buenos Aires. **late 18th century:** Increasing resistance of American Indians and mestizos to Spanish rule; silver production slumped. **1825:** Liberated from Spanish rule by the Venezuelan freedom fighter Simón Bolívar, after whom the country was named, and his general, Antonio José de Sucre; Sucre became Bolivia's first president. **1836–39:** Bolivia became part of a federation with Peru, headed by Bolivian president Andres Santa Cruz, but it dissolved following defeat in war with Chile. **1879–84:** Coastal territory in the Atacama, containing valuable minerals, was lost after defeat in war with Chile. **1903:** Territory was lost to Brazil. **1932–35:** Further territory was lost after defeat by Paraguay in the Chaco War, fought over control of the Chaco Boreal. **1952:** After the military regime was overthrown in the Bolivian National Revolution, Dr Victor Paz Estenssoro of the centrist National Revolutionary Movement (MNR) became president and introduced social reforms. **1964:** An army coup was led by Vice-President Gen René Barrientos. **1967:** There was a peasant uprising, led by Ernesto 'Che' Guevara. The uprising was put down with US help, and Guevara was killed. **1969:** Barrientos was killed in a plane crash, and replaced by Siles Salinas, who was soon deposed in an army coup. **1971:** Col Hugo Banzer Suárez came to power after a military coup. **1974:** An attempted coup prompted Banzer to postpone promised elections and ban political and trade-union activity. **1980:** Inconclusive elections were followed by the country's 189th coup. Allegations of corruption and drug trafficking led to the cancellation of US and European Community (EC) aid. **1982:** With the economy worsening, the military junta handed power over to a civilian administration headed by Siles Zuazo. **1983:** US and EC economic aid resumed as austerity measures were introduced. **1985:** The inflation rate was 23,000%. **1993:** Foreign investment was encouraged as inflation fell to single figures. **1997:** Hugo Banzer was elected president. **2000:** The government lost support due to widespread poverty and the stagnation of the economy. There were violent clashes between security forces and protesters, who called for the resignation of President Banzer.

Practical information

Visa requirements: UK: visa not required for a stay of up to 90 days. USA: visa not required for a stay of up to 90 days
Time difference: GMT –4
Chief tourist attractions: Lake Titicaca; pre-Inca ruins at Tiwanaku; Chacaltaga in the Andes Mountains; UNESCO World Cultural Heritage Sites of Potosí and Sucre; skiing in the Andes
Major holidays: 1 January, 1 May, 6 August, 1 November, 25 December; variable: Carnival, Corpus Christi, Good Friday

BOSNIA-HERZEGOVINA

Located in central Europe, bounded north and west by Croatia, east by the Yugoslavian republic of Serbia, and east and south by the Yugoslavian republic of Montenegro.

National name: *Bosna i Hercegovina/Bosnia-Herzegovina*
Area: 51,129 sq km/19,740 sq mi
Capital: Sarajevo
Major towns/cities: Banja Luka, Mostar, Prijedor, Tuzla, Zenica, Bihac, Goražde
Physical features: barren, mountainous country, part of the Dinaric Alps; limestone gorges; 20 km/12 mi of coastline with no harbour

The stars on a blue field represent Europe. The yellow triangle stands for equality between the three peoples of Bosnia-Herzegovina. Effective date: 4 February 1998.

Government

Head of state: Zivko Radisic from 2000, chairman of the rotating collective presidency
Head of government: Martin Raguz from 2000
Political system: emergent democracy
Political executive: limited presidency
Administrative divisions: ten cantons
Political parties: Party of Democratic Action (PDA), Muslim-oriented; Serbian Renaissance Movement (SPO), Serbian nationalist; Croatian Christian Democratic Union of Bosnia-Herzegovina (CDU), Croatian nationalist; League of Communists (LC) and Socialist Alliance (SA), left wing
Armed forces: 40,000 (1998)
Death penalty: abolished for ordinary crimes in 1997; laws provide for the death penalty for exceptional crimes only
Defence spend: (% GDP) 8.1 (1998)
Health spend: (% GDP) 7.6 (1997)

Economy and resources

Currency: dinar
GDP: (US$) 4.4 billion (1999)
Real GDP growth: (% change on previous year) 10 (1999)
GNP: (US$) 2.3 billion (1996)
GNP per capita (PPP): (US$) 450 (1996 est)
Consumer price inflation: 14.1% (1999)
Unemployment: 28% (1992 est)
Foreign debt: (US$) 3.2 billion (1999)
Major trading partners: Croatia, Yugoslavia, Italy, Slovenia, Germany, Hungary
Resources: copper, lead, zinc, iron ore, coal, bauxite, manganese
Industries: iron and crude steel, armaments, cement, textiles, vehicle assembly, wood products, oil refining, electrical appliances, cigarettes; industrial infrastructure virtually destroyed by war
Exports: coal, domestic appliances (industrial production and mining remain low). Principal market: Croatia 24.9% (1999)
Imports: foodstuffs, basic manufactured goods, processed and semiprocessed goods. Principal source:

Croatia 28.3% (1999)
Arable land: 9.8% (1996)
Agricultural products: before the war, these were maize, wheat, potatoes, rice, tobacco, fruit, olives, grapes; livestock rearing (sheep and cattle); timber reserves

Population and society

Population: 3,972,000 (2000 est)
Population growth rate: 3% (1995–2000)
Population density: (per sq km) 75 (1999 est)
Urban population: (% of total) 43 (2000 est)
Age distribution: (% of total population) 0–14 19%, 15–59 66%, 60+ 15% (2000 est)
Ethnic groups: 40% ethnic Muslim, 38% Serb, 22% Croat. Croats are most thickly settled in southwest Bosnia and western Herzegovina, Serbs in eastern and western Bosnia. Since the start of the civil war in 1992 many Croats and Muslims have fled as refugees to neighbouring states
Language: Serbian, Croat, Bosnian
Religion: 40% Muslim, 31% Serbian Orthodox, 15% Roman Catholic
Education: (compulsory years) 8
Literacy rate: 90% (men); 90% (women) (1992)
Labour force: 2% agriculture, 45% industry, 53% services (1990 est)
Life expectancy: 71 (men); 76 (women) (1995–2000)
Child mortality rate: (under 5, per 1,000 live births) 17 (1995–2000)
TV sets: (per 1,000 people) 41 (1997)
Radios: (per 1,000 people) 248 (1996)
Internet users: (per 10,000 people) 9.1 (1999)

Transport

Airports: international airport: Sarajevo; two smaller civil airports (civil aviation was severely disrupted by fighting in early 1990s; no air services to Sarajevo since 1992)
Railways: total length: 1,021 km/634 mi; total passenger km: 554 million (1991)
Roads: total road network: 21,846 km/13,575 mi (1996); passenger cars: 23 per 1,000 people (1996)

Chronology

1st century AD: Part of Roman province of Illyricum. **395:** On division of Roman Empire, stayed in west, along with Croatia and Slovenia, while Serbia to the east became part of the Byzantine Empire. **7th century:** Settled by Slav tribes. **12–15th centuries:** Independent state. **1463 and 1482:** Bosnia and Herzegovina, in south, successively conquered by Ottoman Turks; many Slavs were converted to Sunni Islam. **1878:** Became an Austrian protectorate, following Bosnian revolt against Turkish rule in 1875–76. **1908:** Annexed by Austrian Habsburgs in wake of Turkish Revolution. **1914:** Archduke Franz Ferdinand, the Habsburg heir, was assassinated in Sarajevo by a Bosnian-Serb extremist, precipitating World War I. **1918:** On the collapse of the Habsburg Empire, the region became part of the Serb-dominated 'Kingdom of Serbs, Croats, and Slovenes', known as Yugoslavia from 1929. **1941:** The region was occupied by Nazi Germany and became 'Greater Croatia', fascist puppet state and the scene of fierce fighting. **1943–44:** Bosnia was liberated by communist Partisans, led by Marshal Tito. **1945:** The region became a republic within the Yugoslav Socialist Federation. **1980:** There was an upsurge in Islamic nationalism. **1990:** Ethnic violence erupted between Muslims and Serbs. Communists were defeated in multiparty elections; a coalition was formed by Serb, Muslim, and Croatian parties. **1991:** The Serb–Croat civil war in Croatia spread unrest into Bosnia. Fears that Serbia planned to annex Serb-dominated parts of the republic led to a declaration of sovereignty by parliament. Serbs within Bosnia established autonomous enclaves. **1992:** Bosnia was admitted into the United Nations (UN). Violent civil war broke out, as an independent 'Serbian Republic of Bosnia-Herzegovina', comprising parts of the east and the west, was proclaimed by Bosnian-Serb militia leader Radovan Karadzic, with Serbian backing. UN forces were drafted into Sarajevo to break the Serb siege of the city; Bosnian Serbs were accused of 'ethnic cleansing', particularly of Muslims. **1993:** A UN–EC peace plan failed. The USA began airdrops of food and medical supplies. Six UN 'safe areas' were created, intended as havens for Muslim civilians. A Croat–Serb partition plan was rejected by Muslims. **1994:** The Serb siege of Sarajevo was lifted after a UN–NATO ultimatum and Russian diplomatic intervention. A Croat–Muslim federation was formed. **1995:** Hostilities resumed. A US-sponsored peace accord, providing for two sovereign states (a Muslim–Croat federation and a Bosnian Serb Republic, the Republika Srpska) as well as a central legislature (House of Representatives, House of Peoples, and three-person presidency), was agreed at Dayton, Ohio. A 60,000-strong NATO peacekeeping force was deployed. **1996:** An International Criminal Tribunal for Former Yugoslavia began in the Hague and an arms-control accord was signed. Full diplomatic relations were established with Yugoslavia. The collective rotating presidency was elected, with Alija Izetbegovic (Muslim), Momcilo Krajisnik (Serb), and Kresimir Zubak (Croat); Izetbegovic was elected overall president. Biljana Plavsic was elected president of the Serb Republic and Gojko Klickovic its prime minister. Edhem Bicakcic became prime minister of the Muslim–Croat Federation. **1997:** Vladimir Soljic was elected president of the Muslim–Croat Federation. The Serb part of Bosnia signed a customs agreement with Yugoslavia. **1998:** A moderate, pro-western government was formed in the Bosnian Serb republic, headed by Milorad Dodik; Nikola Poplasen became president. The first Muslims and Croats were convicted in The Hague for war crimes during 1992. Zivko Radisic and Ante Jelavic replaced Krajisnik and Zubak respectively on the rotating presidency. **2000:** In January, Ejup Ganic became president of the Muslim–Croat federation. Hardline nationalist Mirko Sarovic became president of the Bosnian Serb republic. Izetbegovic was replaced by Halid Genjac as the Muslim member of the rotating presidency. The northeastern town of Brcko, the only territorial dispute outstanding from the Dayton peace accord, was established as a self-governing neutral district in March, to be ruled by an elected alliance. The year saw three changes of prime minister, with Martin Raguz elected in October. **2001:** Biljana Plavsic gave herself up to the war crimes tribunal, but pleaded 'not guilty' to nine counts of war crimes.

Practical information

Visa requirements: UK: visa not required. USA: visa not required

Time difference: GMT +1
Major holidays: 1–2 January, 1 March, 1–2 May, 27 July, 25 November

BOTSWANA FORMERLY BECHUANALAND (UNTIL 1966)

Landlocked country in central southern Africa, bounded south and southeast by South Africa, west and north by Namibia, and northeast by Zimbabwe.

National name: *Republic of Botswana*
Area: 582,000 sq km/224,710 sq mi
Capital: Gaborone
Major towns/cities: Mahalapye, Serowe, Francistown, Selebi-Phikwe, Molepolol, Kange, Maun
Physical features: Kalahari Desert in southwest (70–80% of national territory is desert), plains (Makgadikgadi salt pans) in east, fertile lands and Okavango Delta in north

Blue stands for water and rain. Black and white represent the racial harmony of the people. Effective date: 30 September 1966.

Government

Head of state and government: Festus Mogae from 1998
Political system: liberal democracy
Political executive: limited presidency
Administrative divisions: ten districts and four town councils
Political parties: Botswana Democratic Party (BDP), moderate centrist; Botswana National Front (BNF), moderate left of centre; Botswana Freedom Party (BFP)
Armed forces: 8,500 (1998)
Conscription: military service is voluntary
Death penalty: retained and used for ordinary crimes
Defence spend: (% GDP) 6.5 (1998)
Education spend: (% GNP) 8.6 (1997)
Health spend: (% GDP) 4.2 (1997)

Economy and resources

Currency: franc CFA
GDP: (US$) 5.8 billion (1999 est)
Real GDP growth: (% change on previous year) 8.5 (1999)
GNP: (US$) 5.1 billion (1999)
GNP per capita (PPP): (US$) 6,032 (1999)
Consumer price inflation: 7.1% (1999)
Unemployment: 21.5% (1997)
Foreign debt: (US$) 520 million (1999)
Major trading partners: Lesotho, Namibia, South Africa, Swaziland – all fellow SACU (Southern African Customs Union) members; UK and other European countries, USA
Resources: diamonds (world's third-largest producer), copper-nickel ore, coal, soda ash, gold, cobalt, salt, plutonium, asbestos, chromite, iron, silver, manganese, talc, uranium
Industries: mining, food processing, textiles and clothing, beverages, soap, chemicals, paper, plastics, electrical goods
Exports: diamonds, copper and nickel, beef. Principal market: EU 77% (1998)
Imports: machinery and transport equipment, food, beverages, tobacco, chemicals and rubber products, textiles and footwear, fuels, wood and paper products. Principal source: SACU 76% (1998)
Arable land: 0.6% (1996)
Agricultural products: sorghum, vegetables, pulses; cattle raising (principally for beef production) is main agricultural activity

Population and society

Population: 1,622,000 (2000 est)
Population growth rate: 1.9% (1995–2000)
Population density: (per sq km) 3 (1999 est)
Urban population: (% of total) 50 (2000 est)
Age distribution: (% of total population) 0–14 42%, 15–59 53%, 60+ 5% (2000 est)
Ethnic groups: about 95% Tswana (Butswana) and 4% Kalanga, Basarwa, and Kgalagadi; 1% European
Language: English (official), Setswana (national)
Religion: Christian 50%, animist 50%
Education: not compulsory
Literacy rate: 74% (men); 80% (women) (2000 est)
Labour force: 44% of population: 46% agriculture, 20% industry, 33% services (1990)
Life expectancy: 46 (men); 48 (women) (1995–2000)
Child mortality rate: (under 5, per 1,000 live births) 107 (1995–2000)
Physicians: 1 per 4,130 people (1994)
Hospital beds: 1 per 635 people (1993 est)
TV sets: (per 1,000 people) 27 (1997)
Radios: (per 1,000 people) 156 (1997)

Transport

Airports: international airports: Gaborone (Sir Seretse Khama), Kasane; six domestic airports; total passenger km: 53 million (1995)
Railways: total length: 888 km/552 mi (1995); total passenger km: 96 million (1997)
Roads: total road network: 18,212 km/11,317 mi, of which 23.5% paved (1996); passenger cars: 15.2 per 1,000 people (1996 est)

Chronology

18th century: Formerly inhabited by nomadic hunter-gatherer groups, including the Kung, the area was settled by the Tswana people, from whose eight branches the majority of the people are descended. **1872:** Khama III the Great, a converted Christian, became chief of the Bamangwato, the largest Tswana group. He developed a strong army and greater unity among the Botswana peoples. **1885:** Became the British protectorate of Bechuanaland at the request of Chief Khama, who feared invasion by Boers from the Transvaal (South Africa) following the discovery of gold. **1895:** The southern part of the Bechuanaland Protectorate was annexed by Cape Colony (South Africa). **1960:** A new constitution created a legislative council controlled (until 1963) by a British High Commissioner. **1965:** The capital was transferred from Mafeking to Gaborone. Internal self-government was achieved. **1966:** Independence was achieved from Britain. Name changed to Botswana. **mid-1970s:** The economy grew rapidly as diamond mining expanded. **1985:** South African raid on Gaborone, allegedly in search of African National Congress (ANC) guerrillas. **1993:** Relations with South Africa were fully normalized following the end of apartheid and the establishment of a multiracial government. **1997:** Major constitutional changes reduced the voting age to 18. **1998:** Festus Mogae (BDP) became president.

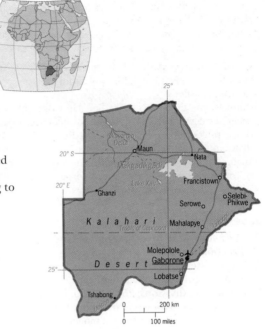

Practical information

Visa requirements: UK: visa not required. USA: visa not required
Time difference: GMT +2
Chief tourist attractions: Kalahari Desert; five game reserves and three national parks, including Chobe National Park, Moremi Wildlife Reserve, and Kalahari Gemsbok National Park
Major holidays: 1–2 January, 30 September, 25–26 December; variable: Ascension Thursday, Good Friday, Easter Monday, Holy Saturday, President's Day (July), July Holiday, October Holiday

BRAZIL

Largest country in South America (almost half the continent), bounded southwest by Uruguay, Argentina, Paraguay and Bolivia; west by Peru and Colombia; north by Venezuela, Guyana, Suriname, and French Guiana; and northeast and southeast by the Atlantic Ocean.

National name: *República Federativa do Brasil/Federative Republic of Brazil*
Area: 8,511,965 sq km/3,286,469 sq mi
Capital: Brasília
Major towns/cities: São Paulo, Belo Horizonte, Nova Iguaçu, Rio de Janeiro, Belém, Recife, Porto Alegre, Salvador, Curitiba, Manaus, Fortaleza
Major ports: Rio de Janeiro, Belém, Recife, Porto Alegre, Salvador
Physical features: the densely forested Amazon basin covers the northern half of the country with a network of rivers; south is fertile; enormous energy resources, both hydroelectric (Itaipú Reservoir on the Paraná, and Tucuruí on the Tocantins) and nuclear (uranium ores); mostly tropical climate

Yellow and the diamond shape represent Brazil's mineral wealth. The motto Ordem e Progresso means 'Order and Progress'. Green stands for the vast forests. Effective date: 15 November 1889.

Government

Head of state and government: Fernando Henrique Cardoso from 1995
Political system: liberal democracy
Political executive: limited presidency
Administrative divisions: 26 states and one federal district
Political parties: Workers' Party (PT), left of centre; Social Democratic Party (PSDB), moderate, left of centre; Brazilian Democratic Movement Party (PMDB), left of centre; Liberal Front Party (PFL), right wing; National Reconstruction Party (PRN), right of centre
Armed forces: 313,000; public security forces under army control 385,600 (1998)
Conscription: 12 months
Death penalty: for exceptional crimes only; last execution 1855
Defence spend: (% GDP) 3.2 (1998)
Education spend: (% GNP) 5.2 (1996)
Health spend: (% GDP) 6.5 (1997)

Economy and resources

Currency: real
GDP: (US$) 760.3 billion (1999)
Real GDP growth: (% change on previous year) 0.8 (1999)
GNP: (US$) 742.8 billion (1999)
GNP per capita (PPP): (US$) 6,317 (1999)
Consumer price inflation: 4.9% (1999)
Unemployment: 7.6% (1998)
Foreign debt: (US$) 212.5 billion (1999 est)
Major trading partners: USA, Argentina, Germany, Japan, the Netherlands, Italy
Resources: iron ore (world's second-largest producer), tin (world's fourth-largest producer), aluminium (world's fourth-largest producer), gold, phosphates, platinum, bauxite, uranium, manganese, coal, copper, petroleum, natural gas, hydroelectric power, forests
Industries: mining, steel, machinery and transport equipment, food processing, textiles and clothing, chemicals, petrochemicals, cement, lumber
Exports: steel products, transport equipment, coffee, iron ore and concentrates, aluminium, iron, tin, soybeans, orange juice (85% of world's concentrates), tobacco, leather footwear, sugar, beef, textiles. Principal market: USA 22.6% (1999)
Imports: mineral fuels, machinery and mechanical appliances, chemical products, foodstuffs, coal, wheat, fertilizers, cast iron and steel. Principal source: USA 24.1% (1999)
Arable land: 6.3% (1996)
Agricultural products: soybeans, coffee (world's largest producer), tobacco, sugar cane (world's third-largest producer), cocoa beans (world's second-largest producer), maize, rice, cassava, oranges; livestock (beef and poultry)

Population and society

Population: 170,115,000 (2000 est)
Population growth rate: 1.3% (1995–2000)
Population density: (per sq km) 20 (1999 est)
Urban population: (% of total) 81 (2000 est)
Age distribution: (% of total population) 0–14 29%, 15–59 63%, 60+ 8% (2000 est)
Ethnic groups: wide range of ethnic groups, including 55% of European origin (mainly Portuguese, Italian, and German), 38% of mixed parentage, 6% of African origin, as well as American Indians and Japanese
Language: Portuguese (official), Spanish, English, French, 120 Indian languages
Religion: Roman Catholic 70%; Indian faiths
Education: (compulsory years) 8
Literacy rate: 85% (men); 85% (women) (2000 est)
Labour force: 44% of population: 23% agriculture, 23% industry, 54% services (1990)
Life expectancy: 63 (men); 71 (women) (1995–2000)
Child mortality rate: (under 5, per 1,000 live births) 48 (1995–2000)
Physicians: 1 per 844 people (1993 est)
Hospital beds: 1 per 299 people (1993 est)
TV sets: (per 1,000 people) 316 (1997)
Radios: (per 1,000 people) 444 (1997)
Internet users: (per 10,000 people) 208.3 (1999)
Personal computer users: (per 100 people) 3.6 (1999)

Transport

Airports: principal international airports: Rio de Janeiro (Galeão), São Paulo (Guarulhos, Viracopos, and Congonhas), Manaus (Eduardo Gomes), Salvador (Dois de Julho); 27 domestic airports; total passenger km: 41,714 million (1997 est)

Railways: total length: 29,099 km/18,082 mi; total passenger km 12,688 million (1997)

Roads: total road network: 1,980,000 km/1,230,372 mi, of which 9.3% paved (1996 est); passenger cars: 122.1 per 1,000 people (1997 est)

Chronology

1500: Originally inhabited by South American Indians. Portuguese explorer Pedro Alvares Cabral sighted and claimed Brazil for Portugal. **1530:** Start of Portuguese colonization; Portugal monopolized trade but colonial government was decentralized. **1580–1640:** Brazil came under Spanish rule along with Portugal. **17th century:** Sugar-cane plantations were established with slave labour in coastal regions, making Brazil the world's largest supplier of sugar; cattle ranching was developed inland. **1695:** Gold was discovered in the central highlands. **1763:** The colonial capital moved from Bahía to Rio de Janeiro. **1770:** Brazil's first coffee plantations were established in Rio de Janeiro. **18th century:** Population in 1798 totalled 3.3 million, of which around 1.9 million were slaves, mainly of African origin; significant growth of gold-mining industry. **1808:** The Portuguese regent, Prince John, arrived in Brazil and established his court at Rio de Janeiro; Brazilian trade opened to foreign merchants. **1815:** The United Kingdom of Portugal, Brazil, and Algarve made Brazil co-equal with Portugal. **1821:** Crown Prince Pedro took over the government of Brazil. **1822:** Pedro defied orders to return to Portugal; he declared Brazil's independence to avoid reversion to colonial status. **1825:** King John VI recognized his son as Emperor Pedro I of Brazil. **1831:** Pedro I abdicated in favour of his infant son, Pedro II. **1847:** The first prime minister was appointed, but the emperor retained many powers. **1865–70:** Brazilian efforts to control Uruguay led to the War of the Triple Alliance with Paraguay. **1888:** Slavery was abolished in Brazil. **1889:** The monarch was overthrown by a liberal revolt; a federal republic was established, with a central government controlled by the coffee planters. **1902:** Brazil produced 65% of the world's coffee. **1915–19:** Lack of European imports during World War I led to rapid industrialization. **1930:** A revolution against the coffee planter oligarchy placed Getúlio Vargas in power; he introduced social reforms. **1937:** Vargas established an authoritarian corporate state. **1942:** Brazil entered World War II as an ally of the USA. **1945–54:** Vargas was ousted by a military coup. In 1951 he was elected president and continued to extend the state control of the economy. In 1954 he committed suicide. **1960:** The capital moved to Brasília. **1964:** A bloodless coup established a technocratic military regime; free political parties were abolished; intense concentration on industrial growth was aided by foreign investment and loans. **1970s:** Economic recession and inflation undermined public support for the military regime. **1985:** After gradual democratization from 1979, Tancredo Neves became the first civilian president in 21 years. **1988:** A new constitution reduced the powers of the president. **1989:** Fernando Collor (PRN) was elected president. Brazil suspended its foreign debt payments. **1992:** Collor was charged with corruption and replaced by Vice-President Itamar Franco. **1994:** A new currency was introduced, the third in eight years. Fernando Henrique Cardoso (PSDB) won presidential elections. Collor was cleared of corruption charges. **1997:** The constitution was amended to allow the president to seek a second term of office. **1998:** President Cardoso was re-elected. A stock market crash weakened Brazil's currency, and although an International Monetary Fund (IMF) rescue package was announced, the government had to devalue the real as foreign capital was withdrawn from the economy. **1999:** The economy began to recover and economic reforms were put in place.

Practical information

Visa requirements: UK: visa not required for tourist visits. USA: visa required
Time difference: GMT –2/5
Chief tourist attractions: Rio de Janeiro and beaches; Iguaçu Falls; tropical forests of Amazon basin; wildlife

of the Pantanal; Mato Grosso
Major holidays: 1 January, 21 April, 1 May, 7 September, 12 October, 2, 15 November, 25 December; variable: Carnival (2 days), Corpus Christi, Good Friday, Holy Saturday, Holy Thursday

BRUNEI

Comprises two enclaves on the northwest coast of the island of Borneo, bounded to the landward side by Sarawak and to the northwest by the South China Sea.

National name: *Negara Brunei Darussalam/State of Brunei*
Area: 5,765 sq km/2,225 sq mi
Capital: Bandar Seri Begawan (and chief port)
Major towns/cities: Seria, Kuala Belait
Physical features: flat coastal plain with hilly lowland in west and mountains in east (Mount Pagon 1,850 m/6,070 ft); 75% of the area is forested; the Limbang valley splits Brunei in two, and its cession to Sarawak in 1890 is disputed by Brunei; tropical climate; Temburong, Tutong, and Belait rivers

The four-feathered wing, the Sayap, symbolizes the protection of justice, tranquillity, prosperity, and peace. The scroll reads 'Brunei City of Peace'. The flag and umbrella are based on ancient royal regalia while the mast represents the state. Effective date: c. 1984.

Government

Head of state and government: Sultan Hassanal Bolkiah Mu'izzaddin Waddaulah from 1967
Political system: absolutist
Political executive: absolute
Administrative divisions: four districts
Political parties: Brunei National Democratic Party (BNDP) and Brunei People's Party (BPP) (both banned); Brunei National United Party (BNUP) (inactive)
Armed forces: 5,000 (1998); plus paramilitary forces of 4,100
Conscription: military service is voluntary
Death penalty: retains the death penalty for ordinary crimes but can be considered abolitionist in practice (last execution 1957)
Defence spend: (% GDP) 6.9 (1998)
Education spend: (% GNP) 3.1 (1996)
Health spend: (% GDP) 5.4 (1997)

Economy and resources

Currency: Bruneian dollar, although the Singapore dollar is also accepted
GDP: (US$) 4.9 billion (1999 est)
Real GDP growth: (% change on previous year) 2.5 (1999)
GNP: (US$) 7.8 billion (1999 est)
GNP per capita (PPP): (US$) 24,824 (1999 est)
Consumer price inflation: 1% (1999 est)
Unemployment: 4.3% (1997)
Foreign debt: (US$) 1.2 billion (1997)
Major trading partners: Singapore, Japan, USA, EU countries, Malaysia, South Korea, Thailand
Resources: petroleum, natural gas
Industries: petroleum refining, textiles, cement, mineral water, canned foods, rubber
Exports: crude petroleum, natural gas and refined products. Principal market: Japan 50.9% (1998)
Imports: machinery and transport equipment, basic manufactures, food and live animals, chemicals. Principal source: Singapore 32.3% (1998)

Arable land: 0.6% (1995)
Agricultural products: rice, cassava, bananas, pineapples, vegetables; fishing; forest resources

Population and society

Population: 328,000 (2000 est)
Population growth rate: 2.2% (1995–2000)
Population density: (per sq km) 56 (1999 est)
Urban population: (% of total) 72 (2000 est)
Age distribution: (% of total population) 0–14 32%, 15–59 64%, 60+ 4% (2000 est)
Ethnic groups: 73% indigenous Malays, predominating in government service and agriculture; more than 15% Chinese, predominating in the commercial sphere
Language: Malay (official), Chinese (Hokkien), English
Religion: Muslim 66%, Buddhist 14%, Christian 10%
Education: (compulsory years) 12
Literacy rate: 95% (men); 88% (women) (2000 est)
Labour force: 1% agriculture, 25% industry, 74% services (1997)
Life expectancy: 73 (men); 78 (women) (1995–2000)
Child mortality rate: (under 5, per 1,000 live births) 11 (1995–2000)
Physicians: 1 per 1,522 people (1994 est)
Hospital beds: 1 per 310 people (1994)
TV sets: (per 1,000 people) 250 (1997)
Radios: (per 1,000 people) 302 (1997)
Internet users: (per 10,000 people) 777.2 (1999)
Personal computer users: (per 100 people) 6.2 (1999)

Transport

Airports: international airports: Bandar Seri Begawan (Brunei International); total passenger km: 2,972 million (1997 est)
Railways: none
Roads: total road network: 1,712 km/1,064 mi, of which 75% paved (1996); passenger cars: 575 per 1,000 people (1996)

Chronology

15th century: An Islamic monarchy was established, ruling Brunei and north Borneo, including the Sabah and Sarawak states of Malaysia. **1841:** Control of Sarawak was lost. **1888:** Brunei became a British protectorate. **1906:** Brunei became a British dependency. **1929:** Oil was discovered. **1941–45:** Brunei was occupied by Japan. **1959:** A written constitution made Britain responsible for defence and external affairs. **1962:** The sultan began rule by decree after a plan to join the Federation of Malaysia was opposed by a rebellion organized by the Brunei People's Party (BPP). **1967:** Hassanal Bolkiah became sultan. **1971:** Brunei was given full internal self-government. **1975:** A United Nations (UN) resolution called for independence for Brunei. **1984:** Independence from Britain was achieved, with Britain maintaining a small force to protect the oil and gas fields. **1985:** The Brunei National Democratic Party (BNDP) was legalized. **1986:** The multiethnic Brunei National United Party (BNUP) was formed; nonroyals were given key cabinet posts for the first time. **1988:** The BNDP and the BNUP were banned. **1991:** Brunei joined the nonaligned movement. **1998:** Prince Billah was proclaimed heir to the throne.

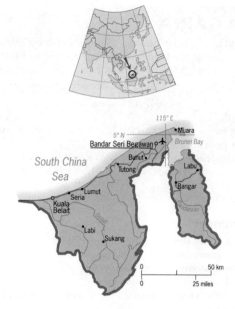

Practical information

Visa requirements: UK: visa not required for visits of up to 30 days. USA: visa not required
Time difference: GMT +8
Chief tourist attractions: tropical rainforest

Major holidays: 1 January, 23 February, 31 May, 15 July, 25 December; variable: Eid-ul-Adha, end of Ramadan, Good Friday, New Year (Chinese), New Year (Muslim), Prophet's Birthday, first day of Ramadan, Meraj (March/April), Revelation of the Koran (May)

BULGARIA

Located in southeast Europe, bounded north by Romania, west by Yugoslavia and the Former Yugoslav Republic of Macedonia, south by Greece, southeast by Turkey, and east by the Black Sea.

National name: *Republika Bulgaria/Republic of Bulgaria*
Area: 110,912 sq km/42,823 sq mi
Capital: Sofia
Major towns/cities: Plovdiv, Varna, Ruse, Burgas, Stara Zagora, Pleven
Major ports: Burgas, Varna
Physical features: lowland plains in north and southeast separated by mountains (Balkan and Rhodope) that cover three-quarters of the country; River Danube in north

White represents a desire for peace and liberty. Green symbolizes freedom and agricultural wealth. Red stands for the courage of spilt blood of the freedom fighters. Effective date: 22 November 1990.

Government

Head of state: Petar Stoyanov from 1997
Head of government: Ivan Kostov from 1997
Political system: emergent democracy
Political executive: parliamentary
Administrative divisions: 28 regions divided into 278 municipalities
Political parties: Union of Democratic Forces (UDF), right of centre; Bulgarian Socialist Party (BSP), left wing, ex-communist; Movement for Rights and Freedoms (MRF), Turkish-oriented, centrist; Civic Alliances for the Republic (CAR), left of centre; Real Reform Movement (DESIR)
Armed forces: 101,500 (1998)
Conscription: compulsory for 12 months
Death penalty: abolished in 1998
Defence spend: (% GDP) 3.7 (1998)
Education spend: (% GNP) 3.2 (1997)
Health spend: (% GDP) 4.8 (1997)

Economy and resources

Currency: lev
GDP: (US$) 12.1 billion (1999)
Real GDP growth: (% change on previous year) 4 (1999)
GNP: (US$) 11.3 billion (1999)
GNP per capita (PPP): (US$) 4,914 (1999)
Consumer price inflation: 9.8% (2000 est)
Unemployment: 13.7% (1997)
Foreign debt: (US$) 10 billion (1999)
Major trading partners: EU countries (principally Germany, Greece, Italy), former USSR (principally Russia), Macedonia, USA
Resources: coal, iron ore, manganese, lead, zinc, petroleum
Industries: food products, petroleum and coal products, metals, mining, paper, beverages and tobacco, electrical machinery, textiles
Exports: base metals, chemical and rubber products, processed food, beverages, tobacco, chemicals, textiles, footwear. Principal market: Italy 13.9% (1999)
Imports: mineral products and fuels, chemical and rubber products, textiles, footwear, machinery and transport equipment, medicines. Principal source: Russia 20.6% (1999)
Arable land: 38% (1996)

Agricultural products: wheat, maize, barley, sunflower seeds, grapes, potatoes, tobacco, roses; viticulture (world's fourth-largest exporter of wine in 1989); forest resources

Population and society

Population: 8,225,000 (2000 est)
Population growth rate: −0.7% (1995–2000)
Population density: (per sq km) 75 (1999 est)
Urban population: (% of total) 70 (2000 est)
Age distribution: (% of total population) 0–14 16%, 15–59 63%, 60+ 21% (2000 est)
Ethnic groups: Southern Slavic Bulgarians constitute around 85% of the population; 9% are ethnic Turks, who during the later 1980s were subjected to government pressure to adopt Slavic names and to resettle elsewhere; 3% Gypsy, 3% Macedonian
Language: Bulgarian (official), Turkish
Religion: Eastern Orthodox Christian, Muslim, Jewish, Roman Catholic, Protestant
Education: (compulsory years) 8
Literacy rate: 99% (men); 98% (women) (2000 est)
Labour force: 24.3% agriculture, 32.2% industry, 43.5% services (1997)
Life expectancy: 68 (men); 75 (women) (1995–2000)
Child mortality rate: (under 5, per 1,000 live births) 20 (1995–2000)
Physicians: 1 per 298 people (1995)
Hospital beds: 1 per 102 people (1995)
TV sets: (per 1,000 people) 366 (1997)
Radios: (per 1,000 people) 543 (1997)
Internet users: (per 10,000 people) 241.6 (1999)
Personal computer users: (per 100 people) 2.7 (1999)

Transport

Airports: international airports: Sofia, Varna, Burgas; seven domestic airports; total passenger km: 1,796 million (1997 est)
Railways: total length: 4,292 km/2,667 mi; total passenger km: 5,059 million (1994)
Roads: total road network: 36,724 km/22,820 mi, of which 92% paved (1997); passenger cars: 208.9 per 1,000 people (1997)

Chronology

c. **3500 BC onwards:** Semi-nomadic pastoralists from the central Asian steppes settled in the area and formed the Thracian community. **mid-5th century BC:** The Thracian state was formed; it was to extend over Bulgaria, northern Greece, and northern Turkey. **4th century BC:** Phillip II and Alexander the Great of Macedonia waged largely unsuccessful campaigns against the Thracian Empire. **AD 50:** The Thracians were subdued and incorporated within the Roman Empire as the province of Moesia Inferior. **3rd–6th centuries:** The Thracian Empire was successively invaded and devastated by the Goths, Huns, Bulgars, and Avars. **681:** The Bulgars, an originally Turkic group that had merged with earlier Slav settlers, revolted against the Avars and established, south of the River Danube, the first Bulgarian kingdom, with its capital at Pliska. **864:** Orthodox Christianity was adopted by Boris I. **1018:** Subjugated by the Byzantines, whose empire had its capital at Constantinople; led to Bulgarian Church breaking with Rome in 1054. **1185:** Second independent Bulgarian Kingdom formed. **mid-13th century:** Bulgarian state destroyed by Mongol incursions. **1396:** Bulgaria became the first European state to be absorbed into the Turkish Ottoman Empire; the imposition of a harsh feudal system and the sacking of the monasteries followed. **1859:** The Bulgarian Catholic Church re-established links with Rome. **1876:** A Bulgarian

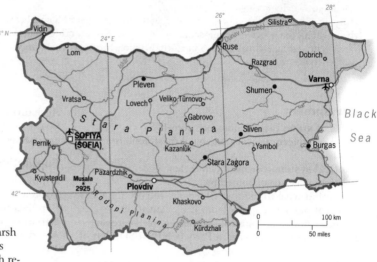

nationalist revolt against Ottoman rule was crushed brutally by Ottomans, with 15,000 massacred at Plovdiv ('Bulgarian Atrocities'). **1878:** At the Congress of Berlin, concluding a Russo-Turkish war in which Bulgarian volunteers had fought alongside the Russians, the area south of the Balkans, Eastern Rumelia, remained an Ottoman province, but the area to the north became the autonomous Principality of Bulgaria, with a liberal constitution and Alexander Battenberg as prince. **1885:** Eastern Rumelia annexed by the Principality; Serbia defeated in war. **1908:** Full independence proclaimed from Turkish rule, with Ferdinand I as tsar. **1913:** Following defeat in the Second Balkan War, King Ferdinand I abdicated and was replaced by his son Boris III. **1919:** Bulgarian Agrarian Union government, led by Alexander Stamboliiski, came to power and redistributed land to poor peasants. **1923:** Agrarian government was overthrown in right-wing coup and Stamboliiski murdered. **1934:** A semifascist dictatorship was established by King Boris III. **1944:** Soviet invasion of German-occupied Bulgaria. **1946:** The monarchy was abolished and a communist-dominated people's republic proclaimed following a plebiscite. **1947:** Gained South Dobruja in the northeast, along the Black Sea, from Romania; Soviet-style constitution established a one-party state; industries and financial institutions were nationalized and cooperative farming introduced. **1954:** Bulgaria became a loyal and cautious satellite of the USSR. **1968:** Bulgaria participated in the Soviet-led invasion of Czechoslovakia. **1971:** A new constitution was introduced. **1985–89:** Haphazard administrative and economic reforms, known as *preustroistvo* ('restructuring'), were introduced under the stimulus of the reformist Soviet leader Mikhail Gorbachev. **1989:** A programme of enforced 'Bulgarianization' resulted in a mass exodus of ethnic Turks to Turkey. Opposition parties were tolerated. **1991:** A new liberal-democratic constitution was adopted. The first noncommunist government was formed. **1993:** A voucher-based 'mass privatization' programme was launched. **1996:** Radical economic and industrial reforms were imposed. There was mounting inflation and public protest at the state of the economy. **1997:** There was a general strike. The UDF leader Ivan Kostov became prime minister. The Bulgarian currency was pegged to the Deutschmark in return for support from the International Monetary Fund. A new political group, the Real Reform Movement (DESIR), was formed. **1999:** Bulgaria joined the Central European Free Trade Agreement (CEFTA).

Practical information

Visa requirements: UK: visa required. USA: visa not required for tourist visits of up to 30 days
Time difference: GMT +2

Chief tourist attractions: Black Sea coastal resorts; mountain scenery; historic towns; skiing resorts
Major holidays: 1 January, 3 March, 1, 24 May, 24–25 December; variable: Easter Monday

BURKINA FASO FORMERLY UPPER VOLTA (UNTIL 1984)

Landlocked country in west Africa, bounded east by Niger, northwest and west by Mali, and south by Côte d'Ivoire, Ghana, Togo, and Benin.

Area: 274,122 sq km/105,838 sq mi
Capital: Ouagadougou
Major towns/cities: Bobo-Dioulasso, Koudougou, Banfora, Ouahigouya, Tenkodogo
Physical features: landlocked plateau with hills in west and southeast; headwaters of the River Volta; semiarid in north, forest and farmland in south; linked by rail to Abidjan in Côte d'Ivoire, Burkina Faso's only outlet to the sea

Red symbolizes the revolution of 1984. The five-pointed star is said to signify the revolution or freedom. Green stands for the country's natural resources. Effective date: 4 August 1984.

Government

Head of state: Blaise Compaoré from 1987
Head of government: Paramanga Ernest Yonli from 2000
Political system: emergent democracy
Political executive: limited presidency
Administrative divisions: 10 regions and 45 provinces
Political parties: Popular Front (FP), centre-left coalition grouping; National Convention of Progressive Patriots–Democratic Socialist Party (CNPP–PSD), left of centre
Armed forces: 5,800 (1998); includes gendarmerie of 4,200
Conscription: military service is voluntary
Death penalty: retained and used for ordinary crimes
Defence spend: (% GDP) 2.5 (1998)
Education spend: (% GNP) 1.5 (1996)
Health spend: (% GDP) 4.2 (1997)

Economy and resources

Currency: franc CFA
GDP: (US$) 2.64 billion (1999)
Real GDP growth: (% change on previous year) 5.8 (1999)
GNP: (US$) 2.6 billion (1999)
GNP per capita (PPP): (US$) 898 (1999 est)
Consumer price inflation: –1.1% (1999)
Unemployment: 8.1% (1994 est)
Foreign debt: (US$) 1.39 million (1998)
Major trading partners: France, Italy, Côte d'Ivoire, Portugal, USA, Thailand, Nigeria
Resources: manganese, zinc, limestone, phosphates, diamonds, gold, antimony, marble, silver, lead
Industries: food processing, textiles, cotton ginning, brewing, processing of hides and skins
Exports: cotton, gold, livestock and livestock products. Principal market: Italy 10.1% (1998)
Imports: machinery and transport equipment, miscellaneous manufactured articles, food products (notably cereals), refined petroleum products, chemicals. Principal source: France 32.8% (1998)
Arable land: 12.4% (1996)

Agricultural products: cotton, sesame seeds, sheanuts (karité nuts), millet, sorghum, maize, sugar cane, rice, groundnuts; livestock rearing (cattle, sheep, and goats)

Population and society

Population: 11,937,000 (2000 est)
Population growth rate: 2.7% (1995–2000)
Population density: (per sq km) 42 (1999 est)
Urban population: (% of total) 19 (2000 est)
Age distribution: (% of total population) 0–14 47%, 15–59 48%, 60+ 5% (2000 est)
Ethnic groups: over 50 ethnic groups, including the nomadic Mossi (48%), Fulani (8%), Gourma (7%), and Bisa-Samo (6%). Settled tribes include: in the north the Lobi-Dagari (4%) and the Mande (7%); in the southeast the Bobo (7%); and in the southwest the Senoufu (6%) and Gourounsi (6%)
Language: French (official), 50 Sudanic languages (90%)
Religion: animist 40%, Sunni Muslim 50%, Christian (mainly Roman Catholic) 10%
Education: (compulsory years) 6
Literacy rate: 34% (men); 14% (women) (2000 est)
Labour force: 54% of population: 92% agriculture, 2% industry, 6% services (1990)
Life expectancy: 44 (men); 45 (women) (1995–2000)
Child mortality rate: (under 5, per 1,000 live births) 171 (1995–2000)
Physicians: 1 per 34,804 people (1993 est)
Hospital beds: 1 per 3,300 people (1993 est)
TV sets: (per 1,000 people) 6 (1997)
Radios: (per 1,000 people) 33 (1997)
Internet users: (per 10,000 people) 3.4 (1999)
Personal computer users: (per 100 people) 0.1 (1999)

Transport

Airports: international airports: Ouagadougou, Bobo-Dioulasso; total passenger km: 256 million (1995)
Railways: total length: 622 km/389 mi; total passenger km: 405 million (1995)
Roads: total road network: 12,100 km/7,519 mi, of which 16% paved (1996 est); passenger cars: 3.6 per 1,000 people (1996 est)

Chronology

13th–14th centuries: Formerly settled by Bobo, Lobi, and Gurunsi peoples, east and centre were conquered by Mossi and Gurma peoples, who established powerful warrior kingdoms, some of which survived until late 19th century. **1895–1903:** France secured protectorates over the Mossi kingdom of Yatenga and the Gurma region, and annexed the Bobo and Lobi lands, meeting armed resistance. **1904:** The French-controlled region, known as Upper Volta, was attached administratively to French Sudan; tribal chiefs were maintained in their traditional seats and the region was to serve as a labour reservoir for more developed colonies to the south. **1919:** Made a separate French colony. **1932:** Partitioned between French Sudan, the Côte d'Ivoire, and Niger. **1947:** Became a French overseas territory. **1960:** Independence was achieved, with Maurice Yaméogo as the first president. **1966:** A military coup was led by Lt-Col Sangoulé Lamizana, and a supreme council of the armed forces established. **1977:** The ban on political activities was removed. A referendum approved a new constitution based on civilian rule. **1978–80:** Lamizana was elected president. In 1980 he was overthrown in a bloodless coup led by Col Saye Zerbo, as the economy deteriorated. **1982–83:** Maj Jean-Baptiste Ouedraogo became president and Capt Thomas Sankara prime minister. In 1983 Sankara seized complete power. **1984:** Upper Volta was renamed Burkina Faso ('land of upright men') to signify a break with the colonial past; literacy and afforestation campaigns were instigated by Sankara, who established links with Libya, Benin, and Ghana. **1987:** Capt Blaise Compaoré became president. **1991:** A new constitution was approved. **1992:** Multiparty elections were won by the pro-Compaoré Popular Front (FP). **1996:** Kadre Desire Ouedraogo was appointed prime minister. **1997:** The CDP won assembly elections. Ouedraogo was reappointed prime minister. **1998:** President Blaise Compaoré was re-elected with an overwhelming majority. **2000:** Prime Minister Ouedraogo resigned and was replaced by Paramanga Ernest Yonli.

Practical information

Visa requirements: UK: visa required. USA: visa required
Time difference: GMT+/–0
Chief tourist attractions: big-game hunting in east and southwest, and along banks of Mouhoun (Black Volta) River; wide variety of wildlife; biennial Ouagadougou film festival
Major holidays: 1, 3 January, 1 May, 4, 15 August, 1 November, 25 December; variable: Ascension Thursday, Eid-ul-Adha, Easter Monday, end of Ramadan, Prophet's Birthday, Whit Monday

BURUNDI FORMERLY URUNDI (UNTIL 1962)

Located in east central Africa, bounded north by Rwanda, west by the Democratic Republic of Congo, southwest by Lake Tanganyika, and southeast and east by Tanzania.

National name: *Republika y'Uburundi/République du Burundi/Republic of Burundi*
Area: 27,834 sq km/10,746 sq mi
Capital: Bujumbura
Major towns/cities: Gitega, Bururi, Ngozi, Muyinga, Ruyigi, Kayanza
Physical features: landlocked grassy highland straddling watershed of Nile and Congo; Lake Tanganyika, Great Rift Valley

Green expresses hope. White symbolizes peace. It is said that the saltire may have been based on the former flag of Belgian airline, Sabena. Red represents the blood shed in the struggle for independence.
Effective date: 27 September 1982.

Government

Head of state and government: Pierre Buyoya from 1996
Political system: military
Political executive: military
Administrative divisions: 15 provinces
Political parties: Front for Democracy in Burundi (FRODEBU), left of centre; Union for National Progress (UPRONA), nationalist socialist; Socialist Party of Burundi (PSB); People's Reconciliation Party (PRP)
Armed forces: 40,000 (1998); plus paramilitary forces of 3,500
Conscription: military service is voluntary
Death penalty: retained and used for ordinary crimes
Defence spend: (% GDP) 7.2 (1998)
Education spend: (% GNP) 3.2 (1996)
Health spend: (% GDP) 4 (1997)

Economy and resources

Currency: Burundi franc
GDP: (US$) 701 million (1999)
Real GDP growth: (% change on previous year) –1 (1999)
GNP: (US$) 800 million (1999)
GNP per capita (PPP): (US$) 553 (1999 est)
Consumer price inflation: 26% (1999 est)
Unemployment: 7.3% (1992)
Foreign debt: (US$) 1.12 billion (1999)
Major trading partners: Belgium, Germany, Zambia, France, Switzerland, Kenya, South Africa
Resources: nickel, gold, tungsten, phosphates, vanadium, uranium, peat; petroleum deposits have been detected
Industries: textiles, leather, food and agricultural products
Exports: coffee, tea, glass products, hides and skins. Principal market: Germany 17.2% (1999)
Imports: machinery and transport equipment, petroleum and petroleum products, cement, malt (and malt flour). Principal source: Belgium 20.2%

(1999)
Arable land: 30% (1996)
Agricultural products: coffee, tea, cassava, sweet potatoes, bananas, beans; cattle rearing

Population and society

Population: 6,695,000 (2000 est)
Population growth rate: 1.7% (1995–2000)
Population density: (per sq km) 236 (1999 est)
Urban population: (% of total) 9 (2000 est)
Age distribution: (% of total population) 0–14 46%, 15–59 50%, 60+ 4% (2000 est)
Ethnic groups: two main groups: the agriculturalist Hutu, comprising about 85% of the population, and the predominantly pastoralist Tutsi, about 14%. There is a small Pygmy (Twa) minority, comprising about 1% of the population, and a few Europeans and Asians
Language: Kirundi, French (both official), Kiswahili
Religion: Roman Catholic 62%, Pentecostalist 5%, Anglican 1%, Muslim 1%, animist
Education: (compulsory years) 6
Literacy rate: 56% (men); 41% (women) (2000 est)
Labour force: 90.7% agriculture, 2.1% industry, 7.2% services (1997 est)
Life expectancy: 41 (men); 44 (women) (1995–2000)
Child mortality rate: (under 5, per 1,000 live births) 179 (1995–2000)
Physicians: 1 per 17,210 people (1995 est)
Hospital beds: 1 per 1,526 people (1995 est)
TV sets: (per 1,000 people) 10 (1997)
Radios: (per 1,000 people) 71 (1997)
Internet users: (per 10,000 people) 3.1 (1999)

Transport

Airports: international airports: Bujumbura; total passenger km: 2 million (1995)
Railways: none
Roads: total road network: 14,480 km/8,998 mi, of which 7.1% paved (1996); passenger cars: 2.8 per 1,000 people (1996 est)

Chronology

10th century: Originally inhabited by the hunter-gatherer Twa Pygmies. Hutu peoples settled in the region and became peasant farmers. **13th century:** Taken over by Banu Hutus. **15th–17th centuries:** The majority Hutu community came under the dominance of the cattle-owning Tutsi peoples, immigrants from the east, who became a semi-aristocracy; the minority Tutsis developed a feudalistic political system, organized around a nominal king, with royal princes in control of local areas. **1890:** Known as Urundi, the Tutsi kingdom, along with neighbouring Rwanda, came under nominal German control as Ruanda-Urundi. **1916:** Occupied by Belgium during World War I. **1923:** Belgium was granted a League of Nations mandate to administer Ruanda-Urundi; it was to rule 'indirectly' through the Tutsi chiefs. **1962:** Burundi was separated from Ruanda-Urundi, and given independence as a monarchy under Tutsi King Mwambutsa IV. **1965:** The king refused to appoint a Hutu prime minister after an election in which Hutu candidates were victorious; an attempted coup by Hutus was brutally suppressed. **1966:** The king was deposed by his teenage son Charles, who became Ntare V; he was in turn deposed by his Tutsi prime minister Col Michel Micombero, who declared Burundi a republic; the Tutsi-dominated Union for National Progress (UPRONA) was declared the only legal political party. **1972:** Ntare V was killed, allegedly by Hutus, provoking a massacre of 150,000 Hutus by Tutsi soldiers; 100,000 Hutus fled to Tanzania. **1976:** An army coup deposed Micombero and appointed the Tutsi Col Jean-Baptiste Bagaza as president. He launched a drive against corruption and a programme of land reforms and economic development. **1987:** Bagaza was deposed in a coup by the Tutsi Maj Pierre Buyoya. **1988:** About 24,000 Hutus were killed by Tutsis and 60,000 fled to Rwanda. **1992:** A new multiparty constitution was adopted following a referendum. **1993:** Melchior Ndadaye, a Hutu, was elected president in the first-ever democratic contest, but was killed in a coup by the Tutsi-dominated army; 100,000 people died in the massacres that followed. **1994:** Cyprien Ntaryamira, a Hutu, became president but was later killed in an air crash along with the Rwandan president Juvenal Habyarimana. There was an eruption of ethnic violence; 750,000 Hutus fled to Rwanda. Hutu Sylvestre Ntibantunganya became head of state, serving with a Tutsi prime minister, as part of a four-year power-sharing agreement. **1995:** Renewed ethnic violence erupted in the capital, Bujumbura, following a massacre of Hutu refugees. **1996:** The former Tutsi president Pierre Buyoya seized power amid renewed ethnic violence; the coup provoked economic sanctions by other African countries. A 'government of national unity' was appointed, with Pascal-Firmin Ndimira as premier. Bujumbura was shelled by Hutu rebels. **1998:** There was renewed fighting between Tutsi-led army and Hutu rebels. A ceasefire was agreed between the warring political factions. The position of head of government was abolished, with President Buyoya assuming the position's authority. **2000:** With the civil war worsening, Nelson Mandela, former president of South Africa, and the new mediator for Burundi, met government, opposition, and rebel leaders in Tanzania. A power-sharing peace agreement was reached by most political factions, though three Tutsi parties declined to sign. Despite the agreement, the war continued. .

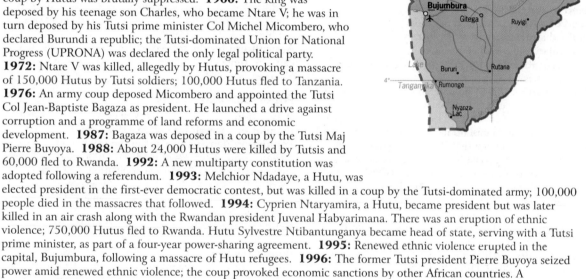

Practical information

Visa requirements: UK: visa required. USA: visa required
Time difference: GMT +2
Chief tourist attractions: tourism is relatively undeveloped

Major holidays: 1 January, 1 May, 1 July, 15 August, 18 September, 1 November, 25 December; variable: Ascension Thursday

CAMBODIA FORMERLY KHMER REPUBLIC (1970–76), KAMPUCHEA (1976–89)

Located in southeast Asia, bounded north and northwest by Thailand, north by Laos, east and southeast by Vietnam, and southwest by the Gulf of Thailand.

National name: *Preah Réaché'anachâkr Kâmpuchéa/Kingdom of Cambodia*
Area: 181,035 sq km/69,897 sq mi
Capital: Phnom Penh
Major towns/cities: Batdambang, Kâmpong Chhnang, Siĕmréab, Prey Vêng, Preah Seihânu
Major ports: Kâmpong Chhnang
Physical features: mostly flat, forested plains with mountains in southwest and north; Mekong River runs north–south; Lake Tonle Sap

The temple of Angkor Wat had five towers but often only three are depicted. Red and blue recall the earlier flags of Cambodia. Effective date: 20 June 1993.

Government

Head of state: King Norodom Sihanouk from 1991
Head of government: Hun Sen from 1998
Political system: emergent democracy
Political executive: dual executive
Administrative divisions: 20 provinces and three municipalities
Political parties: United Front for an Independent, Neutral, Peaceful, and Cooperative Cambodia (FUNCINPEC), nationalist, monarchist; Liberal Democratic Party (BLDP), republican, anticommunist (formerly the Khmer People's National Liberation Front (KPNLF)); Cambodian People's Party (CPP), reform socialist (formerly the communist Kampuchean People's Revolutionary Party (KPRP)); Cambodian National Unity Party (CNUP) (political wing of the Khmer Rouge), ultranationalist communist
Armed forces: 139,000 (1998)
Conscription: military service is compulsory for five years between ages 18 and 35
Death penalty: abolished in 1989
Defence spend: (% GDP) 4.2 (1998)
Education spend: (% GNP) 2.9 (1996)
Health spend: (% GDP) 7.2 (1997)

Economy and resources

Currency: Cambodian riel
GDP: (US$) 3.12 billion (1999)
Real GDP growth: (% change on previous year) 4.3 (1999)
GNP: (US$) 3 billion (1999)
GNP per capita (PPP): (US$) 1,286 (1999 est)
Consumer price inflation: 4.1% (1999)
Foreign debt: (US$) 829 million (1999 est)
Major trading partners: Singapore, Thailand, Vietnam, Japan, Hong Kong, USA, China
Resources: phosphates, iron ore, gemstones, bauxite, silicon, manganese
Industries: rubber processing, seafood processing, rice milling, textiles and garments, pharmaceutical products, cigarettes
Exports: timber, rubber, fishery products, garments. Principal market: Vietnam 18% (1997)

Imports: cigarettes, construction materials, petroleum products, motor vehicles, alcoholic beverages, consumer electronics. Principal source: Thailand 15.9% (1997)
Arable land: 21.1% (1996)
Agricultural products: rice, maize, sugar cane, cassava, bananas; timber and rubber (the two principal export commodities); fishing

Population and society

Population: 11,168,000 (2000 est)
Population growth rate: 2.2% (1995–2000)
Population density: (per sq km) 66 (1999 est)
Urban population: (% of total) 16 (2000 est)
Age distribution: (% of total population) 0–14 41%, 15–59 54%, 60+ 5% (2000 est)
Ethnic groups: 90% Khmer, 5% Vietnamese, 1% Chinese
Language: Khmer (official), French
Religion: Theravada Buddhist 95%, Muslim, Roman Catholic
Education: (compulsory years) 6
Literacy rate: 48% (men); 65% (women) (1995 est)
Labour force: 71.2% agriculture, 7.6% industry, 21.2% services (1997 est)
Life expectancy: 52 (men); 55 (women) (1995–2000)
Child mortality rate: (under 5, per 1,000 live births) 134 (1995–2000)
Physicians: 1 per 9,374 people (1993 est)
Hospital beds: 1 per 453 people (1993 est)
TV sets: (per 1,000 people) 124 (1997)
Radios: (per 1,000 people) 127 (1997)
Internet users: (per 10,000 people) 3.7 (1999)
Personal computer users: (per 100 people) 0.1 (1999)

Transport

Airports: international airports: Phnom Penh (Pochentong); five domestic airports
Railways: total length: 1,370 km/851 mi (1994); total passenger km: 38 million (1995)
Roads: total road network: 35,769 km/22,227 mi, of which 7.5% paved (1997); passenger cars: 4.8 per 1,000 people (1997)

Chronology

1st century AD: Part of the kingdom of Hindu-Buddhist Funan (Fou Nan), centred on Mekong delta region. **6th century:** Conquered by the Chenla kingdom. **9th century:** Establishment by Jayavarman II of extensive and sophisticated Khmer Empire, supported by an advanced irrigation system and architectural achievements. **14th century:** Theravada Buddhism replaced Hinduism. **15th century:** Came under the control of Siam (Thailand), which made Phnom Penh the capital and, later, Champa (Vietnam). **1863:** Became a French protectorate. **1887:** Became part of French Indo-China Union, which included Laos and Vietnam. **1941:** Prince Norodom Sihanouk was elected king. **1941–45:** Occupied by Japan during World War II. **1946:** Recaptured by France; parliamentary constitution adopted. **1949:** Guerrilla war for independence secured semi-autonomy within the French Union. **1953:** Independence was achieved from France as the Kingdom of Cambodia. **1955:** Norodom Sihanouk abdicated as king and became prime minister, representing the Popular Socialist Community mass movement. His father, Norodom Suramarit, became king. **1960:** On the death of his father, Norodom Sihanouk became head of state. **later 1960s:** There was mounting guerrilla insurgency, led by the communist Khmer Rouge, and civil war in neighbouring Vietnam. **1970:** Sihanouk was overthrown by US-backed Lt-Gen Lon Nol in a right-wing coup; the new name of Khmer Republic was adopted; Sihanouk, exiled in China, formed his own guerrilla movement. **1975:** Lon Nol was overthrown by the Khmer Rouge, which was backed by North Vietnam and China; Sihanouk became head of state. **1976:** The Khmer Republic was renamed Democratic Kampuchea. **1976–78:** The Khmer Rouge, led by Pol Pot, introduced an extreme Maoist communist

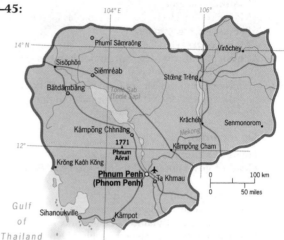

programme, forcing urban groups into rural areas and resulting in over 2.5 million deaths from famine, disease, and maltreatment; Sihanouk was removed from power. **1978–79:** Vietnam invaded and installed a government headed by Heng Samrin, an anti-Pol Pot communist. **1979:** Democratic Kampuchea was renamed the People's Republic of Kampuchea. **1980–82:** Faced by guerrilla resistance from Pol Pot's Chinese-backed Khmer Rouge and Sihanouk's Association of South East Asian Nations (ASEAN) and US-backed nationalists, more than 300,000 Cambodians fled to refugee camps in Thailand and thousands of soldiers were killed. **1985:** The reformist Hun Sen was appointed prime minister and more moderate economic and cultural policies were pursued. **1987–89:** Vietnamese troops were withdrawn. **1989:** The People's Republic of Kampuchea was renamed the State of Cambodia and Buddhism was re-established as the state religion. **1991:** There was a ceasefire, and a United Nations Transitional Authority in Cambodia (UNTAC) agreed to administer the country in conjunction with an all-party Supreme National Council; communism was abandoned. Sihanouk returned as head of state. **1992:** Political prisoners were released, refugees resettled, and freedom of speech restored. However, the Khmer Rouge refused to disarm. **1993:** FUNCINPEC won general elections (boycotted by the Khmer Rouge, who continued fighting); a new constitution was adopted. Sihanouk was reinstated as constitutional monarch; his son Prince Norodom Ranariddh, FUNCINPEC leader, was appointed prime minister, with CPP leader Hun Sen as deputy premier. **1994:** An antigovernment coup was foiled. Seven thousand Khmer Rouge guerrillas surrendered in response to an amnesty. **1995:** Prince Norodom Sirivudh, FUNCINPEC leader and half-brother of King Sihanouk, was exiled for allegedly plotting to assassinate Hun Sen and topple the government. **1996:** There were heightened tensions between Hun Sen's CPP and the royalist FUNCINPEC. **1997:** Pol Pot was sentenced to life imprisonment. FUNCINPEC troops were routed by the CPP, led by Hun Sen. Prime Minister Prince Norodom Ranariddh was deposed and replaced by Ung Huot. There was fighting between supporters of Hun Sen and Ranariddh. **1998:** Ranariddh was found guilty of arms smuggling and colluding with the Khmer Rouge, but was pardoned by the king. Pol Pot died and thousands of Khmer Rouge guerrillas defected. The CPP won elections, and political unrest followed. A new CPP–FUNCINPEC coalition was formed, with Hun Sen as prime minister and Prince Norodom Ranariddh as president. Cambodia re-occupied its UN seat.

Practical information

Visa requirements: UK: visa required. USA: visa required
Time difference: GMT +7
Chief tourist attractions: ancient Khmer ruins and monuments, including great temples of Angkor Thom and Prasat Lingpoun; tropical vegetation and mangrove forests
Major holidays: 9 January, 17 April, 1, 20 May, 22 September; variable: New Year (April)

CAMEROON FORMERLY KAMERUN (UNTIL 1916)

Located in west Africa, bounded northwest by Nigeria, northeast by
Chad, east by the Central African Republic, south by the Republic of the
Congo, Gabon, and Equatorial Guinea, and west by the Atlantic.

National name: *République du Cameroun/Republic of Cameroon*
Area: 475,440 sq km/183,567 sq mi
Capital: Yaoundé
Major towns/cities: Garoua, Douala, Nkongsamba, Maroua, Bamenda,
Bafoussam, Ngaoundéré
Major ports: Douala
Physical features: desert in far north in the Lake Chad basin, mountains
in west, dry savannah plateau in the intermediate area, and dense tropical
rainforest in south; Mount Cameroon 4,070 m/13,358 ft, an active volcano
on the coast, west of the Adamawa Mountains

The single star represents the unity of the former
French and British territories. Effective date: 20 May
1975.

Government

Head of state: Paul Biya from 1982
Head of government: Peter Musonge Mafani from
1996
Political system: emergent democracy
Political executive: limited presidency
Administrative divisions: ten provinces
Political parties: Cameroon People's Democratic
Movement (RDPC), nationalist, left of centre; Front of
Allies for Change (FAC), left of centre (There are 47
parties in Cameroon and seven parties in parliament)
Armed forces: 13,100 (1998); plus 9,000 paramilitary
forces
Conscription: military service is voluntary;
paramilitary compulsory training programme in force
Death penalty: retained and used for ordinary crimes
Defence spend: (% GDP) 2.9 (1998)
Education spend: (% GNP) 2.9 (1996)
Health spend: (% GDP) 5 (1997)

Economy and resources

Currency: franc CFA
GDP: (US$) 9.8 billion (1999)
Real GDP growth: (% change on previous year) 4.4
(1999)
GNP: (US$) 8.5 billion (1999)
GNP per capita (PPP): (US$) 1,444 (1999)
Consumer price inflation: 2.6% (1999)
Unemployment: N/A
Foreign debt: (US$) 10.5 billion (1999)
Major trading partners: France, Spain, Italy, Germany,
the Netherlands, Belgium, Nigeria
Resources: petroleum, natural gas, tin ore, limestone,
bauxite, iron ore, uranium, gold
Industries: petroleum refining, aluminium smelting,
cement, food processing, footwear, beer, cigarettes
Exports: crude petroleum and petroleum products,
timber and timber products, coffee, aluminium, cotton,
bananas. Principal market: France 22% (1999)
Imports: machinery and transport equipment, basic

manufactures, chemicals, fuel. Principal source: France
35% (1999)
Arable land: 12.8% (1996)
Agricultural products: coffee, cocoa, cotton, cassava,
sorghum, millet, maize, plantains, palm (oil and
kernels), rubber, bananas; livestock rearing (cattle and
sheep); forestry; fishing

Population and society

Population: 15,085,000 (2000 est)
Population growth rate: 2.7% (1995–2000)
Population density: (per sq km) 31 (1999 est)
Urban population: (% of total) 49 (2000 est)
Age distribution: (% of total population) 0–14 43%,
15–59 52%, 60+ 5% (2000 est)
Ethnic groups: main groups include the Cameroon
Highlanders (31%), Equatorial Bantu (19%), Kirdi
(11%), Fulani (10%), Northwestern Bantu (8%), and
Eastern Nigritic (7%)
Language: French, English (both official; often spoken
in pidgin), Sudanic languages (in the north), Bantu
languages (elsewhere); there has been some discontent
with the emphasis on French – there are 163 indigenous
peoples with their own African languages
Religion: animist 50%, Christian 33%, Muslim 16%
Education: (compulsory years) 6 in Eastern Cameroon;
7 in Western Cameroon
Literacy rate: 82% (men); 70% (women) (2000 est)
Labour force: 62.7% agriculture, 8.7% industry, 28.6%
services (1997 est)
Life expectancy: 53 (men); 56 (women) (1995–2000)
Child mortality rate: (under 5, per 1,000 live births)
114 (1995–2000)
Physicians: 1 per 16,181 people (1995 est)
Hospital beds: 1 per 357 people (1995 est)
TV sets: (per 1,000 people) 81 (1997)
Radios: (per 1,000 people) 163 (1997)
Internet users: (per 10,000 people) 13.6 (1999)
Personal computer users: (per 100 people) 0.3
(1999)

Transport

Airports: international airports: Douala, Garoua, Yaoundé; eight domestic airports; total passenger km: 681 million (1997 est)
Railways: total length: 1,104 km/686 mi; total passenger

km: 352 million (1993)
Roads: total road network: 34,300 km/21,314 mi, of which 12.5% paved (1995 est); passenger cars: 7.4 per 1,000 people (1996 est)

Chronology

1472: First visited by the Portuguese, who named it the Rio dos Camaroes ('River of Prawns') after the giant shrimps they found in the Wouri River estuary, and later introduced slave trading. **early 17th century:** The Douala people migrated to the coastal region from the east and came to serve as intermediaries between Portuguese, Dutch, and English traders and interior tribes. **1809–48:** The northern savannahs were conquered by the Fulani, Muslim pastoral nomads from the southern Sahara. **1856:** Douala chiefs signed a commercial treaty with Britain and invited British protection. **1884:** A treaty was signed establishing German rule as the protectorate of Kamerun; cocoa, coffee, and banana plantations were developed. **1916:** Captured by Allied forces in World War I. **1919:** Divided under League of Nations' mandates between Britain, which administered the southwest and north (adjoining Nigeria), and France, which administered the east and south. **1946:** The French Cameroon and British Cameroons were made UN trust territories. **1955:** The French crushed a revolt by the Union of the Cameroon Peoples (UPC), southern-based radical nationalists. **1960:** French Cameroon became the independent Republic of Cameroon, with the Muslim Ahmadou Ahidjo as president; a UPC rebellion in the southwest was crushed, and a state of emergency declared. **1961:** Following a UN plebiscite, the northern part of the British Cameroons merged with Nigeria, and the southern part joined the Republic of Cameroon to become the Federal Republic of Cameroon. **1966:** An autocratic one-party regime was introduced; government and opposition parties merged to form the Cameroon National Union (UNC). **1970s:** Petroleum exports made successful investment in education and agriculture possible. **1972:** A new constitution made Cameroon a unitary state. **1982:** President Ahidjo resigned; he was succeeded by his prime minister Paul Biya, a Christian. **1983–84:** Biya began to remove the northern Muslim political 'barons' close to Ahidjo, who went into exile in France. Biya defeated a plot by Muslim officers from the north to overthrow him. **1985:** The UNC adopted the name RDPC. **1990:** There was widespread public disorder as living standards declined; Biya granted an amnesty to political prisoners. **1992:** The ruling RDPC won the first multiparty elections in 28 years, with Biya as president. **1995:** Cameroon was admitted to the Commonwealth. **1996:** Peter Musonge Mafani became prime minister. **1997:** RDPC won assembly elections; Biya was re-elected.

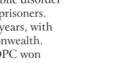

Practical information

Visa requirements: UK: visa not required. USA: visa not required
Chief tourist attractions: national parks; game

reserves; sandy beaches
Major holidays: 1 January, 11 February, 1, 20 May, 15 August, 25 December; variable: Ascension Thursday, Eid-ul-Adha, end of Ramadan, Good Friday

CANADA

Occupies the northern part of the North American continent; bounded to the south by the USA, north by the Arctic Ocean, northwest by Alaska, east by the Atlantic Ocean, and west by the Pacific Ocean.

Area: 9,970,610 sq km/3,849,652 sq mi
Capital: Ottawa
Major towns/cities: Toronto, Montréal, Vancouver, Edmonton, Calgary, Winnipeg, Québec, Hamilton, Saskatoon, Halifax, London, Kitchener, Mississauga, Laval, Surrey
Physical features: mountains in west, with low-lying plains in interior and rolling hills in east; St Lawrence Seaway, Mackenzie River; Great Lakes; Arctic Archipelago; Rocky Mountains; Great Plains or Prairies; Canadian Shield; Niagara Falls; climate varies from temperate in south to arctic in north; 45% of country forested

The maple leaf is a traditional Canadian emblem. Red recalls Canadian lives lost during World War I. White stands for snow. Effective date: 15 February 1965.

Government

Head of state: Queen Elizabeth II from 1952, represented by Governor General Adrienne Clarkson from 1999
Head of government: Jean Chrétien from 1993
Political system: liberal democracy
Political executive: parliamentary
Administrative divisions: ten provinces and three territories
Political parties: Liberal Party, nationalist, centrist; Bloc Québécois, Québec-based, separatist; Reform Party, populist, right wing; New Democratic Party (NDP), moderate left of centre; Progressive Conservative Party (PCP), free enterprise, right of centre; Confederation of Regions (COR); Party of New Brunswick; Green Party of Canada
Armed forces: 60,600 (1998)
Conscription: military service is voluntary
Death penalty: abolished 1998
Defence spend: (% GDP) 1.1 (1998)
Education spend: (% GNP) 7 (1996)
Health spend: (% GDP) 9.1 (1997)

Economy and resources

Currency: Canadian dollar
GDP: (US$) 612.05 billion (1999)
Real GDP growth: (% change on previous year) 4.5 (1999)
GNP: (US$) 591.4 billion (1999)
GNP per capita (PPP): (US$) 23,725 (1999)
Consumer price inflation: 1.7% (1999)
Unemployment: 8.3% (1998)
Major trading partners: USA, EU countries, Japan, China, Mexico, South Korea
Resources: petroleum, natural gas, coal, copper (world's third-largest producer), nickel (world's second-largest producer), lead (world's fifth-largest producer), zinc (world's largest producer), iron, gold, uranium, timber
Industries: transport equipment, food products, paper and related products, wood industries, chemical products, machinery

Exports: motor vehicles and parts, lumber, wood pulp, paper and newsprint, crude petroleum, natural gas, aluminium and alloys, petroleum and coal products. Principal market: USA 85.8% (1999)
Imports: motor vehicle parts, passenger vehicles, computers, foodstuffs, telecommunications equipment. Principal source: USA 76.3% (1999)
Arable land: 4.9% (1996)
Agricultural products: wheat, barley, maize, oats, rapeseed, linseed; livestock production (cattle and pigs)

Population and society

Population: 31,147,000 (2000 est)
Population growth rate: 1% (1995–2000)
Population density: (per sq km) 3 (1999 est)
Urban population: (% of total) 77 (2000 est)
Age distribution: (% of total population) 0–14 19%, 15–59 64%, 60+ 17% (2000 est)
Ethnic groups: about 40% of British Irish origin, 27% French, 20% of other European descent, about 2% American Indians and Inuit, and 11% other, mostly Asian
Language: English (60%), French (24%) (both official), American Indian languages, Inuktitut (Inuit)
Religion: Roman Catholic 45%, various Protestant denominations
Education: (compulsory years) 10
Literacy rate: 99% (men); 99% (women) (2000 est)
Labour force: 3.9% agriculture, 23.2% industry, 72.9% services (1997)
Life expectancy: 76 (men); 82 (women) (1995–2000)
Child mortality rate: (under 5, per 1,000 live births) 7 (1995–2000)
Physicians: 1 per 476 people (1996)
Hospital beds: 1 per 185 people (1994)
TV sets: (per 1,000 people) 708 (1997)
Radios: (per 1,000 people) 1,077 (1997)
Internet users: (per 10,000 people) 3,607.6 (1999)
Personal computer users: (per 100 people) 36.1 (1999)

Transport

Airports: international airports: Calgary, Edmonton, Gander, Halifax, Hamilton, Montréal (Dorval, Mirabel), Ottawa (Uplands), St John's, Saskatoon, Toronto (Lester B Pearson), Vancouver, Winnipeg; domestic services to all major cities/towns; total passenger km: 61,862 million (1997 est) **Railways:** total length: 70,739 km/43,957 mi; total passenger km: 1,519 million (1996) **Roads:** total road network: 901,902 km/560,442 mi, of which 35.3% paved (1995); passenger cars: 458.2 per 1,000 people (1996)

Chronology

35,000 BC: First evidence of people reaching North America from Asia by way of Beringia. *c.* **2000 BC:** Inuit (Eskimos) began settling the Arctic coast from Siberia eastwards to Greenland. *c.* **AD 1000:** Vikings, including Leif Ericsson, established Vinland, a settlement in northeast America that did not survive. **1497:** John Cabot, an Italian navigator in the service of English king Henry VII, landed on Cape Breton Island and claimed the area for England. **1534:** French navigator Jacques Cartier reached the Gulf of St Lawrence and claimed the region for France. **1608:** Samuel de Champlain, a French explorer, founded Québec; French settlers developed fur trade and fisheries. **1663:** French settlements in Canada formed the colony of New France, which expanded southwards. **1670:** Hudson's Bay Company established trading posts north of New France, leading to Anglo-French rivalry. **1689–97:** King William's War: Anglo-French conflict in North America arising from the 'Glorious Revolution' in Europe. **1702–13:** Queen Anne's War: Anglo-French conflict in North America arising from the War of the Spanish Succession in Europe; Britain gained Newfoundland. **1744–48:** King George's War: Anglo-French conflict in North America arising from the War of Austrian Succession in Europe. **1756–63:** Seven Years' War: James Wolfe captured Québec in 1759; France ceded Canada to Britain by the Treaty of Paris. **1775–83:** American Revolution caused an influx of 40,000 United Empire Loyalists, who formed New Brunswick in 1784. **1791:** Canada was divided into Upper Canada (much of modern Ontario) and Lower Canada (much of modern Québec). **1793:** British explorer Alexander Mackenzie crossed the Rocky Mountains to reach the Pacific coast. **1812–14:** War of 1812 between Britain and USA; US invasions repelled by both provinces. **1820s:** Start of large-scale immigration from British Isles caused resentment among French Canadians. **1837:** Rebellions were led by Louis Joseph Papineau in Lower Canada and William Lyon Mackenzie in Upper Canada. **1841:** Upper and Lower Canada united as Province of Canada; achieved internal self-government in 1848. **1867:** British North America Act united Ontario, Québec, Nova Scotia, and New Brunswick in Dominion of Canada. **1869:** Red River Rebellion of Métis (people of mixed French and American Indian descent), led by Louis Riel, against British settlers in Rupert's Land. **1870:** Manitoba (part of Rupert's Land) formed the fifth province of Canada; British Columbia became the sixth in 1871, and Prince Edward Island became the seventh in 1873. **1885:** The Northwest Rebellion was crushed and Riel hanged. The Canadian Pacific Railway was completed. **1905:** Alberta and Saskatchewan were formed from the Northwest Territories and became provinces of Canada. **1914–18:** Half a million Canadian troops fought for the British Empire on the western front in World War I. **1931:** The Statute of Westminster affirmed equality of status between Britain and the Dominions. **1939–45:** World War II: Canadian participation in all theatres. **1949:** Newfoundland became the tenth province of Canada; Canada was a founding member of the North Atlantic Treaty

Organization (NATO). **1960:** The Québec Liberal Party of Jean Lesage launched a 'Quiet Revolution' to re-assert French-Canadian identity. **1970:** Pierre Trudeau invoked the War Measures Act to suppress separatist terrorists of the Front de Libération du Québec. **1976:** The Parti Québécois won control of the Québec provincial government; a referendum rejected independence in 1980. **1982:** 'Patriation' of the constitution removed Britain's last legal control over Canada. **1987:** Meech Lake Accord: a constitutional amendment was proposed to increase provincial powers (to satisfy Québec); it failed to be ratified in 1990. **1992:** A self-governing homeland for the Inuit was approved. **1994:** Canada formed the North American Free Trade Area with USA and Mexico. **1995:** A Québec referendum narrowly rejected a sovereignty proposal. **1997:** The Liberals were re-elected by a narrow margin. **1999:** The government passed a bill making secession by Québec more difficult to achieve. **2000:** The Liberals were elected for a third term.

Practical information

Visa requirements: UK: visa not required. USA: visa not required
Time difference: GMT –3.5/9
Chief tourist attractions: forests; lakes; rivers; Rockies of British Columbia; St Lawrence Seaway; Niagara Falls; fjords of Newfoundland and Labrador; historic cities of Montréal and Québec; museums and art galleries of Toronto, Vancouver, and Ottawa
Major holidays: 1 January, 1 July (except Newfoundland), 11 November, 25–26 December; variable: Good Friday, Easter Monday, Labour Day (September), Thanksgiving (October), Victoria (May), additional days vary between states

CAPE VERDE

A group of islands in the Atlantic, west of Senegal (West Africa).

National name: *República de Cabo Verde/Republic of Cape Verde*
Area: 4,033 sq km/1,557 sq mi
Capital: Praia
Major towns/cities: Mindelo, Santa Marta
Major ports: Mindelo
Physical features: archipelago of ten volcanic islands 565 km/350 mi
west of Senegal; the windward (Barlavento) group includes Santo Antão,
São Vicente, Santa Luzia, São Nicolau, Sal, and Boa Vista; the leeward
(Sotovento) group comprises Maio, São Tiago, Fogo, and Brava; all but
Santa Luzia are inhabited

Blue symbolizes the ocean. The stars represent the ten main islands. The red stripe stands for the road to progress. Effective date: 25 September 1992.

Government

Head of state: Antonio Mascarenhas Monteiro from
1991
Head of government: Gualberto do Rosário from 2000
Political system: emergent democracy
Political executive: limited presidency
Administrative divisions: 14 districts
Political parties: African Party for the Independence of
Cape Verde (PAICV), African nationalist; Movement for
Democracy (MPD), moderate, centrist; Party for
Democratic Convergence (PCD), centrist; Party of Work
and Solidarity (PTS)
Armed forces: 1,100 (1998)
Conscription: selective conscription
Death penalty: abolished in 1981
Defence spend: (% GDP) 1.6 (1998)
Education spend: (% GNP) 4.4 (1993–94)
Health spend: (% GDP) 2.8 (1997)

Economy and resources

Currency: Cape Verde escudo
GDP: (US$) 581 million (1999)
Real GDP growth: (% change on previous year) 8
(1999)
GNP: (US$) 569 million (1999)
GNP per capita (PPP): (US$) 3,497 (1999 est)
Consumer price inflation: 4.3% (1999)
Unemployment: N/A
Foreign debt: (US$) 250 million (1999)
Major trading partners: Portugal, Germany, France,
UK, the Netherlands, Spain, Guinea-Bissau
Resources: salt, pozzolana (volcanic rock), limestone,
basalt, kaolin
Industries: fish processing, machinery and electrical
equipment, transport equipment, textiles, chemicals,
rum
Exports: fish, shellfish and fish products, salt, bananas.
Principal market: Portugal 42.9% (1998)
Imports: food and live animals, machinery and
electrical equipment, transport equipment, mineral
products, metals. Principal source: Portugal 52.4%
(1998)

Arable land: 9.7% (1995)
Agricultural products: maize, beans, potatoes, cassava,
coconuts, sugar cane, bananas, coffee, groundnuts;
fishing (mainly tuna, lobster, shellfish)

Population and society

Population: 428,000 (2000 est)
Population growth rate: 2.3% (1995–2000)
Population density: (per sq km) 104 (1999 est)
Urban population: (% of total) 62 (2000 est)
Age distribution: (% of total population) 0–14 39%,
15–59 55%, 60+ 6% (2000 est)
Ethnic groups: about 70% of mixed descent
(Portuguese and African), known as mestizos or Creoles;
the remainder is mainly African. The European
population is very small
Language: Portuguese (official), Creole
Religion: Roman Catholic 93%, Protestant (Nazarene
Church)
Education: (compulsory years) 6
Literacy rate: 85% (men); 66% (women) (2000 est)
Labour force: 37% of population: 31% agriculture, 30%
industry, 40% services (1990)
Life expectancy: 66 (men); 71 (women) (1995–2000)
Child mortality rate: (under 5, per 1,000 live births)
64 (1995–2000)
Physicians: 1 per 5,280 people (1990 est)
Hospital beds: 1 per 930 people (1990 est)
TV sets: (per 1,000 people) 3.6 (1995)
Radios: (per 1,000 people) 179 (1995)
Internet users: (per 10,000 people) 119.7 (1999)

Transport

Airports: international airports: Sal Island (Amílcar
Cabral), São Tiago; eight domestic airports; total
passenger km: 181 million (1995)
Railways: none
Roads: total road network: 1,100 km/684 mi, of which
78% paved (1996 est); passenger cars: 7.6 per 1,000
people (1996 est)

Chronology

1462: Originally uninhabited; settled by Portuguese, who brought in slave labour from West Africa. **later 19th century:** There was a decline in prosperity as slave trade ended. **1950s:** A liberation movement developed on the islands and the Portuguese African mainland colony of Guinea-Bissau. **1951:** Cape Verde became an overseas territory of Portugal. **1975:** Independence was achieved and a national people's assembly elected, with Aristides of the PAICV as the first executive president; a policy of nonalignment followed. **1981:** The goal of union with Guinea-Bissau was abandoned; Cape Verde became a one-party state. **1988:** There was rising unrest and demand for political reforms. **1991:** In the first multiparty elections, the new Movement for Democracy party (MPD) won a majority and Antonio Mascarenhas Monteiro became president; market-centred economic reforms were introduced. **2000:** Gualberto do Rosário became prime minister.

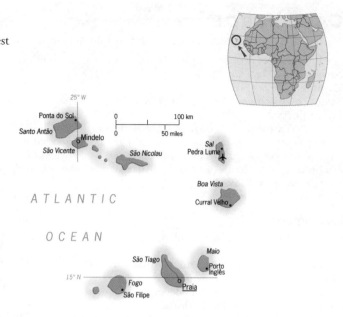

Practical information

Visa requirements: UK: visa required. USA: visa required
Time difference: GMT –1
Chief tourist attractions: mountain scenery; extensive white sandy beaches on islands of São Tiago, Sal, Boa Vista, and Maio
Major holidays: 1, 20 January, 8 March, 1 May, 1 June, 12 September, 24–25 December; variable: Good Friday

CENTRAL AFRICAN REPUBLIC

FORMERLY UBANGI-SHARI (UNTIL 1958), CENTRAL AFRICAN EMPIRE (1976–79)

Landlocked country in Central Africa, bordered northeast and east by
Sudan, south by the Democratic Republic of Congo and the Republic of
the Congo, west by Cameroon, and northwest by Chad.

National name: *République Centrafricaine/Central African Republic*
Area: 622,436 sq km/240,322 sq mi
Capital: Bangui
Major towns/cities: Berbérati, Bouar, Bambari, Bossangoa, Carnot,
Kaga Bandoro
Physical features: landlocked flat plateau, with rivers flowing north and
south, and hills in northeast and southwest; dry in north, rainforest in
southwest; mostly wooded; Kotto and Mbari river falls; the Ubangi River
rises 6 m/20 ft at Bangui during the wet season (June–November)

Red, white, and blue recall the French tricolour. Red,
yellow, and green are the pan-African colours. Red
represents the common blood of mankind which links
African and European nations.
Effective date: 1 December 1958.

Government

Head of state: Ange-Felix Patasse from 1993
Head of government: Anicet Georges Dologuele
from 1999
Political system: emergent democracy
Political executive: limited presidency
Administrative divisions: 14 prefectures and two
economic prefectures
Political parties: Central African People's
Liberation Party (MPLC), left of centre; Central
African Democratic Rally (RDC), nationalist, right
of centre
Armed forces: 2,700 (1998); plus 2,300 in
paramilitary forces
Conscription: selective national service for two-year
period
Death penalty: retains the death penalty for
ordinary crimes but can be considered abolitionist in
practice (last execution 1981)
Defence spend: (% GDP) 4.7 (1998)
Education spend: (% GNP) 2.8 (1993–94)
Health spend: (% GDP) 2.9 (1997)

Economy and resources

Currency: franc CFA
GDP: (US$) 1.05 billion (1999)
Real GDP growth: (% change on previous year) 0.8
(1999)
GNP: (US$) 1.04 billion (1999)
GNP per capita (PPP): (US$) 1,131 (1999 est)
Consumer price inflation: 2.8% (1999)
Unemployment: 5.6% (1993)
Foreign debt: (US$) 790 million (1999)
Major trading partners: France, Belgium,
Luxembourg, Cameroon, China
Resources: gem diamonds and industrial diamonds,
gold, uranium, iron ore, manganese, copper
Industries: food processing, beverages, tobacco,
furniture, textiles, paper, soap

Exports: diamonds, coffee, timber, cotton. Principal
market: Belgium – Luxembourg 63.5% (1999 est)
Imports: machinery, road vehicles and parts, basic
manufactures, food and chemical products. Principal
source: France 34.6% (1999 est)
Arable land: 3.1% (1996)
Agricultural products: cassava, coffee, yams,
maize, bananas, groundnuts; forestry

Population and society

Population: 3,615,000 (2000 est)
Population growth rate: 1.9% (1995–2000)
Population density: (per sq km) 6 (1999 est)
Urban population: (% of total) 41 (2000 est)
Age distribution: (% of total population) 0–14
43%, 15–59 51%, 60+ 6% (2000 est)
Ethnic groups: over 80 ethnic groups, but 66% of
the population falls into one of five: the Baya (34%),
the Banda (27%), the Mandjia (21%), the Sava
(10%), the Mbimu (4%) and the Mbaka (4%). There
are clearly defined ethnic zones; the forest region,
inhabited by Bantu groups, the Mbaka, Lissongo,
Mbimu, and Babinga; the river banks, populated by
the Sango, Yakoma, Baniri, and Buraka; and the
savannah region, where the Banda, Sande, Sara,
Ndle, and Bizao live. Europeans number fewer than
7,000, the majority being French
Language: French (official), Sangho (national),
Arabic, Hunsa, Swahili
Religion: Protestant 25%, Roman Catholic 25%,
animist 24%, Muslim 15%
Education: (compulsory years) 8
Literacy rate: 60% (men); 35% (women) (2000 est)
Labour force: 75.1% agriculture, 3.2% industry,
21.7% services (1997 est)
Life expectancy: 43 (men); 47 (women)
(1995–2000)
Child mortality rate: (under 5, per 1,000 live
births) 157 (1995–2000)

Physicians: 1 per 25,920 people (1993 est)
Hospital beds: 1 per 1,140 people (1993 est)
TV sets: (per 1,000 people) 5 (1997)
Radios: (per 1,000 people) 83 (1997)
Internet users: (per 10,000 people) 2.8 (1999)
Personal computer users: (per 100 people) 0.1 (1999)

Transport

Airports: international airports: Bangui-M'Poko; 37 small airports for international chartered services; total passenger km: 236 million (1995)
Railways: none
Roads: total road network: 24,000 km/14,913 mi, of which 1.8% paved (1996 est); passenger cars: 0.1 per 1,000 people (1996)

Chronology

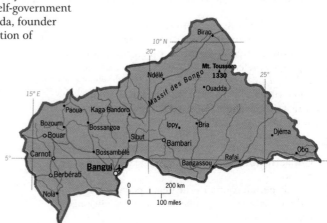

10th century: Immigration by peoples from Sudan to the east and Cameroon to the west. **16th century:** Part of the Gaoga Empire. **16th–18th centuries:** Population reduced greatly by slave raids both by coastal traders and Arab empires in Sudan and Chad. **19th century:** The Zande nation of the Bandia peoples became powerful in the east. Bantu speakers immigrated from Zaire and the Baya from northern Cameroon. **1889–1903:** The French established control over the area, quelling insurrections; a French colony known as Ubangi-Shari was formed and partitioned among commercial concessionaries. **1920–30:** A series of rebellions against forced labour on coffee and cotton plantations were savagely repressed by the French. **1946:** Given a territorial assembly and representation in French parliament. **1958:** Achieved self-government within French Equatorial Africa, with Barthélémy Boganda, founder of the pro-independence Movement for the Social Evolution of Black Africa (MESAN), as prime minister. **1960:** Achieved independence as Central African Republic; David Dacko, the nephew of the late Boganda, was elected president. **1962:** The republic became a one-party state, dominated by MESAN and loyal to the French interest. **1965:** Dacko was ousted in a military coup led by Col Jean-Bedel Bokassa, as the economy deteriorated. **1972:** Bokassa, a violent and eccentric autocrat, declared himself president for life. In 1977 he made himself emperor of the 'Central African Empire'. **1979:** Bokassa was deposed by Dacko in a French-backed bloodless coup, following violent repressive measures including the massacre of 100 children. Bokassa went into exile and the country became known as the Central African Republic again. **1981:** Dacko was deposed in a bloodless coup, led by Gen André Kolingba, and a military government was established. **1983:** A clandestine opposition movement was formed. **1984:** Amnesty for all political party leaders was announced. **1988:** Bokassa, who had returned from exile, was found guilty of murder and embezzlement; he received a death sentence, later commuted to life imprisonment. **1991:** Opposition parties were allowed to form. **1992:** Multiparty elections were promised, but cancelled with Kolingba in last place. **1993:** Kolingba released thousands of prisoners, including Bokassa. Ange-Félix Patassé of the leftist African People's Labour Party (MLPC) was elected president, ending 12 years of military dictatorship. **1996:** There was an army revolt over pay; Patassé was forced into hiding. **1999:** Anicet Georges Dologuele was appointed prime minister.

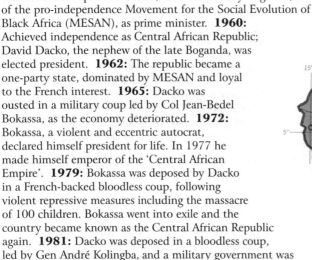

Practical information

Visa requirements: UK: visa required. USA: visa required
Time difference: GMT +1
Chief tourist attractions: waterfalls; forests; wildlife; game reserves; hunting and fishing
Major holidays: 1 January, 29 March, 1 May, 1 June, 13, 15 August, 1 September, 1 November, 1, 25 December; variable: Ascension Thursday, Easter Monday, Whit Monday

CHAD

Landlocked country in central North Africa, bounded north by Libya,
east by Sudan, south by the Central African Republic, and west by
Cameroon, Nigeria, and Niger.

National name: *République du Tchad/Republic of Chad*
Area: 1,284,000 sq km/495,752 sq mi
Capital: Ndjamena (formerly Fort Lamy)
Major towns/cities: Sarh, Moundou, Abéché, Bongor, Doba, Kélo,
Koumra
Physical features: landlocked state with mountains (Tibetsi) and part
of Sahara Desert in north; moist savannah in south; rivers in south flow
northwest to Lake Chad

Blue symbolizes hope, the clear sky, and the streams of
the south. Yellow stands for the sun and the Sahara
Desert. Red represents unity, prosperity, and national
sacrifice. Effective date: 6 November 1959.

Government

Head of state: Idriss Deby from 1990
Head of government: Nagoum Yamassoum from 1999
Political system: emergent democracy
Political executive: limited presidency
Administrative divisions: 14 prefectures
Political parties: Patriotic Salvation Movement (MPS),
left of centre; Alliance for Democracy and Progress
(RDP), left of centre; Union for Democracy and Progress
(UPDT), left of centre; Action for Unity and Socialism
(ACTUS), left of centre; Union for Democracy and the
Republic (UDR), left of centre
Armed forces: 25,400 (1998); plus 9,500 in
paramilitary forces
Conscription: conscription is for three years
Death penalty: retained and used for ordinary crimes
Defence spend: (% GDP) 5.6 (1998)
Education spend: (% GNP) 2.9 (1995)
Health spend: (% GDP) 4.3 (1997)

Economy and resources

Currency: franc CFA
GDP: (US$) 1.6 billion (1999)
Real GDP growth: (% change on previous year) –1.1
(1999)
GNP: (US$) 1.6 billion (1999)
GNP per capita (PPP): (US$) 816 (1999 est)
Consumer price inflation: –6.8% (1999)
Foreign debt: (US$) 1.03 billion (1999)
Major trading partners: France, Portugal, Nigeria,
Cameroon, Germany, Australia, Brazil, India
Resources: petroleum, tungsten, tin ore, bauxite, iron
ore, gold, uranium, limestone, kaolin, titanium
Industries: cotton processing, sugar refinery, beer,
cigarettes, soap, bicycles
Exports: cotton, live cattle, meat, hides and skins.
Principal market: Portugal 38.5% (1999 est)
Imports: petroleum and petroleum products, cereals,
pharmaceuticals, chemicals, machinery and transport
equipment, electrical equipment. Principal source: France
40.5% (1999 est)
Arable land: 2.6% (1996)

Agricultural products: cotton, millet, sugar cane,
sorghum, groundnuts; livestock rearing (cattle, sheep, and
goats)

Population and society

Population: 7,651,000 (2000 est)
Population growth rate: 2.6% (1995–2000)
Population density: (per sq km) 6 (1999 est)
Urban population: (% of total) 24 (2000 est)
Age distribution: (% of total population) 0–14 46%,
15–59 48%, 60+ 6% (2000 est)
Ethnic groups: mainly Arabs in the north, and Pagan, or
Kirdi, groups in the south. There is no single dominant
group in any region, the largest are the Sara, who
comprise about a quarter of the total population.
Europeans, mainly French, constitute a very small
minority
Language: French, Arabic (both official), over 100
African languages
Religion: Muslim 50%, Christian 25%, animist 25%
Education: (compulsory years) 8
Literacy rate: 52% (men); 34% (women) (2000 est)
Labour force: 77.8% agriculture, 6.3% industry, 15.9%
services (1997 est)
Life expectancy: 46 (men); 49 (women) (1995–2000)
Child mortality rate: (under 5, per 1,000 live births)
174 (1995–2000)
Physicians: 1 per 28,570 people (1994 est)
Hospital beds: 1 per 1,565 people (1994 est)
TV sets: (per 1,000 people) 2 (1997)
Radios: (per 1,000 people) 242 (1997)
Internet users: (per 10,000 people) 1.3 (1999)
Personal computer users: (per 100 people) 0.1 (1999)

Transport

Airports: international airports: Ndjamena; 12 small
airports for domestic services; total passenger km: 231
million (1995)
Railways: none
Roads: total road network: 33,400 km/20,775 mi, of
which 0.8% paved (1996 est); passenger cars: 3.2 per
1,000 people (1996 est)

Chronology

7th–9th centuries: Berber pastoral nomads, the Zaghawa, immigrated from the north and became a ruling aristocracy, dominating the Sao people, sedentary black farmers, and establishing the Kanem state. **9th–19th centuries:** The Zaghawa's Saifi dynasty formed the kingdom of Bornu, which stretched to the west and south of Lake Chad, and converted to Islam in the 11th century. At its height between the 15th and 18th centuries, it raided the south for slaves, and faced rivalry from the 16th century from the Baguirmi and Ouadai Arab kingdoms. **1820s:** Visited by British explorers. **1890s–1901:** Conquered by France, who ended slave raiding by Arab kingdoms. **1910:** Became a colony in French Equatorial Africa. Cotton production expanded in the south. **1944:** The pro-Nazi Vichy government signed an agreement giving Libya rights to the Aouzou Strip in northern Chad. **1946:** Became an overseas territory of the French Republic, with its own territorial assembly and representation in the French parliament. **1960:** Independence was achieved, with François Tombalbaye of the Chadian Progressive Party (CPT), dominated by Sara Christians from the south, as president. **1963:** Violent opposition in the Muslim north, led by the Chadian National Liberation Front (Frolinat), backed by Libya, following the banning of opposition parties. **1968:** A revolt of northern militias was quelled with France's help. **1973:** An Africanization campaign was launched by Tombalbaye, who changed his first name to Ngarta. **1975:** Tombalbaye was killed in a military coup led by southerner Gen Félix Malloum. Frolinat continued its resistance. **1978:** Malloum formed a coalition government with former Frolinat leader Hissène Habré, but it soon broke down. **1979:** Malloum was forced to leave the country; an interim government was set up under Gen Goukouni Oueddei (Frolinat). Habré continued his opposition with his Army of the North (FAN), and Libya provided support for Goukouni. **1981–82:** Habré gained control of half the country. Goukouni fled and set up a 'government in exile'. **1983:** Habré's regime was recognized by the Organization of African Unity (OAU) and France, but in the north, Goukouni's supporters, with Libya's help, fought on. Eventually a ceasefire was agreed, with latitude 16°north dividing the country. **1987:** Chad, France, and Libya agreed on an OAU ceasefire to end the civil war between the Muslim Arab north and Christian and animist black African south. **1988:** Libya relinquished its claims to the Aozou Strip. **1990:** President Habré was ousted after the army was defeated by Libyan-backed Patriotic Salvation Movement (MPS) rebel troops based in the Sudan and led by Habré's former ally Idriss Deby. **1991–92:** Several antigovernment coups were foiled. **1993:** A transitional charter was adopted, as a prelude to full democracy at a later date. **1997:** A reconciliation agreement was signed with rebel forces. **1998:** A new rebel force, the Movement for Democracy and Justice in Chad (MDJC), led by a former defence minster, began armed rebellion. **1999:** Nagoum Yamassoum was appointed prime minister. **2000:** Former president Hissene Habré was freed after having been charged with torture and murder.

Practical information

Visa requirements: UK: visa required. USA: visa required
Time difference: GMT +1
Chief tourist attractions: varied scenery – desert in north, dense forest in south
Major holidays: 1 January, 1, 25 May, 7 June, 11 August, 1, 28 November, 25 December; variable: Eid-ul-Adha, Easter Monday, end of Ramadan, Prophet's Birthday

CHILE

Located in South America, bounded north by Peru and Bolivia, east by Argentina, and south and west by the Pacific Ocean.

National name: *República de Chile/Republic of Chile*
Area: 756,950 sq km/292,258 sq mi
Capital: Santiago
Major towns/cities: Concepción, Viña del Mar, Valparaíso, Talcahuano, Puente Alto, Temuco, Antofagasta
Major ports: Valparaíso, Antofagasta, Arica, Iquique, Punta Arenas
Physical features: Andes mountains along eastern border, Atacama Desert in north, fertile central valley, grazing land and forest in south
Territories: Easter Island, Juan Fernández Islands, part of Tierra del Fuego, claim to part of Antarctica

White stands for the snowy peaks of the Andes. Red symbolizes the blood shed by the freedom fighters. Blue represents the clear Andean skies.
Effective date: c. 18 October 1817.

Government

Head of state and government: Ricardo Lagos Escobar from 2000
Political system: emergent democracy
Political executive: limited presidency
Administrative divisions: 12 regions and one metropolitan area
Political parties: Christian Democratic Party (PDC), moderate centrist; National Renewal Party (RN), right wing; Socialist Party of Chile (PS), left wing; Independent Democratic Union (UDI), right wing; Party for Democracy (PPD), left of centre; Union of the Centre-Centre (UCC), right wing; Radical Party (PR), left of centre
Armed forces: 94,500 (1998)
Conscription: one year (army) or two years (navy and air force)
Death penalty: retained and used for ordinary crimes
Defence spend: (% GDP) 3.7 (1998)
Education spend: (% GNP) 3.6 (1997)
Health spend: (% GDP) 6.1 (1997)

Economy and resources

Currency: Chilean peso
GDP: (US$) 71.09 billion (1999)
Real GDP growth: (% change on previous year) –1.1 (1999)
GNP: (US$) 71.1 billion (1999)
GNP per capita (PPP): (US$) 8,370 (1999)
Consumer price inflation: 3.3% (1999)
Unemployment: 7.2% (1998 est)
Foreign debt: (US$) 37.7 billion (1999 est)
Major trading partners: USA, Japan, Brazil, Germany, Argentina, UK, Mexico
Resources: copper (world's largest producer), gold, silver, iron ore, molybdenum, cobalt, iodine, saltpetre, coal, natural gas, petroleum, hydroelectric power
Industries: nonferrous metals, food processing, petroleum refining, chemicals, paper products (cellulose, newsprint, paper and cardboard), motor tyres, beer, glass sheets, motor vehicles
Exports: copper, fruits, timber products, fishmeal, vegetables, manufactured foodstuffs and beverages. Principal market: USA 18.6% (1999)
Imports: machinery and transport equipment, wheat,

chemical and mineral products, consumer goods, raw materials. Principal source: USA 20.8% (1999)
Arable land: 4.5% (1996)
Agricultural products: wheat, sugar beet, potatoes, maize, fruit and vegetables; livestock

Population and society

Population: 15,211,000 (2000 est)
Population growth rate: 1.4% (1995–2000); 1.2% (2000–05)
Population density: (per sq km) 20 (1999 est)
Urban population: (% of total) 86 (2000 est)
Age distribution: (% of total population) 0–14 28%, 15–59 62%, 60+ 10% (2000 est)
Ethnic groups: 65% mestizo (mixed American Indian and Spanish descent), 30% European, remainder mainly American Indian
Language: Spanish (official)
Religion: Roman Catholic 80%, Protestant 13%, atheist and nonreligious 6%
Education: (compulsory years) 8
Literacy rate: 96% (men); 95% (women) (2000 est)
Labour force: 14.4% agriculture, 27.3% industry, 58.3% services (1997)
Life expectancy: 72 (men); 78 (women) (1995–2000)
Child mortality rate: (under 5, per 1,000 live births) 15 (1995–2000)
Physicians: 1 per 942 people (1993 est)
Hospital beds: 1 per 320 people (1993 est)
TV sets: (per 1,000 people) 233 (1997)
Radios: (per 1,000 people) 354 (1997)
Internet users: (per 10,000 people) 416.6 (1999)
Personal computer users: (per 100 people) 6.7 (1999)

Transport

Airports: international airports: Santiago (Arturo Merino Benítez), Arica (Chacalluta); domestic services to main towns; total passenger km: 8,597 million (1997 est)
Railways: total length: 6,572 km/4,084 mi; total passenger km: 816 million (1994)
Roads: total road network: 79,800 km/49,588 mi, of which 13.8% paved (1996 est); passenger cars: 70.5 per 1,000 people (1996)

Chronology

1535: The first Spanish invasion of Chile was abandoned in the face of fierce resistance from indigenous Araucanian Indians. **1541:** Pedro de Valdivia began the Spanish conquest and founded Santiago. **1553:** Valdivia was captured and killed by Araucanian Indians, led by Chief Lautaro. **17th century:** The Spanish developed small agricultural settlements ruled by a government subordinate to the viceroy in Lima, Peru. **1778:** The king of Spain appointed a captain-general to govern Chile. **1810:** A Santiago junta proclaimed Chilean autonomy after Napoleon dethroned the king of Spain. **1814:** The Spanish viceroy regained control of Chile. **1817:** The Army of the Andes, led by José de San Martín and Bernardo O'Higgins, defeated the Spanish. **1818:** Chile achieved independence from Spain with O'Higgins as supreme director. **1823–30:** O'Higgins was forced to resign; a civil war between conservative centralists and liberal federalists ended with conservative victory. **1833:** An autocratic republican constitution created a unitary Roman Catholic state with a strong president and limited franchise. **1851–61:** President Manuel Montt bowed to pressure to liberalize the constitution and reduce privileges of landowners and the church. **1879–84:** Chile defeated Peru and Bolivia in the War of the Pacific and increased its territory by a third. **late 19th century:** Mining of nitrate and copper became a major industry; large-scale European immigration followed the 'pacification' of Araucanian Indians. **1891:** A constitutional dispute between president and congress led to civil war; congressional victory reduced the president to figurehead status. **1925:** A new constitution increased presidential powers, separated church and state, and made primary education compulsory. **1927:** A military coup led to the dictatorship of Gen Carlos Ibáñez del Campo. **1931:** A sharp fall in price of copper and nitrate caused dramatic economic and political collapse. **1938:** A Popular Front of Radicals, Socialists, and Communists took power under Pedro Aguirre Cedra, who introduced economic policies based on the US New Deal. **1948–58:** The Communist Party was banned. **1970:** Salvador Allende, leader of the Popular Unity coalition, became the world's first democratically elected Marxist president; he embarked on an extensive programme of nationalization and radical social reform. **1973:** Allende was killed in a CIA-backed military coup; Gen Augusto Pinochet established a dictatorship combining severe political repression with free-market economics. **1981:** Pinochet began an eight-year term as president under a new constitution described as a 'transition to democracy'. **1983:** Economic recession provoked growing opposition to the governing regime. **1988:** A referendum on whether Pinochet should serve a further term resulted in a clear 'No' vote. **1990:** The military regime ended, with a Christian Democrat (Patricio Aylwin) as president, with Pinochet as commander in chief of the army. An investigation was launched into over 2,000 political executions during the military regime. **1995:** Dante Cordova was appointed prime minister. **1998:** Pinochet retired from the army and was made life senator. Pinochet was placed under arrest in the UK; proceedings began to extradite him to Spain on murder charges. **1999:** The ruling on the extradition of Pinochet to Spain was left to the British government. **2000:** Ricardo Lagos was elected president. Pinochet was found unfit for trial by British doctors and allowed to return to Chile. However, in Chile, Pinochet was stripped of immunity from prosecution. **2001:** Pinochet was arrested and charged organizing the killings of left-wing activists and union leaders during his time in power.

Practical information

Visa requirements: UK: visa not required. USA: visa not required
Time difference: GMT –4
Chief tourist attractions: beaches; Andean skiing resorts; lakes, rivers, desert scenery; Easter Island Neolithic sites
Major holidays: 1 January, 1, 21 May, 29 June, 15 August, 11, 18–19 September, 12 October, 1 November, 8, 25, 31, December; variable: Good Friday, Holy Saturday

CHINA

The largest country in East Asia, bounded to the north by Mongolia; to the
northwest by Tajikistan, Kyrgyzstan, Kazakhstan, and Afghanistan; to the
southwest by India, Nepal, and Bhutan; to the south by Myanmar, Laos, and
Vietnam; to the southeast by the South China Sea; to the east by the East
China Sea, North Korea, and Yellow Sea; and to the northeast by Russia.

National name: *Zhonghua Renmin Gongheguo (Zhongguo)/People's Republic
of China*
Area: 9,572,900 sq km/3,696,000 sq mi
Capital: Beijing (or Peking)
Major towns/cities: Shanghai, Hong Kong, Chongqing, Tianjin,
Guangzhou (English Canton), Shenyang (formerly Mukden), Wuhan,
Nanjing, Harbin, Chengdu, Xi'an
Major ports: Tianjin, Shanghai, Hong Kong, Qingdao, Guangzhou
Physical features: two-thirds of China is mountains or desert (north and west); the low-lying east is irrigated by
rivers Huang He (Yellow River), Chang Jiang (Yangtze-Kiang), Xi Jiang (Si Kiang)
Territories: Paracel Islands

The large star symbolizes the common programme of
the Communist Party. The small stars represent the four
economic classes: peasants, workers, petty bourgeoisie,
and 'patriotic capitalists'. Effective date: 1 October 1949.

Government

Head of state: Jiang Zemin from 1993
Head of government: Zhu Rongji from 1998
Political system: communist
Political executive: communist
Administrative divisions: 23 provinces, 5 autonomous
regions, and 4 municipalities
Political party: Chinese Communist Party (CCP),
Marxist-Leninist-Maoist; eight registered small parties
controlled by the CCP
Armed forces: 2,820,000 (1998); reserves
approximately 1.2 million
Conscription: selective: 3 years (army and marines), 4
years (air force and navy)
Death penalty: retained and used for ordinary crimes
Defence spend: (% GDP) 5.3 (1998)
Education spend: (% GNP) 2.3 (1996)
Health spend: (% GDP) 2.7 (1997)

Economy and resources

Currency: yuan
GDP: (US$) 991.2 billion (1999)
Real GDP growth: (% change on previous year) 7.2
(1999)
GNP: (US$) 980.2 billion (1999)
GNP per capita (PPP): (US$) 3,291 (1999)
Consumer price inflation: –1.3% (1999)
Unemployment: 3.1% (1998)
Foreign debt: (US$) 164.8 billion (1999 est)
Major trading partners: Japan, USA, Taiwan, South
Korea, Germany, the Netherlands, UK, Singapore, Russia
Resources: coal, graphite, tungsten, molybdenum,
antimony, tin (world's largest producer), lead (world's
fifth-largest producer), mercury, bauxite, phosphate rock,
iron ore (world's largest producer), diamonds, gold,
manganese, zinc (world's third-largest producer),
petroleum, natural gas, fish

Industries: raw cotton and cotton cloth, cement, paper,
sugar, salt, plastics, aluminium ware, steel, rolled steel,
chemical fertilizers, silk, woollen fabrics, bicycles,
cameras, electrical appliances; tourism is growing
Exports: basic manufactures, miscellaneous
manufactured articles (particularly clothing and toys),
crude petroleum, machinery and transport equipment,
fishery products, cereals, canned food, tea, raw silk,
cotton cloth. Principal market: USA 21.5% (1999)
Imports: machinery and transport equipment, basic
manufactures, chemicals, wheat, rolled steel, fertilizers.
Principal source: Japan 20.4% (1999)
Arable land: 13.3% (1996)
Agricultural products: sweet potatoes, wheat, maize,
soybeans, rice, sugar cane, tobacco, cotton, jute; world's
largest fish catch

Population and society

Population: 1,277,558,000 (2000 est)
Population growth rate: 0.9% (1995–2000)
Population density: (per sq km) 133 (1999 est)
Urban population: (% of total) 32 (2000 est)
Age distribution: (% of total population) 0–14 25%,
15–59 65%, 60+ 10% (2000 est)
Ethnic groups: 92% Han Chinese, the remainder being
Zhuang, Uygur, Hui (Muslims), Yi, Tibetan, Miao,
Manchu, Mongol, Buyi, or Korean; numerous lesser
nationalities live mainly in border regions
Language: Chinese (dialects include Mandarin (official),
Yue (Cantonese), Wu (Shanghaiese), Minbai, Minnah,
Xiang, Gan, and Hakka)
Religion: Taoist, Confucianist, and Buddhist; Muslim
2–3%; Christian about 1% (divided between the
'patriotic' church established in 1958 and the 'loyal'
church subject to Rome); Protestant 3 million
Education: (compulsory years) 9
Literacy rate: 92% (men); 76% (women) (2000 est)

Labour force: 59% of population: 72% agriculture, 15% industry, 13% services (1990)
Life expectancy: 68 (men); 72 (women) (1995–2000)
Child mortality rate: (under 5, per 1,000 live births) 47 (1995–2000)
Physicians: 1 per 627 people (1996)
Hospital beds: 1 per 417 people (1996)
TV sets: (per 1,000 people) 270 (1997)
Radios: (per 1,000 people) 333 (1997)
Internet users: (per 10,000 people) 70.6 (1999)
Personal computer users: (per 100 people) 1.2 (1999)

Transport

Airports: international airports: Beijing (Capital International Central), Guangzhou (Baiyun), Shanghai (Hongqiao), Hong Kong (Kai Tak); 59 domestic airports; total passenger km: 77,352 million (1997)
Railways: total length: 54,000 km/33,556 mi; total passenger km: 354,825 million (1997)
Roads: total road network: 1,526,389 km/948,498 mi, of which 89.7% paved (1996); passenger cars: 3.2 per 1,000 people (1996 est)

Chronology

c. **3000 BC:** Yangshao culture reached its peak in the Huang He Valley; displaced by Longshan culture in eastern China. *c.* **1766–*c.* 1122 BC:** First major dynasty, the Shang, arose from Longshan culture; writing and calendar developed. *c.* **1122–256 BC:** Zhou people of western China overthrew Shang and set up new dynasty; development of money and written laws. *c.* **500 BC:** Confucius expounded the philosophy which guided Chinese government and society for the next 2,000 years. **403–221 BC:** 'Warring States Period': Zhou Empire broke up into small kingdoms. **221–206 BC:** Qin kingdom defeated all rivals and established first empire with strong central government; emperor Shi Huangdi built the Great Wall of China. **202 BC–AD 220:** Han dynasty expanded empire into central Asia; first overland trade with Europe; art and literature flourished; Buddhism introduced from India.

AD 220–581: Large-scale rebellion destroyed the Han dynasty; the empire split into three competing kingdoms; several short-lived dynasties ruled parts of China. **581–618:** Sui dynasty reunified China and repelled Tatar invaders. **618–907:** Tang dynasty enlarged and strengthened the empire; great revival of culture; major rebellion (875–84). **907–60:** 'Five Dynasties and Ten Kingdoms': disintegration of the empire amid war and economic decline; development of printing. **960–1279:** Song dynasty reunified China and restored order; civil service examinations introduced; population reached 100 million; Manchurians occupied northern China in 1127. **1279:** Mongols conquered all China, which became part of the vast empire of Kublai Khan, founder of the Yuan dynasty; the Venetian traveller Marco Polo visited China (1275–92). **1368:** Rebellions drove out the Mongols; Ming dynasty expanded the empire; architecture flourished in the new capital of Beijing. **1516:** Portuguese explorers reached Macau. Other European traders followed, with the first Chinese porcelain arriving in Europe in 1580. **1644:** Manchurian invasion established the Qing (or Manchu) dynasty; Manchurians were assimilated and Chinese trade and culture continued to thrive. **1796–1804:** Anti-Manchu revolt weakened the Qing dynasty; a population increase in excess of food supplies led to falling living standards and cultural decline. **1839–42:** First Opium War; Britain forced China to cede Hong Kong and open five ports to European trade; Second

Opium War extracted further trade concessions (1856–60). **1850–64:** Millions died in the Taiping Rebellion; Taipings combined Christian and Chinese beliefs and demanded land reform. **1894–95:** Sino-Japanese War: Chinese driven out of Korea. **1897–98:** Germany, Russia, France, and Britain leased ports in China. **1898:** Hong Kong was secured by Britain on a 99-year lease. **1900:** Anti-Western Boxer Rebellion crushed by foreign intervention; jealousy between the Great Powers prevented partition. **1911:** Revolution broke out; Republic of China proclaimed by Sun Zhong Shan (Sun Yat-sen) of Guomindang (National People's Party). **1912:** Abdication of infant emperor Pu-i; Gen Yuan Shih-K'ai became dictator. **1916:** The power of the central government collapsed on the death of Yuan Shih-K'ai; northern China dominated by local warlords. **1919:** Beijing students formed the 4th May movement to protest at the transfer of German possessions in China to Japan. **1921:** Sun Zhong Shan elected president of nominal national government; Chinese Communist Party founded; communists worked with Guomindang to reunite China from 1923. **1925:** Death of Sun Zhong Shan; leadership of Guomindang gradually passed to military commander Jiang Jie Shi (Chiang Kai-shek). **1926–28:** Revolutionary Army of Jiang Jie Shi reunified China; Guomindang broke with communists and tried to suppress them in civil war. **1932:** Japan invaded Manchuria and established the puppet state of Manchukuo. **1934–35:** Communists undertook Long March from Jiangxi and Fujian in south to Yan'an in north to escape encirclement by Guomindang. **1937–45:** Japan renewed invasion of China; Jiang Jie Shi received help from USA and Britain from 1941. **1946:** Civil war resumed between Guomindang and communists led by Mao Zedong. **1949:** Victorious communists proclaimed People's Republic of China under Chairman Mao; Guomindang fled to Taiwan. **1950–53:** China intervened heavily in Korean War. **1958:** 'Great Leap Forward': extremist five-year plan to accelerate output severely weakened the economy. **1960:** Sino-Soviet split: China accused USSR of betraying communism. **1962:** Economic recovery programme under Liu Shaoqi caused divisions between 'rightists' and 'leftists'; brief border war with India. **1966–69:** 'Great Proletarian Cultural Revolution'; leftists overthrew Liu Shaoqi with support of Mao; Red Guards disrupted education, government, and daily life in attempt to enforce revolutionary principles. **1970:** Mao supported the efforts of Prime Minister Zhou Enlai to restore order. **1971:** People's Republic of China admitted to United Nations. **1976:** Deaths of Zhou Enlai and Mao Zedong led to a power struggle between rightists and leftists; Hua Guofeng became leader. **1977–81:** Rightist Deng Xiaoping emerged as supreme leader; pragmatic economic policies introduced market incentives and encouraged foreign trade. **1979:** Full diplomatic relations with USA established **1987:** Deng Xiaoping retired from Politburo but remained a dominant figure. **1989:** Over 2,000 people were killed when the army crushed prodemocracy student demonstrations in Tiananmen Square, Beijing; international sanctions were imposed. **1991:** China and the USSR reached an agreement on their disputed border. **1993:** Jiang Zemin became head of state **1996:** Reunification with Taiwan was declared a priority. **1997:** A border agreement was signed with Russia. Hong Kong was returned to Chinese sovereignty. **1998:** Zhu Rongji became prime minister. The Yangtze in Hubei province flooded, causing widespread devastation. Dissident Xu Wenli was jailed for trying to set up an opposition party. **1999:** The USA and China announced a deal to allow for China's entry into the World Trade Organization (WTO), in exchange for opening China's markets to foreign firms. Macau was returned to China, with the promise that it would have an independent political system for fifty years. The religious sect Falun Gong was banned and its leaders arrested. **2000:** A drive against corruption convicted a number of high-ranking government officials, including a former deputy chairman of the National People's Congress. The first verdicts in trials of at least 200 officials accused of evading tariffs on the importing of US$6.6 billion worth of goods resulted in 14 people being sentenced to death. **2001:** Five members of the Falun Gong set themselves alight in Tiananmen Square, Beijing, in protest at the continued government crackdown on the sect.

Practical information

Visa requirements: UK: visa required. USA: visa required

Time difference: GMT +8

Chief tourist attractions: scenery; historical sites such as the Great Wall, Temple of Heaven, Forbidden City (Beijing), Ming tombs, terracotta warriors (Xian); Buddhist monasteries and temples in Tibet (Xizang)

Major holidays: 1 January, 8 March, 1 May, 1 August, 9 September, 1–2 October; variable: Spring Festival (January/February, 4 days)

COLOMBIA

Located in South America, bounded north by the Caribbean Sea, west by the Pacific Ocean, northwestern corner by Panama, east and northeast by Venezuela, southeast by Brazil, and southwest by Peru and Ecuador.

National name: *República de Colombia/Republic of Colombia*
Area: 1,141,748 sq km/440,828 sq mi
Capital: Bogotá
Major towns/cities: Medellín, Cali, Barranquilla, Cartagena, Bucaramanga, Cúcuta, Ibagué
Major ports: Barranquilla, Cartagena, Buenaventura
Physical features: the Andes mountains run north–south; flat coastland in west and plains (llanos) in east; Magdalena River runs north to Caribbean Sea; includes islands of Providencia, San Andrés, and Mapelo; almost half the country is forested

Yellow represents the golden land of South America. Blue stands for the ocean separating the country from Spain. Red symbolizes the blood and courage of the people resisting the tyrants.
Effective date: 26 November 1861.

Government

Head of state and government: Andres Pastrana from 1998
Political system: liberal democracy
Political executive: limited presidency
Administrative divisions: 32 departments and one capital district
Political parties: Liberal Party (PL), centrist; Conservative Party (PSC), right of centre; M-19 Democratic Alliance (ADM-19), left of centre; National Salvation Movement (MSN), right-of-centre coalition grouping
Armed forces: 146,300 (1998); plus a paramilitary police force of 87,000
Conscription: selective conscription for 1–2 years
Death penalty: abolished in 1910
Defence spend: (% GDP) 3.2 (1998)
Education spend: (% GNP) 4.4 (1996)
Health spend: (% GDP) 9.3 (1997)

Economy and resources

Currency: Colombian peso
GDP: (US$) 88.6 billion (1999)
Real GDP growth: (% change on previous year) –4.5 (1999)
GNP: (US$) 93.6 billion (1999)
GNP per capita (PPP): (US$) 5,709 (1999 est)
Consumer price inflation: 10.9% (1999)
Unemployment: 13% (1998)
Foreign debt: (US$) 32.2 billion (1999 est)
Major trading partners: USA, Venezuela, Germany, Japan, Peru
Resources: petroleum, natural gas, coal, nickel, emeralds (accounts for about half of world production), gold, manganese, copper, lead, mercury, platinum, limestone, phosphates
Industries: food processing, chemical products, textiles, beverages, transport equipment, cement
Exports: coffee, petroleum and petroleum products, coal, gold, bananas, cut flowers, cotton, chemicals, textiles, paper. Principal market: USA 48.5% (1999). Illegal trade in cocaine in 1995; it was estimated that approximately

$3.5 billion (equivalent to about 4% of GDP) was entering Colombia as the proceeds of drug-trafficking
Imports: machinery and transport equipment, chemicals, minerals, food, metals. Principal source: USA 38.3% (1999)
Arable land: 1.9% (1996)
Agricultural products: coffee (world's second-largest producer), cocoa, sugar cane, bananas, tobacco, cotton, cut flowers, rice, potatoes, maize; timber; beef production

Population and society

Population: 42,321,000 (2000 est)
Population growth rate: 1.9% (1995–2000)
Population density: (per sq km) 36 (1999 est)
Urban population: (% of total) 74 (2000 est)
Age distribution: (% of total population) 0–14 33%, 15–59 60%, 60+ 7% (2000 est)
Ethnic groups: 58% mestizo (mixed Spanish and American Indian descent), 20% European, 14% mulatto, 4% black, 3% black American Indian, 1% American Indian
Language: Spanish (official) (95%)
Religion: Roman Catholic
Education: (compulsory years) 5
Literacy rate: 92% (men); 92% (women) (2000 est)
Labour force: 40% of population: 27% agriculture, 23% industry, 50% services (1990)
Life expectancy: 67 (men); 74 (women) (1995–2000)
Child mortality rate: (under 5, per 1,000 live births) 39 (1995–2000)
Physicians: 1 per 1,105 people (1993 est)
Hospital beds: 1 per 732 people (1993 est)
TV sets: (per 1,000 people) 217 (1997)
Radios: (per 1,000 people) 581 (1997)
Internet users: (per 10,000 people) 144.4 (1999)
Personal computer users: (per 100 people) 3.4 (1999)

Transport

Airports: international airports: Santa Fe de Bogotá, DC (El Dorado International), Medellín, Cali, Barranquilla, Bucaramanga, Cartagena, Cúcuta, Leticia, Pereira, San Andrés, Santa Marta; over 80 smaller airports serving

domestic flights; total passenger km: 6,733 million (1997 est)
Railways: total length: 3,380 km/2,100 mi (1995); total
passenger km: 16 million (1992)

Roads: total road network: 106,600 km/66,241 mi, of
which 11.9% paved (1996 est); passenger cars: 21 per
1,000 people (1996 est)

Chronology

late 15th century: Southern Colombia became part of Inca Empire, whose core lay in Peru.
1522: Spanish conquistador Pascual de Andagoya reached the San Juan River. **1536–38:**
Spanish conquest by Jimenez de Quesada overcame powerful Chibcha Indian chiefdom, which
had its capital in the uplands at Bogotá and was renowned for its gold crafts; became part of
Spanish Viceroyalty of Peru, which covered much of South America. **1717:** Bogotá became
capital of the new Spanish Viceroyalty of Nueva (New) Granada, which also ruled
Ecuador and Venezuela. **1809:** Struggle for independence from Spain began. **1819:**
Venezuelan freedom fighter Simón Bolívar, 'the Liberator', who had withdrawn to
Colombia in 1814, raised a force of 5,000 British mercenaries and defeated the
Spanish at the battle of Boyacá, establishing Colombia's independence; Gran
Colombia formed, also comprising Ecuador, Panama, and Venezuela. **1830:**
Became a separate state, which included Panama, on the dissolution of the
Republic of Gran Colombia. **1863:** Became major coffee exporter. Federalizing,
anti-clerical Liberals came to power, with the country divided into nine largely
autonomous 'sovereign' states; the church was disestablished. **1885:**
Conservatives came to power, beginning 45 years of political
dominance; power was recentralized and the church restored to
influence. **1899–1903:** Civil war between Liberals and
Conservatives, ending with Panama's separation as an independent
state. **1930:** Liberals returned to power at the time of the economic
depression; social legislation introduced and a labour movement
encouraged. **1946:** Conservatives returned to power. **1948:** The left-
wing mayor of Bogotá was assassinated to a widespread outcry. **1949–57:**
Civil war, 'La Violencia', during which over 250,000 people died. **1957:**
Hoping to halt violence, Conservatives and Liberals agreed to form National
Front, sharing the presidency. **1970:** National Popular Alliance (ANAPO) formed as
left-wing opposition to National Front. **1974:** National Front accord temporarily ended.
1975: Civil unrest due to disillusionment with government. **1978:** Liberals, under Julio
Turbay, revived the accord and began an intensive fight against drug dealers. **1982:** The Liberals
maintained their control of congress but lost the presidency. Conservative president Belisario Betancur
granted guerrillas an amnesty and freed political prisoners. **1984:** The minister of justice was assassinated by
drug dealers; the campaign against them was stepped up. **1989:** A drug cartel assassinated the leading presidential
candidate and an antidrug war was declared by the president; a bombing campaign by drug traffickers killed hundreds;
the police killed José Rodriguez Gacha, one of the most wanted cartel leaders. **1991:** A new constitution prohibited the
extradition of Colombians wanted for trial in other countries. Several leading drug traffickers were arrested. Many guerrillas
abandoned the armed struggle, but the Colombian Revolutionary Armed Forces (FARC) and the National Liberation Army
remained active. **1993:** Medellín drug-cartel leader Pablo Escobar was shot while attempting to avoid arrest. **1995:**
President Samper came under pressure to resign over corruption allegations; a state of emergency was declared. Leaders of
the Cali drug cartel were imprisoned. **1998:** There were clashes between the army and left-wing guerrillas. The
conservative Andres Pastrana won presidential elections. Peace talks were held with rebels. **1999:** Formal peace talks were
broken off after violence, but later resumed. **2000:** US president Clinton announced US$1.3 billion in aid for Colombia
in January and a further US$1.3 billion in August, most of which was to go to the armed forces. Pastrana's government
sank to just 20% public support. Talks with one rebel group secured the release of 42 hostages in December.

Practical information

Visa requirements: UK: visa not required for a stay of
up to 90 days. USA: visa not required for a stay of up to
90 days
Time difference: GMT –5
Chief tourist attractions: Caribbean coast; 16th-
century walled city of Cartagena; Amazonian town of
Leticia; Andes Mountains; forest and rainforest; pre-
Columbian relics and colonial architecture
Major holidays: 1, 6 January, 29 June, 20 July, 7, 15
August, 12 October, 1, 15 November, 8, 25, 30–31
December; variable: Ascension Thursday, Corpus
Christi, Good Friday, Holy Thursday, St Joseph
(March), Sacred Heart (June)

COMOROS

Located in the Indian Ocean between Madagascar and the east coast of Africa.

National name: *Jumhuriyyat al-Qumur al-Itthadiyah al-Islamiyah* (Arabic), *République fédérale islamique des Comores* (French)/*Federal Islamic Republic of the Comoros*
Area: 1,862 sq km/718 sq mi
Capital: Moroni
Major towns/cities: Mutsamudu, Domoni, Fomboni, Mitsamiouli
Physical features: comprises the volcanic islands of Njazídja, Nzwani, and Mwali (formerly Grande Comore, Anjouan, Mohéli); at northern end of Mozambique Channel in Indian Ocean between Madagascar and coast of Africa

Monogram of Muhammed. Monogram of Allah. The crescent is a symbol of Islam. The four stars represent the islands. Effective date: 3 October 1996.

Government

Head of state: Azali Assoumani from 1999
Head of government: Hamada Madi from 2000
Political system: military
Political executive: military
Administrative divisions: three prefectures (each of the three main islands is a prefecture)
Political parties: National Union for Democracy in the Comoros (UNDC), Islamic, nationalist; Rally for Democracy and Renewal (RDR), left of centre
Armed forces: 800 (1995)
Conscription: military service is voluntary
Death penalty: retained and used for ordinary crimes
Education spend: (% GNP) 3.9 (1995)
Health spend: (% GDP) 4.5 (1997)

Economy and resources

Currency: Comorian franc
GDP: (US$) 203 million (1999 est)
Real GDP growth: (% change on previous year) 0.5 (1999 est)
GNP: (US$) 189 million (1999)
GNP per capita (PPP): (US$) 1,360 (1999 est)
Consumer price inflation: 4% (1999)
Unemployment: 20% (1996 est)
Foreign debt: (US$) 134.3 million (1998)
Major trading partners: France, Germany, Kenya, Pakistan, South Africa
Industries: sawmilling, processing of vanilla and copra, printing, soft drinks, plastics
Exports: vanilla, cloves, ylang-ylang, essences, copra, coffee. Principal market: France 50% (1998 est)
Imports: rice, petroleum products, transport equipment, meat and dairy products, cement, iron and steel, clothing and footwear. Principal source: France 37.5% (1998 est)
Arable land: 35% (1995)
Agricultural products: vanilla, ylang-ylang, cloves, basil, cassava, sweet potatoes, rice, maize, pulses, coconuts, bananas

Population and society

Population: 694,000 (2000 est)
Population growth rate: 2.7% (est 1995–2025)
Population density: (per sq km) 363 (1999 est)
Urban population: (% of total) 33 (2000 est)
Age distribution: (% of total population) 0–14 42%, 15–59 53%, 60+ 5% (2000 est)
Ethnic groups: population of mixed origin, with Africans, Arabs, and Malaysians predominating; the principal ethnic group is the Antalaotra; others are the Catre, the Makoa, the Oimatsaha, and the Sakalava
Language: Arabic, French (both official), Comorian (a Swahili and Arabic dialect), Makua
Religion: Muslim; Islam is the state religion
Education: (compulsory years) 9
Literacy rate: 66% (men); 53% (women) (2000 est)
Labour force: 44% of population: 77% agriculture, 9% industry, 13% services (1990)
Life expectancy: 57 (men); 60 (women) (1995–2000)
Child mortality rate: (under 5, per 1,000 live births) 106 (1995–2000)
Physicians: 1 per 7,500 people (1990)
Hospital beds: 1 per 342 people (1990)
TV sets: (per 1,000 people) 2 (1997)
Radios: (per 1,000 people) 141 (1997)

Transport

Airports: international airport: Moroni-Hahaya, on Njazídja; each of the three other islands has a small airfield; total passenger km: 3 million (1995)
Railways: none
Roads: total road network: 900 km/559 mi, of which 76.5% paved (1996 est); passenger cars: 13.3 per 1,000 people (1996 est)

Chronology

5th century AD: First settled by Malay-Polynesian immigrants. **7th century:** Converted to Islam by Arab seafarers and fell under the rule of local sultans. **late 16th century:** First visited by European navigators. **1886:** Mohéli island in south became a French protectorate. **1904:** Slave trade abolished, ending influx of Africans. **1912:** Grande Comore and Anjouan, the main islands, joined Mohéli to become a French colony, which was attached to Madagascar from 1914. **1947:** Became a French Overseas Territory separate from Madagascar. **1961:** Internal self-government achieved. **1975:** Independence achieved from France, but island of Mayotte to the southeast voted to remain part of France. Joined the United Nations. **1976:** President Ahmed Abdallah was overthrown in a coup by Ali Soilih; relations deteriorated with France as a Maoist-Islamic socialist programme was pursued. **1978:** Soilih was killed by French mercenaries. A federal Islamic republic was proclaimed, with exiled Abdallah restored as president; diplomatic relations re-established with France. **1979:** The Comoros became a one-party state; powers of the federal government increased. **1989:** Abdallah killed by French mercenaries who, under French and South African pressure, turned authority over to French administration; Said Muhammad Djohar became president in a multiparty democracy. **1995:** Djohar was overthrown in a coup led by Denard, who was persuaded to withdraw by French troops. **1997:** Secessionist rebels took control of the island of Anjouan. **1999:** The government was overthrown by an army coup, after granting greater autonomy to the islands of Anjouan and Mohéli. The new president was Colonel Azali Assoumani. **2000:** A coup against the military government was foiled. Hamada Madi was appointed prime minister.

Practical information

Visa requirements: UK: visa required. USA: visa required
Time difference: GMT +3
Chief tourist attractions: rich marine life; beaches; underwater fishing; mountain scenery

Major holidays: 6 July, 27 November; variable: Eid-ul-Adha, Arafa, Ashora, first day of Ramadan, end of Ramadan, New Year (Muslim), Prophet's Birthday

CONGO, DEMOCRATIC REPUBLIC OF

OR CONGO (KINSHASA); FORMERLY REPUBLIC OF CONGO (1960–64), ZAIRE (1971–97)

Located in central Africa, bounded west by the Republic of the Congo, north by the Central African Republic and Sudan, east by Uganda, Rwanda, Burundi, and Tanzania, southeast by Zambia, and southwest by Angola.

National name: *République Démocratique du Congo/Democratic Republic of Congo*
Area: 2,344,900 sq km/905,366 sq mi
Capital: Kinshasa
Major towns/cities: Lubumbashi, Kananga, Mbuji-Mayi, Kisangani, Kolwezi, Likasi, Boma
Major ports: Matadi, Kalemie
Physical features: Congo River basin has tropical rainforest (second-largest remaining in world) and savannah; mountains in east and west; lakes Tanganyika, Albert, Edward; Ruwenzori Range

The small stars stand for the original provinces of Congo at independence in 1960. The single gold star was said to represent the light of civilization. Effective date: 17 May 1997.

Government

Head of state and government: Joseph Kabila from 2001
Political system: military
Political executive: military
Administrative divisions: 11 provinces
Political parties: Popular Movement of the Revolution (MPR), African socialist; Democratic Forces of Congo–Kinshasa (formerly Sacred Union, an alliance of some 130 opposition groups), moderate, centrist; Union for Democracy and Social Progress (UPDS), left of centre; Congolese National Movement–Lumumba (MNC), left of centre
Armed forces: 50,000 (1998); plus paramilitary forces of 37,000
Conscription: military service is compulsory
Death penalty: retained and used for ordinary crimes
Defence spend: (% GDP) 6.6 (1998)
Education spend: (% GNP) N/A
Health spend: (% GDP) 3.7 (1997)

Economy and resources

Currency: congolese franc
GDP: (US$) 7 billion (1999 est)
Real GDP growth: (% change on previous year) –14.5 (1999 est)
GNP: (US$) 5.4 billion (1999 est)
GNP per capita (PPP): (US$) 731 (1999 est)
Consumer price inflation: 333% (1999 est)
Unemployment: N/A
Foreign debt: (US$) 12.93 billion (1998)
Major trading partners: Belgium–Luxembourg, South Africa, USA, Nigeria, Finland, Italy, Kenya
Resources: petroleum, copper, cobalt (65% of world's reserves), manganese, zinc, tin, uranium, silver, gold, diamonds (one of the world's largest producers of industrial diamonds)
Industries: textiles, cement, food processing, tobacco, rubber, engineering, wood products, leather, metallurgy and metal extraction, electrical equipment, transport vehicles
Exports: mineral products (mainly copper, cobalt, industrial diamonds, and petroleum), agricultural products (chiefly coffee). Principal market: Belgium, Luxembourg 62% (1999 est)
Imports: manufactured goods, food and live animals, machinery and transport equipment, chemicals, mineral fuels and lubricants. Principal source: South Africa 28.4% (1999 est)
Arable land: 3.1% (1996)
Agricultural products: coffee, palm oil, palm kernels, sugar cane, cassava, plantains, maize, groundnuts, bananas, yams, rice, rubber, seed cotton; forest resources

Population and society

Population: 51,654,000 (2000 est)
Population growth rate: 2.6% (1995–2000)
Population density: (per sq km) 21 (1999 est)
Urban population: (% of total) 30 (2000 est)
Age distribution: (% of total population) 0–14 48%, 15–59 48%, 60+ 4% (2000 est)
Ethnic groups: almost entirely of African descent, distributed among over 200 ethnic groups, the most numerous being the Kongo, Luba, Lunda, Mongo, and Zande
Language: French (official), Swahili, Lingala, Kikongo, Tshiluba (all national languages), over 200 other languages
Religion: Roman Catholic 41%, Protestant 32%, Kimbanguist 13%, animist 10%, Muslim 1–5%
Education: (compulsory years) 6
Literacy rate: 73% (men); 50% (women) (2000 est)
Labour force: 43% of population: 68% agriculture, 13% industry, 19% services (1990)
Life expectancy: 50 (men); 52 (women) (1995–2000)
Child mortality rate: (under 5, per 1,000 live births) 139 (1995–2000)
Physicians: 1 per 15,150 people (1993 est)
Hospital beds: 1 per 702 people (1993 est)
TV sets: (per 1,000 people) 43 (1997)
Radios: (per 1,000 people) 375 (1997)
Internet users: (per 10,000 people) 0.1 (1999)

Transport

Airports: international airports: Kinshasa (N'djili), Luano (near Lubumbashi), Bukava, Goma, Kisangani; over 40 domestic airports and 150 landing strips; total passenger km: 480 million (1994)
Railways: total length: 4,772 km/2,965 mi; total passenger km: 469 million (1990)
Roads: total road network: 157,000 km/97,560 mi (1996 est); passenger cars: 16.9 per 1,000 people (1996 est)

Chronology

13th century: Rise of Kongo Empire, centred on banks of the Congo River. **1483:** First visited by the Portuguese, who named the area Zaire (from Zadi, 'big water') and converted local rulers to Christianity. **16th–17th centuries:** Great development of slave trade by Portuguese, Dutch, British, and French merchants, initially supplied by Kongo intermediaries. **18th century:** Rise of Luba state, in southern copper belt of north Katanga, and Lunda, in Kasai region in central south. **mid-19th century:** Eastern Zaire invaded by Arab slave traders from East Africa. **1874–77:** Welsh-born US explorer Henry Morton Stanley navigated Congo River to Atlantic Ocean. **1879–87:** Stanley engaged by King Leopold II of Belgium to sign protection treaties with local chiefs and the 'Congo Free State' was awarded to Leopold by 1884–85 Berlin Conference; great expansion in rubber export, using forced labour. **1908:** Leopold was forced to relinquish personal control of Congo Free State, after international condemnation of human-rights abuses. Became a colony of the Belgian Congo and important exporter of minerals. **1959:** Riots in Kinshasa (Léopoldville) persuaded Belgium to decolonize rapidly. **1960:** Independence achieved as Republic of the Congo. Civil war broke out between central government based in Kinshasa (Léopoldville) with Joseph Kasavubu as president, and rich mining province of Katanga. **1961:** Former prime minister Patrice Lumumba was murdered in Katanga; fighting between mercenaries engaged by Katanga secessionist leader Moise Tshombe, and United Nations (UN) troops; Kasai and Kivu provinces also sought (briefly) to secede. **1963:** Katanga secessionist war ended; Tshombe forced into exile. **1964:** Tshombe returned from exile to become prime minister; pro-Marxist groups took control of eastern Zaire. The country was renamed the Democratic Republic of Congo. **1965:** Western-backed Col Sese Seko Mobutu seized power in coup, ousting Kasavubu and Tshombe. **1971:** Country renamed Republic of Zaire, with Mobutu as president as *authenticité* (Africanization) policy launched. **1972:** Mobutu's Popular Movement of the Revolution (MPR) became the only legal political party. Katanga province was renamed Shaba. **1974:** Foreign-owned businesses and plantations seized by Mobutu and given to his political allies. **1977:** Zairean guerrillas invaded Shaba province from Angola, but were repulsed by Moroccan, French, and Belgian paratroopers. **1980s:** The collapse in world copper prices increased foreign debts, and international creditors forced a series of austerity programmes. **1991:** After antigovernment riots, Mobutu agreed to end the ban on multiparty politics and share power with the opposition. **1993:** Rival pro- and anti-Mobutu governments were created. **1994:** There was an influx of Rwandan refugees. **1995:** There was secessionist activity in Shaba and Kasai provinces and interethnic warfare in Kivu, adjoining Rwanda in the east. **1996:** Thousands of refugees were allowed to return to Rwanda. **1997:** Mobutu was ousted by the rebel forces of Laurent Kabila, who declared himself president and changed the name of Zaire back to the Democratic Republic of the Congo. There was fighting between army factions. **1998:** There was a rebellion by Tutsi-led forces, backed by Rwanda and Uganda, against President Kabila; government troops aided by Angola and Zimbabwe put down the rebellion. A constituent assembly was appointed prior to a general election. UN-urged peace talks and a ceasefire agreed by rebel forces failed. **1999:** A peace deal, signed by both the government and rebel factions, was broken in November with fighting in the north of the country, and a reported bombing by the government of the centre in an attempt to free 700 Zimbabwean troops besieged by rebels. **2000:** The war between government and rebel soldiers intensified. President Kabila walked out of a peace conference held in Lusaka, Zambia and called for a summit with Uganda, Rwanda, and Burundi. Ugandan-backed rebels made gains in the northwest of the country. Kabila allowed some United Nations troops into the country, but hindered the operation. **2001:** In January President Kabila was assassinated in politically suspicious circumstances, allegedly by a bodyguard. He was succeeded by his son, Joseph.

Practical information

Visa requirements: UK: visa required. USA: visa required
Time difference: GMT +1/2

Chief tourist attractions: lake and mountain scenery; extensive tropical rainforests along Congo River
Major holidays: 1, 4 January, 1, 20 May, 24, 30 June, 1 August, 14, 27 October, 17, 24 November, 25 December

CONGO, REPUBLIC OF OR CONGO (BRAZZAVILLE)

Located in west-central Africa, bounded north by Cameroon and the
Central African Republic, east and south by the Democratic Republic of
Congo, west by the Atlantic Ocean, and northwest by Gabon.

National name: *République du Congo/Republic of Congo*
Area: 342,000 sq km/132,046 sq mi
Capital: Brazzaville
Major towns/cities: Pointe-Noire, Nkayi, Loubomo, Bouenza,
Mossendjo, Ouesso, Owando
Major ports: Pointe-Noire
Physical features: narrow coastal plain rises to central plateau, then
falls into northern basin; Congo River on the border with the
Democratic Republic of Congo; half the country is rainforest

The flag uses the pan-African colours in a striking
design. Effective date: 10 June 1991.

Government

Head of state and government: Denis Sassou-
Nguessou from 1997
Political system: nationalistic socialist
Political executive: unlimited presidency
Administrative divisions: nine regions and one
capital district
Political parties: Pan-African Union for Social
Democracy (UPADS), moderate, left of centre;
Congolese Movement for Democracy and Integral
Development (MCDDI), moderate, left of centre;
Congolese Labour Party (PCT), left wing
Armed forces: 10,000 (1998); plus a paramilitary
force of 5,000
Conscription: national service is voluntary
Death penalty: retains the death penalty for
ordinary crimes but can be considered abolitionist
in practice (last execution 1982)
Defence spend: (% GDP) 3.9 (1998)
Education spend: (% GNP) 6.2 (1996)
Health spend: (% GDP) 5 (1997)

Economy and resources

Currency: franc CFA
GDP: (US$) 2.27 billion (1999)
Real GDP growth: (% change on previous year)
–0.7 (1999)
GNP: (US$) 1.9 billion (1999)
GNP per capita (PPP): (US$) 897 (1999)
Consumer price inflation: 3.6% (1999)
Foreign debt: (US$) 5.12 billion (1998)
Major trading partners: France, USA, Italy,
South Korea, Germany, China
Resources: petroleum, natural gas, lead, zinc,
gold, copper, phosphate, iron ore, potash, bauxite
Industries: mining, food processing, textiles,
cement, metal goods, chemicals, forest products
Exports: petroleum and petroleum products, saw
logs and veneer logs, veneer sheets. Principal
market: USA 25.9% (1999 est)

Imports: machinery, chemical products, iron and
steel, transport equipment, foodstuffs. Principal
source: France 21.8% (1999 est)
Arable land: 0.4% (1996)
Agricultural products: cassava, plantains, sugar
cane, palm oil, maize, coffee, cocoa; forestry

Population and society

Population: 2,943,000 (2000 est)
Population growth rate: 2.8% (1995–2000);
2.6% (est 2000–05)
Population density: (per sq km) 8 (1999 est)
Urban population: (% of total) 63 (2000 est)
Age distribution: (% of total population) 0–14
46%, 15–59 49%, 60+ 5% (2000 est)
Ethnic groups: predominantly Bantu; population
comprises 15 main ethnic groups and 75 tribes.
The Kongo, or Bakongo, account for about 48% of
the population, then come the Sanga at about
20%, the Bateke, or Teke, at about 17%, and then
the Mboshi, or Boubangui, about 12%
Language: French (official), Kongo, Monokutuba
and Lingala (both patois), and other dialects
Religion: Christian 50%, animist 48%, Muslim 2%
Education: (compulsory years) 10
Literacy rate: 87% (men); 74% (women) (2000
est)
Labour force: 42% of population: 49%
agriculture, 15% industry, 37% services (1990)
Life expectancy: 48 (men); 51 (women)
(1995–2000)
Child mortality rate: (under 5, per 1,000 live
births) 132 (1995–2000)
Physicians: 1 per 3,713 people (1993 est)
Hospital beds: 1 per 306 people (1993 est)
TV sets: (per 1,000 people) 12 (1997)
Radios: (per 1,000 people) 124 (1997)
Internet users: (per 10,000 people) 13.8 (1999)
Personal computer users: (per 100 people) 0.6
(1999)

Transport

Airports: international airports: Brazzaville (Maya-Maya), Pointe-Noire; six domestic airports; total passenger km: 283 million (1995)
Railways: total length: 795 km/494 mi; total passenger km: 302 million (1995)
Roads: total road network: 12,760 km/7,929 mi, of which 9.7% paved (1996 est); passenger cars: 14.1 per 1,000 people (1996 est)

Chronology

late 15th century: First visited by Portuguese explorers, at which time the Bakongo (a six-state confederation centred south of the Congo River in Angola) and Bateke, both Bantu groups, were the chief kingdoms. **16th century:** The Portuguese, in collaboration with coastal peoples, exported slaves from the interior to plantations in Brazil and São Tomé; missionaries spread Roman Catholicism. **1880:** French explorer Pierre Savorgnan de Brazza established French claims to coastal region, with the makoko (king) of the Bateke accepting French protection. **1905:** There was international outrage at revelations of the brutalities of forced labour as ivory and rubber resources were ruthlessly exploited by private concessionaries. **1910:** As Moyen-Congo became part of French Equatorial Africa, which also comprised Gabon and the Central African Republic, with the capital at Brazzaville. **1920s:** More than 17,000 were killed as forced labour was used to build the Congo-Ocean railway; first Bakongo political organization founded. **1940–44:** Supported the 'Free French' anti-Nazi resistance cause during World War II, Brazzaville serving as capital for Gen Charles de Gaulle's forces. **1946:** Became autonomous, with a territorial assembly and representation in French parliament. **1960:** Achieved independence from France, with Abbé Fulbert Youlou, a moderate Catholic Bakongo priest, as the first president. **1963:** Alphonse Massamba-Débat became president and a single-party state was established under the socialist National Revolutionary Movement (MNR). **1968:** A military coup, led by Capt Marien Ngouabi, ousted Massamba-Débat. **1970:** A Marxist People's Republic declared, with Ngouabi's PCT the only legal party. **1977:** Ngouabi was assassinated in a plot by Massamba-Débat, who was executed. **early 1980s:** Petroleum production increased fivefold. **1990:** The PCT abandoned Marxist-Leninism and promised multiparty politics and market-centred reforms in an economy crippled by foreign debt. **1992:** Multiparty elections gave the coalition dominated by the Pan-African Union for Social Democracy (UPADS) an assembly majority, with Pascal Lissouba elected president. **1995:** A new broad-based government was formed, including opposition groups; market-centred economic reforms were instigated, including privatization. **1997:** Violence between factions continued despite the unity government. Sassou-Nguesso took over the presidency.

Practical information

Visa requirements: UK: visa required. USA: visa required
Time difference: GMT +1

Major holidays: 1 January, 18 March, 1 May, 31 July, 13–15 August, 1 November, 25, 31 December; variable: Good Friday, Easter Monday

COSTA RICA

Located in Central America, bounded north by Nicaragua, southeast by Panama, east by the Caribbean Sea, and west by the Pacific Ocean.

National name: *República de Costa Rica/Republic of Costa Rica*
Area: 51,100 sq km/19,729 sq mi
Capital: San José
Major towns/cities: Alajuela, Cartago, Limón, Puntarenas, San Isidro, Desamparados
Major ports: Limón, Puntarenas
Physical features: high central plateau and tropical coasts; Costa Rica was once entirely forested, containing an estimated 5% of the Earth's flora and fauna

The sun of freedom rises from the Caribbean Sea. The stars represent Costa Rica's seven provinces. Red was added to the blue and white flag to reflect the French tricolour. Effective date: 29 September 1848.

Government

Head of state and government: Miguel Angel Rodriguez Echeverria from 1998
Political system: liberal democracy
Political executive: limited presidency
Administrative divisions: seven provinces
Political parties: National Liberation Party (PLN), left of centre; Christian Socialist Unity Party (PUSC), centrist coalition; ten minor parties
Armed forces: army abolished in 1948; 4,300 civil guards and 3,200 rural guards
Death penalty: abolished in 1877
Defence spend: (% GDP) 0.7 (1998)
Education spend: (% GNP) 5.3 (1996)
Health spend: (% GDP) 8.7 (1997)

Economy and resources

Currency: colón
GDP: (US$) 11.08 billion (1999)
Real GDP growth: (% change on previous year) 8 (1997)
GNP: (US$) 9.8 billion (1999)
GNP per capita (PPP): (US$) 5,770 (1999 est)
Consumer price inflation: 10.1% (1999)
Unemployment: 5.7% (1997)
Foreign debt: (US$) 4.2 billion (1999 est)
Major trading partners: USA, EU, Japan, Venezuela, Mexico, Guatemala
Resources: gold, salt, hydro power
Industries: food processing, chemical products, beverages, paper and paper products, textiles and clothing, plastic goods, electrical equipment
Exports: bananas, coffee, sugar, cocoa, textiles, seafood, meat, tropical fruit. Principal market: USA 51.4% (1999)
Imports: raw materials for industry and agriculture, consumer goods, machinery and transport equipment, construction materials. Principal source: USA 56.4% (1999)
Arable land: 5.6% (1996)
Agricultural products: bananas, coffee, sugar cane, maize, potatoes, tobacco, tropical fruit; livestock rearing (cattle and pigs); fishing

Population and society

Population: 4,023,000 (2000 est)
Population growth rate: 2.5% (1995–2000)
Population density: (per sq km) 77 (1999 est)
Urban population: (% of total) 48 (2000 est)
Age distribution: (% of total population) 0–14 32%, 15–59 60%, 60+ 8% (2000 est)
Ethnic groups: about 96% of the population is of European descent, mostly Spanish, and about 2% is of African origin, 1% Amerindian, 1% Chinese
Language: Spanish (official)
Religion: Roman Catholic 95% (state religion)
Education: (compulsory years) 9
Literacy rate: 95% (men); 96% (women) (2000 est)
Labour force: 21.8% agriculture, 23.4% industry, 54.8% services (1997)
Life expectancy: 74 (men); 79 (women) (1995–2000)
Child mortality rate: (under 5, per 1,000 live births) 15 (1995–2000)
Physicians: 1 per 979 people (1995)
Hospital beds: 1 per 566 people (1995)
TV sets: (per 1,000 people) 403 (1997)
Radios: (per 1,000 people) 271 (1997)
Internet users: (per 10,000 people) 381.4 (1999)
Personal computer users: (per 100 people) 4.5 (1999)

Transport

Airports: international airports: San José (Juan Santamaría), Liberia (Daniel Oduber Quirós); 11 domestic airports as well as charter services to provincial towns and villages); total passenger km: 1,920 million (1997 est)
Railways: total length: 950 km/590 mi; total passenger journeys: 335,276 (1994); rail system ceased operating in 1995
Roads: total road network: 35,597 km/22,120 mi, of which 17% paved (1996); passenger cars: 85 per 1,000 people (1996)

Chronology

1502: Visited by Christopher Columbus, who named the area Costa Rica (the rich coast), observing the gold decorations worn by the Guaymi American Indians. **1506:** Colonized by Spain, but there was fierce guerrilla resistance by the indigenous population. Many later died from exposure to European diseases. **18th century:** Settlements began to be established in the fertile central highlands, including San José and Alajuela. **1808:** Coffee was introduced from Cuba and soon became the staple crop. **1821:** Independence achieved from Spain, and was joined initially with Mexico. **1824:** Became part of United Provinces (Federation) of Central America, also embracing El Salvador, Guatemala, Honduras, and Nicaragua. **1838:** Became fully independent when it seceded from the Federation. **later 19th century:** Immigration by Europeans to run and work small coffee farms. **1940–44:** Liberal reforms, including recognition of workers' rights and minimum wages, were introduced by President Rafael Angel Calderón Guradia, founder of the United Christian Socialist Party (PUSC). **1948:** Brief civil war following a disputed presidential election. **1949:** New constitution adopted, giving women and blacks the vote. National army abolished and replaced by civil guard. José Figueres Ferrer, cofounder of the PLN, elected president; he embarked on an ambitious socialist programme, nationalizing the banks and introducing a social security system. **1958–73:** Mainly conservative administrations. **1978:** Sharp deterioration in the state of the economy. **1982:** A harsh austerity programme was introduced. **1985:** Following border clashes with Nicaraguan Sandinista forces, a US-trained antiguerrilla guard was formed. **1986:** Oscar Arias Sanchez (PLN) won the presidency on a neutralist platform. **1987:** Arias won the Nobel Prize for Peace for devising a Central American peace plan signed by the leaders of Nicaragua, El Salvador, Guatemala, and Honduras. **1998:** Miguel Angel Rodriguez Echeverria (PUSC) was elected president.

Practical information

Visa requirements: UK: visa not required. USA: visa not required
Time difference: GMT –6
Chief tourist attractions: nature reserves and national parks make up one-third of the country; Irazú and Poás volcanoes; Orosí valley; colonial ruins at Ujarras; railway through rainforest to Limón; San José (the capital); Pacific beaches; Caribbean beaches at Limón
Major holidays: 1 January, 19 March, 11 April, 1 May, 29 June, 25 July, 2, 15 August, 15 September, 12 October, 8, 25 December; variable: Corpus Christi, Good Friday, Holy Saturday, Holy Thursday

CÔTE D'IVOIRE

Located in West Africa, bounded north by Mali and Burkina Faso, east by Ghana, south by the Gulf of Guinea, and west by Liberia and Guinea.

National name: *République de la Côte d'Ivoire/Republic of the Ivory Coast*
Area: 322,463 sq km/124,502 sq mi
Capital: Yamoussoukro
Major towns/cities: Abidjan, Bouaké, Daloa, Man, Korhogo, Gagnoa
Major ports: Abidjan, San Pedro
Physical features: tropical rainforest (diminishing as exploited) in south; savannah and low mountains in north; coastal plain; Vridi canal, Kossou dam, Monts du Toura

Orange stands for the savannah. White symbolizes the country's rivers. Green represents the forests.
Effective date: 3 December 1959.

Government

Head of state: Laurent Gbagbo from 2000
Head of government: Affi N'Guessan from 2000
Political system: emergent democracy
Political executive: limited presidency
Administrative divisions: 16 regions, comprising 49 departments
Political parties: Democratic Party of Côte d'Ivoire (PDCI), nationalist, free enterprise; Rally of Republicans (RDR), nationalist; Ivorian Popular Front (FPI), left of centre; Ivorian Labour Party (PIT), left of centre; over 20 smaller parties
Armed forces: 8,400 (1998); plus paramilitary forces numbering 7,000
Conscription: selective conscription for six months
Death penalty: retains the death penalty for ordinary crimes but can be considered abolitionist in practice
Defence spend: (% GDP) 0.9 (1998)
Education spend: (% GNP) 5 (1997)
Health spend: (% GDP) 3.2 (1997)

Economy and resources

Currency: franc CFA
GDP: (US$) 11.22 billion (1999)
Real GDP growth: (% change on previous year) 1.5 (1999)
GNP: (US$) 10.4 billion (1999)
GNP per capita (PPP): (US$) 1,546 (1999)
Consumer price inflation: 0.8% (1999)
Unemployment: N/A
Foreign debt: (US$) 14.7 billion (1999)
Major trading partners: France, Nigeria, Germany, the Netherlands, Italy, USA, China
Resources: petroleum, natural gas, diamonds, gold, nickel, reserves of manganese, iron ore, bauxite
Industries: agro-processing (dominated by cocoa, coffee, cotton, palm kernels, pineapples, fish), petroleum refining, tobacco
Exports: cocoa beans and products, petroleum products, timber, coffee, cotton, tinned tuna. Principal market: France 15.2% (1998 est)
Imports: crude petroleum, machinery and vehicles, pharmaceuticals, fresh fish, plastics, cereals. Principal source: France 26.6% (1998 est)
Arable land: 9.1% (1996)

Agricultural products: cocoa (world's largest producer), coffee (world's fifth-largest producer), cotton, rubber, palm kernels, bananas, pineapples, yams, cassava, plantains; fishing; forestry

Population and society

Population: 14,786,000 (2000 est)
Population growth rate: 1.8% (1995–2000)
Population density: (per sq km) 45 (1999 est)
Urban population: (% of total) 46 (2000 est)
Age distribution: (% of total population) 0–14 43%, 15–59 52%, 60+ 5% (2000 est)
Ethnic groups: five principal ethnic groups: the Akan (41%), in the east and centre, the Voltaic (16%), based in the north, the Malinke (15%) and Southern Mande (11%) in the west, and the Kron (4%), based in the centre and the west
Language: French (official), over 60 ethnic languages
Religion: animist 17%, Muslim 39% (mainly in north), Christian 26% (mainly Roman Catholic in south)
Education: (compulsory years) 6
Literacy rate: 55% (men); 39% (women) (2000 est)
Labour force: 37% of population: 52.5% agriculture, 11.8% industry, 35.7% services (1997 est)
Life expectancy: 46 (men); 47 (women) (1995–2000)
Child mortality rate: (under 5, per 1,000 live births) 136 (1995–2000)
Physicians: 1 per 18,000 people (1994 est)
Hospital beds: 1 per 1,670 people (1994 est)
TV sets: (per 1,000 people) 70 (1998)
Radios: (per 1,000 people) 164 (1997)
Internet users: (per 10,000 people) 13.8 (1999)
Personal computer users: (per 100 people) 0.6 (1999)

Transport

Airports: international airports: Abidjan (Port Bouet), Bouaké, Yamoussoukro (San Pedro); domestic services to all major towns; total passenger km: 302 million (1995)
Railways: total length: 660 km/410 mi; total passenger km: 173 million (1993)
Roads: total road network: 50,400 km/31,318 mi, of which 9.7% paved (1996 est); passenger cars: 18.1 per 1,000 people (1996 est)

Chronology

1460s: Portuguese navigators arrived. **16th century:** Ivory export trade developed by Europeans and slave trade, though to a lesser extent than neighbouring areas; Krou people migrated from Liberia to the west and Senoufo and Lubi from the north. **late 17th century:** French coastal trading posts established at Assini and Grand Bassam. **18th–19th centuries:** Akan peoples, including the Baoulé, immigrated from the east and Malinke from the northwest. **1840s:** French began to conclude commercial treaties with local rulers. **1893:** Colony of Côte d'Ivoire created by French, after war with Mandinkas; Baoulé resistance continued until 1917. **1904:** Became part of French West Africa; cocoa production encouraged. **1940–42:** Under pro-Nazi French Vichy regime. **1946:** Became overseas territory in French Union, with own territorial assembly and representation in French parliament: Felix Houphoüet-Boigny, a Western-educated Baoulé chief who had formed the Democratic Party (PDCI) to campaign for autonomy, was elected to the French assembly. **1947:** A French-controlled area to the north, which had been added to Côte d'Ivoire in 1932, separated to create new state of Upper Volta (now Burkina Faso). **1950–54:** Port of Abidjan constructed. **1958:** Achieved internal self-government. **1960:** Independence secured, with Houphouët-Boigny as president of a one-party state. **1960s–1980s:** Political stability, close links maintained with France and economic expansion of 10% per annum, as the country became one of the world's largest coffee producers. **1986:** The country's name was officially changed from Ivory Coast to Côte d'Ivoire. **1987–93:** Per capita incomes fell by 25% owing to an austerity programme promoted by the International Monetary Fund. **1990:** There were strikes and student unrest. Houphoüet-Boigny was re-elected as president as multiparty politics were re-established. **1993:** Houphouët-Boigny died and was succeeded by parliamentary speaker and Baoulé Henri Konan Bedie. **1999:** After a largely bloodless coup over Christmas 1999 Bedie was replaced by a new military leader, General Robert Guei. **2000:** Guei announced suspension of the country's foreign debt repayments in January. A new constitution for the return of civilian rule was approved by referendum. Mutinous soldiers launched three unsuccessful coups against Guei, and a state of emergency was imposed. Guei attempted to sabotage the presidential elections held in October, but was forced to flee. The elections were won by Laurent Gbagbo, and marked by violence against supporters of Alassane Outtara, who had been excluded from standing in the contest. Violence also surrounded the parliamentary elections held in December.

Practical information

Visa requirements: UK: visa required. USA: visa not required for a stay of less than 90 days
Time difference: GMT +/–0
Chief tourist attractions: game reserves; lagoons; forests; Abidjan
Major holidays: 1 January, 1 May, 15 August, 1 November, 7, 24–25, 31 December; variable: Ascension Thursday, Eid-ul-Adha, Good Friday, Easter Monday, Whit Monday, end of Ramadan

CROATIA

Located in central Europe, bounded north by Slovenia and Hungary, west by the Adriatic Sea, and east by Bosnia-Herzegovina and the Yugoslavian republic of Serbia.

National name: *Republika Hrvatska/Republic of Croatia*
Area: 56,538 sq km/21,829 sq mi
Capital: Zagreb
Major towns/cities: Osijek, Split, Dubrovnik, Rijeka, Zadar, Pula
Major ports: chief port: Rijeka (Fiume); other ports: Zadar, Sibenik, Split, Dubrovnik
Physical features: Adriatic coastline with large islands; very mountainous, with part of the Karst region and the Julian and Styrian Alps; some marshland

Red, white, and blue are the pan-Slav colours. The small shields represent Croatia Ancient, Dubrovnik, Dalmatia, Istria, and Slavonia. The flag is based on the tricolour used during World War II. Effective date: 22 December 1990.

Government

Head of state: Stipe Mesic from 2000
Head of government: Ivica Racan from 2000
Political system: emergent democracy
Political executive: limited presidency
Administrative divisions: 21 counties
Political parties: Croatian Democratic Union (CDU), Christian Democrat, right of centre, nationalist; Croatian Social-Liberal Party (CSLP), centrist; Social Democratic Party of Change (SDP), reform socialist; Croatian Party of Rights (HSP), Croat-oriented, ultranationalist; Croatian Peasant Party (HSS), rural-based; Serbian National Party (SNS), Serb-oriented
Armed forces: 56,200 (1998); plus 40,000 in paramilitary forces
Conscription: compulsory for ten months
Death penalty: abolished in 1990
Defence spend: (% GDP) 8.3 (1998)
Education spend: (% GDP) 5.3 (1996)
Health spend: (% GDP) 8.1 (1997)

Economy and resources

Currency: kuna
GDP: (US$) 20.2 billion (1999)
Real GDP growth: (% change on previous year) –0.3 (1999)
GNP: (US$) 20.4 billion (1999)
GNP per capita (PPP): (US$) 6,915 (1999)
Consumer price inflation: 4.2% (1999)
Unemployment: 17.6% (1998)
Foreign debt: (US$) 8.3 billion (1999 est)
Major trading partners: Germany, Italy, Slovenia, Austria, Bosnia-Herzegovina, Russia, Norway
Resources: petroleum, natural gas, coal, lignite, bauxite, iron ore, salt
Industries: food processing, textiles, chemicals, ship-building, metal processing, construction materials. Tourism was virtually eliminated during hostilities, but a revival began in 1992
Exports: machinery and transport equipment, chemicals, foodstuffs, miscellaneous manufactured items (mainly clothing). Principal market: Italy 18% (1999)

Imports: machinery and transport equipment, basic manufactures, mineral fuels, miscellaneous manufactured articles. Principal source: Germany 18.5% (1999)
Arable land: 22.2% (1996)
Agricultural products: wheat, maize, potatoes, plums, sugar beet; livestock rearing (cattle and pigs); dairy products

Population and society

Population: 4,473,000 (2000 est)
Population growth rate: –0.1% (1995–2000); –0.1% (est 2000–05)
Population density: (per sq km) 79 (1999 est)
Urban population: (% of total) 58 (2000 est)
Age distribution: (% of total population) 0–14 17%, 15–59 63%, 60+ 20% (2000 est)
Ethnic groups: in 1991, 77% of the population were ethnic Croats, 12% were ethnic Serbs, and 1% were Slovenes. The civil war that began in 1992 displaced more than 300,000 Croats from Serbian enclaves within the republic, and created some 500,000 refugees from Bosnia in the republic. Serbs are most thickly settled in areas bordering Bosnia-Herzegovina, and in Slavonia, although more than 150,000 fled from Krajina to Bosnia-Herzegovina and Serbia following the region's recapture by the Croatian army in August 1995.
Language: Croat (official), Serbian
Religion: Roman Catholic (Croats) 76.5%; Orthodox Christian (Serbs) 11%, Protestant 1.4%, Muslim 1.2%
Education: (compulsory years) 8
Literacy rate: 99% (men); 97% (women) (2000 est)
Labour force: 10.3% agriculture, 40.7% industry, 49.0% services (1997 est)
Life expectancy: 69 (men); 77 (women) (1995–2000)
Child mortality rate: (under 5, per 1,000 live births) 12 (1995–2000)
Physicians: 1 per 518 people (1993 est)
Hospital beds: 1 per 167 people (1993 est)
TV sets: (per 1,000 people) 267 (1997)
Radios: (per 1,000 people) 336 (1997)
Internet users: (per 10,000 people) 446.7 (1999)
Personal computer users: (per 100 people) 6.7 (1999)

Transport

Airports: international airports: Zagreb (Pleso), Dubrovnik; three domestic airports; total passenger km: 546 million (1997)
Railways: total length: 2,452 km/1,524 mi; total

passenger km: 981 million (1997)
Roads: total road network: 27,840 km/17,300 mi, of which 81.5% paved (1997); passenger cars: 204 per 1,000 people (1997)

Chronology

early centuries AD: Part of Roman region of Pannonia. **AD 395:** On division of Roman Empire, stayed in western half, along with Slovenia and Bosnia. **7th century:** Settled by Carpathian Croats, from northeast; Christianity adopted. **924:** Formed by Tomislav into independent kingdom, which incorporated Bosnia from 10th century. **12th–19th centuries:** Autonomy under Hungarian crown, following dynastic union in 1102. **1526–1699:** Slavonia, in east, held by Ottoman Turks, while Serbs were invited by Austria to settle along the border with Ottoman-ruled Bosnia, in Vojna Krajina (military frontier). **1797–1815:** Dalmatia, in west, ruled by France. **19th century:** Part of Austro-Hungarian Habsburg Empire. **1918:** On dissolution of Habsburg Empire, joined Serbia, Slovenia, and Montenegro in 'Kingdom of Serbs, Croats, and Slovenes', under Serbian Karageorgevic dynasty. **1929:** The Kingdom became Yugoslavia. Croatia continued its campaign for autonomy. **1930s:** Ustasa, a Croat terrorist organization, began a campaign against dominance of Yugoslavia by the non-Catholic Serbs. **1941–44:** Following German invasion, a 'Greater Croatia' Nazi puppet state, including most of Bosnia and western Serbia, formed under Ustasa leader, Ante Pavelic; more than half a million Serbs, Jews, and members of the Romany community were massacred in extermination camps. **1945:** Became constituent republic of Yugoslavia Socialist Federation after communist partisans, led by Croat Marshal Tito, overthrew Pavelic. **1970s:** Separatist demands resurfaced, provoking a crackdown. **late 1980s:** Spiralling inflation and a deterioration in living standards sparked industrial unrest and a rise in nationalist sentiment, which affected the local communist party. **1989:** The formation of opposition parties was permitted. **1990:** The communists were defeated by the conservative nationalist CDU led by ex-Partisan Franjo Tudjman in the first free election since 1938. Sovereignty was declared. **1991:** The Serb-dominated region of Krajina in the southwest announced its secession from Croatia. Croatia declared independence, leading to military conflict with Serbia, and civil war ensued. **1992:** A United Nations (UN) peace accord was accepted; independence was recognized by the European Community (EC) and the USA; Croatia joined the UN. A UN peacekeeping force was stationed in Croatia. Tudjman was elected president. **1993:** A government offensive was launched to retake parts of Serb-held Krajina, violating the 1992 UN peace accord. **1994:** There was an accord with Muslims and ethnic Croats within Bosnia, to the east, to link the recently formed Muslim–Croat federation with Croatia. **1995:** Serb-held western Slavonia and Krajina were captured by government forces; there was an exodus of Croatian Serbs. The offensive extended into Bosnia-Herzegovina to halt a Bosnian Serb assault on Bihac in western Bosnia. Serbia agreed to cede control of eastern Slavonia to Croatia over a two-year period. Zlatko Matesa was appointed prime minister. **1996:** Diplomatic relations between Croatia and Yugoslavia were restored. Croatia entered the Council of Europe. **1997:** The opposition was successful in local elections. The constitution was amended to prevent the weakening of Croatia's national sovereignty. **1998:** Croatia resumed control over East Slavonia. **2000:** In parliamentary elections, the ruling Croatian Democratic Union lost heavily to a centre-left coalition. The reformist Stipe Mesic was elected president. The Social Democrat leader, Ivica Rajan, became prime minister. Constitutional changes reduced the powers of the president and turned Croatia into a parliamentary democracy.

Practical information

Visa requirements: UK: visa not required. USA: visa required
Time difference: GMT +1
Chief tourist attractions: Adriatic coast with 1,185 islands. Owing to civil conflict which began in 1991,

tourist activity has been greatly reduced; historic cities, notably Dubrovnik, have been severely damaged
Major holidays: 1, 6 January, 1, 30 May, 22 June, 15 August, 1 November, 25–26 December; variable: Good Friday, Easter Monday

CUBA

Island in the Caribbean Sea, the largest of the West Indies, off the south coast of Florida and to the east of Mexico.

National name: *República de Cuba/Republic of Cuba*
Area: 110,860 sq km/42,803 sq mi
Capital: Havana
Major towns/cities: Santiago de Cuba, Camagüey, Holguín, Guantánamo, Santa Clara, Bayamo, Cienfuegos
Physical features: comprises Cuba and smaller islands including Isle of Youth; low hills; Sierra Maestra mountains in southeast; Cuba has 3,380 km/2,100 mi of coastline, with deep bays, sandy beaches, coral islands and reefs

The flag is known as the 'Lone Star' banner. The red triangle symbolizes the blood shed in the fight for freedom from Spain. The blue stripes stand for Cuba's three provinces. Effective date: 20 May 1902.

Government

Head of state and government: Fidel Castro Ruz from 1959
Political system: communist
Political executive: communist
Administrative divisions: 14 provinces and the special municipality of the Isle of Youth (Isla de la Juventud)
Political party: Communist Party of Cuba (PCC), Marxist-Leninist
Armed forces: 60,000 (1998)
Conscription: compulsory for two years
Death penalty: retained and used for ordinary crimes
Defence spend: (% GDP) 5.3 (1998)
Education spend: (% GNP) 6.7 (1997)
Health spend: (% GDP) 6.3 (1997)

Economy and resources

Currency: Cuban peso
GDP: (US$) 22.1 billion (1999)
Real GDP growth: (% change on previous year) 6.2 (1999)
GNP: (US$) N/A
GNP per capita (PPP): (US$) N/A
Consumer price inflation: –0.5% (1999 est)
Unemployment: 8% (1996 est)
Foreign debt: (US$) 12 billion (1999 est)
Major trading partners: Canada, Spain, Russia, China, the Netherlands, France, Venezuela
Resources: iron ore, copper, chromite, gold, manganese, nickel, cobalt, silver, salt
Industries: mining, textiles and footwear, cigarettes, cement, food processing (sugar and its by-products), fertilizers
Exports: sugar, minerals, tobacco, citrus fruits, fish products. Principal market: Russia 21.7% (1998)
Imports: mineral fuels, machinery and transport equipment, foodstuffs, beverages. Principal source: Spain 22.3% (1998)
Arable land: 34.3% (1996)
Agricultural products: sugar cane (world's fourth-largest producer of sugar), tobacco, rice, citrus fruits, plantains, bananas; forestry; fishing

Population and society

Population: 11,201,000 (2000 est)
Population growth rate: 0.4% (1995–2000)
Population density: (per sq km) 101 (1999 est)
Urban population: (% of total) 75 (2000 est)
Age distribution: (% of total population) 0–14 21%, 15–59 65%, 60+ 14% (2000 est)
Ethnic groups: predominantly of mixed Spanish and African or Spanish and American Indian origin
Language: Spanish (official)
Religion: Roman Catholic; also Episcopalians and Methodists
Education: (compulsory years) 6
Literacy rate: 97% (men); 97% (women) (2000 est)
Labour force: 15.3% agriculture, 23.0% industry, 61.7% services (1997 est)
Life expectancy: 74 (men); 78 (women) (1995–2000)
Child mortality rate: (under 5, per 1,000 live births) 12 (1995–2000)
Physicians: 1 per 172 people (1998 est)
Hospital beds: 1 per 165 people (1993 est)
TV sets: (per 1,000 people) 239 (1997)
Radios: (per 1,000 people) 352 (1997)
Internet users: (per 10,000 people) 44.8 (1999)
Personal computer users: (per 100 people) 0.7 (1999)

Transport

Airports: international airports: Havana, Santiago de Cuba, Holguín, Camagüey, Varadero; 11 domestic airports; total passenger km: 3,615 million (1997 est)
Railways: total length: 4,807 km/2,987 mi; total passenger km: 2,346 million (1994)
Roads: total road network: 60,858 km/37,817 mi, of which 49% paved (1997); passenger cars: 15.6 per 1,000 people (1997 est)

Chronology

3rd century AD: The Ciboney, Cuba's earliest known inhabitants, were dislodged by the immigration of Taino, Arawak Indians from Venezuela. **1492:** Christopher Columbus landed in Cuba and claimed it for Spain. **1511:** Spanish settlement established at Baracoa by Diego Velazquez. **1523:** Decline of American Indian population and rise of sugar plantations led to import of slaves from Africa.

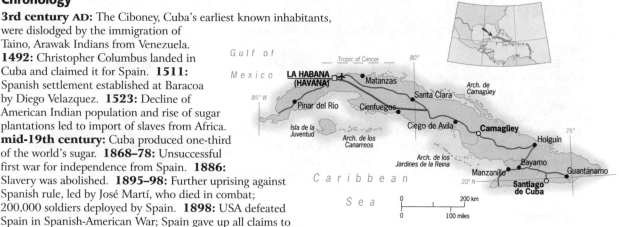

mid-19th century: Cuba produced one-third of the world's sugar. **1868–78:** Unsuccessful first war for independence from Spain. **1886:** Slavery was abolished. **1895–98:** Further uprising against Spanish rule, led by José Martí, who died in combat; 200,000 soldiers deployed by Spain. **1898:** USA defeated Spain in Spanish-American War; Spain gave up all claims to Cuba, which was ceded to the USA. **1901:** Cuba achieved independence; Tomás Estrada Palma became first president of the Republic of Cuba. **1906–09:** Brief period of US administration after Estrada resigned in the face of an armed rebellion by political opponents. **1924–33:** Gerado Machado established a brutal dictatorship. **1925:** Socialist Party founded, from which the Communist Party later developed. **1933:** Army sergeant Fulgencio Batista seized power. **1934:** USA abandoned its right to intervene in Cuba's internal affairs. **1944:** Batista retired and was succeeded by the civilian Ramon Gray San Martin. **1952:** Batista seized power again to begin an oppressive and corrupt regime. **1953:** Fidel Castro Ruz led an unsuccessful coup against Batista. **1956:** Second unsuccessful coup by Castro. **1959:** Batista overthrown by Castro and his 9,000-strong guerrilla army. Constitution was replaced by a 'Fundamental Law', making Castro prime minister, his brother Raúl Castro his deputy, and Argentine-born Ernesto 'Che' Guevara third in command. **1960:** All US businesses in Cuba appropriated without compensation; USA broke off diplomatic relations. **1961:** USA sponsored an unsuccessful invasion by Cuban exiles at the Bay of Pigs. Castro announced that Cuba had become a communist state, with a Marxist-Leninist programme of economic development, and became allied with the USSR. **1962:** Cuban missile crisis: Cuba was expelled from the Organization of American States. Castro responded by tightening relations with the USSR, which installed nuclear missiles in Cuba (subsequently removed at US insistence). US trade embargo imposed. **1965:** Cuba's sole political party renamed Cuban Communist Party (PCC). With Soviet help, Cuba began to make considerable economic and social progress. **1972:** Cuba became a full member of the Moscow-based Council for Mutual Economic Assistance (COMECON). **1976:** New socialist constitution approved; Castro elected president. **1976–81:** Castro became involved in extensive international commitments, sending troops as Soviet surrogates, particularly to Africa. **1982:** Cuba joined other Latin American countries in giving moral support to Argentina in its dispute with Britain over the Falklands. **1984:** Castro tried to improve US–Cuban relations by discussing exchange of US prisoners in Cuba for Cuban 'undesirables' in the USA. **1988:** A peace accord with South Africa was signed, agreeing to the withdrawal of Cuban troops from Angola, as part of a reduction in Cuba's overseas military activities. **1991:** Soviet troops were withdrawn with the collapse of the USSR. **1993:** The US trade embargo was tightened; market-oriented reforms were introduced in the face of a deteriorating economy. **1994:** There was a refugee exodus; US policy on Cuban asylum seekers was revised. **1998:** Castro was confirmed as president for a further five-year term. **1999:** In an immigration dispute with the US, which focused on one child, Cuba demanded the return of illegal immigrants, and condemned the use of the US justice system on such matters. **2000:** Trade talks were cancelled with European Union (EU) officials after EU countries voted in a UN committee to condemn Cuba's human rights record. The USA eased its 40-year-old economic embargo of Cuba, allowing exports of food and medicine.

Practical information

Visa requirements: UK: visa required. USA: visa required
Time difference: GMT –5
Chief tourist attractions: sandy beaches; Sierra Maestra Mountain range, which encircles the port of Santiago; mountain forest of pine and mahogany; Havana city centre, a United Nations (UN) World Heritage site, with colonial fortresses and castles; Santiago, the former capital, with Museum of Colonial Art and Festival de Carib (April)
Major holidays: 1 January, 1 May, 25–26 July, 10 October

CYPRUS

Island in the Mediterranean Sea, off the south coast of Turkey and west coast of Syria.

Although often coloured yellow, the island is intended to be copper, reflecting the country's name, 'Copper Island'. Effective date: c. September 1960.

National name: *Kipriakí Dimokratía/Greek Republic of Cyprus* (south); *Kibris Cumhuriyeti/Turkish Republic of Northern Cyprus* (north)
Area: 9,251 sq km/3,571 sq mi (3,335 sq km/1,287 sq mi is Turkish-occupied)
Capital: Nicosia (divided between Greek and Turkish Cypriots)
Major towns/cities: Limassol, Larnaka, Paphos, Lefkosia, Gazimaĝusa
Major ports: Limassol, Larnaka, and Paphos (Greek); Kyrenia and Famagusta (Turkish)
Physical features: central plain between two east–west mountain ranges

Government

Head of state and government: Glafkos Clerides (Greek) from 1993 and Rauf Denktas (Turkish) from 1976
Political system: liberal democracy
Political executive: limited presidency
Administrative divisions: six districts
Political parties: *Greek zone:* Democratic Party (DEKO), federalist, centre left; Progressive Party of the Working People (AKEL), socialist; Democratic Rally (DISY), centrist; Socialist Party–National Democratic Union of Cyprus (SK–EDEK), socialist; *Turkish zone:* National Unity Party (NUP); Communal Liberation Party (CLP); Republican Turkish Party (RTP); New British Party (NBP)
Armed forces: National Guard of 10,000 (1998); Turkish Republic of Northern Cyprus (TRNC) 4,000, plus 26,000 reserves (1995)
Conscription: is for 26 months
Death penalty: laws provide for the death penalty only for exceptional crimes such as under military law or crimes committed in exceptional circumstances such as wartime; last execution 1962
Defence spend: (% GDP) 5.5 (1998)
Education spend: (% GNP) 4.5 (1997)
Health spend: (% GDP) 5.9 (1997)

Economy and resources

Currency: Cyprus pound and Turkish lira
GDP: (US$) 9.0 billion (1999 est)
Real GDP growth: (% change on previous year) 4.5 (1999)
GNP: (US$) 9.09 billion (1999)
GNP per capita (PPP): (US$) 18,395 (1999 est)
Consumer price inflation: 1.6% (1999 est)
Unemployment: 3.4% (1998)
Foreign debt: (US$) 11.5 billion (1999 est)
Major trading partners: government-controlled area: UK, USA, Arab countries, France, Germany, Greece, Japan, Italy; TRNC area: Turkey, UK, other EU countries
Resources: copper precipitates, beutonite, umber and other ochres
Industries: food processing, beverages, textiles, clothing and leather, chemicals and chemical petroleum, metal products, wood and wood products, tourism, financial services (24 offshore banking units in December 1994)
Exports: clothing, potatoes, pharmaceutical products, manufactured foods, minerals, citrus fruits, industrial products. Principal market: UK 18.2% (1999)
Imports: mineral fuels, textiles, vehicles, metals, tobacco, consumer goods, basic manufactures, machinery and transport equipment, food and live animals. Principal source: UK 11.2% (1999)
Arable land: 10.8% (1995)
Agricultural products: government-controlled area: barley, potatoes, grapes, citrus fruit, olives; TRNC area: wheat, barley, potatoes, citrus fruit, olives; livestock rearing (sheep and goats)

Population and society

Population: 786,000 (2000 est)
Population growth rate: 1.1% (est 1995–2025)
Population density: (per sq km) 84 (1999 est)
Urban population: (% of total) 57 (2000 est)
Age distribution: (% of total population) 0–14 23%, 15–59 62%, 60+ 15% (2000 est)
Ethnic groups: about 80% of the population is of Greek origin, while about 18% are of Turkish descent, and live in the northern part of the island within the self-styled Turkish Republic of Northern Cyprus
Language: Greek, Turkish (both official), English
Religion: Greek Orthodox 78%, Sunni Muslim 18%, Maronite, Armenian Apostolic
Education: (compulsory years) 9
Literacy rate: 99% (men); 95% (women) (2000 est)
Labour force: 9.9% agriculture, 23.5% industry, 66.6% services (1997)
Life expectancy: 76 (men); 80 (women) (1995–2000)
Child mortality rate: (under 5, per 1,000 live births) 9 (1995–2000)
Physicians: 1 per 1,000 people (1994 est)
Hospital beds: 1 per 600 people (1994 est)
TV sets: (per 1,000 people) 325 (1997)
Radios: (per 1,000 people) 406 (1997)
Internet users: (per 10,000 people) 1,027.6 (1999)
Personal computer users: (per 100 people) 16.7 (1999)

Transport

Airports: international airports: Nicosia, Larnaka, Paphos; total passenger km: 2,657 million (1997 est)
Railways: none

Roads: total road network: 10,654 km/6,620 mi, of which 58.3% paved (1997); passenger cars: 357.2 per 1,000 people (1997 est)

Chronology

14th–11th centuries BC: Colonized by Myceneans and Achaeans from Greece. **9th century BC:** Phoenicans settled in Cyprus. **7th century BC:** Several Cypriot kingdoms flourished under Assyrian influence. **414–374 BC:** Under Evagoras of Salamis (in eastern Cyprus) the island's ten city kingdoms were united into one state and Greek culture, including the Greek alphabet, was promoted. **333–58 BC:** Became part of the Greek Hellenistic and then, from 294 BC, the Egypt-based Ptolemaic empire. **58 BC:** Cyprus was annexed by the Roman Empire. **AD 45:** Christianity introduced. **AD 395:** When the Roman Empire divided, Cyprus was allotted to the Byzantine Empire. **7th–10th centuries:** Byzantines and Muslim Arabs fought for control of Cyprus. **1191:** Richard the Lionheart of England conquered Cyprus as a base for the Crusades; he later sold it to a French noble, Guy de Lusignan, who established a feudal monarchy which ruled for three centuries. **1498:** The Venetian Republic took control of Cyprus. **1571:** Conquered by Ottoman Turks, who introduced Turkish Muslim settlers, but permitted Christianity to continue in rural areas. **1821–33:** Period of unrest, following execution of popular Greek Orthodox Archbishop Kyprianos. **1878:** Anglo-Turkish Convention: Turkey ceded Cyprus to British administration in return for defensive alliance. **1914:** Formally annexed by Britain after Turkey entered World War I as a Central Power. **1915:** Greece rejected an offer of Cyprus in return for entry into World War I on Allied side. **1925:** Cyprus became a crown colony. **1931:** Greek Cypriots rioted in support of demand for union with Greece (*enosis*); legislative council suspended. **1948:** Greek Cypriots rejected new constitution because it did not offer links with Greece. **1955:** The National Organization of Cypriot Fighters (EOKA) began a terrorist campaign for *enosis*. **1958:** Britain proposed autonomy for Greek and Turkish Cypriot communities under British sovereignty; plan accepted by Turks, rejected by Greeks; violence increased. **1959:** Britain, Greece, and Turkey agreed to Cypriot independence, with partition and *enosis* both ruled out. **1960:** Cyprus became an independent republic with Archbishop Makarios as president; Britain retained two military bases. **1963:** Makarios proposed major constitutional reforms; Turkish Cypriots withdrew from government and formed separate enclaves; communal fighting broke out. **1964:** United Nations (UN) peacekeeping force installed. **1968:** Intercommunal talks made no progress; Turkish Cypriots demanded federalism; Greek Cypriots insisted on unitary state. **1974:** Coup by Greek officers in Cypriot National Guard installed Nikos Sampson as president; Turkey, fearing *enosis*, invaded northern Cyprus; Greek Cypriot military regime collapsed; President Makarios restored. **1975:** Northern Cyprus declared itself the Turkish Federated State of Cyprus, with Rauf Denktas as president. **1977:** Makarios died; succeeded by Spyros Kyprianou. **1983:** Denktas proclaimed independent Turkish Republic of Cyprus; recognized only by Turkey. **1985:** Summit meeting between Kyprianou and Denktas failed to reach agreement; further peace talks failed in 1989 and 1992. **1988:** Kyprianou was succeeded as Greek Cypriot president by Georgios Vassiliou. **1993:** Glafkos Clerides (DISY) replaced Vassiliou. **1994:** The European Court of Justice declared trade with northern Cyprus illegal. **1996:** Further peace talks were jeopardized by the boundary killing of a Turkish Cypriot soldier; there was mounting tension between north and south. **1997:** UN-mediated peace talks between Clerides and Denktas collapsed. **1998:** President Clerides was re-elected. Denktas refused to meet a British envoy. US mediation failed. Full EU membership negotiations commenced. Greek Cyprus rejected Denktas's confederation proposals. **2000:** Turkish Cypriot President Denktas was re-elected for a fourth five-year term.

Practical information

Visa requirements: UK: visa not required. USA: visa not required
Time difference: GMT +2
Chief tourist attractions: sandy beaches; forested mountains; winter skiing; archaeological and historic sites; the old city of Nicosia, with its Venetian walls and

cathedral (International State Fair and Nicosia Art Festival in May); Limassol, with its 14th-century castle (Spring Carnival, arts festival in July, wine festival in September)
Major holidays: 1, 6 January, 25 March, 1 May, 28–29 October, 25–26 December; variable: Eid-ul-Adha, Good Friday, Easter Monday, end of Ramadan, Holy Saturday, Prophet's Birthday

CZECH REPUBLIC

FORMERLY CZECHOSLOVAKIA (WITH SLOVAKIA) (1918–93)

Landlocked country in east-central Europe, bounded north by Poland, northwest and west by Germany, south by Austria, and east by the Slovak Republic.

National name: *Ceská Republika/Czech Republic*
Area: 78,864 sq km/30,449 sq mi
Capital: Prague
Major towns/cities: Brno, Ostrava, Olomouc, Liberec, Plzen, Hradec Králové, Ceské Budejovice
Physical features: mountainous; rivers: Morava, Labe (Elbe), Vltava (Moldau)

Red and white are the colours of Bohemia, dating back to the 13th century. Blue represents Moravia. Unlike that of the Slovak Republic, the Czech flag is not based on the pan-Slav colours. Effective date: 1 January 1993.

Government

Head of state: Václav Havel from 1993
Head of government: Miloš Zeman from 1998
Political system: liberal democracy
Political executive: parliamentary
Administrative divisions: 72 districts
Political parties: Civic Democratic Party (CDP), right of centre, free-market; Civic Democratic Alliance (CDA), right of centre, free-market; Civic Movement (CM), liberal, left of centre; Communist Party of Bohemia and Moravia (KSCM), reform socialist; Agrarian Party, centrist, rural-based; Liberal National Social Party (LNSP; formerly the Czech Socialist Party (SP)), reform socialist; Czech Social Democratic Party (CSDP), moderate left of centre; Christian Democratic Union–Czech People's Party (CDU–CPP), right of centre; Movement for Autonomous Democracy of Moravia and Silesia (MADMS), Moravian and Silesian-based, separatist; Czech Republican Party, far right
Armed forces: 59,100 (1998)
Conscription: compulsory for 12 months
Death penalty: abolished in 1990
Defence spend: (% GDP) 2.1 (1998)
Education spend: (% GNP) 5.1 (1997)
Health spend: (% GDP) 7.2 (1997)

Economy and resources

Currency: koruna (based on the Czechoslovak koruna)
GDP: (US$) 53.2 billion (1999)
Real GDP growth: (% change on previous year) –0.2 (1999)
GNP: (US$) 52 billion (1999)
GNP per capita (PPP): (US$) 12,289 (1999)
Consumer price inflation: 2.1% (1999)
Unemployment: 6.3% (1998)
Foreign debt: (US$) 23.5 billion (1999 est)
Major trading partners: Germany, Slovak Republic, Austria, Italy, France, Russia
Resources: coal, lignite
Industries: steel, cement, motor cars, textiles, bicycles, beer, trucks and tractors
Exports: basic manufactures, machinery and transport equipment, miscellaneous manufactured articles, chemicals, beer. Principal market: Germany 43% (1999)
Imports: machinery and transport equipment, basic

manufactures, chemicals and chemical products, mineral fuels. Principal source: Germany 37.5% (1999)
Arable land: 40.1% (1996)
Agricultural products: wheat, barley, sugar beet, potatoes, hops; livestock rearing (cattle, pigs, and poultry); dairy farming

Population and society

Population: 10,244,000 (2000 est)
Population growth rate: –0.16% (1995–2000)
Population density: (per sq km) 130 (1999 est)
Urban population: (% of total) 75 (2000 est)
Age distribution: (% of total population) 0–14 17%, 15–69 65%, 60+ 18% (2000 est)
Ethnic groups: predominantly Western Slav Czechs (94%); there is also a sizeable Slovak minority (4%) and small Polish (0.6%), German (0.5%), and Hungarian (0.2%) minorities
Language: Czech (official), Slovak
Religion: Roman Catholic 39%, atheist 30%, Protestant 5%, Orthodox 3%
Education: (compulsory years) 9
Literacy rate: 99% (men); 99% (women) (2000 est)
Labour force: 5.8% agriculture, 41.6% industry, 52.5% services (1997)
Life expectancy: 70 (men); 77 (women) (1995–2000)
Child mortality rate: (under 5, per 1,000 live births) 8 (1995–2000)
Physicians: 1 per 345 people (1996)
Hospital beds: 1 per 109 people (1996)
TV sets: (per 1,000 people) 447 (1997)
Radios: (per 1,000 people) 803 (1997)
Internet users: (per 10,000 people) 682.1 (1999)
Personal computer users: (per 100 people) 10.7 (1999)

Transport

Airports: international airports: Prague (Ruzyne), Brno (Cernovice), Ostrava (International and Mosnov – domestic), Karlovy Vary; total passenger km: 2,442 million (1997 est)
Railways: total length: 9,440 km/5,866 mi; total passenger km: 7,710 million (1997)
Roads: total road network: 55,876 km/34,721 mi, of which 100% paved (1996); passenger cars: 344.4 per 1,000 people (1997)

Chronology

5th century: Settled by West Slavs. **8th century:** Part of Charlemagne's Holy Roman Empire. **9th century:** Kingdom of Greater Moravia, centred around the eastern part of what is now the Czech Republic, founded by the Slavic prince Sviatopluk; Christianity adopted. **906:** Moravia conquered by the Magyars (Hungarians). **995:** Independent state of Bohemia in the northwest, centred around Prague, formed under the Premysl rulers, who had broken away from Moravia; became kingdom in 12th century. **1029:** Moravia became a fief of Bohemia. **1355:** King Charles IV of Bohemia became Holy Roman Emperor.

early 15th century: Nationalistic Hussite religion, opposed to German and papal influence, founded in Bohemia by John Huss. **1526:** Bohemia came under the control of the Austrian Catholic Habsburgs. **1618:** Hussite revolt precipitated the Thirty Years' War, which resulted in the Bohemians' defeat, more direct rule by the Habsburgs, and re-Catholicization. **1867:** With creation of dual Austro-Hungarian monarchy, Bohemia was reduced to a province of Austria, leading to a growth in national consciousness.

1918: Austro-Hungarian Empire dismembered; Czechs joined Slovaks in forming Czechoslovakia as independent democratic nation, with Tomas Masaryk president. **1938:** Under the Munich Agreement, Czechoslovakia was forced to surrender the Sudeten German districts in the north to Germany. **1939:** The remainder of Czechoslovakia annexed by Germany, Bohemia-Moravia being administered as a 'protectorate'; President Eduard Beneš set up a government-in-exile in London; liquidation campaigns against intelligentsia. **1945:** Liberated by Soviet and US troops; communist-dominated government of national unity formed under Beneš; 2 million Sudeten Germans expelled. **1948:** Beneš ousted; communists assumed full control under a Soviet-style single-party constitution. **1950s:** Political opponents purged; nationalization of industries. **1968:** 'Prague Spring' political liberalization programme, instituted by Communist Party leader Alexander Dubcek, crushed by invasion of Warsaw Pact forces to restore the 'orthodox line'. **1969:** New federal constitution, creating a separate Czech Socialist Republic; Gustáv Husák became Communist Party leader. **1977:** The formation of the 'Charter '77' human-rights group by intellectuals encouraged a crackdown against dissidents. **1987:** Reformist Miloš Jakeš replaced Husák as communist leader, and introduced a *prestvaha* ('restructuring') reform programme on the Soviet leader Mikhail Gorbachev's *perestroika* model. **1989:** Prodemocracy demonstrations in Prague; new political parties formed and legalized, including Czech-based Civic Forum under Havel; Communist Party stripped of powers. New 'grand coalition' government formed; Havel appointed state president. Amnesty granted to 22,000 prisoners. **1991:** The Civic Forum split into the centre-right Civic Democratic Party (CDP) and the centre-left Civic Movement (CM), evidence of increasing Czech and Slovak separatism. **1992:** Václav Klaus, leader of the Czech-based CDP, became prime minister; Havel resigned as president following nationalist Slovak gains in assembly elections. The creation of separate Czech and Slovak states and a customs union were agreed. A market-centred economic-reform programme was launched, including mass privatization. **1993:** The Czech Republic became a sovereign state within the United Nations (UN), with Klaus as prime minister. Havel was elected president. **1994:** The Czech Republic joined NATO's 'partnership for peace' programme. Strong economic growth was registered. **1996:** The Czech Republic applied for European Union (EU) membership. **1997:** The former communist leader Miloš Jakeš was charged with treason. The ruling coalition survived a currency crisis. The Czech Republic was invited to begin EU membership negotiations. Klaus resigned after allegations of misconduct. **1998:** Havel was re-elected president. The centre-left Social Democrats won a general election and a minority government was formed by Miloš Zeman, including communist ministers and supported from outside by Václav Klaus, who became parliamentary speaker. Full EU membership negotiations commenced. **1999:** The Czech Republic became a full member of NATO. **2000:** In December, television journalists went on strike, and thousands of protesters demonstrated in Prague, in response to the appointment of a new director general of television who was widely believed to be politically biased. The situation was resolved by the appointment of a new director general in January 2001.

Practical information

Visa requirements: UK: visa not required. USA: visa not required
Time difference: GMT +1
Chief tourist attractions: scenery; winter-sports facilities; historic towns, including Prague, Karlovy Vary, Olomouc, and Cesky; castles and cathedrals; numerous resorts and spas
Major holidays: 1 January, 1 May, 5–6, July, 28 October, 24–26 December; variable: Easter Monday

DENMARK

Peninsula and islands in northern Europe, bounded to the north by the Skagerrak, east by the Kattegat, south by Germany, and west by the North Sea.

National name: *Kongeriget Danmark/Kingdom of Denmark*
Area: 43,075 sq km/16,631 sq mi
Capital: Copenhagen
Major towns/cities: Århus, Odense, Ålborg, Esbjerg, Randers, Kolding, Horsens
Major ports: Århus, Odense, Ålborg, Esbjerg
Physical features: comprises the Jutland peninsula and about 500 islands (100 inhabited) including Bornholm in the Baltic Sea; the land is flat and cultivated; sand dunes and lagoons on the west coast and long inlets on the east; the main island is Sjælland (Zealand), where most of Copenhagen is located (the rest is on the island of Amager)
Territories: the dependencies of Faroe Islands and Greenland

Nordic flags bearing the Scandinavian cross are based on the Danish flag, known as the Dannebrog, 'Danish cloth'.
Effective date: 1 May 1893.

Government

Head of state: Queen Margrethe II from 1972
Head of government: Poul Nyrup Rasmussen from 1993
Political system: liberal democracy
Political executive: parliamentary
Administrative divisions: 14 counties, one city and one borough
Political parties: Social Democrats (SD), left of centre; Conservative People's Party (KF), moderate right-of-centre; Liberal Party (V), left of centre; Socialist People's Party (SF), moderate left wing; Radical Liberals (RV), radical internationalist, left of centre; Centre Democrats (CD), moderate centrist; Progress Party (FP), radical antibureaucratic; Christian People's Party (KrF), interdenominational, family values
Armed forces: 32,100; 100,000 reservists and volunteer Home Guard of 70,500 (1998)
Conscription: 9–12 months (27 months for some ranks)
Death penalty: abolished in 1978
Defence spend: (% GDP) 1.6 (1998)
Education spend: (% GNP) 8.2 (1996)
Health spend: (% GDP) 8 (1997)

Economy and resources

Currency: Danish krone
GDP: (US$) 174.4 billion (1999)
Real GDP growth: (% change on previous year) 2.7 (1998)
GNP: (US$) 170.3 billion (1999)
GNP per capita (PPP): (US$) 24,280 (1999)
Consumer price inflation: 2.5% (1999)
Unemployment: 5.1% (1998)
Major trading partners: EU (principally Germany, Sweden, UK, and France), Norway, USA
Resources: crude petroleum, natural gas, salt, limestone
Industries: mining, food processing, fisheries, machinery, textiles, furniture, electronic goods and transport equipment, chemicals and pharmaceuticals, printing and publishing
Exports: pig meat and pork products, other food products, fish, industrial machinery, chemicals, transport equipment. Principal market: Germany 20% (1999)
Imports: food and live animals, machinery, transport equipment, iron, steel, electronics, petroleum, cereals, paper. Principal source: Germany 21.8% (1999)
Arable land: 54.7% (1996)
Agricultural products: wheat, rye, barley, oats, potatoes, sugar beet, dairy products; livestock production (pigs) and dairy products; fishing

Population and society

Population: 5,293,000 (2000 est)
Population growth rate: 0.2% (1995–2000); 0% (2000–05)
Population density: (per sq km) 123 (1999 est)
Urban population: (% of total) 85 (2000 est)
Age distribution: (% of total population) 0–14 18%, 15–59 62%, 60+ 20% (2000 est)
Ethnic groups: Danish, Inuit (in Greenland), Faroese, German
Language: Danish (official), German
Religion: Evangelical Lutheran 87% (national church), other Protestant and Roman Catholic 3%
Education: (compulsory years) 9
Literacy rate: 99% (men); 99% (women) (2000 est)
Labour force: 53.6% of population: 3.7% agriculture, 26.8% industry, 69.5% services (1997)
Life expectancy: 73 (men); 78 (women) (1995–2000)
Child mortality rate: (under 5, per 1,000 live births) 8 (1995–2000)
Physicians: 1 per 345 people (1994)
Hospital beds: 1 per 204 people (1995)
TV sets: (per 1,000 people) 585 (1998)
Radios: (per 1,000 people) 1,141 (1997)
Internet users: (per 10,000 people) 2,823.0 (1999)
Personal computer users: (per 100 people) 41.4 (1999)

Transport

Airports: international airports: Copenhagen (Kastrup), Århus; ten major domestic airports; total passenger km: 5,262 million (1995)
Railways: total length: 3,359 km/2,087 mi; total passenger km: 4,988 million (1997)
Roads: total road network: 71,600 km/44,492 mi, of which 100% paved (1996 est); passenger cars: 332.7 per 1,000 people (1996 est)

Chronology

5th–6th centuries: Danes migrated from Sweden. **8th–10th centuries:** Viking raids throughout Europe. *c.* **940–85:** Harald Bluetooth unified Kingdom of Denmark and established Christianity. **1014–35:** King Canute I created an empire embracing Denmark, Norway, and England; the empire collapsed after his death. **12th century:** Denmark re-emerged as dominant Baltic power. **1340–75:** Valdemar IV restored order after a period of civil war and anarchy. **1397:** Union of Kalmar: Denmark, Sweden, and Norway (with Iceland) united under a single monarch. **1449:** Sweden broke away from union. **1536:** Lutheranism established as official religion of Denmark. **1563–70:** Unsuccessful war to recover Sweden. There were two further unsuccessful attempts to reclaim Sweden, 1643–45 and 1657–60. **1625–29:** Denmark sided with Protestants in Thirty Years' War. **1665:** Frederick III made himself absolute monarch. **1729:** Greenland became a Danish province. **1780–81:** Denmark, Russia, and Sweden formed 'Armed Neutrality' coalition to protect neutral shipping during the American revolution. **1788:** Serfdom abolished. **1800:** France persuaded Denmark to revive Armed Neutrality against British blockade. **1801:** First Battle of Copenhagen: much of Danish fleet destroyed by British navy. **1807:** Second Battle of Copenhagen: British seized rebuilt fleet to pre-empt Danish entry into Napoleonic War on French side. **1814:** Treaty of Kiel: Denmark ceded Norway to Sweden as penalty for supporting France in Napoleonic War; Denmark retained Iceland. **1849:** Liberal pressure compelled Frederick VII to grant a democratic constitution. **1914–1919:** Denmark neutral during World War I. **1918:** Iceland achieved full self-government. **1929–40:** Welfare state established under left-wing coalition government dominated by Social Democrat Party. **1940–45:** German occupation. **1944:** Iceland declared independence. **1949:** Denmark became a founding member of the North Atlantic Treaty Organization (NATO). **1960:** Denmark joined the European Free Trade Association (EFTA). **1973:** Denmark withdrew from EFTA and joined the European Economic Community (EEC). **1981:** Greenland achieved full self-government. **1992:** A referendum rejected the Maastricht Treaty on European union; it was approved in 1993 after the government negotiated a series of 'opt-out' clauses. **1993:** Conservative leader Poul Schlüter resigned as prime minister due to a legal scandal. **1994:** Schlüter was succeeded as prime minister by Poul Rasmussen, who, leading a Social Democrat-led coalition, won the general election. **1998:** The government won a slim majority in assembly elections. A referendum endorsed the Amsterdam European Union (EU) treaty. **2000:** A referendum rejected joining Europe's single currency and adopting the euro.

Practical information

Visa requirements: UK: visa not required. USA: visa not required
Time difference: GMT +1
Chief tourist attractions: landscape with woods, small lakes; volcanic Faeroe Islands with ancient culture and customs; Copenhagen, with its Tivoli Amusement Park (May–September), palaces, castle, cathedral, national museum, and Little Mermaid sculpture
Major holidays: 1 January, 5 June, 24–26 December; variable: Ascension Thursday, Good Friday, Easter Monday, Holy Thursday, Whit Monday, General Prayer (April/May)

DJIBOUTI FORMERLY FRENCH SOMALILAND (1888–1967), FRENCH TERRITORY OF THE AFARS AND ISSAS (1966–77)

Located on the east coast of Africa, at the south end of the Red Sea, bounded east by the Gulf of Aden, southeast by Somalia, south and west by Ethiopia, and northwest by Eritrea.

National name: *Jumhouriyya Djibouti/Republic of Djibouti*
Area: 23,200 sq km/8,957 sq mi
Capital: Djibouti (and chief port)
Major towns/cities: Tadjoura, Obock, Dikhil, Ali Sabih
Physical features: mountains divide an inland plateau from a coastal plain; hot and arid

Blue recalls the sea and the sky. Green symbolizes the earth. The red star represents unity. White stands for peace. Effective date: 27 June 1977.

Government

Head of state: Ismail Omar Guelleh from 1999
Head of government: Barkat Gourad Hamadou from 1978
Political system: authoritarian nationalist
Political executive: unlimited presidency
Administrative divisions: five districts
Political parties: People's Progress Assembly (RPP), nationalist; Democratic Renewal Party (PRD), moderate left-of-centre; Democratic National Party (DND)
Armed forces: 9,600 (1998); plus 3,900 French troops
Conscription: military service is voluntary
Death penalty: retains the death penalty for ordinary crimes but can be considered abolitionist in practice (no executions since independence)
Defence spend: (% GDP) 5.1 (1998)
Education spend: (% GNP) 3.8 (1993–94)
Health spend: (% GDP) 2.8 (1994)

Economy and resources

Currency: Djibouti franc
GDP: (US$) 530.1 million (1999)
Real GDP growth: (% change on previous year) 1.5 (1999)
GNP: (US$) 511 million (1999)
GNP per capita (PPP): (US$) 1,200 (1999 est)
Consumer price inflation: 2% (1999 est)
Unemployment: 58% (1996 est)
Foreign debt: (US$) 300 million (1999 est)
Major trading partners: Somalia, France, Ethiopia, Yemen, Saudi Arabia, United Arab Emirates, Italy, UK
Industries: mineral water bottling, dairy products and other small-scale enterprises; an important port serving the regional hinterland
Exports: hides, cattle, coffee (exports are largely re-exports). Principal market: Somalia 53% (1998 est)
Imports: vegetable products, foodstuffs, beverages, vinegar, tobacco, machinery and transport equipment, mineral products. Principal source: France 13% (1998 est)
Arable land: 10% (1995)

Agricultural products: mainly market gardening (for example, tomatoes); livestock rearing (over 50% of the population are pastoral nomads herding goats, sheep, and camels); fishing

Population and society

Population: 638,000 (2000 est)
Population growth rate: 1.2% (1995–2000)
Population density: (per sq km) 27 (1999 est)
Urban population: (% of total) 83 (2000 est)
Age distribution: (% of total population) 0–14 41%, 15–59 53%, 60+ 6% (2000 est)
Ethnic groups: population divided mainly into two Hamitic groups; the Issas (Somalis) (60%) in the south, and the minority Afars (or Danakil) (35%) in the north and west. There are also minorities of Europeans (mostly French), as well as Arabs, Sudanese, and Indians
Language: French (official), Issa (Somali), Afar, Arabic
Religion: Sunni Muslim
Education: (compulsory years) 6
Literacy rate: 76% (men); 54% (women) (2000 est)
Life expectancy: 49 (men); 52 (women) (1995–2000)
Child mortality rate: (under 5, per 1,000 live births) 174 (1995–2000)
Physicians: 1 per 6,590 people (1993 est)
Hospital beds: 1 per 3,000 people (1990)
TV sets: (per 1,000 people) 45 (1997)
Radios: (per 1,000 people) 84 (1997)
Internet users: (per 10,000 people) 15.9 (1999)
Personal computer users: (per 100 people) 1.0 (1999)

Transport

Airports: international airport: Djibouti (Ambouli); six domestic airports; total passengers: 106,823 (1997)
Railways: total length: 106 km/67 mi; (part within Djibouti of 781-km/488-mi track linking Djibouti with Addis Ababa, Ethiopia); total passenger km: 279 million (1995)
Roads: total road network: 2,890 km/1,796 mi, of which 12.6% paved (1996 est); passenger cars: 17.2 per 1,000 people (1996 est)

Chronology

3rd century BC: The north settled by Able immigrants from Arabia, whose descendants are the Afars (Danakil). **early Christian era:** Somali Issas settled in coastal areas and south, ousting Afars. **825:** Islam introduced by missionaries. **16th century:** Portuguese arrived to challenge trading monopoly of Arabs. **1862:** French acquired a port at Obock. **1888:** Annexed by France as part of French Somaliland. **1900s:** Railroad linked Djibouti port with the Ethiopian hinterland. **1946:** Became overseas territory within French Union, with own assembly and representation in French parliament. **1958:** Voted to become overseas territorial member of French Community. **1967:** French Somaliland renamed the French Territory of the Afars and the Issas. **early 1970s:** Issas (Somali) peoples campaigned for independence, but the minority Afars, of Ethiopian descent, and Europeans sought to remain French. **1977:** Independence was achieved as Djibouti, with Hassan Gouled Aptidon, the leader of the independence movement, elected president. **1981:** A new constitution made the People's Progress Assembly (RPP) the only legal party. Treaties of friendship were signed with Ethiopia, Somalia, Kenya, and Sudan. **1984:** The policy of neutrality was reaffirmed. The economy was undermined by severe drought. **1992:** A new multiparty constitution was adopted; fighting erupted between government forces and Afar Front for Restoration of Unity and Democracy (FRUD) guerrilla movement in the northeast. **1993:** Opposition parties were allowed to operate, but Gouled was re-elected president. **1994:** A peace agreement was reached with Afar FRUD militants, ending the civil war. **1999:** Ismail Omar Guelleh was elected president.

Practical information

Visa requirements: UK: visa required. USA: visa required
Time difference: GMT +3
Chief tourist attractions: desert scenery in interior; water-sports facilites on coast

Major holidays: 1 January, 1 May, 27 June (2 days), 25 December; variable: Eid-ul-Adha (2 days), end of Ramadan (2 days), New Year (Muslim), Prophet's Birthday, Al-Isra Wal-Mira'age (March–April)

DOMINICA

Island in the eastern Caribbean, between Guadeloupe and Martinique, the largest of the Windward Islands, with the Atlantic Ocean to the east and the Caribbean Sea to the west.

National name: *Commonwealth of Dominica*
Area: 751 sq km/290 sq mi
Capital: Roseau
Major towns/cities: Portsmouth, Marigot, Mahaut, Atkinson, Grand Bay
Major ports: Roseau, Portsmouth, Berekua, Marigot, Rosalie
Physical features: second-largest of the Windward Islands, mountainous central ridge with tropical rainforest

The stars symbolize hope and equality between the ten parishes. Green reflects the island's lush vegetation. The red disc has socialist connotations. Effective date: 3 November 1990.

Government

Head of state: Vernon Shaw from 1998
Head of government: Pierre Charles from 2000
Political system: liberal democracy
Political executive: parliamentary
Administrative divisions: ten parishes
Political parties: Dominica Freedom Party (DFP), centrist; Labour Party of Dominica (LPD), left-of-centre coalition (before 1985 the DLP); Dominica United Workers' Party (DUWP), left of centre
Armed forces: defence force disbanded in 1981; police force of approximately 300
Death penalty: retained and used for ordinary crimes
Education spend: (% GNP) N/A
Health spend: (% GDP) 6 (1997)

Economy and resources

Currency: East Caribbean dollar, although the pound sterling and French franc are also accepted
GDP: (US$) 260.4 million (1999 est)
Real GDP growth: (% change on previous year) 0.9 (1999 est)
GNP: (US$) 231 million (1999)
GNP per capita (PPP): (US$) 4,825 (1999)
Consumer price inflation: 1.5% (1999)
Unemployment: 23% (1995 est)
Foreign debt: (US$) 109 million (1999 est)
Major trading partners: USA, UK, the Netherlands, South Korea, Belgium, Japan, Trinidad and Tobago
Resources: pumice, limestone, clay
Industries: banana packaging, vegetable oils, soap, canned juice, cigarettes, rum, beer, furniture, paint, cardboard boxes, candles, tourism
Exports: bananas, soap, coconuts, grapefruit, galvanized sheets. Principal market: UK 32.8% (1997)
Imports: food and live animals, basic manufactures, machinery and transport equipment, mineral fuels. Principal source: USA 38.3% (1997)
Arable land: 4% (1995)

Agricultural products: bananas, coconuts, mangoes, avocados, papayas, ginger, citrus fruits, vegetables; livestock rearing; fishing

Population and society

Population: 71,000 (2000 est)
Population growth rate: –0.06 (1995–2000)
Population density: (per sq km) 100 (1999 est)
Urban population: (% of total) 71 (2000 est)
Age distribution: (% of total population) 0–14 38%, 15–59 48%, 60+ 14% (2000 est)
Ethnic groups: majority descended from African slaves; a small number of the indigenous Arawaks remain
Language: English (official), a Dominican patois (which reflects earlier periods of French rule)
Religion: Roman Catholic 80%
Education: (compulsory years) 10
Literacy rate: 94% (men); 94% (women) (1994 est)
Labour force: 30.8% agriculture, 21.6% industry, 44.7% services (1992)
Life expectancy: 75 (men); 81 (women) (1998 est)
Child mortality rate: (under 5, per 1,000 live births) 21 (1995)
Physicians: 1 per 225 people (1996)
Hospital beds: 1 per 88 people (1996)
TV sets: (per 1,000 people) 78 (1997)
Radios: (per 1,000 people) 647 (1997)
Internet users: (per 10,000 people) 261.4 (1999)
Personal computer users: (per 100 people) 6.5 (1999)

Transport

Airports: international airports: Roseau (Canefield), Portsmouth/Marigot (Melville Hall); aircraft arrivals and departures: 18,672 (1997)
Railways: none
Roads: total road network: 780 km/485 mi, of which 50.4% paved (1996 est)

Chronology

1493: Visited by the explorer Christopher Columbus, who named the island Dominica ('Sunday Island'). **1627:** Presented by the English King Charles I to the Earl of Carlisle, but initial European attempts at colonization were fiercely resisted by the indigenous Carib community. **later 18th century:** Succession of local British and French conflicts over control of the fertile island. **1763:** British given possession of the island by the Treaty of Paris, but France continued to challenge this militarily until 1805, when there was formal cession in return for the sum of £12,000. **1834:** Slaves, who had been brought in from Africa, were emancipated. **1870:** Became part of the British Leeward Islands federation. **1940:** Transferred to British Windward Islands federation. **1951:** Universal adult suffrage established. **1958–62:** Part of the West Indies Federation. **1960:** Granted separate, semi-independent status, with a legislative council and chief minister. **1961:** Edward le Blanc, leader of the newly formed DLP, became chief minister. **1978:** Independence was achieved as a republic within the Commonwealth, with Patrick John (DLP) as prime minister. **1980:** The DFP won a convincing victory in a general election, and Eugenia Charles became the Caribbean's first woman prime minister. **1983:** A small force participated in the US-backed invasion of Grenada. **1985:** The regrouping of left-of-centre parties resulted in the new Labour Party of Dominica (LPD). **1991:** A Windward Islands confederation comprising St Lucia, St Vincent, Grenada, and Dominica was proposed. **1993:** Charles resigned the DFP leadership, but continued as prime minister. **1995:** DUWP won a general election; Edison James was appointed prime minister and Eugenia Charles retired from politics. **1998:** Vernon Shaw elected president. **2000:** Rosie Douglas was elected prime minister, leading a DLP-DFP coalition, but died in October. He was replaced by Pierre Charles.

Practical information

Visa requirements: UK: visa not required for stays of up to six months. USA: visa not required for stays of up to six months
Time difference: GMT –4
Chief tourist attractions: scenery; nature reserves; marine reserves; rich birdlife, including rare and endangered species such as the imperial parrot
Major holidays: 1 January, 1 May, 3–4 November, 25–26 December; variable: Carnival (2 days), Good Friday, Easter Monday, Whit Monday, August Monday

DOMINICAN REPUBLIC

FORMERLY HISPANIOLA (WITH HAITI) (UNTIL 1844)

Located in the West Indies (eastern Caribbean); occupies the eastern two-thirds of the island of Hispaniola, with Haiti covering the western third; the Atlantic Ocean is to the east and the Caribbean Sea to the west.

National name: *República Dominicana/Dominican Republic*
Area: 48,442 sq km/18,703 sq mi
Capital: Santo Domingo
Major towns/cities: Santiago, La Romana, San Pedro de Macoris, San Francisco de Macoris, La Vega, San Juan, San Cristóbal
Physical features: comprises eastern two-thirds of island of Hispaniola; central mountain range with fertile valleys; Pico Duarte 3,174 m/10,417 ft, highest point in Caribbean islands

The arms show a Bible open at the Gospel of St John, a Trinitarian symbol. The white cross symbolizes faith. The arms appear on national and state flags. Effective date: 14 September 1863.

Government

Head of state and government: Hipólito Mejía from 2000
Political system: liberal democracy
Political executive: limited presidency
Administrative divisions: 29 provinces and a national district (Santo Domingo)
Political parties: Dominican Revolutionary Party (PRD), moderate, left of centre; Christian Social Reform Party (PRSC), independent socialist; Dominican Liberation Party (PLD), nationalist
Armed forces: 24,500 (1998); plus a paramilitary force of 15,000
Conscription: military service is voluntary
Death penalty: abolished in 1966
Defence spend: (% GDP) 1.1 (1998)
Education spend: (% GNP) 2.3 (1997)
Health spend: (% GDP) 4.9 (1997)

Economy and resources

Currency: Dominican Republic peso
GDP: (US$) 17.13 billion (1999)
Real GDP growth: (% change on previous year) 8.3 (1999)
GNP: (US$) 16.1 billion (1999)
GNP per capita (PPP): (US$) 4,653 (1999 est)
Consumer price inflation: 6.5% (1999)
Unemployment: 15.9% (1997)
Foreign debt: (US$) 4.5 billion (1999 est)
Major trading partners: USA, Venezuela, Belgium, Mexico, Japan, South Korea, Haiti, Panama, Canada
Resources: ferro-nickel, gold, silver
Industries: food processing (including sugar refining), petroleum refining, beverages, chemicals, cement
Exports: raw sugar, molasses, coffee, cocoa, tobacco, ferro-nickel, gold, silver. Principal market: USA 43.9% (1998)
Imports: petroleum and petroleum products, coal, foodstuffs, wheat, machinery. Principal source: USA 45.1% (1998)
Arable land: 27.9% (1996)

Agricultural products: sugar cane, cocoa, coffee, bananas, tobacco, rice, tomatoes

Population and society

Population: 8,495,000 (2000 est)
Population growth rate: 1.7% (1995–2000); 1.4% (2000–05)
Population density: (per sq km) 173 (1999 est)
Urban population: (% of total) 65 (2000 est)
Age distribution: (% of total population) 0–14 33%, 15–59 60%, 60+ 7% (2000 est)
Ethnic groups: about 73% of the population are mulattos, of mixed European and African descent; about 16% European; 11% African
Language: Spanish (official)
Religion: Roman Catholic
Education: (compulsory years) 8
Literacy rate: 84% (men); 84% (women) (2000 est)
Labour force: 44% of population: 12.9% agriculture, 23% industry, 64.1% services (1995)
Life expectancy: 69 (men); 73 (women) (1995–2000)
Child mortality rate: (under 5, per 1,000 live births) 46 (1995–2000)
Physicians: 1 per 949 people (1993)
Hospital beds: 1 per 506 people (1993)
TV sets: (per 1,000 people) 95 (1997)
Radios: (per 1,000 people) 178 (1997)
Internet users: (per 10,000 people) 29.9 (1999)

Transport

Airports: international airports: Santo Domingo (Internacional de las Americas), Puerto Plata (La Union), Punta Cana, La Romana; most main cities have domestic airports; total passenger km: 234 million (1994)
Railways: total length: 517 km/321 mi
Roads: total road network: 12,600 km/7,830 mi, of which 49.4% paved (1996 est); passenger cars: 26.9 per 1,000 people (1996 est)

Chronology

14th century: Settled by Carib Indians, who followed an earlier wave of Arawak Indian immigration. **1492:** Visited by Christopher Columbus, who named it Hispaniola ('Little Spain'). **1496:** At Santo Domingo the Spanish established the first European settlement in the western hemisphere, which became capital of all Spanish colonies in America. **first half of 16th century:** One-third of a million Arawaks and Caribs died, as a result of enslavement and exposure to European diseases; black African slaves were consequently brought in to work the island's gold and silver mines, which were swiftly exhausted. **1697:** Divided between France, which held the western third (Haiti), and Spain, which held the east (Dominican Republic, or Santo Domingo). **1795:** Santo Domingo was ceded to France. **1808:** Following a revolt by Spanish Creoles, with British support, Santo Domingo was retaken by Spain. **1821:** Became briefly independent after uprising against Spanish rule, and then fell under the control of Haiti. **1844:** Separated from Haiti to form Dominican Republic. **1861–65:** Under Spanish protection. **1904:** The USA took over the near-bankrupt republic's debts. **1916–24:** Temporarily occupied by US forces. **1930:** Military coup established personal dictatorship of Gen Rafael Trujillo Molina. **1937:** Army massacred 19,000–20,000 Haitians living in the Dominican provinces adjoining the frontier. **1961:** Trujillo assassinated. **1962:** First democratic elections resulted in Juan Bosch, founder of the left-wing Dominican Revolutionary Party (PRD), becoming president. **1963:** Bosch overthrown in military coup. **1965:** 30,000 US marines intervened to restore order and protect foreign nationals after Bosch had attempted to seize power. **1966:** New constitution adopted. Joaquín Balaguer, protégé of Trujillo and leader of the centre-right Christian Social Reform Party (PRSC), became president. **1978:** PRD returned to power. **1985:** PRD president Jorge Blanco was forced by the International Monetary Fund to adopt austerity measures to save the economy. **1986:** The PRSC returned to power. **1996:** Leonel Fernandez of the left-wing PLD was elected president. **2000:** Presidential elections were won by Hipólito Mejía, a social democrat.

Practical information

Visa requirements: UK: visa not required for stays of up to 90 days. USA: visa not required for stays of up to 60 days
Time difference: GMT –4
Chief tourist attractions: beaches on north, east, and southeast coasts; forested and mountainous landscape

Major holidays: 1, 6, 21, 26 January, 27 February, 1 May, 16 August, 24 September, 25 December; variable: Corpus Christi, Good Friday

ECUADOR

Located in South America, bounded north by Colombia, east and south by Peru, and west by the Pacific Ocean.

National name: *República del Ecuador/Republic of Ecuador*
Area: 270,670 sq km/104,505 sq mi
Capital: Quito
Major towns/cities: Guayaquil, Cuenca, Machala, Portoviejo, Manta, Ambato, Santo Domingo
Major ports: Guayaquil
Physical features: coastal plain rises sharply to Andes Mountains, which are divided into a series of cultivated valleys; flat, low-lying rainforest in the east; Galapagos Islands; Cotopaxi, the world's highest active volcano. Ecuador is crossed by the Equator, from which it derives its name

A condor, poised to attack enemies, protects the nation under its wings. Blue symbolizes independence from Spain. Yellow recalls the Federation of Greater Colombia. Red stands for courage.
Effective date: 7 November 1900.

Government

Head of state and government: Gustavo Noboa from 2000
Political system: liberal democracy
Political executive: limited presidency
Administrative divisions: 21 provinces
Political parties: Social Christian Party (PSC), right wing; Ecuadorean Roldosist Party (PRE), populist, centre left; Popular Democracy (DP), centre right; Democratic Left (ID), moderate socialist; Conservative Party (PCE), right wing; Popular Democratic Movement (MPD), far-left
Armed forces: 57,100 (1998)
Conscription: military service is selective for one year
Death penalty: abolished in 1906
Defence spend: (% GDP) 2.6 (1998)
Education spend: (% GNP) 2.6 (1998)
Health spend: (% GDP) 4.6 (1997)

Economy and resources

Currency: sucre
GDP: (US$) 18.7 billion (1999)
Real GDP growth: (% change on previous year) –7.3 (1999)
GNP: (US$) 16.2 billion (1999)
GNP per capita (PPP): (US$) 2,605 (1999)
Consumer price inflation: 52.2% (1999)
Unemployment: 11.5% (1998 est)
Foreign debt: (US$) 15.7 billion (1999 est)
Major trading partners: USA, Colombia, Germany, Chile, Peru, Japan, Italy, Venezuela
Resources: petroleum, natural gas, gold, silver, copper, zinc, antimony, iron, uranium, lead, coal
Industries: food processing, petroleum refining, cement, chemicals, textiles
Exports: petroleum and petroleum products, bananas, shrimps (a major exporter), coffee, seafood products, cocoa beans and products, cut flowers. Principal market: USA 36.9% (1999)
Imports: machinery and transport equipment, basic manufactures, chemicals, consumer goods. Principal source: USA 29.7% (1999)
Arable land: 5.7% (1996)

Agricultural products: bananas, coffee, cocoa, rice, potatoes, maize, barley, sugar cane; fishing (especially shrimp industry); forestry

Population and society

Population: 12,646,000 (2000 est)
Population growth rate: 2.0% (1995–2000); 1.7% (2000–05)
Population density: (per sq km) 46 (1999 est)
Urban population: (% of total) 65 (2000 est)
Age distribution: (% of total population) 0–14 34%, 15–59 59%, 60+ 7% (2000 est)
Ethnic groups: about 55% mestizo (of Spanish-American and American Indian parentage), 25% American Indian, 10% Spanish, 10% African
Language: Spanish (official), Quechua, Jivaro, other indigenous languages
Religion: Roman Catholic
Education: (compulsory years) 6
Literacy rate: 93% (men); 89% (women) (2000 est)
Labour force: 35% of population: 28% agriculture, 20% industry, 52% services (1996 est)
Life expectancy: 67 (men); 73 (women) (1995–2000)
Child mortality rate: (under 5, per 1,000 live births) 60 (1995–2000)
Physicians: 1 per 904 people (1996)
Hospital beds: 1 per 602 people (1993)
TV sets: (per 1,000 people) 294 (1997)
Radios: (per 1,000 people) 419 (1997)
Internet users: (per 10,000 people) 16.1 (1999)
Personal computer users: (per 100 people) 2.0 (1999)

Transport

Airports: international airports: Quito (Mariscal Sucre), Guayaquil (Simón Bolívar); six domestic airports; total passenger km: 1,946 million (1997 est)
Railways: total length: 966 km/600 mi; total passenger km: 27 million (1994)
Roads: total road network: 43,197 km/26,843 mi, of which 18.9% paved (1997); passenger cars: 39.7 per 1,000 people (1996)

Chronology

1450s: The Caras people, whose kingdom had its capital at Quito, conquered by Incas of Peru. **1531:** Spanish conquistador Francisco Pizarro landed on Ecuadorean coast, en route to Peru, where Incas were defeated. **1534:** Conquered by Spanish. Quito, which had been destroyed by American Indians, was refounded by Sebastian de Belalcazar; the area became part of the Spanish Viceroyalty of Peru, which covered much of South America, with its capital at Lima (Peru).

later 16th century: Spanish established large agrarian estates, owned by Europeans and worked by American Indian labourers. **1739:** Became part of new Spanish Viceroyalty of Nueva Granada, which included Colombia and Venezuela, with its capital in Bogotá (Colombia). **1809:** With the Spanish monarchy having been overthrown by Napoleon Bonaparte, the Creole middle class began to press for independence. **1822:** Spanish Royalists defeated by Field Marshal Antonio José de Sucre, fighting for Simón Bolívar, 'The Liberator', at battle of Pichincha, near Quito; became part of independent Gran Colombia, which also comprised Colombia, Panama, and Venezuela. **1830:** Became fully independent state, after leaving Gran Colombia.

1845–60: Political instability, with five presidents holding power, increasing tension between conservative Quito and liberal Guayaquil on the coast, and minor wars with Peru and Colombia. **1860–75:** Power held by Gabriel García Moreno, an autocratic theocrat-Conservative who launched education and public-works programmes. **1895–1912:** Dominated by Gen Eloy Alfaro, a radical, anticlerical Liberal from the coastal region, who reduced the power of the church. **1925–48:** Great political instability; no president completed his term of office. **1941:** Lost territory in Amazonia after defeat in war with Peru. **1948–55:** Liberals in power. **1956:** Camilo Ponce became first conservative president in 60 years. **1960:** Liberals in power, with José María Velasco Ibarra as president. **1962:** Military junta installed. **1968:** Velasco returned as president. **1970s:** Ecuador emerged as significant oil producer. **1972:** Coup put military back in power. **1979:** New democratic constitution; Liberals in power but opposed by right- and left-wing parties. **1981:** Border dispute with Peru flared up again. **1982:** The deteriorating economy and austerity measures provoked strikes, demonstrations, and a state of emergency. **1988:** Unpopular austerity measures were introduced. **1992:** PUR leader Sixto Duran Ballen was elected president; PSC became the largest party in congress. Ecuador withdrew from OPEC to enable it to increase its oil exports. **1994:** There was mounting opposition to Duran's economic liberalization and privatization programme. **1998:** A 157-year border dispute was settled with Peru. **2000:** After the currency lost 65% of its value in 1999, President Mahuad declared a state of emergency, froze all bank accounts valued at over £100, and said that he would introduce the dollar in favour of the sucre. After a bloodless coup in protest to the measures, Gustavo Noboa was sworn in as president. He nevertheless continued with the introduction of the dollar after positive international response to the plans.

Practical information

Visa requirements: UK: visa not required (except for business visits of three–six months). USA: visa not required (except for business visits of three–six months) **Time difference:** GMT –5 **Chief tourist attractions:** Andes Mountains; rainforests of upper Amazon basin; colonial churches and palaces of Guayaquil; Quito, the former Inca capital, with its cathedral, churches, and palaces – the old city is a designated UN World Heritage site; Galapagos Islands, with marine iguanas and giant tortoises **Major holidays:** 1 January, 1, 24 May, 30 June, 24 July, 10 August, 9, 12 October, 2–3 November, 6, 25, 31 December; variable: Carnival (2 days), Good Friday, Holy Thursday

EGYPT

Located in northeast Africa, bounded to the north by the Mediterranean Sea, east by the Palestinian-controlled Gaza Strip, Israel, and the Red Sea, south by Sudan, and west by Libya.

National name: *Jumhuriyyat Misr al-'Arabiyya/Arab Republic of Egypt*
Area: 1,001,450 sq km/386,659 sq mi
Capital: Cairo
Major towns/cities: El Gîza, Shubra Al Khayma, Alexandria, Port Said, El-Mahalla el-Koubra, Tanta, El-Mansoura, Suez
Major ports: Alexandria, Port Said, Suez, Damietta, Shubra Al Khayma
Physical features: mostly desert; hills in east; fertile land along Nile valley and delta; cultivated and settled area is about 35,500 sq km/13,700 sq mi; Aswan High Dam and Lake Nasser; Sinai (the peninsula)

Red, white, and black are the colours of Arab nationalism. Effective date: 4 October 1984.

Government

Head of state: Hosni Mubarak from 1981
Head of government: Atef Obeid from 1999
Political system: liberal democracy
Political executive: limited presidency
Administrative divisions: 26 governates
Political parties: National Democratic Party (NDP), moderate, left of centre; Socialist Labour Party (SLP), right of centre; Liberal Socialist Party, free enterprise; New Wafd Party, nationalist; National Progressive Unionist Party, left wing
Armed forces: 450,000 (1998)
Conscription: 3 years (selective)
Death penalty: retained and used for ordinary crimes
Defence spend: (% GDP) 4.1 (1998)
Education spend: (% GNP) 5.6 (1995)
Health spend: (% GDP) 3.7 (1997)

Economy and resources

Currency: Egyptian pound
GDP: (US$) 92.4 billion (1999)
Real GDP growth: (% change on previous year) 6 (1999)
GNP: (US$) 87.5 billion (1999)
GNP per capita (PPP): (US$) 3,303 (1999)
Consumer price inflation: 3.1% (1999)
Unemployment: 11.8% (1998)
Foreign debt: (US$) 31.6 billion (1999 est)
Major trading partners: EU, USA, Turkey, Japan
Resources: petroleum, natural gas, phosphates, manganese, uranium, coal, iron ore, gold
Industries: petroleum and petroleum products, food processing, petroleum refining, textiles, metals, cement, tobacco, sugar crystal and refined sugar, electrical appliances, fertilizers
Exports: petroleum and petroleum products, textiles, clothing, food, live animals. Principal market: EU 35% (1999)
Imports: wheat, maize, dairy products, machinery and transport equipment, wood and wood products, consumer goods. Principal source: EU 35.9% (1999)

Arable land: 2.8% (1996)
Agricultural products: wheat, cotton, rice, corn, beans

Population and society

Population: 68,470,000 (2000 est)
Population growth rate: 1.9% (1995–2000)
Population density: (per sq km) 67 (1999 est)
Urban population: (% of total) 45 (2000 est)
Age distribution: (% of total population) 0–14 35%, 15–59 58%, 60+ 7% (2000 est)
Ethnic groups: 99% Eastern Hamitic stock (Egyptians, Bedouins, and Berbers)
Language: Arabic (official), Coptic (derived from ancient Egyptian), English, French
Religion: Sunni Muslim 90%, Coptic Christian and other Christian 6%
Education: (compulsory years) 5
Literacy rate: 67% (men); 44% (women) (2000 est)
Labour force: 35% of population: 34% agriculture, 21% industry, 45% services (1995)
Life expectancy: 65 (men); 68 (women) (1995–2000)
Child mortality rate: (under 5, per 1,000 live births) 65 (1995–2000)
Physicians: 1 per 1,316 people (1993 est)
Hospital beds: 1 per 523 people (1997 est)
TV sets: (per 1,000 people) 127 (1997)
Radios: (per 1,000 people) 324 (1997)
Internet users: (per 10,000 people) 29.8 (1999)
Personal computer users: (per 100 people) 1.1 (1999)

Transport

Airports: international airports: Cairo (two), Alexandria (El Nouzha), Luxor; eight domestic airports; total passenger km: 9,018 million (1997 est)
Railways: total length: 4,751 km/2,952 mi; total passenger km: 52,928 million (1997)
Roads: total road network: 64,000 km/39,770 mi, of which 78.1% paved (1996 est); passenger cars: 22.5 per 1,000 people (1996 est)

Chronology

1st century BC–7th century AD: Conquered by Augustus in AD 30, Egypt passed under rule of Roman, and later Byzantine, governors. **AD 639–42:** Arabs conquered Egypt, introducing Islam and Arabic; succession of Arab dynasties followed. **1250:** Mamelukes seized power. **1517:** Became part of Turkish Ottoman Empire. **1798–1801:** Invasion by Napoleon followed by period of French occupation. **1801:** Control regained by Turks. **1869:** Opening of Suez Canal made Egypt strategically important. **1881–82:** Nationalist revolt resulted in British occupation.

1914: Egypt became a British protectorate. **1922:** Achieved nominal independence under King Fuad I. **1936:** Full independence from Britain achieved. King Fuad succeeded by his son Farouk. **1946:** Withdrawal of British troops except from Suez Canal zone. **1952:** Farouk overthrown by army in bloodless coup. **1953:** Egypt declared a republic, with Gen Neguib as president. **1956:** Neguib replaced by Col Gamal Nasser. Nasser announced nationalization of Suez Canal; Egypt attacked by Britain, France, and Israel. Ceasefire agreed following US intervention. **1958:** Short-lived merger of Egypt and Syria as United Arab Republic (UAR). **1967:** Six-Day War with Israel ended in Egypt's defeat and Israeli occupation of Sinai and Gaza Strip. **1970:** Nasser died suddenly; succeeded by Anwar Sadat. **1973:** An attempt to regain territory lost to Israel led to the Yom Kippur War; ceasefire arranged by US secretary of state Henry Kissinger. **1978–79:** Camp David talks in USA resulted in a peace treaty between Egypt and Israel. Egypt expelled from Arab League. **1981:** Sadat was assassinated by Muslim fundamentalists and succeeded by Hosni Mubarak. **1983:** Relations between Egypt and the Arab world improved; only Libya and Syria maintained a trade boycott. **1987:** Egypt was readmitted to the Arab League. **1989:** Relations with Libya improved; diplomatic relations with Syria were restored. **1991:** Egypt participated in the Gulf War on the US-led side and was a major force in convening a Middle East peace conference in Spain. **1994:** The government cracked down on Islamic militants. **1997:** Islamic extremists killed and injured tourists at Luxor. **1999:** President Mubarak was awarded a fourth term as president, and Atef Obeid was appointed as prime minister. **2000:** At least 20 people were killed in clashes between Christians and Muslims in southern Egypt. In parliamentary elections, opposition parties did much better than usual and the banned Muslim Brotherhood won 17 seats, re-establishing their presence in parliament for the first time in a decade.

Practical information

Visa requirements: UK: visa required. USA: visa required

Time difference: GMT +2

Chief tourist attractions: beaches and coral reefs on coast south of Suez; Western Desert, containing the Qattara Depression, the world's largest and lowest depression; ancient pyramids and temples, including those at Saqqara, El Giza, and Karnak

Major holidays: 7 January, 25 April, 1 May, 18 June, 1, 23 July, 6 October; variable: Eid-ul-Adha (2 days), Arafa, end of Ramadan (2 days), New Year (Muslim), Prophet's Birthday, Palm Sunday and Easter Sunday (Eastern Orthodox), Sham-el-Nessim (April/May)

EL SALVADOR

Located in Central America, bounded north and east by Honduras, south and southwest by the Pacific Ocean, and northwest by Guatemala.

National name: *República de El Salvador/Republic of El Salvador*
Area: 21,393 sq km/8,259 sq mi
Capital: San Salvador
Major towns/cities: Soyapango, Santa Ana, San Miguel, Nueva San Salvador, Mejicanos, Apopa, Delgado
Physical features: narrow coastal plain, rising to mountains in north with central plateau

The triangle represents equality and the rainbow signifies peace. The arms may be replaced by the motto Dios, Union, Libertad, 'God, Union, Liberty'.
Effective date: September 1972.

Government

Head of state and government: Francisco Guillermo Flores Pérez from 1999
Political system: emergent democracy
Political executive: limited presidency
Administrative divisions: 14 departments
Political parties: Christian Democrats (PDC), anti-imperialist; Farabundo Martí Liberation Front (FMLN), left wing; National Republican Alliance (ARENA), extreme right wing; National Conciliation Party (PCN), right wing
Armed forces: 24,600 (1998); plus 12,000 in paramilitary forces
Conscription: selective conscription for two years
Death penalty: laws provide for the death penalty only for exceptional crimes such as crimes under military law or crimes committed in exceptional circumstances such as wartime (last known execution in 1973)
Defence spend: (% GDP) 1.7 (1998)
Education spend: (% GNP) 2.5 (1997)
Health spend: (% GDP) 7 (1997)

Economy and resources

Currency: US dollar (replaced Salvadorean colón in 2001)
GDP: (US$) 12.2 billion (1999)
Real GDP growth: (% change on previous year) 3.4 (1999)
GNP: (US$) 11.8 billion (1999)
GNP per capita (PPP): (US$) 4,048 (1999 est)
Consumer price inflation: 0.6% (1999)
Unemployment: 8% (1997)
Foreign debt: (US$) 3.8 billion (1999 est)
Major trading partners: USA, Guatemala, Costa Rica, Honduras, Mexico, Japan, Germany, Venezuela
Resources: salt, limestone, gypsum
Industries: food processing, beverages, petroleum products, textiles, tobacco, paper products, chemical products
Exports: coffee, textiles and garments, sugar, shrimp, footwear, pharmaceuticals. Principal market: USA 63.1% (1999)

Imports: petroleum and other minerals, cereals, chemicals, iron and steel, machinery and transport equipment, consumer goods. Principal source: USA 51.7% (1999)
Arable land: 30.7% (1996)
Agricultural products: coffee, sugar cane, cotton, maize, beans, rice, sorghum; fishing (shrimp)

Population and society

Population: 6,276,000 (2000 est)
Population growth rate: 2% (1995–2000)
Population density: (per sq km) 288 (1999 est)
Urban population: (% of total) 47 (2000 est)
Age distribution: (% of total population) 0–14 36%, 15–59 56%, 60+ 8% (2000 est)
Ethnic groups: about 92% of the population are mestizos, 6% Indians, and 2% of European origin
Language: Spanish (official), Nahuatl
Religion: about 75% Roman Catholic, Protestant
Education: (compulsory years) 9
Literacy rate: 82% (men); 76% (women) (2000 est)
Labour force: 36% of population: 36% agriculture, 21% industry, 43% services (1990)
Life expectancy: 67 (men); 73 (women) (1995–2000)
Child mortality rate: (under 5, per 1,000 live births) 41 (1995–2000)
Physicians: 1 per 1,515 people (1993 est)
Hospital beds: 1 per 680 people (1993 est)
TV sets: (per 1,000 people) 250 (1996)
Radios: (per 1,000 people) 464 (1997)
Internet users: (per 10,000 people) 65.0 (1999)
Personal computer users: (per 100 people) 1.6 (1999)

Transport

Airports: international airport: San Salvador (El Salvador International); three domestic airports; total passenger km: 2,290 million (1997 est)
Railways: total length: 602 km/374 mi; total passenger km: 5 million (1995)
Roads: total road network: 10,029 km/6,232 mi, of which 19.8% paved (1997); passenger cars: 30 per 1,000 people (1997)

Chronology

11th century: Pipils, descendants of the Nahuatl-speaking Toltec and Aztec peoples of Mexico, settled in the country and came to dominate El Salvador until the Spanish conquest. **1524:** Conquered by the Spanish adventurer Pedro de Alvarado and made a Spanish colony, with resistance being crushed by 1540. **1821:** Independence achieved from Spain; briefly joined with Mexico. **1823:** Became part of United Provinces (Federation) of Central America, also embracing Costa Rica, Guatemala, Honduras, and Nicaragua. **1833:** Unsuccessful rebellion against Spanish control of land led by Anastasio Aquino. **1840:** Became fully independent when the Federation was dissolved. **1859–63:** Coffee growing introduced by President Gerardo Barrios. **1932:** Peasant uprising, led by Augustín Farabundo Martí, suppressed by military at a cost of the lives of 30,000, virtually eliminating American Indian Salvadoreans. **1961:** Following a coup, the right-wing National Conciliation Party (PCN) established and in power. **1969:** Brief 'Football War' with Honduras, in which El Salvador attacked Honduras, at the time of a football competition between the two states, following evictions of thousands of Salvadoran illegal immigrants from Honduras. **1977:** Allegations of human-rights violations; growth of left-wing Farabundo Martí National Liberation Front (FMLN) guerrilla activities. Gen Carlos Romero elected president. **1979:** A coup replaced Romero with a military-civilian junta. **1980:** The archbishop of San Salvador and human-rights champion, Oscar Romero, was assassinated; the country was on the verge of civil war. José Napoleón Duarte (PDC) became the first civilian president since 1931. **1979–81:** 30,000 people were killed by right-wing death squads. **1981:** Mexico and France recognized the FMLN guerrillas as a legitimate political force, but the USA actively assisted the government in its battle against them. **1982:** Assembly elections were boycotted by left-wing parties. Held amid considerable violence, they were won by far-right National Republican Alliance (ARENA). **1986:** Duarte sought a negotiated settlement with the guerrillas. **1989:** Alfredo Cristiani (ARENA) became president in rigged elections; rebel attacks intensified. **1991:** A peace accord sponsored by the United Nations (UN) was signed by representatives of the government and the socialist guerrilla group, the FMLN, which became a political party. **1993:** A UN-sponsored commission published a report on war atrocities; there was a government amnesty for those implicated; top military leaders officially retired. **1999:** Francisco Guillermo Flores Pérez was elected president. **2000:** The FMLN displaced the ruling ARENA as the largest party in Congress, but did not win an overall majority. The government signed a free-trade agreement with Mexico. **2001:** El Salvador adopted the US dollar as its currency, phasing out the colón. A powerful earthquake in January killed over 1,000 people and left a million homeless.

Practical information

Visa requirements: UK: visa not required for a stay of up to 90 days. USA: visa required (Tourist Card)
Time difference: GMT –6
Chief tourist attractions: Mayan temples and other remains; upland scenery with lakes and volcanoes; Pacific beaches
Major holidays: 1 January, 1 May, 29–30 June, 15 September, 12 October, 2, 5 November, 24–25, 30–31 December; variable: Good Friday, Holy Thursday, Ash Wednesday, San Salvador (4 days)

EQUATORIAL GUINEA

Located in west-central Africa, bounded north by Cameroon, east and south by Gabon, and west by the Atlantic Ocean; also five offshore islands including Bioco, off the coast of Cameroon.

National name: *República de Guinea Ecuatorial/Republic of Equatorial Guinea*
Area: 28,051 sq km/10,830 sq mi
Capital: Malabo
Major towns/cities: Bata, Mongomo, Ela Nguema, Mbini, Campo Yaunde, Los Angeles
Physical features: comprises mainland Río Muni, plus the small islands of Corisco, Elobey Grande and Elobey Chico, and Bioco (formerly Fernando Po) together with Annobón (formerly Pagalu); nearly half the land is forested; volcanic mountains on Bioco

Blue stands for the sea. Green represents agriculture and natural wealth. White symbolizes peace. Red recalls the struggle for independence.
Effective date: 21 August 1979.

Government

Head of state: Teodoro Obiang Nguema Mbasogo from 1979
Head of government: Angel Serafin Seriche Dougan from 1996
Political system: authoritarian nationalist
Political executive: unlimited presidency
Administrative divisions: seven provinces
Political parties: Democratic Party of Equatorial Guinea (PDGE), nationalist, right of centre, militarily controlled; People's Social Democratic Convention (CSDP), left of centre; Democratic Socialist Union of Equatorial Guinea (UDS), left of centre; Liberal Democratic Convention (CLD)
Armed forces: 1,300 (1998)
Conscription: military service is voluntary
Death penalty: retained and used for ordinary crimes
Defence spend: (% GDP) 1.5 (1998)
Education spend: (% GNP) 1.8 (1996)
Health spend: (% GDP) 3.5 (1997)

Economy and resources

Currency: franc CFA
GDP: (US$) 751 million (1999 est)
Real GDP growth: (% change on previous year) 15 (1999 est)
GNP: (US$) 516 million (1999)
GNP per capita (PPP): (US$) 3,545 (1999 est)
Consumer price inflation: 6% (1999 est)
Foreign debt: (US$) 290 million (1999 est)
Major trading partners: USA, Spain, Italy, China, France, Cameroon, UK
Resources: petroleum, natural gas, gold, uranium, iron ore, tantalum, manganese
Industries: wood processing, food processing
Exports: timber, re-exported ships and boats, textile fibres and waste, cocoa, coffee. Principal market: USA 62% (1998 est)
Imports: ships and boats, petroleum and related products, food and live animals, machinery and transport equipment, beverages and tobacco, basic manufactures. Principal source: USA 35% (1998 est)
Arable land: 4.6% (1995)

Agricultural products: cocoa, coffee, cassava, sweet potatoes, bananas, palm oil, palm kernels; exploitation of forest resources (principally of *okoumé* and *akoga* timber)

Population and society

Population: 453,000 (2000 est)
Population growth rate: 2.5% (1995–2000); 2.4% (2000–05)
Population density: (per sq km) 16 (1999 est)
Urban population: (% of total) 48 (2000 est)
Age distribution: (% of total population) 0–14 43%, 15–59 51%, 60+ 6% (2000 est)
Ethnic groups: the Fang ethnic group, of Bantu origin (80–90%); most other groups have been pushed to the coast by Fang expansion; the Bubi are the indigenous ethnic group of Bioco (island)
Language: Spanish (official), pidgin English, a Portuguese patois (on Annobón, whose people were formerly slaves of the Portuguese), Fang and other African patois (on Río Muni)
Religion: Roman Catholic, Protestant, animist
Education: (compulsory years) 8
Literacy rate: 92% (men); 74% (women) (2000 est)
Labour force: 77% agriculture, 2% industry, 21% services (1990)
Life expectancy: 48 (men); 52 (women) (1995–2000)
Child mortality rate: (under 5, per 1,000 live births) 177 (1995–2000)
Physicians: 1 per 3,889 people (1993)
Hospital beds: N/A
TV sets: (per 1,000 people) 10 (1997)
Radios: (per 1,000 people) 428 (1997)
Internet users: (per 10,000 people) 11.3 (1999)
Personal computer users: (per 100 people) 0.2 (1999)

Transport

Airports: international airports: Malabo, Bata; domestic services operate between major towns; total passenger km: 7 million (1995)
Railways: none
Roads: total road network: 2,880 km/1,790 mi (1996 est); passenger cars: 3.3 per 1,000 people (1996 est)

Chronology

1472: First visited by Portuguese explorers. **1778:** Bioco (formerly known as Fernando Po) Island ceded to Spain, which established cocoa plantations there in the late 19th century, importing labour from West Africa. **1885:** Mainland territory of Mbini (formerly Rio Muni) came under Spanish rule, the whole colony being known as Spanish Guinea, with the capital at Malabu on Bioco Island. **1920s:** League of Nations special mission sent to investigate the forced, quasi-slave labour conditions on the Bioco cocoa plantations, then the largest in the world. **1959:** Became a Spanish Overseas Province; African population finally granted full citizenship. **early 1960s:** On the mainland, the Fang people spearheaded a nationalist movement directed against Spanish favouritism towards Bioco Island and its controlling Bubi tribe. **1963:** Achieved internal autonomy. **1968:** Independence achieved from Spain. Macias Nguema, a nationalist Fang, became first president, discriminating against the Bubi community. **1970s:** The economy collapsed as Spanish settlers and other minorities fled in the face of intimidation by Nguema's brutal, dictatorial regime, which was marked by the murder, torture, and imprisonment of tens of thousands of political opponents and rivals. **1979:** Nguema was overthrown, tried, and executed.

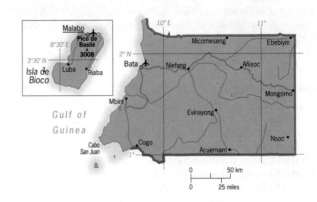

He was replaced by his nephew, Teodoro Obiang Nguema Mbasogo, who established a military regime, but released political prisoners and imposed restrictions on the Catholic Church. **1992:** A new pluralist constitution was approved by referendum. **1993:** Obiang's PDGE won the first multiparty elections on low turnout. **1996:** Obiang was re-elected amid claims of fraud by opponents. Angel Serafin Seriche Dougan became prime minister, and was reappointed in 1998.

Practical information

Visa requirements: UK: visa required. USA: visa required
Time difference: GMT +1
Chief tourist attractions: beaches around small offshore islands; tourism remains undeveloped
Major holidays: 1 January, 1 May, 5 June, 3 August, 12 October, 10, 25 December; variable: Corpus Christi, Good Friday, Constitution (August)

ERITREA

Located in East Africa, bounded north by Sudan, south by Ethiopia, southeast by Djibouti, and east by the Red Sea.

National name: *Hagere Eretra al-Dawla al-Iritra/State of Eritrea*
Area: 125,000 sq km/48,262 sq mi
Capital: Asmara
Major towns/cities: Assab, Keren, Massawa, Adi Ugri, Ed
Major ports: Assab, Massawa
Physical features: coastline along the Red Sea 1,000 km/620 mi; narrow coastal plain that rises to an inland plateau; Dahlak Islands

Green, red, and blue were the colours of the Eritrean People's Liberation Front (EPLF) flag which bore a yellow star at the hoist. Effective date: late 1995.

Government

Head of state and government: Issaias Afwerki from 1993
Political system: nationalistic socialist
Political executive: unlimited presidency
Administrative divisions: eight provinces
Political parties: People's Front for Democracy and Justice (PFDJ) (formerly Eritrean People's Liberation Front: EPLF), left of centre, the only party recognised by the government; Eritrean National Pact Alliance (ENPA), moderate, centrist
Armed forces: 47,100 (1998)
Conscription: compulsory for 18 months
Death penalty: retained and used for ordinary crimes
Defence spend: (% GDP) 35.8 (1998)
Education spend: (% GNP) 1.8 (1995–97)
Health spend: (% GDP) 3.4 (1997 est)

Economy and resources

Currency: Ethiopian nakfa
GDP: (US$) 670 million (1999)
Real GDP growth: (% change on previous year) 0 (1999 est)
GNP: (US$) 800 million (1999)
GNP per capita (PPP): (US$) 1,012 (1999 est)
Consumer price inflation: 12% (1999 est)
Unemployment: 50% (1997 est)
Foreign debt: (US$) 225 million (1999 est)
Major trading partners: Ethiopia, Saudi Arabia, Italy, Sudan, Japan, United Arab Emirates, UK, Korea
Resources: gold, silver, copper, zinc, sulphur, nickel, chrome, potash, basalt, limestone, marble, sand, silicates
Industries: food processing, textiles, leatherwear, building materials, glassware, petroleum products
Exports: textiles, leather and leather products, beverages, petroleum products, basic household goods. Principal market: Sudan 27.2% (1998)
Imports: machinery and transport equipment, petroleum, food and live animals, basic

manufactures. Principal source: Italy 17.4% (1998)
Arable land: 4.4% (1996)
Agricultural products: sorghum, teff (an indigenous grain), maize, wheat, millet; livestock rearing (goats and camels); fisheries

Population and society

Population: 3,850,000 (2000 est)
Population growth rate: 3.8% (1995–2000)
Population density: (per sq km) 30 (1999 est)
Urban population: (% of total) 19 (2000 est)
Age distribution: (% of total population) 0–14 44%, 15–59 52%, 60+ 4% (2000 est)
Ethnic groups: ethnic Tigrinya 50%, Tigre and Kunama 40%, Afar 4%, Saho 3%
Language: Tigre, Tigrinya, Arabic, English, Afar, Amharic, Kunama, Italian
Religion: mainly Sunni Muslim and Coptic Christian, some Roman Catholic, Protestant, and animist
Education: (compulsory years) 7
Literacy rate: 67% (men); 44% (women) (2000 est)
Life expectancy: 49 (men); 52 (women) (1995–2000)
Child mortality rate: (under 5, per 1,000 live births) 146 (1995–2000)
Physicians: 1 per 45,588 people (1993 est)
TV sets: (per 1,000 people) 14 (1998)
Radios: (per 1,000 people) 91 (1997)
Internet users: (per 10,000 people) 1.3 (1999)

Transport

Airports: international airport: Asmara (Yohannes IV); two domestic airports
Railways: none
Roads: total road network: 4,010 km/2,492 mi, of which 21.8% paved (1996 est); passenger cars: 1.5 per 1,000 people (1996 est)

Chronology

4th–7th centuries AD: Part of Ethiopian Aksum kingdom. **8th century:**
Islam introduced to coastal areas by Arabs. **12th–16th centuries:** Under
influence of Ethiopian Abyssinian kingdoms. **mid-16th century:** Came under
control of Turkish Ottoman Empire. **1882:** Occupied by Italy.
1889: Italian colony of Eritrea created out of Ottoman areas
and coastal districts of Ethiopia. **1920s:** Massawa
developed into the largest port in East Africa. **1935–36:**
Used as base for Italy's conquest of Ethiopia and
became part of Italian East Africa. **1941:** Became
British protectorate after Italy removed from
North Africa. **1952:** Federation formed with
Ethiopia by United Nations (UN). **1958:**
Eritrean People's Liberation Front (EPLF) was
formed to fight for independence after a
general strike was brutally suppressed by
Ethiopian rulers. **1962:** Annexed by Ethiopia,
sparking a secessionist rebellion which was to
last 30 years and claim 150,000 lives. **1974:**
Ethiopian emperor Haile Selassie was deposed by
the military; the EPLF continued the struggle for
independence. **1977–78:** The EPLF cleared the
territory of Ethiopian forces, but the position was
soon reversed by the Soviet-backed Marxist Ethiopian
government of Col Mengistu Haile Mariam. **mid-1980s:**
There was severe famine in Eritrea and a refugee crisis as the
Ethiopian government sought forcible resettlement. **1990:** The strategic port of Massawa was
captured by Eritrean rebel forces. **1991:** Ethiopian president Mengistu was overthrown. The
EPLF secured the whole of Eritrea and a provisional government was formed under Issaias Afwerki.
1993: Independence was approved in a regional referendum and recognized by Ethiopia. A transitional
government was established, with Afwerki elected president; 500,000 refugees outside Eritrea began to return.
1998: Border disputes with Ethiopia escalated, with bombing raids from both sides. **1999:** The border dispute
with Ethiopia erupted into war in February. Peace was agreed the following month, but fighting was renewed.
2000: An Ethiopian military offensive caused Eritrea to pull its forces back to the line it held before the
escalation of violence in 1998. A ceasefire was agreed in June and a peace agreement signed in December that
provided for UN troops to keep the peace along the border.

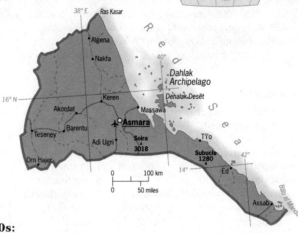

Practical information

Visa requirements: UK: visa required. USA: visa
required
Time difference: GMT +3
Chief tourist attractions: tourism remains largely
undeveloped; Dahlak Islands (a coral archipelago rich

in marine life) near Massawa; unique ecosystem on
escarpment that rises from coastal plain
Major holidays: 1, 6 January, 24 May, 20 June, 1
September, 25 December; variable: Eid-ul-Adha, Arafa,
end of Ramadan

ESTONIA

Located in northern Europe, bounded east by Russia, south by Latvia, and north and west by the Baltic Sea.

National name: *Eesti Vabariik/Republic of Estonia*
Area: 45,000 sq km/17,374 sq mi
Capital: Tallinn
Major towns/cities: Tartu, Narva, Kohtla-Järve, Pärnu
Physical features: lakes and marshes in a partly forested plain; 774 km/481 mi of coastline; mild climate; Lake Peipus and Narva River forming boundary with Russian Federation; Baltic islands, the largest of which is Saaremaa

Blue stands for faith and loyalty, the sea, lakes, and the sky. Black represents past suffering and is the colour of the traditional peasant's jacket. White symbolizes virtue and enlightenment and is the colour of snow, birch bark, and the midnight sun. Effective date: 16 November 1990.

Government

Head of state: Lennart Meri from 1992
Head of government: Mart Laar from 1999
Political system: emergent democracy
Political executive: dual executive
Administrative divisions: 15 counties and six towns
Political parties: Coalition Party (KMU), ex-communist, left of centre, 'social market'; Isamaa (National Fatherland Party, or Pro Patria), right wing, nationalist, free market; Estonian Reform Party (ERP), free market; Centre Party (CP), moderate nationalist (formerly the Estonian Popular Front (EPF; Rahvarinne); Estonian National Independence Party (ENIP), radical nationalist; Communist Party of Estonia (CPE); Our Home is Estonia; Estonian Social Democratic Party (ESDP) (last three draw much of their support from ethnic Russian community)
Armed forces: 4,300 (1998); plus 14,000 reservists and paramilitary border guard of 2,800
Conscription: compulsory for 12 months (men and women)
Death penalty: abolished in 1998
Defence spend: (% GDP) 1.3 (1998)
Education spend: (% GNP) 7.2 (1997)
Health spend: (% GDP) 6.4 (1997)

Economy and resources

Currency: kroon
GDP: (US$) 5.1 billion (1999)
Real GDP growth: (% change on previous year) –1.4 (1999)
GNP: (US$) 5 billion (1999)
GNP per capita (PPP): (US$) 7,826 (1999)
Consumer price inflation: 3.5% (1999)
Unemployment: 3.6% (1997)
Foreign debt: (US$) 1.5 billion (1999)
Major trading partners: Finland, Russia, Sweden, Germany, Latvia, the Netherlands, Lithuania, UK, Japan
Resources: oilshale, peat, phosphorite ore, superphosphates
Industries: machine building, electronics, electrical engineering, textiles, fish and food processing, consumer goods
Exports: foodstuffs, animal products, textiles, timber products, base metals, mineral products, machinery. Principal market: Finland 19.4% (1999)

Imports: machinery and transport equipment, food products, textiles, mineral products. Principal source: Finland 22.8% (1999)
Arable land: 26.7% (1996)
Agricultural products: wheat, rye, barley, potatoes, other vegetables; livestock rearing (cattle and pigs); dairy farming

Population and society

Population: 1,396,000 (2000 est)
Population growth rate: –1.2% (1995–2000)
Population density: (per sq km) 31 (1999 est)
Urban population: (% of total) 69 (2000 est)
Age distribution: (% of total population) 0–14 17%, 15–59 63%, 60+ 20% (2000 est)
Ethnic groups: 65% Finno-Ugric ethnic Estonians, 28% Russian, 2.5% Ukrainian, 1.5% Belorussian, 1% Finnish
Language: Estonian (official), Russian
Religion: Eastern Orthodox, Evangelical Lutheran, Russian Orthodox, Muslim, Judaism
Education: (compulsory years) 9
Literacy rate: 99% (men); 99% (women) (2000 est)
Labour force: 12.2% agriculture, 31.4% industry, 56.4% services (1997 est)
Life expectancy: 63 (men); 75 (women) (1995–2000)
Child mortality rate: (under 5, per 1,000 live births) 25 (1995–2000)
Physicians: 1 per 318 people (1997)
Hospital beds: 1 per 99.1 people (1997)
TV sets: (per 1,000 people) 479 (1997)
Radios: (per 1,000 people) 693 (1997)
Internet users: (per 10,000 people) 1,383.5 (1999)
Personal computer users: (per 100 people) 13.5 (1999)

Transport

Airports: international airports: Tallinn; three domestic airports; total passenger km: 209 million (1997)
Railways: total length: 1,018 km/633 mi; total passenger km: 262 million (1997)
Roads: total road network: 16,438 km/10,215 mi, of which 50.8% paved (1997); passenger cars: 294.1 per 1,000 people (1997)

Chronology

1st century AD: First independent state formed. **9th century:** Invaded by Vikings. **13th century:** Tallinn, in the Danish-controlled north, joined Hanseatic League, a northern European union of commercial towns; Livonia, comprising southern Estonia and Latvia, came under control of German Teutonic Knights and was converted to Christianity. **1561:** Sweden took control of northern Estonia. **1629:** Sweden took control of southern Estonia from Poland. **1721:** Sweden ceded the country to tsarist Russia. **late 19th century:** Estonian nationalist movement developed in opposition to Russian political and cultural repression and German economic control. **1914:** Occupied by German troops. **1918–19:** Estonian nationalists, led by Konstantin Pats, proclaimed and achieved independence, despite efforts by the Russian Red Army to regain control. **1920s:** Land reforms and cultural advances under democratic regime. **1934:** Pats overthrew parliamentary democracy in a quasi-fascist coup at a time of economic depression; Baltic Entente mutual defence pact signed with Latvia and Lithuania. **1940:** Estonia incorporated into Soviet Union (USSR); 100,000 Estonians deported to Siberia or killed. **1941–44:** German occupation during World War II. **1944:** USSR regained control; 'Sovietization' followed, including agricultural collectivization and immigration of ethnic Russians. **late 1980s:** Beginnings of nationalist dissent, encouraged by *glasnost* initiative of reformist Soviet leader Mikhail Gorbachev. **1988:** Popular Front (EPF) established to campaign for democracy. Sovereignty declaration issued by state assembly rejected by USSR as unconstitutional. **1989:** Estonian replaced Russian as the main language. **1990:** The CPE monopoly of power was abolished; pro-independence candidates secured a majority after multiparty elections; a coalition government was formed with EPF leader Edgar Savisaar as prime minister; Arnold Rüütel became president. The prewar constitution was partially restored. **1991:** Independence was achieved after an attempted anti-Gorbachev coup in Moscow; the CPE was outlawed. Estonia joined the United Nations (UN). **1992:** Savisaar resigned over food and energy shortages; Isamaa leader Lennart Meri became president and free-marketer Mart Laar prime minister. **1993:** Estonia joined the Council of Europe and signed a free-trade agreement with Latvia and Lithuania. **1994:** The last Russian troops were withdrawn. A radical economic reform programme was introduced; a controversial law on 'aliens' was passed, requiring non-ethnic Estonians to apply for residency. Laar resigned. **1995:** Former communists won the largest number of seats in a general election; a left-of-centre coalition was formed under Tiit Vahi. **1996:** President Meri was re-elected. The ruling coalition collapsed; Prime Minister Tiit Vahi continued with a minority government. **1997:** Vahi, accused of corruption, resigned and was replaced by Mart Siimann. Estonia was invited to begin European Union (EU) membership negotiations. **1998:** The legislature voted to ban electoral alliances in future elections. **1999:** Mart Laar became prime minister.

Practical information

Visa requirements: UK: visa not required. USA: visa not required
Time difference: GMT +2

Chief tourist attractions: historic towns of Tallinn and Tartu; nature reserves; coastal resorts
Major holidays: 1 January, 24 February, 1 May, 23–24 June, 25–26 December; variable: Good Friday

ETHIOPIA FORMERLY ABYSSINIA (UNTIL THE 1920S)

Located in East Africa, bounded north by Eritrea, northeast by Djibouti, east and southeast by Somalia, south by Kenya, and west and northwest by Sudan.

National name: *Ya'Ityopya Federalawi Dimokrasiyawi Repeblik/Federal Democratic Republic of Ethiopia*
Area: 1,096,900 sq km/423,513 sq mi
Capital: Addis Ababa
Major towns/cities: Jima, Dire Dawa, Harer, Nazret, Dese, Gonder, Mek'ele, Bahir Dar
Physical features: a high plateau with central mountain range divided by Rift Valley; plains in east; source of Blue Nile River; Danakil and Ogaden deserts

Blue stands for peace. Red represents power and faith. Yellow stands for the church, peace, natural wealth, and love. Green symbolizes the land and hope. Effective date: 6 February 1996.

Government

Head of state: Negasso Ghidada from 1995
Head of government: Meles Zenawi from 1995
Political system: emergent democracy
Political executive: limited presidency
Administrative divisions: nine states and two chartered cities
Political parties: Ethiopian People's Revolutionary Democratic Front (EPRDF), nationalist, left of centre; Tigré People's Liberation Front (TPLF); Ethiopian People's Democratic Movement (EPDM); United Oromo Liberation Front, Islamic nationalist
Armed forces: 120,000 (1998)
Conscription: military service is voluntary
Death penalty: retained and used for ordinary crimes
Defence spend: (% GDP) 6.0 (1998)
Education spend: (% GNP) 4 (1996)
Health spend: (% GDP) 3.8 (1997)

Economy and resources

Currency: Ethiopian birr
GDP: (US$) 6.53 billion (1999)
Real GDP growth: (% change on previous year) 0 (1999)
GNP: (US$) 6.6 billion (1999)
GNP per capita (PPP): (US$) 599 (1999)
Consumer price inflation: 4% (1999)
Unemployment: N/A
Foreign debt: (US$) 10.3 billion (1999 est)
Major trading partners: Germany, Russia, Saudi Arabia, USA, Japan, UK, Italy, France
Resources: gold, salt, platinum, copper, potash. Reserves of petroleum have not been exploited
Industries: food processing, petroleum refining, beverages, textiles
Exports: coffee, hides and skins, petroleum products, fruit and vegetables. Principal market: Germany 25% (1998)
Imports: machinery, aircraft and other vehicles, petroleum and petroleum products, basic manufactures, chemicals and related products. Principal source: Russia 11% (1998)
Arable land: 11.3% (1996)
Agricultural products: coffee, teff (an indigenous grain), barley, maize, sorghum, sugar cane; livestock rearing (cattle and sheep) and livestock products (hides, skins, butter, and ghee)

Population and society

Population: 62,565,000 (2000 est)
Population growth rate: 2.5% (1995–2000)
Population density: (per sq km) 56 (1999 est)
Urban population: (% of total) 18 (2000 est)
Age distribution: (% of total population) 0–14 46%, 15–59 49%, 60+ 5% (2000 est)
Ethnic groups: over 80 different ethnic groups, the two main ones are the Amhara and Oromo who comprise about 60% of the population; other groups include Sidamo (9%), Shankella (6%), Somali (16%), and Atar (4%)
Language: Amharic (official), Arabic, Tigrinya, Orominga, about 100 other local languages
Religion: Muslim 45%, Ethiopian Orthodox Church (which has had its own patriarch since 1976) 35%, animist 12%, other Christian 8%
Education: (compulsory years) 6
Literacy rate: 43% (men); 34% (women) (2000 est)
Labour force: 88.6% agriculture, 2% industry, 9.4% services (1995)
Life expectancy: 42 (men); 44 (women) (1995–2000)
Child mortality rate: (under 5, per 1,000 live births) 184 (1995–2000)
Physicians: 1 per 32,499 people (1993 est)
Hospital beds: 1 per 4,141 people (1993 est)
TV sets: (per 1,000 people) 5 (1997)
Radios: (per 1,000 people) 195 (1997)
Internet users: (per 10,000 people) 1.6 (1999)
Personal computer users: (per 100 people) 0.1 (1999)

Transport

Airports: international airports: Addis Ababa (Bole), Dire Dawa; over 40 small domestic airports or airfields; total passenger km: 1,915 million (1997)
Railways: total length: 781 km/485 mi; total passenger km: 157 million (1997)
Roads: total road network: 28,500 km/17,710 mi, of which 15% paved (1996 est); passenger cars: 0.9 per 1,000 people (1997)

Chronology

1st–7th centuries AD: Founded by Semitic immigrants from Saudi Arabia, the kingdom of Aksum and its capital, northwest of Adwa, flourished. It reached its peak in the 4th century when Coptic Christianity was introduced from Egypt. **7th century onwards:** Islam was spread by Arab conquerors. **11th century:** Emergence of independent Ethiopian kingdom of Abyssinia, which was to remain dominant for nine centuries. **late 15th century:** Abyssinia visited by Portuguese explorers. **1889:** Abyssinia reunited by Menelik II. **1896:** Invasion by Italy defeated by Menelik at Adwa, who went on to annex Ogaden in the southeast and areas to the west. **1916:** Haile Selassie became regent. **1930:** Haile Selassie became emperor. **1936:** Conquered by Italy and incorporated in Italian East Africa. **1941:** Return of Emperor Selassie after liberation by the British. **1952:** Ethiopia federated with Eritrea. **1962:** Eritrea annexed by Selassie; Eritrean People's Liberation front (EPLF) resistance movement began, a rebellion that was to continue for 30 years. **1963:** First conference of Selassie-promoted Organization of African Unity (OAU) held in Addis Ababa. **1973–74:** Severe famine in northern Ethiopia; 200,000 died in Wallo province. **1974:** Haile Selassie deposed and replaced by a military government. **1977:** Col Mengistu Haile Mariam took over the government. Somali forces ejected from the Somali-peopled Ogaden in the southeast. **1977–79:** 'Red Terror' period in which Mengistu's single-party Marxist regime killed thousands of people and promoted collective farming; Tigré People's Liberation Front guerrillas began fighting for regional autonomy in the northern highlands. **1984:** The Workers' Party of Ethiopia (WPE) was declared the only legal political party. **1985:** The worst famine in more than a decade; Western aid was sent and forcible internal resettlement programmes undertaken in Eritrea and Tigré in the north. **1987:** Mengistu Mariam was elected president under a new constitution. There was another famine; food aid was hindered by guerrillas. **1989:** Peace talks with Eritrean rebels were mediated by the former US president Jimmy Carter. **1991:** Mengistu was overthrown; a transitional government was set up by the opposing Ethiopian People's Revolutionary Democratic Front (EPRDF), headed by Meles Zenawi. The EPLF took control of Eritrea. Famine again gripped the country. **1993:** Eritrean independence was recognized after a referendum; private farming and market sector were encouraged by the EPRDF government. **1994:** A new federal constitution was adopted. **1995:** The ruling EPRDF won a majority in the first multiparty elections to an interim parliament. Negasso Ghidada was chosen as president; Zenawi was appointed premier. **1998:** There was a border dispute with Eritrea. **1999:** The border dispute with Eritrea erupted into war. Peace proposals were agreed by Eritrea but fighting continued. **2000:** A ceasefire with Eritrea was agreed in June and a peace agreement signed in December that provided for UN troops to keep the peace along the border. Haile Selassie was ceremoniously reburied in Addis Ababa.

Practical information

Visa requirements: UK: visa required. USA: visa required
Time difference: GMT +3
Chief tourist attractions: early Christian churches and monuments; ancient capitals of Gonder and Aksum; Blue Nile Falls; national parks of Semien and Bale Mountains
Major holidays: 7, 19 January, 2 March, 6 April, 1 May, 12, 27 September; variable: Eid-ul-Adha, end of Ramadan, Ethiopian New Year (September), Prophet's Birthday, Ethiopian Good Friday, and Easter

FIJI ISLANDS

844 islands and islets in the southwest Pacific Ocean.

National name: *Matanitu Ko Viti/Republic of the Fiji Islands*
Area: 18,333 sq km/7,078 sq mi
Capital: Suva
Major towns/cities: Lautoka, Nadi, Ba, Labasa, Nausori, Lami
Major ports: Lautoka, Levuka
Physical features: comprises about 844 Melanesian and Polynesian islands
and islets (about 100 inhabited), the largest being Viti Levu (10,429 sq
km/4,028 sq mi) and Vanua Levu (5,556 sq km/2,146 sq mi); mountainous,
volcanic, with tropical rainforest and grasslands; almost all islands surrounded by coral reefs; high volcanic peaks

The bright blue field stands for the Pacific Ocean. The shield is taken from the coat of arms.
Effective date: 10 October 1970.

Government

Head of state: Ratu Josefa Iloilo from 2000
Head of government: Laisenia Qarase from 2000
Political system: military
Political executive: military
Administrative divisions: 14 provinces
Political parties: National Federation Party (NFP),
moderate left of centre, Indian; Fijian Labour Party
(FLP), left of centre, Indian; United Front, Fijian;
Fijian Political Party (FPP), Fijian centrist
Armed forces: 3,500 (1998)
Conscription: military service is voluntary
Death penalty: laws provide for the death penalty
only for exceptional crimes such as crimes under
military law or crimes committed in exceptional
cricumstances such as wartime (last execution 1964)
Defence spend: (% GDP) 1.6 (1998)
Education spend: (% GNP) N/A
Health spend: (% GDP) 4.2 (1997)

Economy and resources

Currency: Fiji dollar
GDP: (US$) 1.7 billion (1999)
Real GDP growth: (% change on previous year) 7.8
(1999 est)
GNP: (US$) 1.8 billion (1999)
GNP per capita (PPP): (US$) 4,536 (1999)
Consumer price inflation: 1.7% (1999)
Unemployment: 6% (1997 est)
Foreign debt: (US$) 193 million (1998)
Major trading partners: Australia, New Zealand,
Japan, UK, USA, Singapore
Resources: gold, silver, copper
Industries: food processing (sugar, molasses, and
copra), ready-made garments, animal feed, cigarettes,
cement, tourism
Exports: sugar, gold, fish and fish products, clothing,
re-exported petroleum products, timber, ginger,
molasses. Principal market: Australia 29.4% (1998)
Imports: basic manufactured goods, machinery and
transport equipment, food, mineral fuels. Principal
source: Australia 40.1% (1998)
Arable land: 10.9% (1995)

Agricultural products: sugar cane, coconuts, ginger,
rice, tobacco, cocoa; forestry (for timber)

Population and society

Population: 817,000 (2000 est)
Population growth rate: 1.2% (1995–2000)
Population density: (per sq km) 44 (1999 est)
Urban population: (% of total) 49 (2000 est)
Age distribution: (% of total population) 0–14 31%,
15–59 61%, 60+ 8% (2000 est)
Ethnic groups: 51% Fijians (of Melanesian and
Polynesian descent), 44% Indian
Language: English (official), Fijian, Hindi
Religion: Methodist 37%, Hindu 38%, Muslim 8%,
Roman Catholic 8%, Sikh
Education: not compulsory
Literacy rate: 95% (men); 89% (women) (2000 est)
Labour force: 43.5% agriculture, 25.1% industry,
30.4% services (1996)
Life expectancy: 71 (men); 75 (women)
(1995–2000)
Child mortality rate: (under 5, per 1,000 live births)
23 (1995–2000)
Physicians: 1 per 1,785 people (1995)
Hospital beds: 1 per 432 people (1995)
TV sets: (per 1,000 people) 27 (1997)
Radios: (per 1,000 people) 636 (1997)
Internet users: (per 10,000 people) 93.0 (1999)

Transport

Airports: international airports: Nadi; 16 domestic
airports and airfields; total passenger km: 1,979
million (1997 est)
Railways: no passenger railway system; Fiji Sugar
Cane Corporation operates a 595-km/370-mi railway
Roads: total road network: 3,440 km/2,138 mi, of
which 49.2% paved (1996 est); passenger cars: 37.3
per 1,000 people (1996 est)

Chronology

***c*. 1500 BC:** Peopled by Polynesian and, later, by Melanesian settlers. **1643:** The islands were visited for the first time by a European, the Dutch navigator Abel Tasman. **1830s:** Arrival of Western Christian missionaries. **1840s–50s:** Western Fiji came under dominance of a Christian convert prince, Cakobau, ruler of Bau islet, who proclaimed himself Tui Viti (King of Fiji), while the east was controlled by Ma'afu, a Christian prince from Tonga. **1857:** British consul appointed, encouraging settlers from Australia and New Zealand to set up cotton farms in Fiji. **1874:** Fiji became a British crown colony after a deed of cession was signed by King Cakobau. **1875–76:** A third of the Fijian population were wiped out by a measles epidemic; a rebellion against the British was suppressed with the assitance of Fijian chiefs. **1877:** Fiji became the headquarters of the British Western Pacific High Commission (WPHC), which controlled other British protectorates in the Pacific region. **1879–1916:** Indian labourers brought in, on ten-year indentured contracts, to work sugar plantations. **1904:** Legislative Council formed, with elected Europeans and nominated Fijians, to advise the British governor. **1963:** Legislative Council enlarged; women and Fijians were enfranchised. The predominantly Fijian Alliance Party (AP) formed. **1970:** Independence was achieved from Britain; Ratu Sir Kamisese Mara of the AP was elected as the first prime minister. **1973:** Ratu Sir George Cakobau, the great-grandson of the chief who had sworn allegiance to the British in 1874, became governor general. **1985:** The FLP was formed by Timoci Bavadra, with trade-union backing. **1987:** After a general election had brought to power an Indian-dominated coalition led by Bavadra, Lt-Col Sitiveni Rabuka seized power in a military coup, and proclaimed a Fijian-dominated republic outside the Commonwealth. **1990:** A new constitution, favouring indigenous (Melanese) Fijians, was introduced. Civilian rule was re-established, with resignations from the cabinet of military officers, but Rabuka remained as home affairs minister, with Mara as prime minister. **1992:** A general election produced a coalition government with Rabuka of the FPP as prime minister. **1994:** Ratu Sir Kamisese Mara became president. **1997:** A nondiscriminatory constitution was introduced. Fiji was re-admitted to the Commonwealth. **1999:** President Mara's term in office was renewed for an additional five years. Mahendra Chaudhry became Fiji's first female prime minister and first prime minister of Indian descent. **2000:** A coup led by George Speight took cabinet members hostage and ended Mara's presidency. The head of Fiji's armed forces, Commodore Frank Bainimarama, announced that he was taking power, proclaimed martial law, and revoked the 1997 non-discriminatory constitution (the aim of Speight's coup). When the hostages were released in July, the military handed over executive power to the new president, Ratu Josefa Iloilo, and installed Laisenia Qarase as prime minister. Speight was arrested and charged with treason. The high court later ruled that Chaudhry's deposed government should be reinstated.

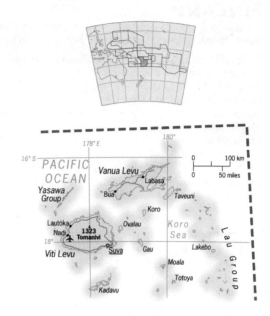

Practical information

Visa requirements: UK: visa not required. USA: visa not required
Time difference: GMT +12
Chief tourist attractions: climate; scenery on Viti Levu; watersports, including diving, white-water rafting, and surfing on Tavarua; fishing; museum in Suva

Major holidays: 1 January, 12 October, 25–26 December; variable: Diwali, Good Friday, Easter Monday, Holy Saturday, Prophet's Birthday, August Bank Holiday, Queen's Birthday (June), Prince Charles's Birthday (November)

FINLAND

Located in Scandinavia, bounded to the north by Norway, east by Russia, south and west by the Baltic Sea, and northwest by Sweden.

National name: *Suomen Tasavalta* (Finnish)/*Republiken Finland* (Swedish)/*Republic of Finland*
Area: 338,145 sq km/130,557 sq mi
Capital: Helsinki (Swedish Helsingfors)
Major towns/cities: Tampere, Turku, Espoo, Vantaa, Oulu
Major ports: Turku, Oulu
Physical features: most of the country is forest, with low hills and about 60,000 lakes; one-third is within the Arctic Circle; archipelago in south includes Åland Islands; Helsinki is the most northerly national capital on the European continent. At the 70th parallel there is constant daylight for 73 days in summer and 51 days of uninterrupted night in winter.

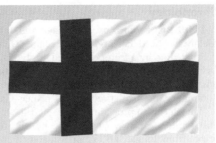

Blue represents Finland's 60,000 lakes. White stands for the snow which covers the ground for 5–7 months each year. Effective date: 1 June 1978.

Government

Head of state: Tarja Halonen from 2000
Head of government: Paavo Lipponen from 1995
Political system: liberal democracy
Political executive: dual executive
Administrative divisions: 12 provinces
Political parties: Finnish Social Democratic Party (SSDP), moderate left of centre; National Coalition Party (KOK), moderate right of centre; Finnish Centre Party (KESK), radical centrist, rural-oriented; Swedish People's Party (SFP), independent Swedish-oriented; Finnish Rural Party (SMP), farmers and small businesses; Left-Wing Alliance (VL), left wing; Finnish Christian League (SKL), centre-right
Armed forces: 31,700 (1998)
Conscription: up to 11 months, followed by refresher training of 40–100 days (before age 50)
Death penalty: abolished in 1972
Defence spend: (% GDP) 1.5 (1998)
Education spend: (% GNP) 7.5 (1997)
Health spend: (% GDP) 7.4 (1997)

Economy and resources

Currency: markka
GDP: (US$) 126.1 billion (1999)
Real GDP growth: (% change on previous year) 5 (1998)
GNP: (US$) 122.9 billion (1999)
GNP per capita (PPP): (US$) 21,209 (1999)
Consumer price inflation: 1.2% (1999)
Unemployment: 11.4% (1998)
Major trading partners: Germany, Sweden, UK, USA, Russia, Denmark, Norway, the Netherlands
Resources: copper ore, lead ore, gold, zinc ore, silver, peat, hydro power, forests
Industries: food processing, paper and paper products, machinery, printing and publishing, wood products, metal products, shipbuilding, chemicals, clothing and footwear
Exports: metal and engineering products, gold, paper and paper products, machinery, ships, wood and pulp, clothing and footwear, chemicals. Principal market: Germany 11.8% (1998)
Imports: mineral fuels, machinery and transport equipment, food and live animals, chemical and related products, textiles, iron and steel. Principal source: Germany 15.2% (1998)
Arable land: 8.1% (1996)
Agricultural products: oats, sugar beet, potatoes, barley, hay; forestry and animal husbandry

Population and society

Population: 5,176,000 (2000 est)
Population growth rate: 0.3% (1995–2000); 0.3% (2000–05)
Population density: (per sq km) 15 (1999 est)
Urban population: (% of total) 67 (2000 est)
Age distribution: (% of total population) 0–14 18%, 15–59 62%, 60+ 20% (2000 est)
Ethnic groups: predominantly Finnish; significant Swedish minority (6% of population); small minorities of native Saami (Lapp) and Russians
Language: Finnish (93%), Swedish (6%) (both official), Saami (Lapp), Russian
Religion: Evangelical Lutheran 87%, Greek Orthodox 1%
Education: (compulsory years) 9
Literacy rate: 99% (men); 99% (women) (2000 est)
Labour force: 49.4% of population: 7.1% agriculture, 27.5% industry, 65.4% services (1997)
Life expectancy: 73 (men); 81 (women) (1995–2000)
Child mortality rate: (under 5, per 1,000 live births) 6 (1995–2000)
Physicians: 1 per 345 people (1996)
Hospital beds: 1 per 107 people (1996)
TV sets: (per 1,000 people) 640 (1998)
Radios: (per 1,000 people) 1,496 (1997)
Internet users: (per 10,000 people) 3,227.4 (1999)
Personal computer users: (per 100 people) 36.01 (1999)

Transport

Airports: international airports: Helsinki (Vantaa); 20 domestic airports; total passenger km: 11,924 million (1997) **Railways:** total length: 5,859 km/3,641 mi; total

passenger km: 3,376 million (1997) **Roads:** total road network: 77,796 km/48,342 mi, of which 64% paved (1997); passenger cars: 378 per 1,000 people (1997)

Chronology

1st century: Occupied by Finnic nomads from Asia who drove out native Saami (Lapps) to the far north. **12th–13th centuries:** A series of Swedish crusades conquered Finns and converted them to Christianity. **16th–17th centuries:** Finland was a semi-autonomous Swedish duchy with Swedish landowners ruling Finnish peasants; Finland was allowed relative autonomy, becoming a grand duchy in 1581. **1634:** Finland fully incorporated into Swedish kingdom. **1700–21:** Great Northern War between Sweden and Russia; half of Finnish population died in famine and epidemics. **1741–43 and 1788–90:** Further Russo-Swedish wars; much of the fighting took place in Finland. **1808:** Russia invaded Sweden (with support of Napoleon). **1809:** Finland ceded to Russia as grand duchy with Russian tsar as grand duke; Finns retained their own legal system and Lutheran religion and were exempt from Russian military service. **1812:** Helsinki became capital of grand duchy. **19th century:** Growing prosperity was followed by rise of national feeling among new Finnish middle class. **1904–05:** Policies promoting Russification of Finland provoked a national uprising; Russians imposed military rule. **1917:** Finland declared independence. **1918:** Bitter civil war between Reds (supported by Russian Bolsheviks) and Whites (supported by Germany); Baron Carl Gustaf Mannerheim led the Whites to victory. **1919:** Republican constitution adopted with Kaarlo Juho Ståhlberg as first president. **1927:** Land reform broke up big estates and created many small peasant farms. **1939–40:** Winter War: USSR invaded Finland after a demand for military bases was refused. **1940:** Treaty of Moscow: Finland ceded territory to USSR. **1941:** Finland joined the German attack on USSR in the hope of regaining lost territory. **1944:** Finland agreed separate armistice with USSR; German troops withdrawn. **1947:** Finno-Soviet peace treaty: Finland forced to cede 12% of its total area and to pay $300 million in reparations. **1948:** Finno-Soviet Pact of Friendship, Cooperation, and Mutual Assistance (YYA treaty): Finland pledged to repel any attack on USSR through its territories. **1950s:** Unstable centre-left coalitions excluded communists from government and adopted strict neutrality in foreign affairs. **1955:** Finland joined the United Nations (UN) and the Nordic Council. **1956:** There was a general strike as a result of unemployment and inflation. **1973:** Trade agreements were signed with the European Economic Community (EEC) and Comecon. **1991:** There was a swing towards the Centre Party in a general election. **1995:** Finland joined the European Union (EU); the Social Democrats won a general election, with Paavo Lipponen becoming prime minister. **2000:** Tarja Halonen was elected president. A former foreign minister, she became the first woman president of Finland.

Practical information

Visa requirements: UK: visa not required. USA: visa not required
Time difference: GMT +2
Chief tourist attractions: scenery: forests, lakes (Europe's largest inland water system)

Major holidays: 1 January, 1 May, 31 October, 1 November, 1, 24–26, 31 December; variable: Ascension Thursday, Good Friday, Easter Monday, Midsummer Eve and Day (June), Twelfthtide (January), Whitsuntide (May/June)

FRANCE

Located in western Europe, bounded to the northeast by Belgium, Luxembourg, and Germany, east by Germany, Switzerland, and Italy, south by the Mediterranean Sea, southwest by Spain and Andorra, and west by the Atlantic Ocean.

Red and blue were taken from the arms of Paris. White was the colour of the Bourbon dynasty.
Effective date: 5 March 1848.

National name: *République Française/French Republic*
Area: (including Corsica) 543,965 sq km/210,024 sq mi
Capital: Paris
Major towns/cities: Lyon, Lille, Bordeaux, Toulouse, Nantes, Marseille, Nice, Strasbourg, Montpellier, Rennes, Le Havre
Major ports: Marseille, Nice, Le Havre
Physical features: rivers Seine, Loire, Garonne, Rhône; mountain ranges Alps, Massif Central, Pyrenees, Jura, Vosges, Cévennes; Auvergne mountain region; Mont Blanc (4,810 m/15,781 ft); Ardennes forest; Riviera; caves of Dordogne with relics of early humans; the island of Corsica
Territories: Guadeloupe, French Guiana, Martinique, Réunion, St Pierre and Miquelon, Southern and Antarctic Territories, New Caledonia, French Polynesia, Wallis and Futuna, Mayotte, Bassas da India, Clipperton Island, Europa Island, Glorioso Islands, Juan de Nova Island, Tromelin Island

Government

Head of state: Jacques Chirac from 1995
Head of government: Lionel Jospin from 1997
Political system: liberal democracy
Political executive: dual executive
Administrative divisions: 22 regions containing 96 departments, four overseas departments, two territorial collectivities, and four overseas territories
Political parties: Rally for the Republic (RPR), neo-Gaullist conservative; Union for French Democracy (UDF), centre right; Socialist Party (PS), left of centre; Left Radical Movement (MRG), left of centre; French Communist Party (PCF), Marxist-Leninist; National Front, far right; Greens, fundamentalist-ecologist; Génération Ecologie, pragmatic ecologist; Movement for France, right wing, anti-Maastricht
Armed forces: 358,800; paramilitary gendarmerie 93,400 (1998)
Conscription: military service is compulsory for 10 months
Death penalty: abolished in 1981
Defence spend: (% GDP) 2.8 (1998)
Education spend: (% GNP) 6.1 (1996)
Health spend: (% GDP) 9.6 (1997)

Economy and resources

Currency: franc
GDP: (US$) 1,410.3 billion (1999)
Real GDP growth: (% change on previous year) 2.9 (1999)
GNP: (US$) 1,427.2 billion (1999)
GNP per capita (PPP): (US$) 21,897 (1999)
Consumer price inflation: 0.6% (1999)
Unemployment: 11.9% (1998)
Major trading partners: EU (principally Germany, Italy, Belgium–Luxembourg, UK), USA

Resources: coal, petroleum, natural gas, iron ore, copper, zinc, bauxite
Industries: mining, quarrying, food products, transport equipment, non-electrical machinery, electrical machinery, weapons, metals and metal products, yarn and fabrics, wine, tourism, aircraft
Exports: machinery and transport equipment, food and live animals, chemicals, beverages and tobacco, textile yarn, fabrics and other basic manufactures, clothing and accessories, perfumery and cosmetics. Principal market: Germany 15.8% (1999)
Imports: food and live animals, mineral fuels, machinery and transport equipment, chemicals and chemical products, basic manufactures. Principal source: Germany 17.2% (1999)
Arable land: 33.2% (1996)
Agricultural products: wheat, sugar beet, maize, barley, vine fruits, potatoes, fruit, vegetables; livestock and dairy products

Population and society

Population: 59,080,000 (2000 est)
Population growth rate: 0.4% (1995–2000); 0.2% (2000–05)
Population density: (per sq km) 108 (1999 est)
Urban population: (% of total) 76 (2000 est)
Age distribution: (% of total population) 0–14 19%, 15–59 60%, 60+ 21% (2000 est)
Ethnic groups: predominantly French ethnic, of Celtic and Latin descent; Basque minority in southwest; 7% of the population are immigrants – a third of these are from Algeria and Morocco and live mainly in the Marseille Midi region and in northern cities, 20% originate from Portugal, and 10% each from Italy and Spain
Language: French (official; regional languages include Basque, Breton, Catalan, Corsican, and Provençal)
Religion: Roman Catholic, about 90%; also Muslim,

Protestant, and Jewish minorities
Education: (compulsory years) 10
Literacy rate: 99% (men); 99% (women) (2000 est)
Labour force: 44% of population: 5.1% agriculture,
27.7% industry, 67.2% services (1993)
Life expectancy: 74 (men); 82 (women) (1995–2000)
Child mortality rate: (under 5, per 1,000 live births) 8
(1995–2000)
Physicians: 1 per 345 people (1996)
Hospital beds: 1 per 112 people (1996)
TV sets: (per 1,000 people) 606 (1997)
Radios: (per 1,000 people) 937 (1997)
Internet users: (per 10,000 people) 961.2 (1999)
Personal computer users: (per 100 people) 22.1
(1999)

Transport

Airports: international airports: Paris (Orly, Roissy-
Charles de Gaulle, Le Bourget), Bordeaux (Merignac),
Lille (Lesquin), Lyon, Marseille, Nice, Strasbourg,
Toulouse (Blagnac); 45 domestic airports; total passenger
km: 80,153 million (1997 est)
Railways: total length: 34,322 km/21,328 mi; total
passenger km: 55,560 million (1995)
Roads: total road network: 892,900 km/554,848 mi, of
which 100% paved (1997); passenger cars: 442 per 1,000
people (1997)

Chronology

5th century BC: Celtic
peoples invaded the region.
58–51 BC: Romans
conquered Celts and formed
province of Gaul. **5th
century AD:** Gaul
overrun by Franks and
other Germanic tribes.
481–511: Frankish
chief Clovis accepted
Christianity and formed a
kingdom based at Paris; under
his successors, the Merovingian
dynasty, the kingdom
disintegrated. **751–68:** Pepin
the Short usurped the Frankish
throne, reunified the kingdom,
and founded the Carolingian
dynasty. **768–814:**
Charlemagne conquered much of
Western Europe and created the
Holy Roman Empire. **843:**
Treaty of Verdun divided the Holy
Roman Empire into three, with the
western portion corresponding to modern
France. **9th–10th centuries:** Weak
central government allowed the great nobles to
become virtually independent. **987:** Frankish crown
passed to House of Capet; the Capets ruled the district around
Paris, but were surrounded by vassals more powerful than themselves. **1180–1223:** Philip II doubled
the royal domain and tightened control over the nobles; the power of the Capets gradually extended with
support of church and towns. **1328:** When Charles IV died without an heir, Philip VI established the House of
Valois. **1337:** Start of the Hundred Years' War: Edward III of England disputed the Valois succession and claimed the
throne. English won victories at Crécy in 1346 and Agincourt in 1415. **1429:** Joan of Arc raised the siege of Orléans;
Hundred Years' War ended with Charles VII expelling the English in 1453. **1483:** France annexed Burgundy and
Brittany after Louis XI had restored royal power. **16th–17th centuries:** French kings fought the Habsburgs (of
Holy Roman Empire and Spain) for supremacy in Western Europe. **1562–98:** Civil wars between nobles were fought
under religious slogans, Catholic versus Protestant (or Huguenot). **1589–1610:** Henry IV, first king of Bourbon
dynasty, established peace, religious tolerance, and absolute monarchy. **1634–48:** The ministers Richelieu and

Mazarin, by intervening in the Thirty Years' War, secured Alsace and made France the leading power in Europe. **1701–14:** War of the Spanish Succession: England, Austria, and allies checked expansionism of France under Louis XIV. **1756–63:** Seven Years' War: France lost most of its colonies in India and Canada to Britain. **1789:** French Revolution abolished absolute monarchy and feudalism; First Republic proclaimed and revolutionary wars began in 1792. **1799:** Napoleon Bonaparte seized power in coup; crowned himself emperor in 1804; France conquered much of Europe. **1814:** Defeat of France; restoration of Bourbon monarchy; comeback by Napoleon defeated at Waterloo in 1815. **1830:** Liberal revolution deposed Charles X in favour of his cousin Louis Philippe, the 'Citizen King'. **1848:** Revolution established Second Republic; conflict between liberals and socialists; Louis Napoleon, nephew of Napoleon I, elected president. **1852:** Louis Napoleon proclaimed Second Empire, taking title Napoleon III. **1870–71:** Franco-Prussian War: France lost Alsace-Lorraine; Second Empire abolished; Paris Commune crushed; Third Republic founded. **late 19th century:** France colonized Indo-China, much of North Africa, and South Pacific. **1914–18:** France resisted German invasion in World War I; Alsace-Lorraine recovered in 1919. **1936–37:** Left-wing 'Popular Front' government introduced many social reforms. **1939:** France entered World War II. **1940:** Germany invaded and occupied northern France; Marshal Pétain formed right-wing puppet regime at Vichy; resistance maintained by Maquis and Free French; Germans occupied all France in 1942. **1944:** Allies liberated France; provisional government formed by Gen Charles de Gaulle, leader of Free French. **1946:** Fourth Republic proclaimed. **1949:** Became a member of NATO; withdrew from military command structure in 1966. **1954:** French withdrew from Indo-China after eight years of war; start of guerrilla war against French rule in Algeria. **1957:** France was a founder member of the European Economic Community. **1958:** Algerian crisis caused collapse of Fourth Republic; de Gaulle took power, becoming president of the Fifth Republic in 1959. **1962:** Algeria achieved independence. **1968:** Revolutionary students rioted in Paris; there was a general strike throughout France. **1981:** François Mitterrand was elected the Fifth Republic's first socialist president. **1995:** Jacques Chirac (RPR) was elected president. There was widespread condemnation of the government's decision to resume nuclear tests in the South Pacific, and this was stopped in 1996 . **1996:** Spending cuts were agreed to meet European Monetary Union entry criteria. Unemployment was at a post-war high. **1997:** A general election was called by President Chirac, with victory for Socialists; Lionel Jospin (PS) was appointed prime minister. **1998:** There were protests by the unemployed. **1999:** Two-thirds of France was declared a disaster zone after powerful storms struck Europe and caused widespread damage. **2000:** An Air France Concorde aircraft crash killed 113 people. Disruptive protests over fuel prices forced the government to make tax concessions on fuel. Fears over BSE in French cattle rose and President Chirac called for an immediate ban on cattle remains in all French animal feed.

Practical information

Visa requirements: UK: visa not required. USA: visa not required

Time difference: GMT +1

Chief tourist attractions: Paris, with its boulevards, historic buildings, art treasures, theatres, restaurants, and night clubs; resorts on Mediterranean and Atlantic coasts; many ancient towns; châteaux of the Loire valley; theme parks (Futuroscope and EuroDisney)

Major holidays: 1 January, 1, 8 May, 14 July, 14, 15 August, 31 October, 1, 11 November, 24–25, 31 December; variable: Ascension Eve, Ascension Thusday, Good Friday, Easter Monday, Holy Saturday, Whit Holiday Eve, Whit Monday, Law of 20 December 1906, Law of 23 December 1904

GABON

Located in central Africa, bounded north by Cameroon, east and south by
the Congo, west by the Atlantic Ocean, and northwest by Equatorial Guinea.

National name: *République Gabonaise/Gabonese Republic*
Area: 267,667 sq km/103,346 sq mi
Capital: Libreville
Major towns/cities: Port-Gentil, Franceville (or Masuku), Lambaréné,
Mouanda, Oyem, Mouila
Major ports: Port-Gentil and Owendo
Physical features: virtually the whole country is tropical rainforest;
narrow coastal plain rising to hilly interior with savannah in east and
south; Ogooué River flows north–west

The flag's unusual 3:4 proportions are laid down by
law. Green represents the forests. Blue symbolizes the
Atlantic Ocean. Effective date: 9 August 1960.

Government

Head of state: Omar Bongo from 1967
Head of government: Jean-François Ntoutoume-Emane
from 1999
Political system: emergent democracy
Political executive: limited presidency
Administrative divisions: nine provinces
Political parties: Gabonese Democratic Party (PDG),
nationalist; Gabone Progress Party (PGP), left of centre;
National Lumberjacks Rally (RNB), left of centre
Armed forces: 4,700; plus a paramilitary force of 4,800
(1998)
Conscription: military service is voluntary
Death penalty: retained and used for ordinary crimes
Defence spend: (% GDP) 2.2 (1998)
Education spend: (% GNP) 2.8 (1996)
Health spend: (% GDP) 3 (1997)

Economy and resources

Currency: franc CFA
GDP: (US$) 4.6 billion (1999 est)
Real GDP growth: (% change on previous year) 1.8
(1999 est)
GNP: (US$) 4.04 billion (1999)
GNP per capita (PPP): (US$) 5,325 (1999)
Consumer price inflation: 0.7% (1999 est)
Unemployment: 20% (1996 est)
Foreign debt: (US$) 4.2 billion (1999 est)
Major trading partners: France, USA, Japan, the
Netherlands, Cameroon, China
Resources: petroleum, natural gas, manganese (one of
world's foremost producers and exporters), iron ore,
uranium, gold, niobium, talc, phosphates
Industries: mining, food processing (particularly sugar),
petroleum refining, processing of other minerals, timber
preparation, chemicals
Exports: petroleum and petroleum products, manganese,
timber and wood products, uranium. Principal market:
USA 56% (1998)
Imports: machinery and apparatus, transport
equipment, food products, metals and metal products.
Principal source: France 39% (1998)

Arable land: 1.3% (1996)
Agricultural products: cassava, sugar cane, cocoa,
coffee, plantains, maize, groundnuts, bananas, palm oil;
forestry (forests cover approximately 75% of the land)

Population and society

Population: 1,226,000 (2000 est)
Population growth rate: 2.6% (1995–2000)
Population density: (per sq km) 4 (1999 est)
Urban population: (% of total) 81 (2000 est)
Age distribution: (% of total population) 0–14 40%,
15–59 51%, 60+ 9% (2000 est)
Ethnic groups: 40 Bantu peoples in four main groupings:
the Fang, Eshira, Mbede, and Okande; there are also
Pygmies and about 10% Europeans (mainly French)
Language: French (official), Fang (in the north), Bantu
languages, and other local dialects
Religion: Christian 60% (mostly Roman Catholic),
animist about 4%, Muslim 1%
Education: (compulsory years) 10
Literacy rate: 80% (men); 62% (women) (2000 est)
Labour force: 64.2% agriculture, 10.8% industry, 25%
services (1994)
Life expectancy: 51 (men); 54 (women) (1995–2000)
Child mortality rate: (under 5, per 1,000 live births)
135 (1995–2000)
Physicians: 1 per 1,987 people (1993 est)
Hospital beds: 1 per 305 people (1993 est)
TV sets: (per 1,000 people) 55 (1997)
Radios: (per 1,000 people) 183 (1997)
Internet users: (per 10,000 people) 41.8 (1999)
Personal computer users: (per 100 people) 1.0 (1999)

Transport

Airports: international airports: Port-Gentil, Masuku,
Libreville; 65 public domestic-services airfields; total
passenger km: 764 million (1997 est)
Railways: total length: 668 km/415 mi; total passenger
journeys: 180,000 (1994)
Roads: total road network: 7,670 km/4,766 mi, of which
8.2% paved (1996 est); passenger cars: 17.3 per 1,000
people (1996 est)

Chronology

12th century: Immigration of Bantu speakers into an area previously peopled by Pygmies. **1472:** Gabon Estuary first visited by Portuguese navigators, who named it Gabao ('hooded cloak'), after the shape of the coastal area. **17th–18th centuries:** Fang, from Cameroon in the north, and Omiene peoples colonized the area, attracted by the presence in coastal areas of European traders, who developed the ivory and slave trades, which lasted until the mid-19th century. **1839–42:** Mpongwe coastal chiefs agreed to transfer sovereignty to France; Catholic and Protestant missionaries attracted to the area. **1849:** Libreville ('Free Town') formed by slaves from a slave ship liberated by the French. **1889:** Became part of French Congo, with Congo. **1910:** Became part of French Equatorial Africa, which also comprised Congo, Chad, and Central African Republic. **1890s–1920s:** Human and natural resources exploited by private concessionary companies. **1940–44:** Supported the 'Free French' anti-Nazi cause during World War II. **1946:** Became overseas territory within the French Community, with its own assembly. **1960:** Independence achieved; Léon M'ba, a Fang of the pro-French Gabonese Democratic Block (BDG), became the first president. **1967:** M'ba died and was succeeded by his protégé Albert Bernard Bongo, drawn from the Teke community. **1968:** A One-party state established, with the BDG dissolved and replaced by Gabonese Democratic Party (PDG). **1973:** Bongo converted to Islam and changed his first name to Omar, but continued to follow a pro-Western policy course and exploit rich mineral resources to increase prosperity. **1989:** A coup attempt against Bongo was defeated; the economy deteriorated. **1990:** The PDG won the first multiparty elections since 1964. French troops were sent in to maintain order following antigovernment riots. **1993:** A national unity government was formed, including some opposition members. **1998:** A new party, Rassemblement des Gaullois, was recognized. President Bongo was re-elected. **1999:** Jean-François Ntoutoume-Emane was appointed prime minister

Practical information

Visa requirements: UK: visa required. USA: visa required
Time difference: GMT +1
Chief tourist attractions: national parks; tropical vegetation
Major holidays: 1 January, 12 March, 1 May, 17 August, 1 November, 25 December; variable: Eid-ul-Adha, Easter Monday, end of Ramadan, Whit Monday

GAMBIA, THE

Located in west Africa, bounded north, east, and south by Senegal and west by the Atlantic Ocean.

National name: *Republic of the Gambia*
Area: 10,402 sq km/4,016 sq mi
Capital: Banjul
Major towns/cities: Serekunda, Brikama, Bakau, Farafenni, Sukuta, Gunjur, Basse
Physical features: consists of narrow strip of land along the River Gambia; river flanked by low hills

Red represents the sun. Blue stands for the Gambia river. Green symbolizes agriculture.
Effective date: 18 February 1965.

Government

Head of state and government: Yahya Jammeh from 1994
Political system: transitional
Political executive: transitional
Administrative divisions: five divisions and one city (Banjul)
Political parties: Alliance for Patriotic Reorientation and Construction (APRC), authoritarian, anti-democratic; National Reconciliation Party (NRP), reformist, pro-democratic; People's Democratic Organization for Independence and Socialism (PDOIS), socialist; United Democratic Party (UDP), reformist. The Progressive People's Party (PPP), moderate centrist; National Convention Party (NCP), left of centre; and the Gambian People's Party (GPP) were all banned in 1996
Armed forces: 800 (1998)
Conscription: military service is mainly voluntary
Death penalty: retains the death penalty for ordinary crimes but can be considered abolitionist in practice (last execution 1981)
Defence spend: (% GDP) 3.6 (1998)
Education spend: (% GNP) 4.9 (1997 est)
Health spend: (% GDP) 4.5 (1997)

Economy and resources

Currency: dalasi
GDP: (US$) 437 million (1999)
Real GDP growth: (% change on previous year) 5.6% (1999 est)
GNP: (US$) 430 million (1999)
GNP per capita (PPP): (US$) 1,492 (1999)
Consumer price inflation: 3.8% (1999)
Unemployment: 26% (1995 est)
Foreign debt: (US$) 421 million (1999 est)
Major trading partners: UK, Belgium, Luxembourg, Italy, Hong Kong, China, Japan
Resources: ilmenite, zircon, rutile, petroleum (discovered, but not exploited)
Industries: food processing (fish, fish products, and vegetable oils), beverages, construction materials
Exports: groundnuts and related products, cotton lint, fish and fish preparations, hides and skins. Principal market: Belgium–Luxembourg 58.7% (1999 est)
Imports: food and live animals, basic manufactures, machinery and transport equipment, mineral fuels and lubrications, miscellaneous manufactured articles, chemicals. Principal source: China (including Hong Kong) 49.1% (1999 est)
Arable land: 17.5% (1996)
Agricultural products: groundnuts, cotton, rice, citrus fruits, avocados, sesame seed, millet, sorghum, maize; livestock rearing (cattle); fishing

Population and society

Population: 1,305,000 (2000 est)
Population growth rate: 3.2% (1995–2000)
Population density: (per sq km) 122 (1999 est)
Urban population: (% of total) 33 (2000 est)
Age distribution: (% of total population) 0–14 40%, 15–59 55%, 60+ 5% (2000 est)
Ethnic groups: wide mix of ethnic groups, the largest is the Madinka (about 40%); other main groups are the Fula (13.5%), Wolof (13%), Jola (7%), and Serahuli (7%)
Language: English (official), Mandinka, Fula, Wolof, other indigenous dialects
Religion: Muslim 85%, with animist and Christian minorities
Education: free, but not compulsory
Literacy rate: 44% (men); 29% (women) (2000 est)
Labour force: 50% of population: 79.6% agriculture, 4.2% industry, 16.2% services (1994)
Life expectancy: 45 (men); 49 (women) (1995–2000)
Child mortality rate: (under 5, per 1,000 live births) 203 (1995–2000)
Physicians: 1 per 14,530 people (1991)
Hospital beds: 1 per 1,642 people (1993 est)
TV sets: (per 1,000 people) 4 (1997)
Radios: (per 1,000 people) 165 (1997)
Internet users: (per 10,000 people) 31.6 (1999)
Personal computer users: (per 100 people) 0.4 (1999)

Transport

Airports: international airport: Banjul (Yundum); total passenger km: 50 million (1994)
Railways: none

Roads: total road network: 2,700 km/1,678 mi, of which 35.4% paved (1996 est); passenger cars: 8.2 per 1,000 people (1996 est)

Chronology

13th century: Wolof, Malinke (Mandingo), and Fulani tribes settled in the region from east and north. **14th century:** Became part of the great Muslim Mali Empire, which, centred to northeast, also extended across Senegal, Mali, and southern Mauritania. **1455:** The Gambia River was first sighted by the Portuguese. **1663 and 1681:** The British and French established small settlements on the river at Fort James and Albreda. **1843:** The Gambia became a British crown colony, administered with Sierra Leone until 1888. **1965:** Independence was achieved as a constitutional monarchy within the Commonwealth, with Dawda K Jawara of the People's Progressive Party (PPP) as prime minister at the head of a multiparty democracy. **1970:** The Gambia became a republic, with Jawara as president. **1982:** The Gambia formed the Confederation of Senegambia with Senegal, which involved the integration of military forces, economic and monetary union, and coordinated foreign policy. **1994:** Jawara was ousted in a military coup, and fled to Senegal; Yahya Jammeh was named acting head of state. **1996:** A civilian constitution was adopted.

Practical information

Visa requirements: UK: visa not required for visits of up to 90 days. USA: visa required
Time difference: GMT +/–0
Chief tourist attractions: beaches; coastal resorts; birdlife

Major holidays: 1 January, 1, 18 February, 1 May, 15 August, 25 December; variable: Eid-ul-Adha, Ashora, end of Ramadan (2 days), Good Friday, Prophet's Birthday

GEORGIA

Located in the Caucasus of southeastern Europe, bounded north by Russia, east by Azerbaijan, south by Armenia, and west by the Black Sea.

National name: *Sak'art'velo/Georgia*
Area: 69,700 sq km/26,911 sq mi
Capital: Tbilisi
Major towns/cities: Kutaisi, Rustavi, Batumi, Zugdidi, Gori
Physical features: largely mountainous with a variety of landscape from the subtropical Black Sea shores to the ice and snow of the crest line of the Caucasus; chief rivers are Kura and Rioni

Black recalls the country's tragic past. White reflects the Georgians' hopes for the future. Dark red is the national colour and is said to represent happiness.
Effective date: 14 November 1990.

Government

Head of state: Eduard Shevardnadze from 1992
Head of government: Giorgi Arsenishvili from 2000
Political system: emergent democracy
Political executive: limited presidency
Administrative divisions: 53 regions, nine cities and two autonomous republics
Political parties: Citizens' Union of Georgia (CUG), nationalist, pro-Shevardnadze; National Democratic Party of Georgia (NDPG), nationalist; Round Table/Free Georgia Bloc, nationalist; Georgian Popular Front (GPF), moderate nationalist, prodemocratization; Georgian Communist Party (GCP); National Independence Party (NIP), ultranationalist; Front for the Reinstatement of Legitimate Power in Georgia, strong nationalist
Armed forces: 33,200 (1998)
Conscription: compulsory for two years
Death penalty: abolished in 1997
Defence spend: (% GDP) 2.5 (1998)
Education spend: (% GNP) 5.2 (1996)
Health spend: (% GDP) 4.4 (1997)

Economy and resources

Currency: lari
GDP: (US$) 4.2 billion (1999)
Real GDP growth: (% change on previous year) 3 (1999)
GNP: (US$) 3.4 billion (1999)
GNP per capita (PPP): (US$) 3,606 (1999)
Consumer price inflation: 19.2% (1999)
Unemployment: 2.8% (1996)
Foreign debt: (US$) 1.66 billion (1999 est)
Major trading partners: Russia, Turkey, Azerbaijan, USA, Armenia, Germany, EU
Resources: coal, manganese, barytes, clay, petroleum and natural gas deposits, iron and other ores, gold, agate, marble, alabaster, arsenic, tungsten, mercury
Industries: metalworking, light industrial goods, motor cars, food processing, textiles (including silk), chemicals, construction materials
Exports: metal products, machinery, tea, beverages, food and tobacco products. Principal market: Russia 18.7% (1999)

Imports: mineral fuels, chemical and petroleum products, food products (mainly wheat and flour), light industrial products, beverages. Principal source: Russia 19.2% (1999)
Arable land: 11.1% (1996)
Agricultural products: grain, tea, citrus fruits, wine grapes, flowers, tobacco, almonds, sugar beet; sheep and goat farming; forest resources

Population and society

Population: 4,968,000 (2000 est)
Population growth rate: –1.1% (1995–2000)
Population density: (per sq km) 72 (1999 est)
Urban population: (% of total) 61 (2000 est)
Age distribution: (% of total population) 0–14 22%, 15–59 60%, 60+ 18% (2000 est)
Ethnic groups: 70% ethnic Georgian, 8% Armenian, 7% ethnic Russian, 6% Azeri, 3% Ossetian, 2% Abkhazian, and 2% Greek
Language: Georgian (official), Russian, Abkazian, Armenian, Azeri
Religion: Georgian Orthodox, also Muslim
Education: (compulsory years) 9
Literacy rate: 99% (men); 99% (women) (2000 est)
Labour force: 29.8% agriculture, 20.5% industry, 49.7% services (1995)
Life expectancy: 69 (men); 77 (women) (1995–2000)
Child mortality rate: (under 5, per 1,000 live births) 23 (1995–2000)
Physicians: 1 per 200 people (1994)
Hospital beds: 1 per 105 people (1994)
TV sets: (per 1,000 people) 473 (1997)
Radios: (per 1,000 people) 555 (1997)
Internet users: (per 10,000 people) 36.7 (1999)

Transport

Airports: international airports: Tbilisi; total passenger km: 308 million (1995)
Railways: total length: 1,583 km/984 mi; total passenger km: 1,003 million (1993)
Roads: total road network: 21,700 km/13,484 mi, of which 93.5% paved (1996 est); passenger cars: 79.7 per 1,000 people (1996 est)

Chronology

4th century BC: Georgian kingdom founded. **1st century BC:** Part of Roman Empire. **AD 337:** Christianity adopted. **458:** Tbilisi founded by King Vakhtang Gorgasal. **mid-7th century:** Tbilisi brought under Arab rule and renamed Tiflis. **1121:** Tbilisi liberated by King David II the Builder, of the Gagrationi dynasty. An empire was established across the Caucasus region, remaining powerful until Mongol onslaughts in the 13th and 14th centuries. **1555:** Western Georgia fell to Turkey and Eastern Georgia to Persia (Iran). **1783:** Treaty of Georgievsk established Russian dominance over Georgia. **1804–13:** First Russo-Iranian war fought largely over Georgia. **late 19th century:** Abolition of serfdom and beginnings of industrialization, but Georgian church suppressed. **1918:** Independence established after Russian Revolution. **1921:** Invaded by Red Army; Soviet republic established. **1922–36:** Linked with Armenia and Azerbaijan as the Transcaucasian Federation. **1930s:** Rapid industrial development, but resistance to agricultural collectivization and violent political purges instituted by the Georgian Soviet dictator Joseph Stalin. **1936:** Became separate republic within the USSR. **early 1940s:** 200,000 Meskhetians deported from southern Georgia to Central Asia on Stalin's orders. **1972:** Drive against endemic corruption launched by new Georgian Communist Party (GCP) leader Eduard Shevardnadze. **1978:** Violent demonstrations by nationalists in Tbilisi. **1981–88:** Increasing demands for autonomy were encouraged from 1986 by the *glasnost* initiative of the reformist Soviet leader Mikhail Gorbachev. **1989:** The formation of the nationalist Georgian Popular Front led the minority Abkhazian and Ossetian communities in northwest and central-north Georgia to demand secession, provoking interethnic clashes. A state of emergency was imposed in Abkhazia; 20 pro-independence demonstrators were killed in Tbilisi by Soviet troops; Georgian sovereignty was declared by parliament. **1990:** A nationalist coalition triumphed in elections and Gamsakhurdia became president. The GCP seceded from the Communist Party of the USSR. **1991:** Independence was declared. The GCP was outlawed and all relations with the USSR severed. Demonstrations were held against the increasingly dictatorial Gamsakhurdia; a state of emergency was declared. **1992:** Gamsakhurdia fled to Armenia; Shevardnadze, with military backing, was appointed interim president. Georgia was admitted into the United Nations (UN). Clashes continued in South Ossetia and Abkhazia, where independence had been declared. **1993:** The conflict with Abkhazi separatists intensified, forcing Shevardnadze to seek Russian military help. Otar Patsatsia was appointed prime minister. **1994:** Georgia joined the Commonwealth of Independent States (CIS). A military cooperation pact was signed with Russia. A ceasefire was agreed with the Abkhazi separatists; 2,500 Russian peacekeeping troops were deployed in the region and paramilitary groups disarmed. Inflation exceeded 5,000% per annum. **1996:** A cooperation pact with the European Union (EU) was signed as economic growth resumed. Elections to the secessionist Abkhazi parliament were declared illegal by the Georgian government. **1997:** A new opposition party, Front for the Reinstatement of Legitimate Power in Georgia, was formed. There were talks between the government and the breakaway Abkhazi government. **1998:** There was another outbreak of fighting in Abkhazia. **2000:** President Shevardnadze won a second term as president in elections. Giorgi Arsenishvili became secretary of state (prime minister). The government signed a pact with the Abkhazian prime minister, Vyacheslav Tsugba, both sides repudiating the use of force to settle the conflict.

Practical information

Visa requirements: UK: visa required. USA: visa required
Time difference: GMT +3
Chief tourist attractions: mountain scenery; health spas with mineral waters; waterfalls and caves
Major holidays: 1, 19 January, 3, 26 May, 28 August, 14 October, 23 November; variable: Orthodox Christmas (January), Orthodox Easter (March/April)

GERMANY

Located in central Europe, bounded north by the North and Baltic Seas and Denmark, east by Poland and the Czech Republic, south by Austria and Switzerland, and west by France, Luxembourg, Belgium, and the Netherlands.

National name: *Bundesrepublik Deutschland/Federal Republic of Germany*
Area: 357,041 sq km/137,853 sq mi
Capital: Berlin
Major towns/cities: Cologne, Hamburg, Munich, Essen, Frankfurt am Main, Dortmund, Stuttgart, Düsseldorf, Leipzig, Dresden, Hannover
Major ports: Hamburg, Kiel, Bremerhaven, Rostock
Physical features: flat in north, mountainous in south with Alps; rivers Rhine, Weser, Elbe flow north, Danube flows southeast, Oder and Neisse flow north along Polish frontier; many lakes, including Müritz; Black Forest, Harz Mountains, Erzgebirge (Ore Mountains), Bavarian Alps, Fichtelgebirge, Thüringer Forest

Black and red recall the tunics worn by soldiers during the Napoleonic wars. Gold was added to create a flag similar to the French tricolour, a symbol of revolution. Effective date: 23 May 1949.

Government

Head of state: Johannes Rau from 1999
Head of government: Gerhard Schroeder from 1998
Political system: liberal democracy
Political executive: parliamentary
Administrative divisions: 16 states
Political parties: Christian Democratic Union (CDU), right of centre, 'social market'; Christian Social Union (CSU), right of centre; Social Democratic Party (SPD), left of centre; Free Democratic Party (FDP), liberal; Greens, environmentalist; Party of Democratic Socialism (PDS), reform-socialist (formerly Socialist Unity Party: SED); German People's Union (DVU), far-right; German Communist Party (DKP)
Armed forces: 333,500 (1998)
Conscription: 10 months
Death penalty: abolished in the Federal Republic of Germany in 1949 and in the German Democratic Republic in 1987
Defence spend: (% GDP) 1.5 (1998)
Education spend: (% GNP) 4.8 (1996)
Health spend: (% GDP) 10.7 (1997)

Economy and resources

Currency: Deutschmark
GDP: (US$) 2,081.2 billion (1999)
Real GDP growth: (% change on previous year) 1.4 (1999)
GNP: (US$) 2,079.2 billion (1999)
GNP per capita (PPP): (US$) 22,404 (1999)
Consumer price inflation: 0.6% (1999)
Unemployment: 9.4% (1998)
Major trading partners: EU (particularly France, the Netherlands, and Ireland), USA, Japan, Switzerland
Resources: lignite, hard coal, potash salts, crude oil, natural gas, iron ore, copper, timber, nickel, uranium
Industries: mining, road vehicles, chemical products, transport equipment, nonelectrical machinery, metals and metal products, electrical machinery, electronic goods, cement, food and beverages
Exports: road vehicles, electrical machinery, metals and metal products, textiles, chemicals. Principal market: France 11.4% (1999)
Imports: road vehicles, electrical machinery, food and live animals, clothing and accessories, crude petroleum and petroleum products. Principal source: France 10.3% (1999)
Arable land: 33.9% (1996)
Agricultural products: potatoes, sugar beet, barley, wheat, maize, rapeseed, vine fruits; livestock (cattle, pigs, and poultry) and fishing

Population and society

Population: 82,220,000 (2000 est)
Population growth rate: 0.14% (1995–2000)
Population density: (per sq km) 230 (1999 est)
Urban population: (% of total) 88 (2000 est)
Age distribution: (% of total population) 0–14 16%, 15–59 60%, 60+ 24% (2000 est)
Ethnic groups: predominantly Germanic (92%; notable Danish and Slavonic ethnic minorities in the north; significant population of foreigners (92%), numbering about 7.4 million (1998). The largest community is Turkish (2 million), followed by nationals of the former Yugoslavia (1.4 million)
Language: German (official)
Religion: Protestant (mainly Lutheran) 38%, Roman Catholic 34%
Education: (compulsory years) 12
Literacy rate: 99% (men); 99% (women) (2000 est)
Labour force: 48.2% of population: 3.3% agriculture, 37.5% industry, 59.1% services (1996)
Life expectancy: 74 (men); 80 (women) (1995–2000)
Child mortality rate: (under 5, per 1,000 live births) 6 (1995–2000)
Physicians: 1 per 294 people (1996)
Hospital beds: 1 per 104 people (1996)
TV sets: (per 1,000 people) 580 (1998)

Radios: (per 1,000 people) 948 (1997)
Internet users: (per 10,000 people) 1,934.8 (1999)
Personal computer users: (per 100 people) 29.7 (1999)

Transport

Airports: international airports: Berlin-Tegel (Otto Lilienthal), Berlin-Schönefeld, Berlin-Tempelhof, Leipzig/Halle, Dresden (Klotsche), Bremen (Neuenland), Cologne, Düsseldorf (Lohausen), Frankfurt, Hamburg, Hannover (Langenhagen), Munich (Franz Joseph Strauss), Münster-Osnabrück, Nuremberg, Saarbrucken (Ensheim), Stuttgart (Echterdingen); several domestic airports; total passenger km: 87,983 million (1997 est)
Railways: total length: 40,209 km/24,986 mi; total passenger km: 64,020 million (1997)
Roads: total road network: 656,074 km/407,684 mi, of which 99.1% paved (1997); passenger cars: 506 per 1,000 people (1997)

Chronology

c. **1000 BC:** Germanic tribes from Scandinavia began to settle the region between the rivers Rhine, Elbe, and Danube. **AD 9:** Romans tried and failed to conquer Germanic tribes. **5th century:** Germanic tribes plundered Rome, overran Western Europe, and divided it into tribal kingdoms. **496:** Clovis, King of the Franks, conquered the Alemanni tribe of western Germany. **772–804:** After series of fierce wars, Charlemagne extended Frankish authority over Germany, subjugated Saxons, imposed Christianity, and took title of Holy Roman Emperor. **843:** Treaty of Verdun divided the Holy Roman Empire into three, with eastern portion corresponding to modern Germany; local princes became virtually independent. **919:** Henry the Fowler restored central authority and founded Saxon dynasty. **962:** Otto the Great enlarged the kingdom and revived title of Holy Roman Emperor. **1024–1254:** Emperors of Salian and Hohenstaufen dynasties came into conflict with popes; frequent civil wars allowed German princes to regain independence. **12th century:** German expansion eastwards into lands between rivers Elbe and Oder. **13th–14th centuries:** Hanseatic League of Allied German cities became a great commercial and naval power. **1438:** Title of Holy Roman Emperor became virtually hereditary in the Habsburg family of Austria. **1517:** Martin Luther began the Reformation; Emperor Charles V tried to suppress Protestantism; civil war ensued. **1555:** Peace of Augsburg: Charles V forced to accept that each German prince could choose the religion of his own lands. **1618–48:** Thirty Years' War: bitter conflict, partly religious, between certain German princes and emperor, with foreign intervention; the war wrecked the German economy and reduced the Holy Roman Empire to a name. **1701:** Frederick I, Elector of Brandenburg, promoted to King of Prussia. **1740:** Frederick the Great of Prussia seized Silesia from Austria and retained it through war of Austrian Succession (1740–48) and Seven Years' War (1756–63).

1772–95: Prussia joined Russia and Austria in the partition of Poland. **1792:** Start of French Revolutionary Wars, involving many German states, with much fighting on German soil. **1806:** Holy Roman Empire abolished; France formed puppet Confederation of the Rhine in western Germany and defeated Prussia at Battle of Jena. **1813–15:** National revival enabled Prussia to take part in the defeat of Napoleon at Battles of Leipzig and Waterloo. **1814–15:** Congress of Vienna rewarded Prussia with Rhineland, Westphalia, and much of Saxony; loose German Confederation formed by 39 independent states. **1848–49:** Liberal revolutions in many German states; Frankfurt Assembly sought German unity; revolutions suppressed. **1862:** Otto von Bismarck became prime minister of Prussia. **1866:** Seven Weeks' War: Prussia defeated Austria, dissolved German Confederation, and established North German Confederation under Prussian leadership. **1870–71:** Franco-Prussian War; southern German states agreed to German unification; German Empire proclaimed, with King of Prussia as emperor and Bismarck as chancellor. **1890:** Wilhelm II dismissed Bismarck and sought to make Germany a leading power in world politics. **1914:** Germany encouraged the Austrian attack on Serbia that started World War I; Germany invaded Belgium and France. **1918:** Germany defeated; a revolution overthrew the monarchy. **1919:** Treaty of Versailles: Germany lost land to France, Denmark, and Poland; demilitarization and reparations imposed; Weimar Republic proclaimed. **1922–23:** Hyperinflation: in 1922, one dollar was worth 50 marks; in 1923, one dollar was worth 2.5 trillion marks. **1929:** Start of economic slump caused mass unemployment and brought Germany close to revolution. **1933:** Adolf Hitler, leader of Nazi Party, became chancellor. **1934:** Hitler took title of Führer (leader), murdered rivals, and created one-party state with militaristic and racist ideology; rearmament reduced unemployment. **1938:** Germany annexed Austria and Sudeten; occupied remainder of Czechoslovakia in 1939. **1939:** German invasion of Poland started World War II; Germany defeated France in 1940, attacked USSR in 1941, and pursued extermination of Jews. **1945:** Germany defeated and deprived of its conquests; eastern lands transferred to Poland; USA, USSR, UK, and France established zones of occupation. **1948–49:** Disputes between Western allies and USSR led to Soviet blockade of West Berlin. **1949:** Partition of Germany: US, French, and British zones in West Germany became Federal Republic of Germany with Konrad Adenauer as chancellor; Soviet zone in East Germany became communist German Democratic Republic led by Walter Ulbricht. **1953:** Uprising in East Berlin suppressed by Soviet troops. **1955:** West Germany became a member of NATO; East Germany joined the Warsaw Pact. **1957:** West Germany was a founder member of the European Economic Community. **1960s:** West Germany achieved rapid growth and great prosperity. **1961:** East Germany constructed Berlin Wall to prevent emigration to West Berlin (part of West Germany). **1969:** Willy Brandt, Social Democratic Party chancellor of West Germany, sought better relations with USSR and East Germany. **1971:** Erich Honecker succeeded Ulbricht as Communist Party leader, and became head of state in 1976. **1972:** The Basic Treaty established relations between West Germany and East Germany as between foreign states. **1982:** Helmut Kohl (Christian Democratic Union) became the West German chancellor. **1989:** There was a mass exodus of East Germans to West Germany via Hungary; East Germany opened its frontiers, including the Berlin Wall. **1990:** The communist regime in East Germany collapsed; Germany was reunified with Kohl as chancellor. **1991:** Germany took the lead in pressing for closer European integration in the Maastricht Treaty. **1995:** Unemployment reached 3.8 million. **1996:** There was a public-sector labour dispute over welfare reform plans and the worsening economy. Spending cuts were agreed to meet European Monetary Union entry criteria. **1998:** Unemployment reached a post-war high of 12.6%. The CDU–CSU–FDP coalition was defeated in a general election and a 'Red–Green' coalition government was formed by the SPD and the Greens, with Gerhard Schroeder as chancellor. Kohl was replaced as CDU leader by Wolfgang Schäuble. **1999:** A delay was announced in the planned phasing out of nuclear power. Social Democrat Johannes Rau was elected president. **2000:** Former chancellor Helmut Kohl admitted accepting secret, and therefore illegal, donations to his party, and a criminal investigation was launched as he resigned his honorary leadership of the CDU. Leader Wolfgang Schäuble was also forced to resign, and was replaced by Angela Merkel. The first cases of BSE were discovered in German cattle. **2001:** Kohl was heavily fined for accepting illegal donations to his party, but spared a criminal trial.

Practical information

Visa requirements: UK: visa not required. USA: visa not required

Time difference: GMT +1

Chief tourist attractions: spas; summer and winter resorts; medieval towns and castles; Black Forest; Rhine Valley; North and Baltic Sea coasts; mountains of Thuringia and Bavaria and the Erzgebirge

Major holidays: 1, 6 January, 1 May, 3 October, 1 November, 25–26 December; variable: Good Friday, Easter Monday, Ascension Thursday, Whit Monday, Corpus Christi, Assumption

GHANA FORMERLY THE GOLD COAST (UNTIL 1957)

Located in West Africa, bounded north by Burkina Faso, east by Togo, south by the Gulf of Guinea, and west by Côte d'Ivoire.

National name: *Republic of Ghana*
Area: 238,540 sq km/92,100 sq mi
Capital: Accra
Major towns/cities: Kumasi, Tamale, Tema, Sekondi Takoradi, Cape Coast, Koforidua, Bolgatanga, Obuasi
Major ports: Sekondi, Tema
Physical features: mostly tropical lowland plains; bisected by River Volta

Ghana was the first country to adopt the pan-African colours. The star is known as the 'lode star of African freedom'. Effective date: 28 February 1966.

Government

Head of state and government: John Kufuor from 2001
Political system: emergent democracy
Political executive: limited presidency
Administrative divisions: ten regions
Political parties: National Democratic Congress (NDC), centrist, progovernment; New Patriotic Party (NPP), left of centre
Armed forces: 7,000; plus a paramilitary force of 1,000 (1998)
Conscription: military service is voluntary
Death penalty: retained and used for ordinary crimes
Defence spend: (% GDP) 1.4 (1998)
Education spend: (% GNP) 3.1 (1993–94)
Health spend: (% GDP) 3.1 (1997)

Economy and resources

Currency: cedi
GDP: (US$) 7.6 billion (1999)
Real GDP growth: (% change on previous year) 4.2 (1999)
GNP: (US$) 7.4 billion (1999)
GNP per capita (PPP): (US$) 1,793 (1999 est)
Consumer price inflation: 12.4% (1999)
Unemployment: 20% (1997 est)
Foreign debt: (US$) 7.2 billion (1999)
Major trading partners: Togo, Nigeria, UK, USA, Germany, Japan, Italy
Resources: diamonds, gold, manganese, bauxite
Industries: food processing, textiles, vehicles, aluminium, cement, paper, chemicals, petroleum products, tourism
Exports: gold, cocoa and related products, timber. Principal market: Togo 12% (1998)
Imports: raw materials, machinery and transport equipment, petroleum, food, basic manufactures. Principal source: Nigeria 14% (1998)
Arable land: 12.3% (1996)
Agricultural products: cocoa (world's third-largest producer), coffee, bananas, oil palm, maize, rice, cassava, plantain, yams, coconuts, kola nuts, limes, shea nuts; forestry (timber production)

Population and society

Population: 20,212,000 (2000 est)
Population growth rate: 2.7% (1995–2000)
Population density: (per sq km) 83 (1999 est)
Urban population: (% of total) 38 (2000 est)
Age distribution: (% of total population) 0–14 43%, 15–59 52%, 60+ 5% (2000 est)
Ethnic groups: over 75 ethnic groups; most significant are the Akan in the south and west (44%), the Mole-Dagbani in the north (16%), the Ewe in the south (13%), the Ga in the region of the capital city (8%), and the Fanti in the coastal area
Language: English (official), Ga, other African languages
Religion: Christian 40%, animist 32%, Muslim 16%
Education: (compulsory years) 9
Literacy rate: 80% (men); 63% (women) (2000 est)
Labour force: 47% of population: 47.5% agriculture, 12.8% industry, 39.7% services (1994)
Life expectancy: 58 (men); 62 (women) (1995–2000)
Child mortality rate: (under 5, per 1,000 live births) 101 (1995–2000)
Physicians: 1 per 22,970 people (1993 est)
Hospital beds: 1 per 685 people (1993 est)
TV sets: (per 1,000 people) 99 (1998)
Radios: (per 1,000 people) 238 (1997)
Internet users: (per 10,000 people) 10.2 (1999)
Personal computer users: (per 100 people) 0.3 (1999)

Transport

Airports: international airport: Accra (Koteka); four domestic airports; total passenger km: 611 million (1995)
Railways: total length: 953 km/592 mi; total passenger km: 277 million (1991)
Roads: total road network: 37,800 km/23,489 mi, of which 25% paved (1996 est); passenger cars: 4.7 per 1,000 people (1996 est)

Chronology

5th–12th century: Ghana Empire (from which present-day country's name derives) flourished, with its centre 500 mi/800 km to the northwest, in Mali. **13th century:** In coastal and forest areas Akan peoples founded the first states. **15th century:** Gold-seeking Mande traders entered northern Ghana from the northeast, founding Dagomba and Mamprussi states; Portuguese navigators visited coastal region, naming it the 'Gold Coast', building a fort at Elmina, and slave trading began. **17th century:** Gonja kingdom founded in north by Mande speakers; Ga and Ewe states founded in southeast by immigrants from Nigeria; in central Ghana, controlling gold reserves around Kumasi, the Ashanti, a branch of the Akans, founded what became the most powerful state in precolonial Ghana. **1618:** British trading settlement established on Gold Coast. **18th–19th centuries:** Centralized Ashanti kingdom at its height, dominating between Komoe River in the west and Togo Mountains in the east and active in slave trade; Fante state powerful along coast in the south. **1874:** Britain, after ousting the Danes and Dutch and defeating the Ashanti, made the Gold Coast (the southern provinces) a crown colony. **1898–1901:** After three further military campaigns, Britain finally subdued and established protectorates over Ashanti and the northern territories. **early 20th century:** The colony developed into a major cocoa-exporting region. **1917:** West Togoland, formerly German-ruled, was administered with the Gold Coast as British Togoland. **1949:** Campaign for independence launched by Kwame Nkrumah, who formed the Convention People's Party (CPP) and became prime minister in 1952. **1957:** Independence achieved, within the Commonwealth, as Ghana, which included British Togoland; Nkrumah became prime minister. Policy of 'African socialism' and nonalignment pursued. **1960:** Became a republic, with Nkrumah as president. **1964:** Ghana became a one-party state, dominated by the CCP, and developed links with communist bloc. **1972:** A coup placed Col Ignatius Acheampong at the head of a military government as the economy deteriorated. **1978:** Acheampong was deposed in a bloodless coup. Flight-Lt Jerry Rawlings, a populist soldier who launched a drive against corruption, came to power. **1979:** There was a return to civilian rule. **1981:** Rawlings seized power again. All political parties were banned. **1992:** A pluralist constitution was approved in a referendum, lifting the ban on political parties. Rawlings won presidential elections. **1993:** The fourth republic of Ghana was formally inaugurated. **1994:** Ethnic clashes in the north left more than 6,000 people dead. **1996:** The New Democratic Congress (NDC) won an assembly majority. Rawlings was re-elected as president. **2001:** John Kufuor, leader of the liberal New Patriotic Party, was elected president.

Practical information

Visa requirements: UK: visa required. USA: visa required
Time difference: GMT +/–0
Chief tourist attractions: game reserves; beaches; traditional festivals; old castles and trading posts
Major holidays: 1 January, 6 March, 1 May, 4 June, 1 July, 25–26, 31 December; variable: Good Friday, Easter Monday, Holy Saturday

GREECE

Located in southeast Europe, comprising the southern part of the Balkan peninsula, bounded to the north by the Former Yugoslav Republic of Macedonia and Bulgaria, to the northwest by Albania, to the northeast by Turkey, to the east by the Aegean Sea, to the south by the Mediterranean Sea, and to the west by the Ionian Sea.

The cross represents the Greek Orthodox faith. Blue stands for the sea and sky. The shade has varied over the years. White symbolizes purity. Effective date: 22 December 1978.

National name: *Elliniki Dimokratia/Hellenic Republic*
Area: 131,957 sq km/50,948 sq mi
Capital: Athens
Major towns/cities: Thessaloniki, Peiraias, Patra, Iraklion, Larisa, Peristerio, Kallithéa
Major ports: Peiraias, Thessaloniki, Patra, Iraklion
Physical features: mountainous (Mount Olympus); a large number of islands, notably Crete, Corfu, and Rhodes, and Cyclades and Ionian Islands

Government

Head of state: Costis Stephanopoulos from 1995
Head of government: Costas Simitis from 1996
Political system: liberal democracy
Political executive: parliamentary
Administrative divisions: 13 prefectures and one autonomous region (Mount Athos)
Political parties: Panhellenic Socialist Movement (PASOK), nationalist, democratic socialist; New Democracy Party (ND), right of centre; Democratic Renewal (DIANA), centrist; Communist Party (KKE), left wing; Political Spring, moderate, left of centre
Armed forces: 168,500 (1998)
Conscription: 19–24 months
Death penalty: abolished in 1993
Defence spend: (% GDP) 4.8 (1998)
Education spend: (% GNP) 3 (1996)
Health spend: (% GDP) 8.6 (1997)

Economy and resources

Currency: drachma
GDP: (US$) 123.9 billion (1999)
Real GDP growth: (% change on previous year) 3.5 (1999)
GNP: (US$) 124 billion (1999)
GNP per capita (PPP): (US$) 14,595 (1999)
Consumer price inflation: 2.7% (1999)
Unemployment: 10% (1998 est)
Major trading partners: Germany, Italy, France, the Netherlands, USA, UK
Resources: bauxite, nickel, iron pyrites, magnetite, asbestos, marble, salt, chromite, lignite
Industries: food products, metals and metal products, textiles, petroleum refining, machinery and transport equipment, tourism, wine
Exports: fruit and vegetables, clothing, mineral fuels and lubricants, textiles, iron and steel, aluminium and aluminium alloys. Principal market: Germany 17.7% (1998)
Imports: petroleum and petroleum products, machinery and transport equipment, food and live animals, chemicals and chemical products. Principal source: Italy 15.8% (1998)

Arable land: 22.2% (1996)
Agricultural products: fruit and vegetables, cereals, sugar beet, tobacco, olives; livestock and dairy products

Population and society

Population: 10,645,000 (2000 est)
Population growth rate: 0.3% (1995–2000); 0% (2000–05)
Population density: (per sq km) 81 (1999 est)
Urban population: (% of total) 60 (2000 est)
Age distribution: (% of total population) 0–14 15%, 15–59 61%, 60+ 24% (2000 est)
Ethnic groups: predominantly Greek (98%); main minorities are Turks, Slavs, and Albanians
Language: Greek (official)
Religion: Greek Orthodox, over 96%; about 1% Muslim
Education: (compulsory years) 9
Literacy rate: 98% (men); 96% (women) (2000 est)
Labour force: 20.2% agriculture, 22.9% industry, 56.9% services (1997)
Life expectancy: 76 (men); 81 (women) (1995–2000)
Child mortality rate: (under 5, per 1,000 live births) 9 (1995–2000)
Physicians: 1 per 256 people (1994)
Hospital beds: 1 per 200 people (1994)
TV sets: (per 1,000 people) 466 (1997)
Radios: (per 1,000 people) 477 (1997)
Internet users: (per 10,000 people) 705.8 (1999)
Personal computer users: (per 100 people) 6.0 (1999)

Transport

Airports: international airports: Athens (Athinai), Iraklion/Crete, Thessaloniki (Micra), Corfu (Kerkira), Rhodes (Paradisi); 25 domestic airports, of which 14 are also international; total passenger km: 9,261 million (1997 est)
Railways: total length: 2,474 km/1,537 mi; total passenger km: 1,783 million (1997)
Roads: total road network: 117,000 km/72,704 mi, of which 91.8% paved (1996 est); passenger cars: 223 per 1,000 people (1996)

Chronology

c. **2000–1200 BC:** Mycenaean civilization flourished. *c.* **1500–1100 BC:** Central Greece and Peloponnese invaded by tribes of Achaeans, Aeolians, Ionians, and Dorians. *c.* **1000–500 BC:** Rise of the Greek city states; Greek colonies established around the shores of the Mediterranean. *c.* **490–404 BC:** Ancient Greek culture reached its zenith in the democratic city state of Athens. **357–338 BC:** Philip II of Macedon won supremacy over Greece; cities fought to regain and preserve independence. **146 BC:** Roman Empire defeated Macedon and annexed Greece.

476 AD: Western Roman Empire ended; Eastern Empire continued as Byzantine Empire, based at Constantinople, with essentially Greek culture. **1204:** Crusaders partitioned Byzantine Empire; Athens, Achaea, and Thessaloniki came under Frankish rulers. **late 14th century–1461:** Ottoman Turks conquered mainland Greece and captured Constantinople in 1453; Greek language and culture preserved by Orthodox Church. **1685:** Venetians captured Peloponnese; regained by Turks in 1715. **late 18th century:** Beginnings of Greek nationalism among émigrés and merchant class. **1814:** *Philike Hetairia* ('Friendly Society') formed by revolutionary Greek nationalists in Odessa. **1821:** *Philike Hetairia* raised Peloponnese brigands in revolt against Turks; War of Independence ensued. **1827:** Battle of Navarino: Britain, France, and Russia intervened to destroy Turkish fleet; Count Ioannis Kapodistrias elected president of Greece. **1829:** Treaty of Adrianople: under Russian pressure, Turkey recognized independence of small Greek state. **1832:** Great Powers elected Otto of Bavaria as king of Greece. **1843:** Coup forced King Otto to grant a constitution. **1862:** Mutiny and rebellion led King Otto to abdicate. **1863:** George of Denmark became king of the Hellenes. **1864:** Britain transferred Ionian islands to Greece. **1881:** Following Treaty of Berlin in 1878, Greece was allowed to annex Thessaly and part of Epirus. **late 19th century:** Politics dominated by Kharilaos Trikoupis, who emphasized economic development, and Theodoros Deliyiannis, who emphasized territorial expansion. **1897:** Greco-Turkish War ended in Greek defeat. **1908:** Cretan Assembly led by Eleutherios Venizelos proclaimed union with Greece. **1910:** Venizelos became prime minister and introduced financial, military, and constitutional reforms. **1912–13:** Balkan Wars: Greece annexed a large area of Epirus and Macedonia. **1916:** 'National Schism': Venizelos formed rebel pro-Allied government while royalists remained neutral. **1917–18:** Greek forces fought on Allied side in World War I. **1919–22:** Greek invasion of Asia Minor; after Turkish victory, a million refugees came to Greece. **1924:** Republic declared amid great political instability. **1935:** Greek monarchy restored with George II. **1936:** Gen Ioannia Metaxas established right-wing dictatorship. **1940:** Greece successfully repelled Italian invasion. **1941–44:** German occupation of Greece; rival monarchist and communist resistance groups operated from 1942. **1946–49:** Civil war: communists defeated by monarchists with military aid from Britain and USA. **1952:** Became a member of NATO. **1967:** 'Greek Colonels' seized power under George Papadopoulos; political activity banned; King Constantine II exiled. **1973:** Republic proclaimed with Papadopoulos as president. **1974:** Cyprus crisis caused downfall of military regime; Constantinos Karamanlis returned from exile to form Government of National Salvation and restore democracy. **1981:** Andreas Papandreou was elected Greece's first socialist prime minister; Greece entered the European Community. **1997:** Direct talks with Turkey resulted in an agreement to settle all future disputes peacefully. **2000:** Simitis was re-elected as prime minister, and Stephanopoulos was re-elected as president. Greece and Turkey signed a series of agreements aimed at improving relations between the two countries. **2001:** Greece joined the euro.

Practical information

Visa requirements: UK/USA: visa not required
Time difference: GMT +2

Chief tourist attractions: Aegean islands; historical and archaeological remains – palace of Knossos on Crete, Delphi, the Acropolis in Athens; climate

GRENADA

Island in the Caribbean, the southernmost of the Windward Islands.

Area: (including the southern Grenadine Islands, notably Carriacou and Petit Martinique) 344 sq km/133 sq mi
Capital: St George's
Major towns/cities: Grenville, Sauteurs, Victoria, Gouyave
Physical features: southernmost of the Windward Islands; mountainous; Grand-Anse beach; Annandale Falls; the Great Pool volcanic crater

Yellow represents sunshine, warmth, and wisdom. Green symbolizes the lush vegetation and agriculture. Effective date: 7 February 1974.

Government

Head of state: Queen Elizabeth II from 1974, represented by Governor General Daniel Williams from 1996
Head of government: Keith Mitchell from 1995
Political system: liberal democracy
Political executive: parliamentary
Administrative divisions: six parishes and one dependency
Political parties: Grenada United Labour Party (GULP), nationalist, left of centre; National Democratic Congress (NDC), centrist; National Party (TNP), centrist
Armed forces: no standing army; 730-strong regional security unit (1997)
Death penalty: retained and used for ordinary crimes but can be considered abolitionist in practice (last execution in 1978)
Health spend: (% GDP) 6.3 (1997)

Economy and resources

Currency: East Caribbean dollar
GDP: (US$) 351 million (1999)
Real GDP growth: (% change on previous year) 6.2 (1999 est)
GNP: (US$) 335 million (1999)
GNP per capita (PPP): (US$) 5,847 (1999)
Consumer price inflation: 0.5% (1999 est)
Unemployment: 14% (1997)
Foreign debt: (US$) 182.8 million (1998)
Major trading partners: USA, UK, Trinidad and Tobago, the Netherlands, Germany
Industries: agricultural products (nutmeg oil distillation), rum, beer, soft drinks, cigarettes, clothing, tourism
Exports: cocoa, bananas, cocoa, mace, fresh fruit. Principal market: UK, USA, France 18.5% each (1995)
Imports: foodstuffs, mineral fuels, machinery and transport equipment, basic manufactures, beverages, tobacco. Principal source: USA 30% (1995)
Arable land: 11.8% (1995)

Agricultural products: cocoa, bananas, nutmeg (world's second-largest producer), and mace, sugar cane, fresh fruit and vegetables; livestock productions (for domestic use); fishing

Population and society

Population: 94,000 (2000 est)
Population growth rate: 0.3% (1995–2025)
Population density: (per sq km) 286 (1999 est)
Urban population: (% of total) 38 (2000 est)
Age distribution: (% of total population) 0–14 43%, 15–59 51%, 60+ 6% (2000 est)
Ethnic groups: majority is of black African descent
Language: English (official), some French-African patois
Religion: Roman Catholic 53%, Anglican about 14%, Seventh Day Adventist, Pentecostal, Methodist
Education: (compulsory years) 11
Literacy rate: 90% (1997 est)
Labour force: 17.1% agriculture, 22.4% industry, 60.5% services (1995)
Life expectancy: 69 (men); 74 (women) (1998 est)
Child mortality rate: (under 5, per 1,000 live births) 33 (1995)
Physicians: 1 per 1,253 people (1996)
Hospital beds: 1 per 290 people (1996)
TV sets: (per 1,000 people) 353 (1997)
Radios: (per 1,000 people) 615 (1997)
Internet users: (per 10,000 people) 214.2 (1999)
Personal computer users: (per 100 people) 11.8 (1999)

Transport

Airports: international airports: St George's (Point Salines); total aircraft arrivals: 11,310 (1995)
Railways: none
Roads: total road network: 1,040 km/646 mi, of which 61.3% paved (1996 est)

Chronology

1498: Sighted by the explorer Christopher Columbus; Spanish named it Grenada since its hills were reminiscent of the Andalusian city. **1650:** Colonized by French settlers from Martinique, who faced resistance from the local Carib Indian community armed with poison arrows, before the defeated Caribs performed a mass suicide. **1783:** Ceded to Britain as a colony by the Treaty of Versailles; black African slaves imported to work cotton, sugar, and tobacco plantations. **1795:** Abortive rebellion against British rule led by Julien Fedon, a black planter. **1834:** Slavery abolished. **1950:** Left-wing Grenada United Labour Party (GULP) founded by trade union leader Eric Gairy. **1951:** Universal adult suffrage granted and GULP elected to power in a nonautonomous local assembly. **1958–62:** Part of the Federation of the West Indies. **1967:** Internal self-government achieved. **1974:** Independence achieved within the Commonwealth, with Gairy as prime minister. **1979:** Autocratic Gairy was removed in a bloodless coup led by left-wing Maurice Bishop of the New Jewel Movement. The constitution was suspended and a People's Revolutionary Government established. **1982:** Relations with the USA and Britain deteriorated as ties with Cuba and the USSR strengthened. **1983:** After attempts to improve relations with the USA, Bishop was overthrown by left-wing opponents, precipitating a military coup by Gen Hudson Austin. The USA invaded; there were 250 fatalities. Austin was arrested and the 1974 constitution was reinstated. **1984:** Newly formed centre-left New National Party (NNP) won a general election and its leader became prime minister. **1991:** Integration of the Windward Islands was proposed. **1995:** A general election was won by the NNP, led by Keith Mitchell. A plague of pink mealy bugs caused damage to crops estimated at $60 million. **1999:** The ruling NNP gained a sweeping general election victory.

Practical information

Visa requirements: UK: visa not required. USA: visa not required
Time difference: GMT –4
Chief tourist attractions: white sandy beaches; mountainous interior; rainforest
Major holidays: 1–2 January, 7 February, 1 May, 3–4 August, 25 October, 25–26 December; variable: Corpus Christi, Good Friday, Easter Monday, Whit Monday

GUATEMALA

Located in Central America, bounded north and northwest by Mexico, east by Belize and the Caribbean Sea, southeast by Honduras and El Salvador, and southwest by the Pacific Ocean.

National name: *República de Guatemala/Republic of Guatemala*
Area: 108,889 sq km/42,042 sq mi
Capital: Guatemala City
Major towns/cities: Quezaltenango, Escuintla, Puerto Barrios (naval base), Mixco, Villa Nueva, Chinautla
Physical features: mountainous; narrow coastal plains; limestone tropical plateau in north; frequent earthquakes

The quetzal, the national bird of Guatemala, symbolizes freedom. The blue bands stand for the Caribbean Sea and the Pacific Ocean. The weapons represent the defence of liberty. Effective date: 15 September 1968.

Government

Head of state and government: Alfonso Portillo from 2000
Political system: liberal democracy
Political executive: limited presidency
Administrative divisions: 22 departments
Political parties: Guatemalan Christian Democratic Party (PDCG), Christian, left of centre; Centre Party (UCN), centrist; Revolutionary Party (PR), radical; Movement of National Liberation (MLN), extreme right wing; Democratic Institutional Party (PID), moderate conservative; Solidarity and Action Movement (MAS), right of centre; Guatemalan Republican Front (FRG), right wing; National Advancement Party (PAN), right of centre; Social Democratic Party (PSD), right of centre
Armed forces: 31,400 (1998); plus paramilitary forces of 9,800
Conscription: selective conscription for 30 months
Death penalty: retained and used for ordinary crimes
Defence spend: (% GDP) 1.2 (1998)
Education spend: (% GNP) 1.7 (1996)
Health spend: (% GDP) 2.4 (1997)

Economy and resources

Currency: quetzal
GDP: (US$) 18.02 billion (1999)
Real GDP growth: (% change on previous year) 3.5 (1999)
GNP: (US$) 18.4 billion (1999)
GNP per capita (PPP): (US$) 3,517 (1999 est)
Consumer price inflation: 5.2% (1999)
Unemployment: 5.2% (1997 est)
Foreign debt: (US$) 4.5 billion (1999)
Major trading partners: USA, El Salvador, Mexico, Costa Rica, Venezuela, Germany, Japan, Honduras
Resources: petroleum, antimony, gold, silver, nickel, lead, iron, tungsten
Industries: food processing, textiles, pharmaceuticals, chemicals, tobacco, non-metallic minerals, sugar, electrical goods, tourism
Exports: coffee, bananas, sugar, cardamoms, shellfish, tobacco. Principal market: USA 51.4% (1998)
Imports: raw materials and intermediate goods for industry, consumer goods, mineral fuels and lubricants. Principal source: USA 42.8% (1998)

Arable land: 12.6% (1996)
Agricultural products: coffee, sugar cane, bananas, cardamoms, cotton; one of the largest sources of essential oils (citronella and lemon grass); livestock rearing; fishing (chiefly shrimp); forestry (mahogany and cedar)

Population and society

Population: 11,385,000 (2000 est)
Population growth rate: 2.6% (1995–2000)
Population density: (per sq km) 102 (1999 est)
Urban population: (% of total) 40 (2000 est)
Age distribution: (% of total population) 0–14 44%, 15–59 51%, 60+ 5% (2000 est)
Ethnic groups: two main ethnic groups: American Indians and ladinos (others, including Europeans, black Africans, and mestizos). American Indians are descended from the highland Mayas
Language: Spanish (official), 22 Mayan languages (45%)
Religion: Roman Catholic 70%, Protestant 10%, traditional Mayan
Education: (compulsory years) 6
Literacy rate: 76% (men); 61% (women) (2000 est)
Labour force: 48.2% agriculture, 18.1% industry, 33.7% services (1995 est)
Life expectancy: 61 (men); 67 (women) (1995–2000)
Child mortality rate: (under 5, per 1,000 live births) 61 (1995–2000)
Physicians: 1 per 3,999 people (1993 est)
Hospital beds: 1 per 1,191 people (1993 est)
TV sets: (per 1,000 people) 126 (1997)
Radios: (per 1,000 people) 79 (1997)
Internet users: (per 10,000 people) 58.6 (1999)
Personal computer users: (per 100 people) 1.0 (1999)

Transport

Airports: international airport: Guatemala City (La Aurora); over 380 airstrips serving internal travel; total passenger km: 500 million (1995)
Railways: total length: 782 km/486 mi; total passenger km: 16,580 million (1995)
Roads: total road network: 13,100 km/8,140 mi, of which 27.6% paved (1996 est); passenger cars: 8.8 per 1,000 people (1996 est)

Chronology

c. **AD 250–900:** Part of culturally advanced Maya civilization. **1524:** Conquered by the Spanish adventurer Pedro de Alvarado and became a Spanish colony. **1821:** Independence achieved from Spain, joining Mexico initially. **1823:** Became part of United Provinces (Federation) of Central America, also embracing Costa Rica, El Salvador, Honduras, and Nicaragua. **1839:** Achieved full independence. **1844–65:** Rafael Carrera held power as president. **1873–85:** The country was modernized on liberal lines by President Justo Rufino Barrios, the army was built up, and coffee growing introduced. **1944:** Juan José Arevalo became president, ending a period of rule by dictators. Socialist programme of reform instituted by Arevalo, including establishing a social security system and redistributing land expropriated from large estates to landless peasants. **1954:** Col Carlos Castillo Armas became president in a US-backed coup, after United Fruit Company plantations had been nationalized. Land reform halted. **1966:** Civilian rule was restored. **1970s:** More than 50,000 died in a spate of political violence as the military regime sought to liquidate left-wing dissidents. **1970:** The military were back in power. **1976:** An earthquake killed 27,000 and left more than 1 million homeless. **1981:** Growth of an antigovernment guerrilla movement. Death squads and soldiers killed an estimated 11,000 civilians during the year. **1985:** A new constitution was adopted; PDCG won the congressional elections. **1989:** Over 100,000 people were killed, and 40,000 reported missing, since 1980. **1991:** Diplomatic relations established with Belize, which Guatemala had long claimed. **1994:** Peace talks were held with Guatemalan Revolutionary National Unity (URNG) rebels. Right-wing parties secured a majority in congress after elections. **1995:** The government was criticized by USA and United Nations for widespread human-rights abuses. There was a ceasefire by rebels, the first in 30 years. **1996:** A peace agreement was signed which ended the 36-year war. **2000:** Alfonso Portillo, a right-wing candidate, became president. Guatemala signed a free-trade agreement with Mexico. The US dollar was accepted as a second currency.

Practical information

Visa requirements: UK: visa required for business visits and tourist visits of over 90 days. USA: visa not required for a stay of up to 90 days
Time difference: GMT –6
Chief tourist attractions: mountainous and densely forested landscape; Mayan temples and ruins at Tikal, Palenque, and Kamihal Juyú; Guatemala City, with its archaeological and historical museums
Major holidays: 1 January, 1 May, 30 June, 1 July, 15 September, 12, 20 October, 1 November, 24–25, 31 December; variable: Good Friday, Holy Thursday, Holy Saturday

GUINEA

Located in West Africa, bounded north by Senegal, northeast by Mali, southeast by Côte d'Ivoire, south by Liberia and Sierra Leone, west by the Atlantic Ocean, and northwest by Guinea-Bissau.

National name: *République de Guinée/Republic of Guinea*
Area: 245,857 sq km/94,925 sq mi
Capital: Conakry
Major towns/cities: Labé, Nzérékoré, Kankan, Kindia, Mamou, Siguiri
Physical features: flat coastal plain with mountainous interior; sources of rivers Niger, Gambia, and Senegal; forest in southeast; Fouta Djallon, area of sandstone plateaux, cut by deep valleys

Red represents work. Yellow symbolizes justice. The design is based on the French tricolour. Green stands for solidarity. Effective date: 10 November 1958.

Government

Head of state: Lansana Conté from 1984
Head of government: Lamine Sidime from 1999
Political system: emergent democracy
Political executive: limited presidency
Administrative divisions: eight administrative regions, including Conakry; the country is subdivided into 34 regions, including Conakry (which is divided into three communities)
Political parties: Party of Unity and Progress (PUP), centrist; Rally of the Guinean People (RPG), left of centre; Union of the New Republic (UNR), left of centre; Party for Renewal and Progress (PRP), left of centre
Armed forces: 9,700; plus paramilitary forces of 9,600 (1998)
Conscription: military service is compulsory for two years
Death penalty: retained and used for ordinary crimes
Defence spend: (% GDP) 1.8 (1998)
Education spend: (% GNP) 1.9 (1997)
Health spend: (% GDP) 3.5 (1997)

Economy and resources

Currency: Guinean franc
GDP: (US$) 3.7 billion (1999)
Real GDP growth: (% change on previous year) 3.7 (1999 est)
GNP: (US$) 3.7 billion (1999)
GNP per capita (PPP): (US$) 1,761 (1999)
Consumer price inflation: 4.5% (1999)
Foreign debt: (US$) 3.15 billion (1998)
Major trading partners: France, USA, Belgium, Ukraine, Hong Kong, Spain, Ireland, Côte d'Ivoire
Resources: bauxite (world's top exporter of bauxite and second-largest producer of bauxite ore), alumina, diamonds, gold, granite, iron ore, uranium, nickel, cobalt, platinum
Industries: processing of agricultural products, cement, beer, soft drinks, cigarettes
Exports: bauxite, alumina, diamonds, coffee. Principal market: USA 15.4% (1999)
Imports: foodstuffs, mineral fuels, semi-manufactured

goods, consumer goods, textiles and clothing, machinery and transport equipment. Principal source: France 23.1% (1999)
Arable land: 2.4% (1996)
Agricultural products: cassava, millet, rice, fruits, oil palm, groundnuts, coffee, vegetables, sweet potatoes, yams, maize; livestock rearing (cattle); fishing; forestry

Population and society

Population: 7,430,000 (2000 est)
Population growth rate: 0.8% (1995–2000)
Population density: (per sq km) 30 (1999 est)
Urban population: (% of total) 33 (2000 est)
Age distribution: (% of total population) 0–14 44%, 15–59 52%, 60+ 4% (2000 est)
Ethnic groups: 24 ethnic groups, including the Malinke (30%), Peuhl (30%), and Soussou (16%)
Language: French (official), Susu, Pular (Fulfude), Malinke, and other African languages
Religion: Muslim 85%, Christian 6%, animist
Education: (compulsory years) 6
Literacy rate: 55% (men); 27% (women) (2000 est)
Labour force: 49% of population: 87% agriculture, 2% industry, 11% services (1990)
Life expectancy: 46 (men); 47 (women) (1995–2000)
Child mortality rate: (under 5, per 1,000 live births) 207 (1995–2000)
Physicians: 1 per 7,445 people (1993 est)
Hospital beds: 1 per 1,712 people (1993 est)
TV sets: (per 1,000 people) 41 (1997)
Radios: (per 1,000 people) 47 (1997)
Internet users: (per 10,000 people) 6.4 (1999)
Personal computer users: (per 100 people) 0.4 (1999)

Transport

Airports: international airport: Conakry; eight domestic airports; total passenger km: 52 million (1995)
Railways: total length: 662 km/411 mi; total passenger km: 41 million (1991)
Roads: total road network: 30,500 km/18,953 mi, of which 16.5% paved (1996 est); passenger cars: 2 per 1,000 people (1996 est)

Chronology

c. **AD 900:** The Susi people, a community related to the Malinke, immigrated from the northeast, pushing the indigenous Baga towards the Atlantic coast. **13th century:** Susi kingdoms established, extending their influence to the coast; northeast Guinea was part of Muslim Mali Empire, centred to northeast. **mid-15th century:** Portuguese traders visited the coast and later developed trade in slaves and ivory. **1849:** French protectorate established over coastal region around Nunez River, which was administered with Senegal. **1890:** Separate Rivières du Sud colony formed. **1895:** Renamed French Guinea, the colony became part of French West Africa. **1946:** French Guinea became an overseas territory of France. **1958:** Full independence from France achieved as Guinea, after referendum, rejected remaining within French Community; Sékou Touré of the Democratic Party of Guinea (PDG) elected president. **1960s and 1970s:** Touré established socialist one-party state, leading to deterioration in economy as 200,000 fled abroad. **1979:** Strong opposition to Touré's rigid Marxist policies forced him to accept a return to mixed economy and legalize private enterprise. **1984:** Touré died. A bloodless military coup brought Col Lansana Conté to power; the PDG was outlawed and political prisoners released; and there were market-centred economic reforms. **1991:** Antigovernment general strike and mass protests. **1992:** The constitution was amended to allow for multiparty politics. **1993:** Conté was narrowly re-elected in the first direct presidential election. **1998–99:** President Conté was re-elected, and named his prime minister as Lamine Sidime. **2000–01:** From October 2000, civil wars in Liberia and Sierra Leone began to spill over into Guinea, creating hundreds of thousands of refugees.

Practical information

Visa requirements: UK: visa required. USA: visa required
Time difference: GMT +/–0

Major holidays: 1 January, 3 April, 1 May, 15 August, 2 October, 1 November, 25 December; variable: Eid-ul-Adha, Easter Monday, end of Ramadan, Prophet's Birthday

GUINEA-BISSAU FORMERLY PORTUGUESE GUINEA (UNTIL 1974)

Located in West Africa, bounded north by Senegal, east and southeast by Guinea, and southwest by the Atlantic Ocean.
National name: *República da Guiné-Bissau/Republic of Guinea-Bissau*
Area: 36,125 sq km/13,947 sq mi
Capital: Bissau (and chief port)
Major towns/cities: Bafatá, Bissorã, Bolama, Gabú, Bubaque, Cacheu, Catio, Farim
Physical features: flat coastal plain rising to savannah in east

Red, yellow, and green are the pan-African colours.
Effective date: 24 September 1973.

Government

Head of state: Kumba Ialá from 2000
Head of government: Caetano N'Tchama from 2000
Political system: military
Political executive: military
Administrative divisions: nine regions
Political parties: African Party for the Independence of Portuguese Guinea and Cape Verde (PAIGC), nationalist socialist; Party for Social Renovation (PRS), left of centre; Guinea-Bissau Resistance–Bafatá Movement (PRGB-MB), centrist
Armed forces: 7,300; plus paramilitary gendarmerie of 2,000 (1998)
Conscription: selective conscription
Death penalty: abolished in 1993
Defence spend: (% GDP) 5.5 (1998)
Education spend: (% GNP) N/A
Health spend: (% GDP) 5.7 (1997)

Economy and resources

Currency: Guinean peso
GDP: (US$) 221 million (1999)
Real GDP growth: (% change on previous year) 8.7 (1999 est)
GNP: (US$) 195 million (1999)
GNP per capita (PPP): (US$) 595 (1999)
Consumer price inflation: –0.9% (1999)
Unemployment: N/A
Foreign debt: (US$) 964 million (1998)
Major trading partners: Spain, Thailand, India, Portugal, Côte d'Ivoire, the Netherlands, Japan
Resources: bauxite, phosphate, petroleum (largely unexploited)
Industries: food processing, brewing, cotton processing, fish and timber processing
Exports: cashew nuts, palm kernels, groundnuts, fish and shrimp, timber. Principal market: India 59.1% (1998 est)
Imports: foodstuffs, machinery and transport equipment, fuels, construction materials. Principal source: Portugal 26.1% (1998 est)
Arable land: 10.7% (1996)

Agricultural products: groundnuts, sugar cane, plantains, palm kernels, rice, coconuts, millet, sorghum, maize, cashew nuts; fishing; forest resources

Population and society

Population: 1,213,000 (2000 est)
Population growth rate: 2.2% (1995–2000)
Population density: (per sq km) 33 (1999 est)
Urban population: (% of total) 24 (2000 est)
Age distribution: (% of total population) 0–14 43%, 15–59 51%, 60+ 6% (2000 est)
Ethnic groups: majority originated in Africa, and comprises five main ethnic groups: the Balanta in the central region, the Fulani in the north, the Malinke in the northern central area, and the Mandyako and Pepel near the coast
Language: Portuguese (official), Crioulo (a Cape Verdean dialect of Portuguese), African languages
Religion: animist 58%, Muslim 40%, Christian 5% (mainly Roman Catholic)
Education: (compulsory years) 6
Literacy rate: 60% (men); 19% (women) (2000 est)
Labour force: 48% of population: 85% agriculture, 2% industry, 13% services (1990)
Life expectancy: 44 (men); 47 (women) (1995–2000)
Child mortality rate: (under 5, per 1,000 live births) 203 (1995–2000)
Physicians: 1 per 5,556 people (1993)
Hospital beds: 1 per 741 people (1993)
Radios: (per 1,000 people) 43 (1997)
Internet users: (per 10,000 people) 12.6 (1999)

Transport

Airports: international airport: Bissau (Bissalanca); ten domestic airports; total passenger km: 10 million (1995)
Railways: none
Roads: total road network: 4,400 km/2,734 mi, of which 10.3% paved (1996 est); passenger cars: 5.7 per 1,000 people (1996 est)

Chronology

10th century: Known as Gabu, became a tributary kingdom of the Mali Empire to northeast. **1446:** Portuguese arrived, establishing nominal control over coastal areas and capturing slaves to send to Cape Verde. **1546:** Gabu kingdom became independent of Mali and survived until 1867. **1879:** Portugal, which had formerly administered the area with Cape Verde islands, created the separate colony of Portuguese Guinea. **by 1915:** The interior had been subjugated by the Portuguese. **1956:** African Party for the Independence of Portuguese Guinea and Cape Verde (PAIGC) formed to campaign for independence from Portugal. **1961:** The PAIGC began to wage a guerrilla campaign against Portuguese rule. **1973:** Independence was declared in the two-thirds of the country that had fallen under the control of the PAIGC; heavy losses were sustained by Portuguese troops who tried to put down the uprising. **1974:** Independence separately from Cape Verde accepted by Portugal, with Luiz Cabral (PAIGC) president. **1981:** PAIGC was confirmed as the only legal party, with João Vieira as its secretary general; Cape Verde decided not to form a union. **1984:** A new constitution made Vieira head of both government and state. **1991:** Other parties were legalized in response to public pressure. **1994:** PAIGC secured a clear assembly majority and Vieira narrowly won the first multiparty presidential elections. **1999:** President Vieira was ousted by the army. **2000:** Kumba Yalla became president, and Caetano N'Tchama became prime minister.

Practical information

Visa requirements: UK: visa required. USA: visa required
Time difference: GMT +/–0

Major holidays: 1, 20 January, 8 February, 8 March, 1 May, 3 August, 12, 24 September, 14 November, 25 December

GUYANA

Located in South America, bounded north by the Atlantic Ocean, east by Surináme, south and southwest by Brazil, and northwest by Venezuela.

National name: *Cooperative Republic of Guyana*
Area: 214,969 sq km/82,999 sq mi
Capital: Georgetown (and chief port)
Major towns/cities: Linden, New Amsterdam, Bartica, Corriverton
Major ports: New Amsterdam
Physical features: coastal plain rises into rolling highlands with savannah in south; mostly tropical rainforest; Mount Roraima; Kaietur National Park, including Kaietur Falls on the Potaro (tributary of Essequibo) 250 m/821 ft

The black fimbriation (narrow border) expresses endurance. White stands for Guyana's rivers. Effective date: 26 May 1966.

Government

Head of state: Bharrat Jagdeo from 1999
Head of government: Samuel Hinds from 1999
Political system: liberal democracy
Political executive: limited presidency
Administrative divisions: ten regions
Political parties: People's National Congress (PNC), Afro-Guyanan, nationalist socialist; People's Progressive Party (PPP), Indian-based, left wing
Armed forces: 1,600; plus a paramilitary force of 1,500 (1998)
Conscription: military service is voluntary
Death penalty: retained and used for ordinary crimes
Defence spend: (% GDP) 1 (1998)
Education spend: (% GNP) 5 (1997)
Health spend: (% GDP) 5.1 (1997)

Economy and resources

Currency: Guyanese dollar
GDP: (US$) 649 million (1999)
Real GDP growth: (% change on previous year) 3 (1999)
GNP: (US$) 653 million (1999)
GNP per capita (PPP): (US$) 3,242 (1999 est)
Consumer price inflation: 8.7% (1999)
Unemployment: 12% (1992 est)
Foreign debt: (US$) 1.35 billion (1999 est)
Major trading partners: USA, Canada, UK, Trinidad and Tobago, Italy, France, Japan
Resources: gold, diamonds, bauxite, copper, tungsten, iron, nickel, quartz, molybdenum
Industries: agro-processing (sugar, rice, coconuts, and timber), mining, rum, pharmaceuticals, textiles
Exports: sugar, bauxite, alumina, rice, gold, rum, timber, molasses, shrimp. Principal market: USA 24.1% (1998)
Imports: mineral fuels and lubricants, machinery, capital goods, consumer goods. Principal source: USA 27.9% (1998)
Arable land: 2.4% (1995)
Agricultural products: sugar cane, rice, coffee, cocoa, coconuts, copra, tobacco, fruit and vegetables; forestry (timber production; approximately 76% of total land area was forested 1993)

Population and society

Population: 861,000 (2000 est)
Population growth rate: 0.7% (1995–2000)
Population density: (per sq km) 4 (1999 est)
Urban population: (% of total) 38 (2000 est)
Age distribution: (% of total population) 0–14 30%, 15–59 63%, 60+ 7% (2000 est)
Ethnic groups: about 51% descended from settlers from the subcontinent of India; about 43% Afro-Indian; small minorities of American Indians, Chinese, Europeans, and people of mixed race
Language: English (official), Hindi, American Indian languages
Religion: Christian 57%, Hindu 34%, Sunni Muslim 9%
Education: (compulsory years) 10
Literacy rate: 99% (men); 98% (women) (2000 est)
Labour force: 40% of population (1990): 27% agriculture, 26% industry, 47% services (1993)
Life expectancy: 61 (men); 68 (women) (1995–2000)
Child mortality rate: (under 5, per 1,000 live births) 78 (1995–2000)
Physicians: 1 per 3,360 people (1991)
TV sets: (per 1,000 people) 55 (1997)
Radios: (per 1,000 people) 498 (1997)
Internet users: (per 10,000 people) 35.1 (1999)
Personal computer users: (per 100 people) 2.5 (1999)

Transport

Airports: international airport: Georgetown (Timehri); the larger settlements in the interior have airstrips serving domestic flights; total passenger km: 235 million (1995)
Railways: none
Roads: total road network: 7,970 km/4,952 mi, of which 7.4% paved (1996 est); passenger cars: 30 per 1,000 people (1993 est)

Chronology

1498: The explorer Christopher Columbus sighted Guyana, whose name, 'land of many waters', was derived from a local American Indian word. *c.* **1620:** Settled by Dutch West India Company, who established armed bases and brought in slaves from Africa. **1814:** After a period of French rule, Britain occupied Guyana during the Napoleonic Wars and purchased Demerara, Berbice, and Essequibo. **1831:** Became British colony under name of British Guiana. **1834:** Slavery was abolished, resulting in an influx of indentured labourers from India and China to work on sugar plantations. **1860:** Settlement of the Rupununi Savannah commenced. **1860s:** Gold was discovered. **1899:** International arbitration tribunal found in favour of British Guiana in a long-running dispute with Venezuela over lands west of Essequibo River. **1953:** Assembly elections won by left-wing People's Progressive Party (PPP), drawing most support from the Indian community; Britain suspended constitution and installed interim administration, fearing communist takeover. **1961:** Internal self-government granted; Cheddi Jagan (PPP) became prime minister. **1964:** Racial violence between the Asian- and African-descended communities. **1966:** Independence achieved from Britain as Guyana, with PNC leader Forbes Burnham as prime minister. **1970:** Guyana became a republic within the Commonwealth, with Raymond Arthur Chung as president; Burnham remained as prime minister. **1980:** Burnham became the first executive president under the new constitution, which ended the three-year boycott of parliament by the PPP. **1992:** PPP had a decisive victory in the first completely free assembly elections for 20 years; a privatization programme was launched. **1997:** Cheddi Jagan died. His wife, Janet Jagan, was elected president. **1998:** Violent antigovernment protests. Government and opposition agreed to an independent audit of elections. **1999:** A constitutional reform commission was appointed. Bharrat Jagdeo replaced Janet Jagan as president.

Practical information

Visa requirements: UK: visa not required. USA: visa not required
Time difference: GMT –3
Chief tourist attractions: scenery – Kaietur Falls along Potaro River

Major holidays: 1 January, 23 February, 1 May, 1 August, 25–26 December; variable: Eid-ul-Adha, Diwali, Good Friday, Easter Monday, Prophet's Birthday, Phagwah (March), Caribbean (July)

HAITI

FORMERLY HISPANIOLA (WITH DOMINICAN REPUBLIC) (UNTIL 1844)

Located in the Caribbean, occupying the western part of the island of Hispaniola; to the east is the Dominican Republic.

National name: *République d'Haïti/Republic of Haiti*
Area: 27,750 sq km/10,714 sq mi
Capital: Port-au-Prince
Major towns/cities: Cap-Haïtien, Gonaïves, Les Cayes, St Marc, Carrefour, Delmas, Pétionville
Physical features: mainly mountainous and tropical; occupies western third of Hispaniola Island in Caribbean Sea

Blue represents the black population and links with Africa. Red stands for those of mixed race. The original blue and red flag was based on the French tricolour. Effective date: 25 February 1986.

Government

Head of state: Jean-Bertrand Aristide from 2001
Head of government: Jean-Bertrand Aristide from 2001
Political system: transitional
Political executive: transitional
Administrative divisions: nine departments, subdivided into *arrondissements* and communes
Political parties: National Front for Change and Democracy (FNCD), left of centre; Organization of People in Struggle (OPL), populist; Fanmi Lavalas (FL), personalist
Armed forces: 7,300 (1994); armed forces effectively dissolved in 1995 following restoration of civilian rule in 1994; 5,300 in paramilitary forces (1998)
Conscription: military service is voluntary
Death penalty: abolished in 1987
Defence spend: (% GDP) 2.4 (1998)
Education spend: (% GNP) N/A
Health spend: (% GDP) 4.6 (1997)

Economy and resources

Currency: gourde
GDP: (US$) 3.9 billion (1999 est)
Real GDP growth: (% change on previous year) 2.4 (1999 est)
GNP: (US$) 3.6 billion (1999)
GNP per capita (PPP): (US$) 1,407 (1999 est)
Consumer price inflation: 8.7% (1999)
Unemployment: 68% (1997 est)
Foreign debt: (US$) 1.05 billion (1998)
Major trading partners: USA, the Netherlands, Antilles, France, Italy, Germany, Japan, UK
Resources: marble, limestone, calcareous clay, unexploited copper and gold deposits
Industries: food processing, metal products, machinery, textiles, chemicals, clothing, toys, electronic and electrical equipment, tourism; much of industry closed down during the international embargo imposed by the UN after Aristide was deposed in 1991
Exports: manufactured articles, coffee, essential oils, sisal. Principal market: USA 81.4% (1997)

Imports: food and live animals, mineral fuels and lubricants, textiles, machinery, chemicals, pharmaceuticals, raw materials, vehicles. Principal source: USA 59.4% (1997)
Arable land: 20.3% (1996)
Agricultural products: coffee, sugar cane, rice, maize, sorghum, cocoa, sisal, sweet potatoes, bananas, cotton

Population and society

Population: 8,222,000 (2000 est)
Population growth rate: 1.7% (1995–2000)
Population density: (per sq km) 291 (1999 est)
Urban population: (% of total) 36 (2000 est)
Age distribution: (% of total population) 0–14 41%, 15–59 53%, 60+ 6% (2000 est)
Ethnic groups: about 95% black African descent, the remainder are mulattos or Europeans
Language: French (20%), Creole (both official)
Religion: Christian 95% (of which 70% are Roman Catholic), voodoo 4%
Education: (compulsory years) 6
Literacy rate: 52% (men); 48% (women) (2000 est)
Labour force: 45% of population: 68% agriculture, 9% industry, 23% services (1990)
Life expectancy: 51 (men); 56 (women) (1995–2000)
Child mortality rate: (under 5, per 1,000 live births) 105 (1995–2000)
Physicians: 1 per 10,855 people (1993)
Hospital beds: 1 per 1,251 people (1993)
TV sets: (per 1,000 people) 5 (1997)
Radios: (per 1,000 people) 55 (1997)
Internet users: (per 10,000 people) 7.4 (1999)

Transport

Airports: international airport: Port-au-Prince (Mais Gaté); one domestic airport (Cap-Haïtien) and four smaller airfields
Railways: none
Roads: total road network: 4,160 km/2,585 mi, of which 24.3% paved (1996 est); passenger cars: 4.4 per 1,000 people (1996 est)

Chronology

14th century: Settled by Carib Indians, who followed an earlier wave of Arawak Indian immigration. **1492:** The first landing place of the explorer Christopher Columbus in the New World, who named the island Hispaniola ('Little Spain'). **1496:** At Santo Domingo, now in the Dominican Republic to the east, the Spanish established the first European settlement in the western hemisphere, which became capital of all Spanish colonies in America. **first half of 16th century:** A third of a million Arawaks and Caribs died, as a result of enslavement and exposure to European diseases; black African slaves were consequently brought in to work the island's gold and silver mines, which were swiftly exhausted. **1697:** Spain ceded western third of Hispaniola to France, which became known as Haiti, but kept the east, which was known as Santo Domingo (the Dominican Republic). **1804:** Independence achieved after uprising against French colonial rule led by the former slave Toussaint l'Ouverture, who had died in prison in 1803, and Jean-Jacques Dessalines. **1818–43:** Ruled by Jean-Pierre Boyer, who excluded the blacks from power. **1821:** Santo Domingo fell under the control of Haiti until 1844. **1847–59:** Blacks reasserted themselves under President Faustin Soulouque. **1844:** Hispaniola was split into Haiti and the Dominican Republic. **1915:** Haiti invaded by USA as a result of political instability caused by black-mulatto friction; remained under US control until 1934. **1956:** Dr François Duvalier (Papa Doc), a voodoo physician, seized power in military coup and was elected president one year later. **1964:** Duvalier pronounced himself president for life, establishing a dictatorship based around a personal militia, the Tonton Macoutes. **1971:** Duvalier died, succeeded by his son Jean-Claude (Baby Doc); thousands murdered during Duvalier era. **1988:** A military coup installed Brig-Gen Prosper Avril as president, with a civilian government under military control. **1990:** Left-wing Catholic priest Jean-Bertrand Aristide was elected president. **1991:** Aristide was overthrown in a military coup led by Brig-Gen Raoul Cedras. Sanctions were imposed by the Organization of American States (OAS) and the USA. **1993:** United Nations (UN) embargo was imposed. Aristide's return was blocked by the military. **1994:** The threat of a US invasion led to the regime recognizing Aristide as president. **1995:** UN peacekeepers were drafted in to replace US troops. Assembly elections were won by Aristide's supporters. René Préval was elected to replace Aristide as president. **1998:** Jacques-Edouard Alexis was nominated prime minister and endorsed by the assembly. **1999:** President Préval dissolved parliament. Elections were repeatedly delayed. **2000:** Aristide's Fanmi Lavalas Party won parliamentary elections, which were boycotted by the opposition. **2001:** Aristide became president for the third time.

Practical information

Visa requirements: UK: visa not required. USA: visa not required
Time difference: GMT –5
Chief tourist attractions: beaches; subtropical vegetation

Major holidays: 1–2 January, 14 April, 1 May, 15 August, 17, 24 October, 1–2, 18 November, 5, 25 December; variable: Ascension Thursday, Carnival, Corpus Christi, Good Friday

HONDURAS

Located in Central America, bounded north by the Caribbean Sea, southeast by Nicaragua, south by the Pacific Ocean, southwest by El Salvador, and west and northwest by Guatemala.

National name: *República de Honduras/Republic of Honduras*
Area: 112,100 sq km/43,281 sq mi
Capital: Tegucigalpa
Major towns/cities: San Pedro Sula, La Ceiba, El Progreso, Choluteca, Juticalpa, Danlí
Major ports: La Ceiba, Puerto Cortés
Physical features: narrow tropical coastal plain with mountainous interior, Bay Islands, Caribbean reefs

The blue and white triband is based on the flag of the Central American Federation (CAF), a grouping of African states imposed by the British government in 1953. The stars represent the Federation's five original members. Effective date: 18 January 1949.

Government

Head of state and government: Carlos Flores from 1998
Political system: liberal democracy
Political executive: limited presidency
Administrative divisions: 18 departments
Political parties: Liberal Party of Honduras (PLH), left of centre; National Party of Honduras (PNH), right wing
Armed forces: 8,300; plus paramilitary forces numbering 6,000 (1998)
Conscription: military service is voluntary (conscription abolished in 1995)
Death penalty: abolished in 1956
Defence spend: (% GDP) 2 (1998)
Education spend: (% GNP) 3.6 (1996)
Health spend: (% GDP) 7.5 (1997)

Economy and resources

Currency: lempira
GDP: (US$) 5.3 billion (1999)
Real GDP growth: (% change on previous year) –1.9 (1999 est)
GNP: (US$) 4.8 billion (1999)
GNP per capita (PPP): (US$) 2,254 (1999 est)
Consumer price inflation: 11.6% (1999 est)
Unemployment: 3.2% (1997)
Foreign debt: (US$) 5.5 billion (1999 est)
Major trading partners: USA, Guatemala, Japan, El Salvador, Germany, Nicaragua, Mexico
Resources: lead, zinc, silver, gold, tin, iron, copper, antimony
Industries: food processing, petroleum refining, cement, beverages, wood products, chemical products, textiles, beer, rum
Exports: bananas, lobsters and prawns, zinc, meat. Principal market: USA 35.4% (1999)
Imports: machinery, appliances and electrical equipment, mineral fuels and lubricants, chemical products, consumer goods. Principal source: USA 47.1% (1999)
Arable land: 15.1% (1996)
Agricultural products: coffee, bananas, maize, sorghum, plantains, beans, rice, sugar cane, citrus fruits;

fishing (notably shellfish); livestock rearing (cattle); timber production

Population and society

Population: 6,485,000 (2000 est)
Population growth rate: 2.8% (1995–2000); 2.5% (2000–05)
Population density: (per sq km) 56 (1999 est)
Urban population: (% of total) 53 (2000 est)
Age distribution: (% of total population) 0–14 42%, 15–59 52%, 60+ 6% (2000 est)
Ethnic groups: about 90% of mixed American Indian and Spanish descent (known as ladinos or mestizos); there are also Salvadorean, Guatemalan, American, and European minorities
Language: Spanish (official), English, American Indian languages
Religion: Roman Catholic 97%
Education: (compulsory years) 6
Literacy rate: 74% (men); 75% (women) (2000 est)
Labour force: 34% of population: 37% agriculture, 22% industry, 41% services (1997)
Life expectancy: 68 (men); 72 (women) (1995–2000)
Child mortality rate: (under 5, per 1,000 live births) 49 (1995–2000)
Physicians: 1 per 1,266 people (1993 est)
Hospital beds: 1 per 994 people (1995)
TV sets: (per 1,000 people) 90 (1997)
Radios: (per 1,000 people) 386 (1997)
Internet users: (per 10,000 people) 31.7 (1999)
Personal computer users: (per 100 people) 1.0 (1999)

Transport

Airports: international airports: Tegucigalpa (Toncontín), San Pedro Sula, Roatún, La Ceiba; over 30 smaller airports serving domestic flights; total passenger km: 341 million (1995)
Railways: total length: 939 km/583 mi; passenger journeys: 705,200 (1991)
Roads: total road network: 15,400 km/9,569 mi, of which 20.3% paved (1996 est); passenger cars: 13.1 per 1,000 people (1996 est)

Chronology

c. **AD 250–900:** Part of culturally advanced Maya civilization. **1502:** Visited by Christopher Columbus, who named the country Honduras ('depths') after the deep waters off the north coast. **1525:** Colonized by Spain, who founded the town of Trujillo, but met with fierce resistance from the American Indian population. **17th century onwards:** The northern 'Mosquito Coast' fell under the control of British buccaneers, as the Spanish concentrated on the inland area, with a British protectorate being established over the coast until 1860. **1821:** Achieved independence from Spain and became part of Mexico. **1823:** Became part of United Provinces (Federation) of Central America, also embracing Costa Rica, El Salvador, Guatemala, and Nicaragua, with the Honduran liberal Gen Francisco Morazan, president of the Federation from 1830. **1838:** Achieved full independence when federation dissolved. **1880:** Capital transferred from Comayagua to Tegucigalpa. **later 19th–early 20th centuries:** The USA's economic involvement significant, with banana production, which provided two-thirds of exports in 1913, being controlled by the United Fruit Company; political instability, with frequent changes of constitution and military coups. **1925:** Brief civil war. **1932–49:** Under a right-wing National Party (PNH) dictatorship, led by Gen Tiburcio Carias Andino. **1963–74:** Following a series of military coups, Gen Oswaldo López Arelano held power, before resigning after allegedly accepting bribes from a US company. **1969:** Brief 'Football War' with El Salvador, which attacked Honduras at the time of a football competition between the two states, following evictions of thousands of Salvadoran illegal immigrants from Honduras. **1980:** The first civilian government in more than a century was elected, with Dr Roberto Suazo of the centrist Liberal Party (PLH) as president, but the commander in chief of the army, Gen Gustavo Alvárez, retained considerable power. **1983:** There was close involvement with the USA in providing naval and air bases and allowing Nicaraguan counter-revolutionaries ('Contras') to operate from Honduras. **1989:** The government and opposition declared support for a Central American peace plan to demobilize Nicaraguan Contras (thought to number 55,000 with their dependents) based in Honduras. **1992:** A border dispute with El Salvador dating from 1861 was finally resolved. **1997:** Carlos Flores (PLH) won the presidential elections, beginning his term of office in January 1998.

Practical information

Visa requirements: UK: visa not required with full British passport. USA: visa not required
Time difference: GMT –6
Chief tourist attractions: Mayan ruins at Copán; beaches on north coast; fishing and boating in Trujillo Bay and Lake Yojoa, near San Pedro
Major holidays: 1 January, 14 April, 1 May, 15 September, 3, 12, 21 October, 25, 31 December; variable: Good Friday, Holy Thursday

HUNGARY

Located in central Europe, bounded north by the Slovak Republic, northeast by Ukraine, east by Romania, south by Yugoslavia and Croatia, and west by Austria and Slovenia.

National name: *Magyar Köztársaság/Republic of Hungary*
Area: 93,032 sq km/35,919 sq mi
Capital: Budapest
Major towns/cities: Miskolc, Debrecen, Szeged, Pécs, Györ, Nyíregyháza, Székesfehérvár, Kecskemét
Physical features: Great Hungarian Plain covers eastern half of country; Bakony Forest, Lake Balaton, and Transdanubian Highlands in the west; rivers Danube, Tisza, and Raba; more than 500 thermal springs

Red stands for strength. White symbolizes faithfulness. Green represents hope. Effective date: 1 October 1957.

Government

Head of state: Ferenc Mádl from 2000
Head of government: Viktor Orban from 1998
Political system: liberal democracy
Political executive: parliamentary
Administrative divisions: 19 counties and the capital city (with 22 districts)
Political parties: over 50, including Hungarian Socialist Party (HSP), reform-socialist; Alliance of Free Democrats (AFD), centrist, radical free market; Hungarian Democratic Forum (MDF), nationalist, right of centre; Independent Smallholders Party (ISP), right of centre, agrarian; Christian Democratic People's Party (KDNP), right of centre; Federation of Young Democrats, liberal, anticommunist; Fidesz, right of centre
Armed forces: 43,300 (1998)
Conscription: 12 months (men aged 18–23)
Death penalty: abolished in 1990
Defence spend: (% GDP) 1.4 (1998)
Education spend: (% GNP) 4.7 (1996)
Health spend: (% GDP) 6.5 (1997)

Economy and resources

Currency: forint
GDP: (US$) 48.4 billion (1999)
Real GDP growth: (% change on previous year) 4.5 (1999)
GNP: (US$) 46.8 billion (1999)
GNP per capita (PPP): (US$) 10,479 (1999)
Consumer price inflation: 10% (1999)
Unemployment: 9.6% (1998)
Foreign debt: (US$) 29.7 billion (1999)
Major trading partners: Germany, Italy, Austria, USA, Russia
Resources: lignite, brown coal, natural gas, petroleum, bauxite, hard coal
Industries: food and beverages, tobacco, steel, chemicals, petroleum and plastics, engineering, transport equipment, pharmaceuticals, textiles, cement
Exports: raw materials, semi-finished products, industrial consumer goods, food and agricultural products, transport equipment. Principal market: Germany 38.4% (1999)
Imports: mineral fuels, raw materials, semi-finished products, transport equipment, food products, consumer goods. Principal source: Germany 32.7% (1999)
Arable land: 52.1% (1996)
Agricultural products: wheat, maize, sugar beet, barley, potatoes, sunflowers, grapes; livestock and dairy products

Population and society

Population: 10,036,000 (2000 est)
Population growth rate: –0.4% (1995–2000)
Population density: (per sq km) 108 (1999 est)
Urban population: (% of total) 64 (2000 est)
Age distribution: (% of total population) 0–14 17%, 15–59 63%, 60+ 20% (2000 est)
Ethnic groups: 90% indigenous, or Magyar; there is a large Romany community of around 600,000; other ethnic minorities include Germans, Croats, Romanians, Slovaks, Serbs, and Slovenes
Language: Hungarian (official)
Religion: Roman Catholic 65%, Calvinist 20%, other Christian denominations, Jewish, atheist
Education: (compulsory years) 10
Literacy rate: 99% (men); 99% (women) (2000 est)
Labour force: 39.7% of population: 8% agriculture, 34% industry, 58% services (1997)
Life expectancy: 67 (men); 75 (women) (1995–2000)
Child mortality rate: (under 5, per 1,000 live births) 12 (1995–2000)
Physicians: 1 per 238 people (1995)
Hospital beds: 1 per 109 people (1995)
TV sets: (per 1,000 people) 436 (1997)
Radios: (per 1,000 people) 689 (1997)
Internet users: (per 10,000 people) 587.7 (1999)
Personal computer users: (per 100 people) 7.4 (1999)

Transport

Airports: international airport: Budapest (Ferihegy); six domestic airports; total passenger km: 3,049 million (1997)
Railways: total length: 7,707 km/4,789 mi; total passenger km: 8,669 million (1997)
Roads: total road network: 188,203 km/116,949 mi, of which 43.4% paved (1997); passenger cars: 228 per 1,000 people (1997)

Chronology

1st century AD: Region formed part of Roman Empire. **4th century:** Germanic tribes overran central Europe. *c.* **445:** Attila the Hun established a short-lived empire, including Hungarian nomads living far to the east. *c.* **680:** Hungarians settled between the Don and Dniepr rivers under Khazar rule. **9th century:** Hungarians invaded central Europe; ten tribes united under Árpád, chief of the Magyar tribe, who conquered the area corresponding to modern Hungary in 896. **10th century:** Hungarians colonized Transylvania and raided their neighbours for plunder and slaves. **955:** Battle of Lech: Germans led by Otto the Great defeated the Hungarians. **1001:** St Stephen founded the Hungarian kingdom to replace tribal organization and converted the Hungarians to Christianity. **12th century:** Hungary became a major power when King Béla III won temporary supremacy over the Balkans. **1308–86:** Angevin dynasty ruled after the Arpádian line died out. **1456:** Battle of Belgrade: János Hunyadi defeated Ottoman Turks and saved Hungary from invasion. **1458–90:** Under Mátyás I Corvinus, Hungary enjoyed military success and cultural renaissance. **1526:** Battle of Mohács: Turks under Suleiman the Magnificent decisively defeated the Hungarians. **16th century:** Partition of Hungary between Turkey, Austria, and the semi-autonomous Transylvania. **1699:** Treaty of Karlowitz: Austrians expelled the Turks from Hungary, which was reunified under Habsburg rule. **1707:** Prince Ferenc Rákóczi II led an uprising against the Austrians. **1780–90:** Joseph II's attempts to impose uniform administration throughout the Austrian Empire provoked nationalist reaction among the Hungarian nobility. **early 19th century:** 'National Revival' movement led by Count Stephen Széchenyi and Lajos Kossuth. **1848:** Hungarian Revolution: nationalists proclaimed self-government; Croat minority resisted Hungarian rule. **1849:** Austrians crushed revolution with Russian support. **1867:** Austria conceded equality to Hungary within the dual monarchy of Austria-Hungary. **1918:** Austria-Hungary collapsed in military defeat; Count Mihály Károlyi proclaimed Hungarian Republic. **1920:** Treaty of Trianon: Hungary lost 72% of its territory to Czechoslovakia, Romania, and Yugoslavia; Admiral Miklós Horthy restored the Kingdom of Hungary with himself as regent. **1938–41:** Diplomatic collaboration with Germany allowed Hungary to regain territories lost in 1920; Hungary declared war on USSR in alliance with Germany in 1941. **1944:** Germany occupied Hungary and installed a Nazi regime. **1945:** USSR 'liberated' Hungary. **1947:** Peace treaty restored 1920 frontiers. **1949:** Hungary became a Soviet-style dictatorship; communist leader Mátyás Rákosi pursued Stalinist policies. **1956:** Hungarian uprising: anti-Soviet demonstrations led the USSR to invade, crush dissent, and install János Kádár as communist leader. **1961:** Kádár began to introduce limited liberal reforms. **1988:** The Hungarian Democratic Forum was formed by opposition groups. **1989:** The communist dictatorship was dismantled, and a transitional constitution restored multi-party democracy. The opening of the border with Austria destroyed the 'Iron Curtain'. **1990:** Elections were won by a centre–right coalition. **1991:** The withdrawal of Soviet forces was completed. **1996:** A friendship treaty with the Slovak Republic was signed, as was a cooperation treaty with Romania. **1997:** Hungary was invited to join NATO and to begin negotiations for membership of the European Union. A referendum showed clear support in favour of joining NATO. **1998:** Viktor Orban, leader of right-of-centre Fidesz, became prime minister after the general election. **1999:** Hungary became a full member of NATO. **2000:** Ferenc Mádl was elected president.

Practical information

Visa requirements: UK: visa not required. USA: visa not required
Time difference: GMT +1
Chief tourist attractions: boating and fishing on Lake Balaton; Budapest and other historical cities; Budapest has thermal springs feeding swimming pools equipped with modern physiotherapy facilities
Major holidays: 1 January, 15 March, 1 May, 20 August, 23 October, 25–26 December; variable: Easter Monday

ICELAND

Island in the North Atlantic Ocean, situated south of the Arctic Circle, between Greenland and Norway.

National name: *Lyðveldið Island/Republic of Iceland*
Area: 103,000 sq km/39,768 sq mi
Capital: Reykjavik
Major towns/cities: Akureyri, Kópavogur, Hafnerfjördur, Gardhabaer, Keflavík, Reykjanesbaer, Vestmannaeyjar
Physical features: warmed by the Gulf Stream; glaciers and lava fields cover 75% of the country; active volcanoes (Hekla was once thought the gateway to Hell), geysers, hot springs, and new islands created offshore (Surtsey in 1963); subterranean hot water heats 85% of Iceland's homes; Sidujokull glacier moving at 100 metres a day

Red symbolizes the fire from Iceland's volcanos. White represents ice. Blue stands for the mountains.
Effective date: 19 June 1915.

Government

Head of state: Ólafur Ragnar Grímsson from 1996
Head of government: Davíd Oddsson from 1991
Political system: liberal democracy
Political executive: parliamentary
Administrative divisions: 23 counties and 14 independent towns
Political parties: Independence Party (IP), right of centre; Progressive Party (PP), radical socialist; People's Alliance (PA), socialist; Social Democratic Party (SDP), moderate, left of centre; Citizens' Party, centrist; Women's Alliance, women- and family-oriented
Armed forces: no defence forces of its own; US forces under NATO are stationed there: 2,500 military personnel and a 130-strong coastguard (1998)
Death penalty: abolished in 1928
Education spend: (% GNP) 5.4 (1996)
Health spend: (% GDP) 7.9 (1997)

Economy and resources

Currency: krona
GDP: (US$) 8.8 billion (1999)
Real GDP growth: (% change on previous year) 4.5 (1999)
GNP: (US$) 8.1 billion (1999)
GNP per capita (PPP): (US$) 26,283 (1999)
Consumer price inflation: 3.2% (1999)
Unemployment: 2.8% (1998)
Major trading partners: EU (principally Germany, UK, and Denmark), Norway, USA, Japan, Sweden
Resources: aluminium, diatomite, hydroelectric and thermal power, fish
Industries: mining, fish processing, processed aluminium, fertilizer, construction, cement
Exports: fish products, aluminium, ferrosilicon, diatomite, fertilizer, animal products. Principal market: UK 19.7% (1999)
Imports: machinery and transport equipment, motor vehicles, petroleum and petroleum products, foodstuffs, textiles. Principal source: Germany 11.8% (1999)

Arable land: 0.1% (1995)
Agricultural products: hay, potatoes, turnips; fishing industry, dairy products and livestock (lamb)

Population and society

Population: 281,000 (2000 est)
Population growth rate: 0.9% (1995–2000)
Population density: (per sq km) 3 (1999 est)
Urban population: (% of total) 93 (2000 est)
Age distribution: (% of total population) 0–14 23%, 15–59 62%, 60+ 15% (2000 est)
Ethnic groups: most of the population is descended from Norwegians and Celts
Language: Icelandic (official)
Religion: Evangelical Lutheran about 90%, other Protestant and Roman Catholic about 4%
Education: (compulsory years) 9
Literacy rate: 99% (men); 99% (women) (2000 est)
Labour force: 54.6% of population: 8.5% agriculture, 25.4% industry, 66.1% services (1997)
Life expectancy: 77 (men); 81 (women) (1995–2000)
Child mortality rate: (under 5, per 1,000 live births) 6 (1995–2000)
Physicians: 1 per 333 people (1994)
Hospital beds: 1 per 68 people (1991)
TV sets: (per 1,000 people) 358 (1997)
Radios: (per 1,000 people) 950 (1997)
Internet users: (per 10,000 people) 5,385.6 (1999)
Personal computer users: (per 100 people) 35.9 (1999)

Transport

Airports: international airport: Keflavik (45 km/28 mi southwest of Reykjavik); ten major domestic airports, 12 local airports; total passenger km: 3,216 million (1997 est)
Railways: none
Roads: total road network: 12,691 km/7,886 mi, of which 25.7% paved (1997); passenger cars: 487 per 1,000 people (1997)

Chronology

7th century: Iceland discovered by Irish seafarers. **874:** First Norse settler, Ingólfr Arnarson, founded a small colony at Reykjavik. ***c.* 900:** Norse settlers came in larger numbers, mainly from Norway. **930:** Settlers established an annual parliament, the Althing, to make laws and resolve disputes. **985:** Eric the Red left Iceland to found a settlement in Greenland. **1000:** Icelanders adopted Christianity. **1263:** Icelanders recognized the authority of the king of Norway after a brief civil war. **1397:** Norway and Iceland were united with Denmark and Sweden under a single monarch.

15th century: Norway and Iceland were increasingly treated as appendages of Denmark, especially after Sweden seceded in 1449. **1783:** Poisonous volcanic eruption caused great loss of life. **1814:** Norway passed to the Swedish crown; Iceland remained under Danish rule. **1845:** Althing was re-established in modernized form. **1874:** New constitution gave Iceland limited autonomy. **1918:** Iceland achieved full self-government under the Danish crown. **1940:** British forces occupied Iceland after Germany invaded Denmark; US troops took over in 1941. **1944:** Iceland became an independent republic under President Sveinn Björnsson. **1949:** Iceland became a member of NATO. **1958:** The introduction of an exclusive fishing limit led to the first 'Cod War', when Icelandic patrol boats clashed with British fishing boats. **1972–73:** Iceland extended its fishing limit, renewing confrontations with Britain. **1975–76:** The further extension of the fishing limit caused the third 'Cod War' with the UK. **1985:** Iceland declared itself a nuclear-free zone. **1991:** Davíd Oddsson was appointed prime minister. **1992:** Iceland defied a world ban to resume its whaling industry. **1996:** Ólafur Ragnar Grímsson was elected president.

Practical information

Visa requirements: UK: visa not required. USA: visa not required
Time difference: GMT +/–0
Chief tourist attractions: rugged volcanic landscape with geysers

Major holidays: 1 January, 1 May, 17 June, 25–26 December; variable: Ascension Thursday, Good Friday, Easter Monday, Holy Thursday, First Day of Summer, August Holiday Monday

INDIA

Located in southern Asia, bounded to the north by China, Nepal, and Bhutan; east by Myanmar and Bangladesh; northwest by Pakistan and Afghanistan; and southeast, south, and southwest by the Indian Ocean.

National name: *Bharat* (Hindi)/*India*; *Bharatiya Janarajya* (unofficial)/*Republic of India*
Area: 3,166,829 sq km/1,222,713 sq mi
Capital: New Delhi
Major towns/cities: Mumbai (formerly Bombay), Kolkata (formerly Calcutta), Chennai (formerly Madras), Bangalore, Hyderabad, Ahmadabad, Kanpur, Pune, Nagpur, Bhopal, Jaipur, Lucknow, Surat
Major ports: Kolkata, Mumbai, Chennai

Orange represents Hinduism. Green stands for Islam. White expresses hope for peace and unity between the two religious groups. Effective date: 22 July 1947.

Physical features: Himalayas on northern border; plains around rivers Ganges, Indus, Brahmaputra; Deccan peninsula south of the Narmada River forms plateau between Western and Eastern Ghats mountain ranges; desert in west; Andaman and Nicobar Islands, Lakshadweep (Laccadive Islands)

Government

Head of state: Kocheril Raman Narayanan from 1997
Head of government: Atal Behari Vajpayee from 1998
Political system: liberal democracy
Political executive: parliamentary
Administrative divisions: 28 states and seven centrally administered union territories
Political parties: All India Congress Committee, or Congress, cross-caste and cross-religion coalition, left of centre; Janata Dal (People's Party), secular, left of centre; Bharatiya Janata Party (BJP), radical right wing, Hindu-chauvinist; Communist Party of India (CPI), Marxist-Leninist; Communist Party of India–Marxist (CPI–M), West Bengal-based moderate socialist
Armed forces: 1,175,000 (1998)
Conscription: none, although all citizens are constitutionally obliged to perform national service when called upon
Death penalty: retained and used for ordinary crimes
Defence spend: (% GDP) 3 (1998)
Education spend: (% GNP) 3.2 (1997 est)
Health spend: (% GDP) 5.2 (1997)

Economy and resources

Currency: rupee
GDP: (US$) 459.8 billion (1999)
Real GDP growth: (% change on previous year) 6.4 (1999 est)
GNP: (US$) 442.2 billion (1999)
GNP per capita (PPP): (US$) 2,149 (1999 est)
Consumer price inflation: 4.7% (1999)
Unemployment: 22% (1997 est)
Foreign debt: (US$) 99.8 billion (1999 est)
Major trading partners: USA, Germany, Japan, Belgium, Luxembourg, Saudi Arabia, United Arab Emirates, UK
Resources: coal, iron ore, copper ore, bauxite, chromite, gold, manganese ore, zinc, lead, limestone, crude oil, natural gas, diamonds

Industries: mining (including coal, iron and manganese ores, diamonds, and gold), manufacturing (iron and steel, mineral oils, shipbuilding, chemical products, road transport, cotton cloth, sugar, petroleum refining products)
Exports: tea (world's largest producer), coffee, fish, iron and steel, leather, textiles, clothing, polished diamonds, handmade carpets, engineering goods, chemicals. Principal market: USA 22.3% (1999)
Imports: nonelectrical machinery, mineral fuels and lubricants, pearls, precious and semiprecious stones, chemicals, transport equipment. Principal source: USA 8.9% (1999)
Arable land: 54.7% (1996)
Agricultural products: cotton, tea, wheat, rice, coffee, cashew nuts, jute, spices, sugar cane, oil seeds

Population and society

Population: 1,013,662,000 (2000 est)
Population growth rate: 1.6% (1995–2000)
Population density: (per sq km) 315 (1999 est)
Urban population: (% of total) 28 (2000 est)
Age distribution: (% of total population) 0–14 33%, 15–59 60%, 60+ 7% (2000 est)
Ethnic groups: 72% of Indo-Aryan descent; 25% (predominantly in south) Dravidian; 3% Mongoloid
Language: Hindi, English, Assamese, Bengali, Gujarati, Kannada, Kashmiri, Konkani, Malayalam, Manipuri, Marathi, Nepali, Oriya, Punjabi, Sanskrit, Sindhi, Tamil, Telugu, Urdu (all official), more than 1,650 dialects
Religion: Hindu 80%, Sunni Muslim 10%, Christian 2.5%, Sikh 2%, Buddhist, Jewish
Education: (compulsory years) 8
Literacy rate: 68% (men); 45% (women) (2000 est)
Labour force: 61% agriculture, 17% industry, 22% services (1997 est)
Life expectancy: 62 (men); 63 (women) (1995–2000)
Child mortality rate: (under 5, per 1,000 live births) 89 (1995–2000)

Physicians: 1 per 2,459 people (1993 est)
Hospital beds: 1 per 1,364 people (1993)
TV sets: (per 1,000 people) 69 (1997)
Radios: (per 1,000 people) 121 (1997)
Internet users: (per 10,000 people) 20.0
(1999)
Personal computer users: (per 100
people) 0.3 (1999)

Transport

Airports: international airports: Ahmadabad, Mumbai, Kolkata, Delhi
(Indira Gandhi), Goa, Hyderabad, Chennai, Thiruvanathapuram; over
70 domestic airports; total passenger km: 23,354 million (1997–98)
Railways: total length: 62,660 km/38,937 mi; total passenger km:
370,560 million (1997–98)
Roads: total road network: 2,060,000 km/1,280,084 mi, of which
50.2% paved (1996 est); passenger cars: 4.5 per 1,000 people (1996 est)

Chronology

c. **2500–1500 BC:** The earliest Indian civilization
evolved in the Indus Valley with the city states of
Harappa and Mohenjo Daro. *c.* **1500–1200 BC:**
Aryan peoples from the northwest overran
northern India and the Deccan; Brahmanism (a
form of Hinduism) developed. **321 BC:**
Chandragupta, founder of the Mauryan
dynasty, began to unite northern India
in a Hindu Empire. **268–232 BC:**
Mauryan Empire reached its height
under Asoka, who ruled two-thirds
of India from his capital Pataliputra.
c. **180 BC:** Shunga dynasty
replaced the Mauryans; Hindu
Empire began to break up into
smaller kingdoms. **AD
320–480:** Gupta dynasty
reunified northern India. *c.* **500:**
Raiding Huns from central Asia
destroyed the Gupta dynasty; India
reverted to many warring kingdoms.
11th–12th centuries: Rajput
princes of northern India faced
repeated Muslim invasions by
Arabs, Turks, and Afghans, and in
1206 the first Muslim dynasty was
established at Delhi. **14th–16th
centuries:** Muslim rule extended
over northern India and the
Deccan; south remained
independent under the Hindu
Vijayanagar dynasty. **1498:** Explorer
Vasco da Gama reached India, followed by
Portuguese, Dutch, French, and English
traders. **1526:** Last Muslim invasion: Zahir ud-
din Muhammad (Babur) defeated the Sultan of
Delhi at Battle of Panipat and established the
Mogul Empire, which was consolidated by Akbar
the Great (1556–1605). **1600:** East India Company founded by English merchants, who settled in Madras, Bombay,
and Calcutta. **17th century:** Mogul Empire reached its zenith under Jahangir (1605–27), Shah Jehan (1628–58),
and Aurangzeb (1658–1707). **1739:** Persian king Nadir Shah invaded India and destroyed Mogul prestige; the
British and French supported rival Indian princes in subsequent internal wars. **1757:** Battle of Plassey: Robert Clive
defeated Siraj al-Daulah, nawab of Bengal; Bengal came under control of the British East India Company. **1772–85:**
Warren Hastings, British governor general of Bengal, raised the Indian army and pursued expansionist policies. **early
19th century:** British took control (directly or indirectly) throughout India by defeating powerful Indian states in a
series of regional wars. **1858:** 'Indian Mutiny': mutiny in Bengal army erupted into widespread anti-British revolt;

rebels sought to restore powers of Mogul emperor. **1858:** British defeated the rebels; East India Company dissolved; India came under the British crown. **1885:** Indian National Congress founded in Bombay as a focus for nationalism. **1909:** Morley–Minto Reforms: Indians received the right to elect members of Legislative Councils; Hindus and Muslims formed separate electorates. **1919:** British forces killed 379 Indian demonstrators at Amritsar; India Act (Montagu–Chelmsford Reforms) conceded a measure of provincial self-government. **1920–22:** Mohandas Gandhi won control of the Indian National Congress, which launched a campaign of civil disobedience in support of the demand for complete self-rule. **1935:** India Act provided for Indian control of federal legislature, with defence and external affairs remaining the viceroy's responsibility. **1940:** Muslim League called for India to be partitioned along religious lines. **1947:** British India partitioned into two independent dominions of India (mainly Hindu) and Pakistan (mainly Muslim) amid bloody riots; Jawaharlal Nehru of Congress Party became prime minister. **1950:** India became a republic within the Commonwealth. **1962:** India lost a brief border war with China; retained Kashmir in war with Pakistan in 1965. **1966:** Indira Gandhi, daughter of Nehru, became prime minister. **1971:** India defeated Pakistan in a war and helped East Pakistan become independent as Bangladesh. **1975:** Found guilty of electoral corruption, Mrs Gandhi declared a state of emergency and arrested opponents. **1977–79:** The Janata Party formed a government under Morarji Desai. **1980:** Mrs Gandhi, heading a Congress Party splinter group, Congress (I) ('I' for Indira), was returned to power. **1984:** Troops cleared Sikh separatists from the Golden Temple, Amritsar; Mrs Gandhi was assassinated by Sikh bodyguards; her son Rajiv Gandhi became prime minister. **1989:** After financial scandals, Congress lost elections; V P Singh formed a Janata Dal minority government. **1990:** Direct rule was imposed on Jammu and Kashmir after an upsurge in Muslim separatist violence; rising interethnic and religious conflict was seen in the Punjab and elsewhere. **1992:** The destruction of a mosque at Ayodhya, northern India, by Hindu extremists resulted in widespread violence. **1995:** Bombay was renamed Mumbai. **1996:** Madras was renamed Chennai. Direct central rule was imposed on Uttar Pradesh after inconclusive assembly elections. **1997:** Kocheril Raman Narayanan became the first 'untouchable' to be elected president. **1998:** Atal Behari Vajpayee, leader of the Bharatiya Janata party, was elected prime minister. The creation of three new states was proposed. India carried out five underground nuclear explosions, meeting with international condemnation. There were floods in Uttar Pradesh. **1999:** The Indian government renounced further nuclear weapons testing and promised to sign the Comprehensive Test Ban Treaty. India used air power to attack 'infiltrators' in Kashmir. In June Kashmir peace talks were offered to Pakistan. **2000:** Relations with Pakistan worsened after India accused Pakistan of involvement (which it denied) of the hijacking of an Indian airliner by Kashmiri militants. Three new states were created: Uttaranchal was carved out of Uttar Pradesh, Jharkhand out of Bihar, and Chattisgarh out of Madhya Pradesh. Former premier Narasimha Rao was convicted on corruption charges relating to 1993. India declared a unilateral ceasefire in Kashmir in November, renewing it twice over the following months. **2001:** Over 30,000 people were killed in an earthquake in Gujarat.

Practical information

Visa requirements: UK: visa required. USA: visa required
Time difference: GMT +5.5
Chief tourist attractions: historic palaces, forts, and temples; Taj Mahal; varied scenery; wildlife
Major holidays: 1 (some states), 26 January, 1 May (some states), 30 June, 15 August, 2 October, 25, 31 December; variable: New Year (Parsi, some states)

INDONESIA FORMERLY DUTCH EAST INDIES (UNTIL 1949)

Located in southeast Asia, made up of 13,677 islands situated on or near the Equator, between the Indian and Pacific oceans.

National name: *Republik Indonesia/Republic of Indonesia*
Area: 1,904,569 sq km/735,354 sq mi
Capital: Jakarta
Major towns/cities: Surabaya, Bandung, Medan, Semarang, Palembang, Tangerang, Bandar Lampung, Ujung Pandang, Malang
Major ports: Tanjung Priok, Surabaya, Semarang (Java), Ujung Pandang (Sulawesi)

Red represents the body as well as gallantry and freedom. White stands for the soul, purity, and justice. Effective date: 17 August 1945.

Physical features: comprises 13,677 tropical islands (over 6,000 of them are inhabited): the Greater Sundas (including Java, Madura, Sumatra, Sulawesi, and Kalimantan (part of Borneo)), the Lesser Sunda Islands/Nusa Tenggara (including Bali, Lombok, Sumbawa, Flores, Sumba, Alor, Lomblen, Timor, Roti, and Savu), Maluku/Moluccas (over 1,000 islands including Ambon, Ternate, Tidore, Tanimbar, and Halmahera), and Irian Jaya (part of New Guinea); over half the country is tropical rainforest; it has the largest expanse of peatlands in the tropics

Government

Head of state and government: Abdurrahman Wahid from 1999
Political system: emergent democracy
Political executive: limited presidency
Administrative divisions: 27 provinces
Political parties: Sekber Golkar, ruling military-bureaucrat-farmers' party; United Development Party (PPP), moderate Islamic; Indonesian Democratic Party (PDI), nationalist Christian
Armed forces: 299,000; paramilitary forces 200,000 (1998)
Conscription: 2 years (selective)
Death penalty: retained and used for ordinary crimes
Defence spend: (% GDP) 2.6 (1998)
Education spend: (% GNP) 1.4 (1996)
Health spend: (% GDP) 1.7 (1997)

Economy and resources

Currency: rupiah
GDP: (US$) 140.9 billion (1999)
Real GDP growth: (% change on previous year) 0.3 (1999)
GNP: (US$) 119.5 billion (1999)
GNP per capita (PPP): (US$) 2,439 (1999)
Consumer price inflation: 20.5% (1999)
Unemployment: 15.5% (1998 est)
Foreign debt: (US$) 141.2 billion (1999 est)
Major trading partners: Japan, Singapore, USA, Hong Kong, Australia, Germany, the Netherlands, South Korea
Resources: petroleum (principal producer of petroleum in the Far East), natural gas, bauxite, nickel (world's third-largest producer), copper, tin (world's second-largest producer), gold, coal, forests
Industries: petroleum refining, food processing, textiles, wood products, tobacco, chemicals, fertilizers, rubber, cement
Exports: petroleum and petroleum products, natural and manufactured gas, textiles, rubber, palm oil, wood and wood products, electrical and electronic products, coffee, fishery products, coal, copper, tin, pepper, tea. Principal market: Japan 21.4% (1999)
Imports: machinery, transport and electrical equipment,

manufactured goods, chemical and mineral products. Principal source: Japan 12.1% (1999)
Arable land: 9.9% (1996)
Agricultural products: rice, cassava, maize, coffee, spices, tea, cocoa, tobacco, sugar cane, sweet potatoes, palm, rubber, coconuts, nutmeg; fishing

Population and society

Population: 212,107,000 (2000 est)
Population growth rate: 1.4% (1995–2000)
Population density: (per sq km) 110 (1999 est)
Urban population: (% of total) 41 (2000 est)
Age distribution: (% of total population) 0–14 31%, 15–59 61%, 60+ 8% (2000 est)
Ethnic groups: comprises more than 300 ethnic groups, the majority of which are of Malay descent; important Malay communities include Javanese (about 45% of the population), Sundanese (14%), and Madurese (7%); the largest non-Malay community is the Chinese (2%); substantial numbers of Indians, Melanesians, Micronesians, and Arabs
Language: Bahasa Indonesia (closely related to Malay; official), Javanese, Dutch, over 550 regional languages and dialects
Religion: Muslim 87%, Protestant 6%, Roman Catholic 3%, Hindu 2%, and Buddhist 1% (the continued spread of Christianity, together with an Islamic revival, have led to greater religious tensions)
Education: (compulsory years) 6
Literacy rate: 92% (men); 82% (women) (2000 est)
Labour force: 44% of population: 55% agriculture, 14% industry, 31% services (1990)
Life expectancy: 63 (men); 67 (women) (1995–2000)
Child mortality rate: (under 5, per 1,000 live births) 63 (1995–2000)
Physicians: 1 per 6,688 people (1996)
Hospital beds: 1 per 1,667 people (1996)
TV sets: (per 1,000 people) 136 (1998)
Radios: (per 1,000 people) 156 (1997)
Internet users: (per 10,000 people) 19.1 (1999)
Personal computer users: (per 100 people) 0.9 (1999)

Transport

Airports: international airports: Jakarta (Sukarno-Hatta), Irian Jaya (Frans Kaisepo), Bali (Ngurah Rai), Surabaya, Manado (Sam Ratulangi); over 60 domestic airports; total passenger km: 26,516 million (1997 est)

Railways: total length: 6,583 km/4,091 mi; total passenger km: 15,812 million (1996)
Roads: total road network: 393,000 km/244,210 mi, of which 45.5% paved (1996 est); passenger cars: 10.8 per 1,000 people (1996 est)

Chronology

3000–500 BC: Immigrants from southern China displaced original Melanesian population. **6th century AD:** Start of Indian cultural influence; small Hindu and Buddhist kingdoms developed. **8th century:** Buddhist maritime empire of Srivijaya expanded to include all Sumatra and Malay peninsula. **13th century:** Islam introduced to Sumatra by Arab merchants; spread throughout the islands over next 300 years. **14th century:** Eastern Javanese kingdom of Majapahit destroyed Srivijaya and dominated the region. *c.* **1520:** Empire of Majapahit disintegrated. **16th century:** Portuguese merchants broke the Muslim monopoly on the spice trade. **1602:** Dutch East India Company founded; it displaced the Portuguese and monopolized trade with the Spice Islands. **1619:** Dutch East India Company captured the port of Jakarta in Java and renamed it Batavia. **17th century:** Dutch introduced coffee plants

and established informal control over central Java through divide-and-rule policy. **1799:** The Netherlands took over interests of bankrupt Dutch East India Company. **1808:** French forces occupied Java; British expelled them in 1811 and returned Java to the Netherlands in 1816. **1824:** Anglo-Dutch Treaty: Britain recognized entire Indonesian archipelago as Dutch sphere of influence. **1825–30:** Java War: Prince Dipo Negoro led unsuccessful revolt against Dutch rule; further revolt 1894–96. **19th century:** Dutch formalized control over Java and conquered other islands; cultivation of coffee and sugar under tight official control made the Netherlands Indies one of the richest colonies in the world. **1908:** Dutch completed conquest of Bali. **1927:** Communist revolts suppressed; Achmed Sukarno founded Indonesian Nationalist Party (PNI) to unite diverse anti-Dutch elements. **1929:** Dutch imprisoned Sukarno and tried to suppress PNI. **1942–45:** Japanese occupation; PNI installed as anti-Western puppet government. **1945:** When Japan surrendered, President Sukarno declared an independent republic, but the Dutch set about restoring colonial rule by force. **1949:** Under US pressure, the Dutch agreed to transfer sovereignty of the Netherlands Indies (except Dutch New Guinea or Irian Jaya) to the Republic of the United States of Indonesia. **1950:** President Sukarno abolished federalism and proclaimed unitary Republic of Indonesia dominated by Java; revolts in Sumatra and South Moluccas. **1959:** To combat severe political instability, Sukarno imposed authoritarian 'guided democracy'. **1963:** The Netherlands ceded Irian Jaya to Indonesia. **1963–66:** Indonesia tried to break up Malaysia by means of blockade and guerrilla attacks. **1965–66:** Clashes between communists and army; Gen Raden Suharto imposed emergency administration and massacred up to 700,000 alleged communists. **1968:** Suharto formally replaced Sukarno as president and proclaimed 'New Order' under strict military rule. **1970s:** Rising oil exports brought significant agricultural and industrial growth. **1975:** Indonesia invaded East Timor when Portuguese rule collapsed; 200,000 died in ensuing war. **1986:** After suppressing a revolt on Irian Jaya, Suharto introduced a programme to settle 65,000 Javanese there and on outer islands. **1997:** Forest fires in Borneo and Sumatra caused catastrophic environmental damage. **1998:** Following mass riots, Suharto stepped down as president. The repressive legislation of the Suharto era was repealed and political parties were legalized. **1999:** Ethnic violence continued in Borneo. The government held a referendum on independence for East Timor in August, but after an overwhelming vote in favour, pro-Indonesian militias killed hundreds and displaced thousands of citizens. **2000:** A UN transitional government was established in East Timor. Irian Jaya unilaterally declared independence, while violence continued in Aceh.

Practical information

Visa requirements: UK: visa not required. USA: visa not required
Time difference: GMT +7/9
Chief tourist attractions: Java, with its temples and volcanic scenery; Bali, with its Hindu-Buddhist temples and religious festivals; Lombok, Sumatra, and Celebes
Major holidays: 1 January, 17 August, 25 December; variable: Ascension Thursday, Eid-ul-Adha, end of Ramadan (2 days), Good Friday, New Year (Icaka, March), New Year (Muslim), Prophet's Birthday, Ascension of the Prophet (March/April), Waisak (May)

IRAN FORMERLY PERSIA (UNTIL 1935)

Located in southwest Asia, bounded north by Armenia, Azerbaijan, the Caspian Sea, and Turkmenistan; east by Afghanistan and Pakistan; south and southwest by the Gulf of Oman and the Persian Gulf; west by Iraq; and northwest by Turkey.

National name: *Jomhûrî-ye Eslâmi-ye Îrân/Islamic Republic of Iran*
Area: 1,648,000 sq km/636,292 sq mi
Capital: Tehran
Major towns/cities: Esfahan, Mashhad, Tabriz, Shiraz, Ahvaz, Kermanshah, Qom, Karaj
Major ports: Abadan
Physical features: plateau surrounded by mountains, including Elburz and Zagros; Lake Rezayeh; Dasht-e-Kavir desert; occupies islands of Abu Musa, Greater Tunb and Lesser Tunb in the Gulf

Green represents Islam. White symbolizes peace. Red stands for courage. Effective date: 29 July 1980.

Government

Head of state and government: Ali Akbar Mohtashami from 2000
Leader of the Islamic Revolution: Seyed Ali Khamenei from 1989
Political system: Islamic nationalist
Political executive: unlimited presidency
Administrative divisions: 28 provinces
Political parties: since President Khatami's election (1997), several political parties have been licensed including Executives of Construction, Islamic Iran Solidarity Party, and Islamic Partnership Front
Armed forces: 540,000; plus 350,000 army reserves and 280,000 paramilitary forces (1998)
Conscription: military service is compulsory for two years
Death penalty: retained and used for ordinary crimes
Defence spend: (% GDP) 6.5 (1998)
Education spend: (% GNP) 4.0 (1995)
Health spend: (% GDP) 4.4 (1997)

Economy and resources

Currency: rial
GDP: (US$) 101.1 billion (1999)
Real GDP growth: (% change on previous year) 2.5 (1999)
GNP: (US$) 110.5 billion (1999)
GNP per capita (PPP): (US$) 5,163 (1999)
Consumer price inflation: 21% (1999)
Unemployment: 11% (1998 est)
Foreign debt: (US$) 11.3 billion (1999 est)
Major trading partners: Germany, Japan, Italy, United Arab Emirates, China, France
Resources: petroleum, natural gas, coal, magnetite, gypsum, iron ore, copper, chromite, salt, bauxite, decorative stone
Industries: mining, petroleum refining, textiles, food processing, transport equipment
Exports: crude petroleum and petroleum products, agricultural goods, carpets, metal ores. Principal market: Japan 20.5% (1998)
Imports: machinery and motor vehicles, paper, textiles, iron and steel and mineral products, chemicals and chemical products. Principal source: Germany 11% (1998)

Arable land: 10.9% (1996)
Agricultural products: wheat, barley, sugar beet, sugar cane, rice, fruit, tobacco, livestock (cattle, sheep, and chickens) for meat and wool production

Population and society

Population: 67,702,000 (2000 est)
Population growth rate: 1.7% (1995–2000)
Population density: (per sq km) 41 (1999 est)
Urban population: (% of total) 62 (2000 est)
Age distribution: (% of total population) 0–14 36%, 15–59 58%, 60+ 6% (2000 est)
Ethnic groups: about 66% of Persian origin, 25% Turkic, 5% Kurdish, and 4% Arabic
Language: Farsi (official), Kurdish, Turkish, Arabic, English, French
Religion: Shiite Muslim (official) 91%, Sunni Muslim 8%; Zoroastrian, Christian, Jewish, and Baha'i comprise about 1%
Education: (compulsory years) 5
Literacy rate: 83% (men); 70% (women) (2000 est)
Labour force: 29% of population: 39% agriculture, 23% industry, 39% services (1990)
Life expectancy: 69 (men); 70 (women) (1995–2000)
Child mortality rate: (under 5, per 1,000 live births) 52 (1995–2000)
Physicians: 1 per 1,273 people (1997 est)
Hospital beds: 1 per 656 people (1997 est)
TV sets: (per 1,000 people) 157 (1998)
Radios: (per 1,000 people) 265 (1997)
Internet users: (per 10,000 people) 15.0 (1999)
Personal computer users: (per 100 people) 5.2 (1999)

Transport

Airports: international airports: Tehran (Mehrabad), Abadan, Esfahan; over 20 domestic airports; total passenger km: 7,380 million (1997 est)
Railways: total length: 5,093 km/3,165 mi; total passenger km: 7,044 million (1996)
Roads: total road network: 162,000 km/100,667 mi, of which 50% paved (1996 est); passenger cars: 25.6 per 1,000 people (1996 est)

Chronology

***c.* 2000 BC:** Migration from southern Russia of Aryans, from whom Persians claim descent. **612 BC:** The Medes, from northwest Iran, destroyed Iraq-based Assyrian Empire to the west and established their own empire which extended into central Anatolia (Turkey-in-Asia). **550 BC:** Cyrus the Great overthrew Medes' empire and founded the First Persian Empire, the Achaemenid, conquering much of Asia Minor, including Babylonia (Palestine and Syria) in 539 BC. Expansion continued into Afghanistan under Darius I, who ruled 521–486 BC. **499–449 BC:** The Persian Wars with Greece ended Persian domination of the ancient world. **330 BC:** Collapse of Achaemenid Empire following defeat by Alexander the Great. **AD 224:** Sassanian Persian Empire founded by Ardashir, with its capital at Ctesiphon, in the northeast. **637:** Sassanian Empire destroyed by Muslim Arabs at battle of Qadisiya; Islam replaced Zoroastrianism. **750–1258:** Dominated by the Persianized Abbasid dynasty, who reigned as caliphs (Islamic civil and religious leaders), with a capital in Baghdad (Iraq). **1380s:** Conquered by the Mongol leader, Tamerlane. **1501:** Emergence of Safavids; the arts and architecture flourished, particularly under Abbas I, 'the Great', who ruled 1588–1629. **1736:** The Safavids were deposed by the warrior Nadir Shah Afshar, who ruled until 1747. **1790:** Rise of the Qajars, who transferred the capital from Esfahan in central Iran to Tehran, further north. **19th century:** Increasing influence in the north of tsarist Russia, which took Georgia and much of Armenia 1801–28. Britain exercised influence in the south and east, and fought Iran 1856–57 over claims to Herat (western Afghanistan). **1906:** Parliamentary constitution adopted after a brief revolution. **1925:** Qajar dynasty overthrown, with some British official help, in a coup by Col Reza Khan, a nationalist Iranian Cossack military officer, who was crowned shah ('king of kings'), with the title Reza Shah Pahlavi. **1920s onwards:** Economic modernization, Westernization, and secularization programme launched, which proved unpopular with traditionalists. **1935:** Name changed from Persia to Iran. **1941:** Pahlavi Shah was forced to abdicate during World War II by Allied occupation forces and was succeeded by his son Muhammad Reza Pahlavi, who continued the modernization programme. **1946:** British, US, and Soviet occupation forces left Iran. **1951:** Oilfields nationalized by radical prime minister Muhammad Mossadeq as anti-British and US sentiment increased. **1953:** Mossadeq deposed, the nationalization plan changed, and the US-backed shah, Muhammad Reza Shah Pahlavi, took full control of the government. **1963:** Hundreds of protesters, who demanded the release of the arrested fundamentalist Shiite Muslim leader Ayatollah Ruhollah Khomeini, were killed by troops. **1970s:** Spiralling world oil prices brought rapid economic expansion. **1975:** The shah introduced a single-party system. **1977:** The mysterious death in An Najaf of Mustafa, eldest son of the exiled Ayatollah Ruhollah Khomeini, sparked demonstrations by students, which were suppressed with the loss of six lives. **1978:** Opposition to the shah was organized from France by Ayatollah Ruhollah Khomeini, who demanded a return to the principles of Islam. Hundreds of demonstrators were killed by troops in Jaleh Square, Tehran. **1979:** Amid mounting demonstrations by students and clerics, the shah left the country; Khomeini returned to create a nonparty theocratic Islamic state. Revolutionaries seized 66 US hostages at the embassy in Tehran; US economic boycott. **1980:** Iraq invaded Iran, provoking a bitter war. The exiled shah died. **1981:** US hostages were released. **1985–87:** Fighting intensified in the Iran–Iraq War, with heavy loss of life. **1989:** Khomeini issued a fatwa (public order) for the death of British writer Salman Rushdie for blasphemy against Islam. **1990:** Generous peace terms with Iraq were accepted to close the Iran–Iraq war. **1991:** Nearly 1 million Kurds arrived from northwest Iraq, fleeing persecution by Saddam Hussein after the Gulf War between Iraq and UN forces. **1993:** Free-market economic reforms were introduced. **1997:** Reformist politician Seyyed Muhammad Khatami was elected president. **1998:** There were signs of rapprochement with the West. There was increased tension with Afghanistan, after the murder of Iranian civilians by the Talibaan. **1999:** Diplomatic relations with the UK were to be restored. **2000:** Ali Akbar Mohtashami, a former radical, was elected to lead the reforming majority in Iran's parliament. The conservative judiciary closed 15 pro-democracy newspapers, and the minister for culture resigned in the face of strong opposition to his policies. **2001:** Eight of Khatami's prominent supporters were convicted of crimes relating to expression and thought.

Practical information

Visa requirements: UK/USA: visa required
Time difference: GMT +3.5
Chief tourist attractions: wealth of historical sites, notably Esfahan, Tabriz, Rasht, Persepolis, and Susa
Major holidays: 11 February, 20–25 March, 1–2 April, 5 June; variable: Eid-ul-Adha, Ashora, end of Ramadan, Prophet's Birthday, Prophet's Mission (April), Birth of the Twelfth Imam (April/May), Martyrdom of Imam Ali (May), Death of Imam Jaffar Sadegh (June/July), Birth of Imam Reza (July), Id-E-Gihadir (August), Death of the Prophet and Martyrdom of Imam Hassan (October/November)

IRAQ

Located in southwest Asia, bounded north by Turkey, east by Iran, southeast by the Persian Gulf and Kuwait, south by Saudi Arabia, and west by Jordan and Syria.

National name: *al-Jumhuriyya al'Iraqiyya/Republic of Iraq*
Area: 434,924 sq km/167,924 sq mi
Capital: Baghdad
Major towns/cities: Mosul, Basra, Kirkuk, Hilla, An Najaf, An Nasiriya, As Sulamaniya, Arbil
Major ports: Basra, Um Qass (closed from 1980)
Physical features: mountains in north, desert in west; wide valley of rivers Tigris and Euphrates running northwest–southeast; canal linking Baghdad and Persian Gulf opened in 1992

Red stands for courage. White represents generosity. Black symbolizes Islamic triumphs.
Effective date: 22 January 1991.

Government

Head of state and government: Saddam Hussein from 1979
Political system: nationalistic socialist
Political executive: unlimited presidency
Administrative divisions: 18 provinces
Political party: Arab Ba'ath Socialist Party, nationalist socialist
Armed forces: 429,500; plus 650,000 army reserves (1998)
Conscription: military service is compulsory for 18–24 months; it is waived on the payment of the equivalent of $800
Death penalty: retained and used for ordinary crimes
Defence spend: (% GDP) 7.3 (1998)
Education spend: (% GNP) 4.6 (1988)
Health spend: (% GDP) 4.2 (1997 est)

Economy and resources

Currency: Iraqi dinar
GDP: (US$) 18.6 billion (1999 est)
Real GDP growth: (% change on previous year) 18 (1999 est)
GNP: (US$) N/A
GNP per capita (PPP): (US$) N/A
Consumer price inflation: 135% (1999 est)
Foreign debt: (US$) 130.5 billion (1999 est)
Major trading partners: USA, France, Japan, the Netherlands, Australia, China
Resources: petroleum, natural gas, sulphur, phosphates
Industries: chemical, petroleum, coal, rubber and plastic products, food processing, nonmetallic minerals, textiles, mining
Exports: crude petroleum (accounting for more than 98% of total export earnings (1980–89), dates and other dried fruits. Principal market: USA 56.4% (1999 est)
Imports: machinery and transport equipment, basic manufactured articles, cereals and other foodstuffs, iron and steel, military goods. Principal source: France 19.2% (1999 est)
Arable land: 12.6% (1996)

Agricultural products: dates, wheat, barley, maize, sugar beet, sugar cane, melons, rice; livestock rearing (notably production of eggs and poultry meat)

Population and society

Population: 23,115,000 (2000 est)
Population growth rate: 2.8% (1995–2000); 2.8% (2000–05)
Population density: (per sq km) 52 (1999 est)
Urban population: (% of total) 77 (2000 est)
Age distribution: (% of total population) 0–14 41%, 15–59 54%, 60+ 5% (2000 est)
Ethnic groups: about 79% Arab, 16% Kurdish (mainly in northeast), 3% Persian, 2% Turkish
Language: Arabic (80%) (official), Kurdish (15%), Assyrian, Armenian
Religion: Shiite Muslim 60%, Sunni Muslim 37%, Christian 3%
Education: (compulsory years) 6
Literacy rate: 65% (men); 46% (women) (2000 est)
Labour force: 12% agriculture, 20% industry, 68% services (1997 est)
Life expectancy: 61 (men); 64 (women) (1995–2000)
Child mortality rate: (under 5, per 1,000 live births) 116 (1995–2000)
Physicians: 1 per 1,659 people (1993 est)
Hospital beds: 1 per 584 people (1993 est)
TV sets: (per 1,000 people) 83 (1997)
Radios: (per 1,000 people) 228 (1996)

Transport

Airports: international airports: Baghdad (Saddam), Basra, Bamerui; at least three domestic airports (many civilian airports sustained heavy damage during the 1991 Gulf War); total passenger km: 20 million (1994)
Railways: total length: 2,389 km/1,485 mi; total passenger km: 1,169 million (1997)
Roads: total road network: 47,400 km/29,454 mi, of which 86% paved (1996 est); passenger cars: 0.6 per 1,000 people (1996 est)

Chronology

c. **3400 BC:** The world's oldest civilization, the Sumerian, arose in the land between the rivers Euphrates and Tigris, known as lower Mesopotamia, which lies in the heart of modern Iraq. Its cities included Lagash, Eridu, Uruk, Kish, and Ur. *c.* **2350 BC:** The confederation of Sumerian city-states was forged into an empire by the Akkadian leader Sargon. **7th century BC:** In northern Mesopotamia, the Assyrian Empire, based around the River Tigris and formerly dominated by Sumeria and Euphrates-centred Babylonia, created a vast empire covering much of the Middle East. **612 BC:** The Assyrian capital of Nineveh was destroyed by Babylon and Mede (in northwest Iran). *c.* **550 BC:** Mesopotamia came under Persian control. **AD 114:** Conquered by the Romans. **266:** Came under the rule of the Persian-based Sassanians. **637:** Sassanian Empire destroyed by Muslim Arabs at battle of Qadisiya, in southern Iraq; Islam spread. **750–1258:** Dominated by Abbasid dynasty, who reigned as caliphs (Islamic civil and religious leaders) in Baghdad. **1258:** Baghdad invaded and burned by Tatars. **1401:** Baghdad destroyed by Mongol ruler Tamerlane. **1533:** Annexed by Suleiman the Magnificent, becoming part of the Ottoman Empire until the 20th century, despite recurrent anti-Ottoman insurrections. **1916:** Occupied by Britain during World War I. **1920:** Iraq became a British League of Nations protectorate. **1921:** Hashemite dynasty established, with Faisal I installed by Britain as king. **1932:** Independence achieved from British protectorate status, with Gen Nuri-el Said as prime minister. **1941–45:** Occupied by Britain during World War II. **1955:** Signed the Baghdad Pact collective security treaty with the UK, Iran, Pakistan, and Turkey. **1958:** Monarchy overthrown in military-led revolution, in which King Faisal was assassinated; Iraq became a republic; joined Jordan in an Arab Federation; withdrew from Baghdad Pact as left-wing military regime assumed power. **1963:** Joint socialist-nationalist Ba'athist-military coup headed by Col Salem Aref and backed by US Central Intelligence Agency; reign of terror launched against the left. **1968:** Ba'athist military coup put Maj-Gen Ahmed Hassan al-Bakr in power. **1979:** Al-Bakr was replaced by Saddam Hussein of the Arab Ba'ath Socialist Party. **1980:** The war between Iraq and Iran broke out. **1985–87:** Fighting in the Iran–Iraq war intensified, with heavy loss of life. **1988:** There was a ceasefire and talks began with Iran. Iraq used chemical weapons against Kurdish rebels seeking greater autonomy in the northwest. **1989:** There was an unsuccessful coup against President Hussein; Iraq successfully launched a ballistic test missile. **1990:** A peace treaty favouring Iran was agreed. Iraq invaded and annexed Kuwait in August. US forces massed in Saudi Arabia at the request of King Fahd. The United Nations (UN) ordered Iraqi withdrawal and imposed a total trade ban; further UN resolution sanctioned the use of force. All foreign hostages were released. **1991:** US-led Allied forces launched an aerial assault on Iraq and destroyed the country's infrastructure; a land–sea–air offensive to free Kuwait was successful. Uprisings of Kurds and Shiites were brutally suppressed by surviving Iraqi troops. Allied troops established 'safe havens' for Kurds in the north prior to the withdrawal, and left a rapid-reaction force near the Turkish border. **1992:** The UN imposed a 'no-fly zone' over southern Iraq to protect Shiites. **1993:** Iraqi incursions into the 'no-fly zone' prompted US-led alliance aircraft to bomb strategic targets in Iraq. There was continued persecution of Shiites in the south. **1994:** Iraq renounced its claim to Kuwait, but failed to fulfil the other conditions required for the lifting of UN sanctions. **1996:** Iraqi-backed attacks on Kurds prompted US retaliation; these air strikes destroyed Iraqi military bases in the south. **1997:** Iraq continued to resist the US and Allied pressure to allow UN weapons inspections. **1998:** Iraq expelled UN weapons inspectors. In April the UN inspectors' report showed that Iraq had failed to meet UN requirements on the destruction of chemical and biological weapons. In December US and UK forces launched Operation Desert Fox which lasted four days; there were further clashes between US–UK forces and Baghdad over the no-fly zone, which continued in to 1999. **1999:** In February US–UK air strikes resumed for a short time. The UK suggested the lifting of sanctions if Iraq resumed cooperation with the UN. **2000:** The UN head of humanitarian aid efforts in Iraq resigned in protest against continuing sanctions. The Iraq–Syria border was re-opened, and Iraq began pumping oil to Syria in contravention of the UN-approved oil-for-food programme. **2001:** Iraq signed free-trade agreements with Egypt and Syria.

Practical information

Visa requirements: UK: visa required. USA: visa required
Time difference: GMT +3

Major holidays: 1, 6 January, 8 February, 21 March, 1 May, 14, 17 July; variable: Eid-ul-Adha (4 days), Ashora, end of Ramadan (3 days), New Year (Muslim), Prophet's Birthday

IRELAND OR ÉIRE

Located in the main part of the island of Ireland, in northwest Europe. It is bounded to the east by the Irish Sea, south and west by the Atlantic Ocean, and northeast by Northern Ireland.

National name: *Poblacht Na hÉireann/Republic of Ireland*
Area: 70,282 sq km/27,135 sq mi
Capital: Dublin
Major towns/cities: Cork, Limerick, Galway, Waterford, Dundalk, Bray
Major ports: Cork, Dun Laoghaire, Limerick, Waterford, Galway
Physical features: central plateau surrounded by hills; rivers Shannon, Liffey, Boyne; Bog of Allen; Macgillicuddy's Reeks, Wicklow Mountains; Lough Corrib, lakes of Killarney; Galway Bay and Aran Islands

Green represents the Catholic people. Orange stands for the Protestant people. White is a symbol of peace.
Effective date: 29 December 1937.

Government

Head of state: Mary McAleese from 1997
Head of government: Bertie Ahern from 1997
Political system: liberal democracy
Political executive: parliamentary
Administrative divisions: 26 counties within four provinces
Political parties: Fianna Fáil (Soldiers of Destiny), moderate right of centre; Fine Gael (Irish Tribe or United Ireland Party), moderate left of centre; Labour Party, moderate left of centre; Progressive Democrats, radical free-enterprise; Sinn Fein
Armed forces: 11,500 (1998)
Conscription: military service is voluntary
Death penalty: abolished in 1990
Defence spend: (% GDP) 1.0 (1998)
Education spend: (% GNP) 5.8 (1996)
Health spend: (% GDP) 6.3 (1997)

Economy and resources

Currency: Irish pound, or punt Eireannach
GDP: (US$) 84.9 billion (1999)
Real GDP growth: (% change on previous year) 9.8 (1999)
GNP: (US$) 71.4 billion (1999)
GNP per capita (PPP): (US$) 19,180 (1999)
Consumer price inflation: 1.6% (1999)
Unemployment: 7.7% (1998)
Major trading partners: UK, USA, Germany, France, Japan, the Netherlands, Singapore
Resources: lead, zinc, peat, limestone, gypsum, petroleum, natural gas, copper, silver
Industries: textiles, machinery, chemicals, electronics, motor vehicle manufacturing and assembly, food processing, beer, tourism
Exports: beef and dairy products, live animals, machinery and transport equipment, electronic goods, chemicals. Principal market: UK 22.2% (1998)
Imports: petroleum products, machinery and transport equipment, chemicals, foodstuffs, animal feed, textiles and clothing. Principal source: UK 33.6% (1998)

Arable land: 19.3% (1996)
Agricultural products: barley, potatoes, sugar beet, wheat, oats; livestock (cattle) and dairy products

Population and society

Population: 3,730,000 (2000 est)
Population growth rate: 0.7% (1995–2000)
Population density: (per sq km) 53 (1999 est)
Urban population: (% of total) 59 (2000 est)
Age distribution: (% of total population) 0–14 21%, 15–59 63%, 60+ 16% (2000 est)
Ethnic groups: most of the population has Celtic origins
Language: Irish Gaelic, English (both official)
Religion: Roman Catholic 92%, Church of Ireland, other Protestant denominations 3%
Education: (compulsory years) 9
Literacy rate: 99% (men); 99% (women) (2000 est)
Labour force: 41.3% of population: 10.4% agriculture, 27.2% industry, 62.3% services (1996)
Life expectancy: 74 (men); 79 (women) (1995–2000)
Child mortality rate: (under 5, per 1,000 live births) 8 (1995–2000)
Physicians: 1 per 578 people (1996)
Hospital beds: 1 per 200 people (1996)
TV sets: (per 1,000 people) 403 (1997)
Radios: (per 1,000 people) 699 (1997)
Internet users: (per 10,000 people) 1,198.4 (1999)
Personal computer users: (per 100 people) 32.4 (1999)

Transport

Airports: international airports: Dublin, Shannon, Cork, Knock (Horan), Galway; five domestic airports; total passengers carried: 27.9 million (1996)
Railways: total length: 1,967 km/1,222 mi; total passenger km: 1,102 million (1994)
Roads: total road network: 92,500 km/57,480 mi, of which 94.1% paved (1996 est); passenger cars: 279 per 1,000 people (1996 est)

Chronology

3rd century BC: The Gaels, a Celtic people, invaded Ireland and formed about 150 small kingdoms. **AD *c*. 432:** St Patrick introduced Christianity. **5th–9th centuries:** Irish Church was a centre of culture and scholarship. **9th–11th centuries:** The Vikings raided Ireland until defeated by High King Brian Bóruma at Clontarf in 1014. **12th–13th centuries:** Anglo-Norman adventurers conquered much of Ireland, but no central government was formed and many became assimilated.

14th–15th centuries: Irish chieftains recovered their lands, restricting English rule to the Pale around Dublin. **1536:** Henry VIII of England made ineffectual efforts to impose the Protestant Reformation on Ireland. **1541:** Irish parliament recognized Henry VIII as king of Ireland; Henry gave peerages to Irish chieftains. **1579:** English suppressed Desmond rebellion, confiscated rebel lands, and tried to 'plant' them with English settlers. **1610:** James I established plantation of Ulster with Protestant settlers from England and Scotland. **1641:** Catholic Irish rebelled against English rule; Oliver Cromwell brutally reasserted English control (1649–50); Irish landowners evicted and replaced with English landowners. **1689–91:** Williamite War: following the 'Glorious Revolution', the Catholic Irish unsuccessfully supported James II against Protestant William III in civil war. Penal laws barred Catholics from obtaining wealth and power. **1720:** Act passed declaring British Parliament's right to legislate for Ireland. **1739–41:** Famine killed one-third of population of 1.5 million. **1782:** Protestant landlords led by Henry Grattan secured end of restrictions on Irish trade and parliament. **1798:** British suppressed revolt by Society of United Irishmen (with French support) led by Wolfe Tone. **1800:** Act of Union abolished Irish parliament and created United Kingdom of Great Britain and Ireland, effective 1801. **1829:** Daniel O'Connell secured Catholic Emancipation Act, which permitted Catholics to enter parliament. **1846–52:** Potato famine reduced population by 20% through starvation and emigration. **1870:** Land Act increased security for tenants but failed to halt agrarian disorder; Isaac Butt formed political party to campaign for Irish home rule (devolution). **1885:** Home-rulers, led by Charles Stewart Parnell, held balance of power in Parliament; first Home Rule Bill rejected in 1886; second Home Rule Bill defeated in 1893. **1905:** Arthur Griffith founded the nationalist movement Sinn Fein ('Ourselves Alone'). **1914:** Ireland came close to civil war as Ulster prepared to resist implementation of Home Rule Act (postponed because of World War I). **1916:** Easter Rising: nationalists proclaimed a republic in Dublin; British crushed revolt and executed 15 leaders. **1919:** Sinn Fein MPs formed Irish parliament in Dublin in defiance of British government.
1919–21: Irish Republican Army (IRA) waged guerrilla war against British forces. **1921:** Anglo-Irish Treaty partitioned Ireland; northern Ireland (Ulster) remained part of the United Kingdom; southern Ireland won full internal self-government with dominion status. **1922:** Irish Free State proclaimed; IRA split over Anglo-Irish Treaty led to civil war (1922–23). **1932:** Anti-Treaty party, Fianna Fáil, came to power under Éamon de Valera. **1937:** New constitution established Eire (Gaelic name for Ireland) as a sovereign state and refused to acknowledge partition. **1949:** After remaining neutral in World War II, Eire left the Commonwealth and became the Republic of Ireland. **1973:** Ireland joined European Economic Community. **1985:** The Anglo-Irish Agreement gave the Republic of Ireland a consultative role, but no powers, in the government of Northern Ireland. **1990:** Mary Robinson was elected as the first female president. **1993:** The Downing Street Declaration, a joint Anglo-Irish peace proposal for Northern Ireland, was issued. **1997:** Mary McAleese was elected president; she appointed Bertie Ahern as her prime minister. **1998:** A multiparty agreement (the Good Friday Agreement) was reached on the future of Northern Ireland. The subsequent referendum showed a large majority in favour of dropping Ireland's claim to Northern Ireland. Strict legislation was passed against terrorism. **1999:** The IRA agreed to begin decommissioning discussions and a coalition government was established, with David Trimble as first minister. Powers were devolved to the province by the British government in December. **2000:** After it was revealed that there had been no arms handover, the British Secretary of State for Northern Ireland suspended the Northern Ireland Assembly and reintroduced direct rule. Within hours of the suspension of the Assembly, the British government announced a new IRA initiative on arms decommissioning.

Practical information

Visa requirements: UK/USA: visa not required
Time difference: GMT +/–0
Chief tourist attractions: scenery, notably the Killarney lakes and the west coast; Dublin, with its many literary associations and famous pub life
Major holidays: 1 January, 17 March, 25–26 December; variable: Good Friday, Easter Monday, June Holiday, August Holiday, October Holiday, Christmas Holiday

ISRAEL

Located in southwest Asia, bounded north by Lebanon, east by Syria and Jordan, south by the Gulf of Aqaba, and west by Egypt and the Mediterranean Sea.

National name: *Medinat Israel/State of Israel*
Area: 20,800 sq km/8,030 sq mi (as at 1949 armistice)
Capital: Jerusalem (not recognized by the United Nations)
Major towns/cities: Tel Aviv-Yafo, Haifa, Bat-Yam, Holon, Ramat Gan, Petah Tiqwa, Rishon le Ziyyon, Beersheba
Major ports: Tel Aviv-Yafo, Haifa, 'Akko (formerly Acre), Elat
Physical features: coastal plain of Sharon between Haifa and Tel Aviv noted since ancient times for its fertility; central mountains of Galilee, Samaria, and Judea; Dead Sea, Lake Tiberias, and River Jordan Rift Valley along the east are below sea level; Negev Desert in the south; Israel occupies Golan Heights, West Bank, East Jerusalem, and Gaza Strip (the last was awarded limited autonomy, with West Bank town of Jericho, in 1993)

The Star of David is a centuries-old symbol of Judaism. Blue and white are traditional Jewish colours. Effective date: 21 November 1948.

Government

Head of state: Moshe Katsav from 2000
Head of government: Ariel Sharon from 2001
Political system: liberal democracy
Political executive: parliamentary
Administrative divisions: six districts
Political parties: Israel Labour Party, moderate, left of centre; Consolidation Party (Likud), right of centre; Meretz (Vitality), left-of-centre alliance
Armed forces: 175,000; 430,000 reservists (1998)
Conscription: voluntary for Christians, Circassians, and Muslims; compulsory for Jews and Druzes (men 36 months, women 21 months)
Death penalty: exceptional crimes only; last execution 1962
Defence spend: (% GDP) 11.6 (1998)
Education spend: (% GNP) 7.2 (1996)
Health spend: (% GDP) 8.2 (1997)

Economy and resources

Currency: shekel
GDP: (US$) 99.1 billion (1999)
Real GDP growth: (% change on previous year) 2.3 (1999)
GNP: (US$) 104 billion (1999)
GNP per capita (PPP): (US$) 16,867 (1999)
Consumer price inflation: 5.2% (1999)
Unemployment: 8.6% (1998)
Foreign debt: 36.4 billion (1999)
Major trading partners: USA, UK, Germany, Belgium–Luxembourg, UK, Italy, Japan, Switzerland, the Netherlands
Resources: potash, bromides, magnesium, sulphur, copper ore, gold, salt, petroleum, natural gas
Industries: food processing, beverages, tobacco, electrical machinery, chemicals, petroleum and coal products, metal products, diamond polishing, transport equipment, tourism
Exports: citrus fruits, worked diamonds, machinery and parts, military hardware, food products, chemical products, textiles and clothing. Principal market: USA 35.5% (1999)
Imports: machinery and parts, rough diamonds, chemicals and related products, crude petroleum and petroleum products, motor vehicles. Principal source: USA 20.3% (1999)

Arable land: 17% (1996)
Agricultural products: citrus fruits, vegetables, potatoes, wheat, melons, pumpkins, avocados; poultry and fish

Population and society

Population: 6,217,000 (2000 est)
Population growth rate: 2.2% (1995–2000)
Population density: (per sq km) 293 (1999 est)
Urban population: (% of total) 91 (2000 est)
Age distribution: (% of total population) 0–14 28%, 15–59 59%, 60+ 13% (2000 est)
Ethnic groups: around 81% of the population is Jewish, the majority of the remainder Arab. Under the Law of Return 1950, 'every Jew shall be entitled to come to Israel as an immigrant'; those from the East and Eastern Europe are Ashkenazim, and those from Mediterranean Europe (Spain, Portugal, Italy, France, Greece) and Arab Africa are Sephardim (over 50% of Sephardic descent); an Israeli-born Jew is a Sabra
Language: Hebrew, Arabic (both official), English, Yiddish, other European and west Asian languages
Religion: Israel is a secular state, but the predominant faith is Judaism 80%; Sunni Muslim (about 15%), Christian, Druze
Education: (compulsory years) 11
Literacy rate: 98% (men); 94% (women) (2000 est)
Labour force: 2.4% agriculture, 19.5% industry, 78.1% services (1997 est)
Life expectancy: 76 (men); 80 (women) (1995–2000)
Child mortality rate: (under 5, per 1,000 live births) 10 (1995–2000)
Physicians: 1 per 350 people (1991)
Hospital beds: 1 per 164 people (1997 est)
TV sets: (per 1,000 people) 318 (1998)
Radios: (per 1,000 people) 520 (1997)
Internet users: (per 10,000 people) 1,639.0 (1999)
Personal computer users: (per 100 people) 24.6 (1999)

Transport

Airports: international airports: Tel Aviv (Ben Gurion), Elat; domestic airports in all major cities; total passenger km: 11,794 million (1997 est)

Railways: total length: 530 km/329 mi; total passenger km: 346 million (1997)

Roads: total road network: 15,464 km/9,609 mi, of which 100% paved (1997); passenger cars: 210.4 per 1,000 people (1997)

Chronology

c. **2000 BC:** Abraham, father of the Jewish people, is believed to have come to Palestine from Mesopotamia. *c.* **1225 BC:** Moses led the Jews out of slavery in Egypt towards the promised land of Palestine. **11th century BC:** Saul established a Jewish kingdom in Palestine; developed by kings David and Solomon. **586 BC:** Jews defeated by Babylon and deported; many returned to Palestine in 539 BC. **333 BC:** Alexander the Great of Macedonia conquered the entire region. **3rd century BC:** Control of Palestine contested by Ptolemies of Egypt and Seleucids of Syria. **142 BC:** Jewish independence restored after Maccabean revolt. **63 BC:** Palestine fell to Roman Empire. **70 AD:** Romans crushed Zealot rebellion and destroyed Jerusalem; start of dispersion of Jews (diaspora). **614:** Persians took Jerusalem from Byzantine Empire. **637:** Muslim Arabs conquered Palestine. **1099:** First Crusade captured Jerusalem; Christian kingdom lasted a century before falling to sultans of Egypt. **1517:** Palestine conquered by the Ottoman Turks. **1897:** Theodor Herzl organized the First Zionist Congress at Basel to publicize Jewish claims to Palestine. **1917:** The Balfour Declaration: Britain expressed support for the creation of a Jewish National Home in Palestine. **1918:** British forces expelled the Turks from Palestine, which became a British League of Nations mandate in 1920. **1929:** Severe violence around Jerusalem caused by Arab alarm at doubling of Jewish population in ten years. **1933:** Jewish riots in protest at British attempts to restrict Jewish immigration. **1937:** The Peel Report, recommending partition, accepted by most Jews but rejected by Arabs. **1939:** Britain postponed independence plans on account of World War II, and increased military presence. **1946:** Resumption of terrorist violence. **1947:** United Nations (UN) voted for partition of Palestine. **1948:** Britain withdrew; Independent State of Israel proclaimed with David Ben-Gurion as prime minister; Israel repulsed invasion by Arab nations; many Palestinian Arabs settled in refugee camps in the Gaza Strip and West Bank. **1952:** Col Gamal Nasser of Egypt stepped up blockade of Israeli ports and support of Arab guerrillas in Gaza. **1956:** War between Israel and Egypt; Israeli invasion of Gaza and Sinai followed by withdrawal in 1957. **1964:** Palestine Liberation Organization (PLO) founded to unite Palestinian Arabs with the aim of overthrowing the state of Israel. **1967:** Israel defeated Egypt, Syria, and Jordan in the Six-Day War; Gaza, West Bank, east Jerusalem, Sinai, and Golan Heights captured. **1969:** Yassir Arafat became chair of the PLO; escalation of terrorism and border raids. **1973:** Yom Kippur War: Israel repulsed surprise attack by Egypt and Syria. **1977:** President Anwar Sadat of Egypt began peace initiative. **1979:** Camp David talks ended with signing of peace treaty between Israel and Egypt; Israel withdrew from Sinai. **1980:** United Jerusalem was declared the capital of Israel. **1982:** Israeli forces invaded southern Lebanon to drive out PLO guerrillas; occupation continued until 1985. **1988:** The Israeli handling of Palestinian uprising (Intifada) in the occupied territories provoked international criticism. **1990:** The PLO formally recognized the state of Israel. **1991:** Iraq launched missile attacks on Israel during the Gulf War. **1992:** A Labour government was elected under Yitzhak Rabin. **1993:** Rabin and Arafat signed a peace accord; Israel granted limited autonomy to Gaza Strip and Jericho. **1994:** Arafat became the head of an autonomous Palestinian authority in Gaza and Jericho; a peace agreement was reached between Israel and Jordan. **1995:** Rabin was assassinated by a Jewish opponent of the peace accord. **1996:** A Likud government was elected under Binjamin Netanyahu, a critic of the peace accord. **1997:** A Jewish settlement in east Jerusalem was widely condemned. There were suicide bombs by Hamas in Jerusalem. **1998:** Violence flared on the West Bank between Palestinians and Israeli troops, again stalling the peace process. The Wye Peace Agreement was signed with the PLO. President Clinton attempted to restart the peace process. **1999:** The South Lebanon 'security zone' was expanded. Yassir Arafat delayed the declaration of an independent state until after the Israeli elections. Ehud Barak (Labour) was elected prime minister and restarted peace negotiations. **2000:** Israel withdrew from the Golan Heights. In September, renewed violence between Palestinians and Israeli security forces broke out and quickly escalated. Repeated efforts to end the violence failed, and Barak announced his resignation in December. **2001:** Ariel Sharon was elected prime minister.

Practical information

Visa requirements: UK/USA: visa not required
Time difference: GMT +2
Chief tourist attractions: resorts along Mediterranean coast, Red Sea coast (Elat), and Dead Sea; ancient city of Jerusalem, with its four quarters (Armenian, Christian, Jewish, and Muslim) and many sites, including Temple Mount, the Cathedral of St James, Dome of the Rock, Al-Aqsa Mosque, and Church of the Holy Sepulchre; historic sites of Bethlehem, Nazareth, Masada, Megiddo, Jericho; caves of the Dead Sea scrolls

ITALY

Located in southern Europe, bounded north by Switzerland and
Austria, east by Slovenia, Croatia, and the Adriatic Sea, south by the
Ionian and Mediterranean seas, and west by the Tyrrhenian and
Ligurian seas and France. It includes the Mediterranean islands of
Sardinia and Sicily.

The colours of the flag date from Napoleon's invasion in
1796. The design is based on the French tricolour.
Effective date: 19 June 1946.

National name: *Repubblica Italiana/Italian Republic*
Area: 301,300 sq km/116,331 sq mi
Capital: Rome
Major towns/cities: Milan, Naples, Turin, Palermo, Genoa, Bologna,
Florence
Major ports: Naples, Genoa, Palermo, Bari, Catania, Trieste
Physical features: mountainous (Maritime Alps, Dolomites,
Apennines) with narrow coastal lowlands; continental Europe's only active volcanoes: Vesuvius, Etna, Stromboli;
rivers Po, Adige, Arno, Tiber, Rubicon; islands of Sicily, Sardinia, Elba, Capri, Ischia, Lipari, Pantelleria; lakes Como,
Maggiore, Garda

Government

Head of state: Carlo Azeglio Ciampi from 1999
Head of government: Giuliano Amato from 2000
Political system: liberal democracy
Political executive: parliamentary
Administrative divisions: 94 provinces within 20
regions (of which five have a greater degree of autonomy)
Political parties: Forza Italia (Go Italy!), free market,
right of centre; Northern League (LN), Milan-based,
federalist, right of centre; National Alliance (AN),
neofascist; Italian Popular Party (PPI), Catholic, centrist;
Italian Renewal Party, centrist; Democratic Party of the
Left (PDS), Pro-European, moderate left wing (ex-
communist); Italian Socialist Party (PSI), moderate
socialist; Italian Republican Party (PRI), social
democratic, left of centre; Democratic Alliance (AD),
moderate left of centre; Christian Democratic Centre
(CCD), Christian, centrist; Olive Tree alliance, left of
centre; Panella List, radical liberal; Union of the
Democratic Centre (UDC), right of centre; Pact for Italy,
reformist; Communist Refoundation (RC), Marxist;
Verdi, environmentalist; La Rete (the Network), anti-
Mafia
Armed forces: 298,400 (1998)
Conscription: 12 months
Death penalty: abolished in 1994
Defence spend: (% GDP) 2 (1998)
Education spend: (% GNP) 4.7 (1996)
Health spend: (% GDP) 7.6 (1997)

Economy and resources

Currency: lira
GDP: (US$) 1,149.9 billion (1999)
Real GDP growth: (% change on previous year) 1.4
(1999)
GNP: (US$) 1,136 billion (1999)
GNP per capita (PPP): (US$) 20,751 (1999)
Consumer price inflation: 1.6% (1999)

Unemployment: 12.3% (1998)
Major trading partners: EU (principally Germany,
France, the Netherlands, Spain, and UK), USA
Resources: lignite, lead, zinc, mercury, potash, sulphur,
fluorspar, bauxite, marble, petroleum, natural gas, fish
Industries: machinery and machine tools, textiles,
leather, footwear, food and beverages, steel, motor
vehicles, chemical products, wine, tourism
Exports: machinery and transport equipment, textiles,
clothing, footwear, wine (leading producer and exporter),
metals and metal products, chemicals, wood, paper and
rubber goods. Principal market: Germany 16.7% (1999)
Imports: mineral fuels and lubricants, machinery and
transport equipment, chemical products, foodstuffs,
metal products. Principal source: Germany 19.2% (1999)
Arable land: 27.6% (1996)
Agricultural products: sugar beet, grapes, wheat, maize,
tomatoes, olives, citrus fruits, vegetables; fishing

Population and society

Population: 57,298,000 (2000 est)
Population growth rate: –0.01% (1995–2000); –0.2%
(2000–05)
Population density: (per sq km) 190 (1999 est)
Urban population: (% of total) 67 (2000 est)
Age distribution: (% of total population) 0–14 14%,
15–59 62%, 60+ 24% (2000 est)
Ethnic groups: mainly Italian; some minorities of
German origin
Language: Italian (official), German and Ladin (in the
north), French (in the Valle d'Aosta region), Greek and
Albanian (in the south)
Religion: Roman Catholic 98% (state religion)
Education: (compulsory years) 8
Literacy rate: 99% (men); 98% (women) (2000 est)
Labour force: 6.8% agriculture, 32% industry, 61.2%
services (1997)
Life expectancy: 75 (men); 81 (women) (1995–2000)

Child mortality rate: (under 5, per 1,000 live births) 8 (1995–2000)
Physicians: 1 per 182 people (1996)
Hospital beds: 1 per 176 people (1996)
TV sets: (per 1,000 people) 486 (1998)
Radios: (per 1,000 people) 878 (1997)
Internet users: (per 10,000 people) 872.0 (1999)
Personal computer users: (per 100 people) 19.2 (1999)

Transport

Airports: international airports: Bologna (G Marconi), Genoa (Cristoforo Colombo), Milan (Linate and Malpensa), Naples (Capodichino), Palermo (Punta Rais), Pisa (Galileo Galilei), Rome (Leonardo da Vinci and Ciampino), Turin, Venice (Marco Polo); over 30 domestic airports; total passenger km: 37,728 million (1997 est)
Railways: total length: 15,942 km/9,906 mi; total passenger km: 50,300 million (1996)
Roads: total road network: 307,682 km/191,194 mi, of which 100% paved (1996); passenger cars: 528 per 1,000 people (1996)

Chronology

4th and 3rd centuries BC: Italian peninsula united under Roman rule. **AD 476:** End of Western Roman Empire. **568:** Invaded by Lombards. **756:** Papal States created in central Italy. **800:** Charlemagne united Italy and Germany in Holy Roman Empire. **12th and 13th centuries:** Papacy and Holy Roman Empire contended for political supremacy; papal power reached its peak under Innocent III (1198–1216). **1183:** Cities of Lombard League (founded in 1164) became independent. **14th century:** Beginnings of Renaissance in northern Italy. **15th century:** Most of Italy ruled by five rival states: the city-states of Milan, Florence, and Venice; the Papal States; and the Kingdom of Naples. **1494:** Charles VIII of France invaded Italy. **1529–59:** Spanish Habsburgs secured dominance in Italy. **17th century:** Italy effectively part of Spanish Empire; economic and cultural decline. **1713:** Treaty of Utrecht gave political control of most of Italy to Austrian Habsburgs. **1796–1814:** France conquered Italy, setting up satellite states and introducing principles of French Revolution. **1815:** Old regimes largely restored; Italy divided between Austria, Papal States, Naples, Sardinia, and four duchies. **1831:** Giuseppe Mazzini founded the 'Young Italy' movement with the aim of creating a unified republic. **1848–49:** Liberal revolutions occurred throughout Italy; reversed everywhere except Sardinia, which became a centre of nationalism under the leadership of Count Camillo di Cavour. **1859:** France and Sardinia forcibly expelled Austrians from Lombardy. **1860:** Sardinia annexed duchies and Papal States (except Rome); Giuseppe Garibaldi overthrew Neapolitan monarchy. **1861:** Victor Emmanuel II of Sardinia proclaimed King of Italy in Turin. **1866:** Italy gained Venetia after defeat of Austria by Prussia. **1870:** Italian forces occupied Rome in defiance of Pope, completing unification of Italy. **1882:** Italy joined Germany and Austria-Hungary in Triple Alliance. **1896:** Attempt to conquer Ethiopia defeated at Battle of Adowa. **1900:** King Umberto I assassinated by an anarchist. **1912:** Annexation of Libya and Dodecanese after Italo-Turkish War. **1915:** Italy entered World War I on side of Allies. **1919:** Peace treaties awarded Trentino, South Tyrol, and Trieste to Italy. **1922:** Mussolini established fascist dictatorship following period of strikes and agrarian revolts. **1935–36:** Conquest of Ethiopia. **1939:** Invasion of Albania. **1940:** Italy entered World War II as ally of Germany. **1943:** Allies invaded southern Italy; Mussolini removed from power; Germans occupied northern and central Italy. **1945:** Allies completed liberation. **1946:** Monarchy replaced by a republic. **1947:** Peace treaty stripped Italy of its colonies. **1948:** New constitution adopted; Christian Democrats emerged as main party of government in political system marked by ministerial instability. **1957:** Italy became a founder member of European Economic Community (EEC). **1963:** Creation of first of long series of fragile centre-left coalition governments. **1976:** Communists attempt to join the coalition, the 'historic compromise', rejected by the Christian Democrats. **1978:** Christian Democrat Aldo Moro, the architect of historic compromise, was murdered by Red Brigade guerrillas infiltrated by Western intelligence agents. **1983–87:** Bettino Craxi, Italy's first Socialist prime minister, led the coalition. The economy improved. **1993:** A major political crisis was triggered by the exposure of government corruption and Mafia links, and governing parties were discredited. A new electoral system replaced proportional representation, with 75% majority voting. **1999:** Carlo Azeglio Ciampi was elected president. Former prime minister Prodi became president of the new European Commission. **2000:** After his centre-left coalition was beaten in regional elections, Massimo d'Alema resigned his position as prime minister, which he had held since 1998, and Giuliano Amato was sworn in as head of Italy's 58th government since 1945. Amato later conceded leadership of the coalition in the next general election to Franceso Rutelli, the mayor of Rome.

Practical information

Visa requirements: UK: visa not required. USA: visa not required
Time difference: GMT +1
Chief tourist attractions: Alpine and Mediterranean scenery; ancient Greek and Roman archaeological remains; medieval, Renaissance, and baroque churches (including St Peter's, Rome); Renaissance towns and palaces; museums; art galleries; opera houses
Major holidays: 1, 6 January, 25 April, 1 May, 14 August (mid-August holiday, 2 days), 1 November, 8, 25–26 December; variable: Easter Monday

JAMAICA

Island in the Caribbean Sea, south of Cuba and west of Haiti.

Area: 10,957 sq km/4,230 sq mi
Capital: Kingston
Major towns/cities: Montego Bay, Spanish Town, St Andrew, Portmore, May Pen
Physical features: mountainous tropical island; Blue Mountains (so called because of the haze over them)

Black, yellow, and green are colours found in many African flags and reflect the islanders' heritage. Effective date: 6 August 1962.

Government

Head of state: Queen Elizabeth II from 1962, represented by Governor General Howard Felix Hanlan Cooke from 1991
Head of government: Percival Patterson from 1992
Political system: liberal democracy
Political executive: parliamentary
Administrative divisions: 14 parishes
Political parties: Jamaica Labour Party (JLP), moderate, centrist; People's National Party (PNP), left of centre; National Democratic Union (NDM), centrist
Armed forces: 3,300 (1998)
Conscription: military service is voluntary
Death penalty: retained and used for ordinary crimes
Defence spend: (% GDP) 0.9 (1998)
Education spend: (% GNP) 7.5 (1996)
Health spend: (% GDP) 6 (1997)

Economy and resources

Currency: Jamaican dollar
GDP: (US$) 6.2 billion (1999)
Real GDP growth: (% change on previous year) –0.4 (1999)
GNP: (US$) 6 billion (1999)
GNP per capita (PPP): (US$) 3,276 (1999)
Consumer price inflation: 6% (1999)
Unemployment: 16.2% (1996)
Foreign debt: (US$) 3.9 billion (1999)
Major trading partners: USA, UK, Mexico, Venezuela, EU, Canada, Norway
Resources: bauxite (one of world's major producers), marble, gypsum, silica, clay
Industries: mining and quarrying, bauxite processing, food processing, petroleum refining, clothing, cement, glass, tourism
Exports: bauxite, alumina, gypsum, sugar, bananas, garments, rum. Principal market: USA 39.5% (1998)
Imports: mineral fuels, machinery and transport equipment, basic manufactures, chemicals, food and live animals, miscellaneous manufactured articles. Principal source: USA 47.7% (1998)

Arable land: 16.6% (1996)
Agricultural products: sugar cane, bananas, citrus fruit, coffee, cocoa, coconuts; livestock rearing (goats, cattle, and pigs)

Population and society

Population: 2,583,000 (2000 est)
Population growth rate: 0.9% (1995–2000)
Population density: (per sq km) 234 (1999 est)
Urban population: (% of total) 56 (2000 est)
Age distribution: (% of total population) 0–14 31%, 15–59 59%, 60+ 10% (2000 est)
Ethnic groups: 76% of African descent, about 15% of mixed African-European origin, plus about 3% Indian, 3% European, and 1% Chinese
Language: English (official), Jamaican Creole
Religion: Protestant 70%, Rastafarian
Education: (compulsory years) 6
Literacy rate: 83% (men); 91% (women) (2000 est)
Labour force: 22% agriculture, 21% industry, 57% services (1996 est)
Life expectancy: 73 (men); 77 (women) (1995–2000)
Child mortality rate: (under 5, per 1,000 live births) 27 (1995–2000)
Physicians: 1 per 6,420 people (1993)
Hospital beds: 1 per 463 people (1993)
TV sets: (per 1,000 people) 182 (1997)
Radios: (per 1,000 people) 480 (1997)
Internet users: (per 10,000 people) 234.4 (1999)
Personal computer users: (per 100 people) 4.3 (1999)

Transport

Airports: international airports: Kingston (Norman Manley), Montego Bay (Donald Sangster); four domestic airports; total passenger km: 2,720 million (1997 est)
Railways: total length: 272 km/169 mi; total passenger km: 1.2 million (1992)
Roads: total road network: 19,000 km/11,807 mi, of which 70.7% paved (1996 est); passenger cars: 39.6 per 1,000 people (1996 est)

Chronology

c. **AD 900:** Settled by Arawak Indians, who gave the island the name Jamaica ('well watered'). **1494:** The explorer Christopher Columbus reached Jamaica. **1509:** Occupied by Spanish; much of Arawak community died from exposure to European diseases; black African slaves brought in to work sugar plantations. **1655:** Captured by Britain and became its most valuable Caribbean colony. **1838:** Slavery abolished. **1870:** Banana plantations established as sugar cane industry declined in face of competition from European beet sugar. **1938:** Serious riots during the economic depression and, as a sign of growing political awareness, the People's National Party (PNP) was formed by Norman Manley. **1944:** First constitution adopted. **1958–62:** Part of West Indies Federation. **1959:** Internal self-government granted. **1962:** Independence achieved within the Commonwealth, with Alexander Bustamante of the centre-right Jamaica Labour Party (JLP) as prime minister. **1981:** Diplomatic links with Cuba were severed, and a free-market economic programme was pursued. **1988:** The island was badly damaged by Hurricane Gilbert. **1992:** Percival Patterson of the PNP became prime minister. **1998:** Violent crime increased as the economy declined.

Practical information

Visa requirements: UK: visa not required. USA: visa not required
Time difference: GMT –5
Chief tourist attractions: climate; beaches; mountains; historic buildings

Major holidays: 1 January, 23 May, 5 August, 20 October, 25–26 December; variable: Ash Wednesday, Good Friday, Easter Monday

JAPAN

A group of islands in northeast Asia situated between the Sea of Japan (to the west) and the north Pacific (to the east), east of North and South Korea.

National name: *Nihon-koku/State of Japan*
Area: 377,535 sq km/145,766 sq mi
Capital: Tokyo
Major towns/cities: Yokohama, Osaka, Nagoya, Fukuoka, Kita-Kyushu, Kyoto, Sapporo, Kobe, Kawasaki, Hiroshima
Major ports: Osaka, Nagoya, Yokohama, Kobe
Physical features: mountainous, volcanic (Mount Fuji, volcanic Mount Aso, Japan Alps); comprises over 1,000 islands, the largest of which are Hokkaido, Honshu, Kyushu, and Shikoku

The mon, the central red disc, is called Hi-no-maru or sun-disc. The disc is set slightly towards the hoist. White symbolizes honesty and purity.
Effective date: 5 August 1854.

Government

Head of state: Emperor Akihito from 1989
Head of government: Yoshiro Mori from 2000
Political system: liberal democracy
Political executive: parliamentary
Administrative divisions: 47 prefectures
Political parties: Liberal Democratic Party (LDP), right of centre; Shinshinto (New Frontier Party) opposition coalition, centrist reformist; Social Democratic Party of Japan (SDPJ, former Socialist Party), left of centre but moving towards centre; Shinto Sakigake (New Party Harbinger), right of centre; Japanese Communist Party (JCP), socialist; Democratic Party of Japan (DPJ), Sakigake and SDPJ dissidents; Komeito (Clean Government Party), reformist, pro-democratic
Armed forces: self-defence forces: 242,600; US forces stationed there: 44,800 (1998)
Conscription: military service is voluntary
Death penalty: retained and used for ordinary crimes
Defence spend: (% GDP) 1 (1998)
Education spend: (% GNP) 3.6 (1996)
Health spend: (% GDP) 7.2 (1997)

Economy and resources

Currency: yen
GDP: (US$) 4,395.1 billion (1999)
Real GDP growth: (% change on previous year) 0.3 (1999)
GNP: (US$) 4,078.9 billion (1999)
GNP per capita (PPP): (US$) 24,041 (1999)
Consumer price inflation: –0.3% (1999)
Unemployment: 4.1% (1998)
Major trading partners: USA, China, Australia, South Korea, Indonesia, Germany, Taiwan
Resources: coal, iron, zinc, copper, natural gas, fish
Industries: motor vehicles, steel, machinery, electrical and electronic equipment, chemicals, textiles
Exports: motor vehicles, electronic goods and components, chemicals, iron and steel products, scientific and optical equipment. Principal market: USA 30.7% (1999)
Imports: mineral fuels, foodstuffs, live animals, bauxite, iron ore, copper ore, coking coal, chemicals, textiles, wood. Principal source: USA 21.7% (1999)
Arable land: 10.5% (1996)

Agricultural products: rice, potatoes, cabbages, sugar cane, sugar beet, citrus fruit; one of the world's leading fishing nations

Population and society

Population: 126,714,000 (2000 est)
Population growth rate: 0.2% (1995–2000); 0.1% (2000–05)
Population density: (per sq km) 335 (1999 est)
Urban population: (% of total) 79 (2000 est)
Age distribution: (% of total population) 0–14 15%, 15–59 62%, 60+ 23% (2000 est)
Ethnic groups: more than 99% of Japanese descent; Ainu (aboriginal people of Japan) in north Japan (Hokkaido, Kuril Islands)
Language: Japanese (official), Ainu
Religion: Shinto, Buddhist (often combined), Christian (less than 1%)
Education: (compulsory years) 9
Literacy rate: 99% (men); 99% (women) (2000 est)
Labour force: 5.3% agriculture, 33.1% industry, 61.6% services (1997)
Life expectancy: 77 (men); 83 (women) (1995–2000)
Child mortality rate: (under 5, per 1,000 live births) 6 (1995–2000)
Physicians: 1 per 542 people (1994)
Hospital beds: 1 per 75 people (1994)
TV sets: (per 1,000 people) 708 (1997)
Radios: (per 1,000 people) 955 (1997)
Internet users: (per 10,000 people) 1,446.6 (1999)
Personal computer users: (per 100 people) 28.7 (1999)

Transport

Airports: international airports: Tokyo (Narita), Fukuoka, Kagoshima, Kansai, Nagoya, Osaka; one principal domestic services airport (Haneda), smaller airports cover connections between major towns and islands; total passenger km: 151,048 million (1997 est)
Railways: total length: 38,125 km/23,690 mi; total passenger km: 402,513 million (1994)
Roads: total road network: 1,147,532 km/713,076 mi, of which 74.3% paved (1996); passenger cars: 376 per 1,000 people (1996)

Chronology

c. **400 AD:** The Yamato, one of many warring clans, unified central Japan. **5th–6th centuries:**
Writing, Confucianism, and Buddhism spread to Japan from China and Korea. **646:** Start of Taika
Reform: Emperor Kotoku organized central government on Chinese model. **794:** Heian became imperial
capital; later called Kyoto. **858:** Imperial court fell under control of Fujiwara clan, who reduced the
emperor to a figurehead. **11th century:** Central government grew ineffectual; real power exercised by
great landowners (daimyo) with private armies of samurai. **1185:** Minamoto clan seized power under
Yoritomo, who established military rule. **1192:** Emperor gave Yoritomo the title of shogun (general); the
shogun ruled in the name of the emperor. **1274:** Mongol conqueror Kublai Khan attempted to invade
Japan, making a second attempt in 1281; on both occasions Japan was saved by a typhoon. **1336:**
Warlord Takauji Ashikaga overthrew Minamoto shogunate; emperor recognized Ashikaga shogunate in
1338. **16th century:** Power of Ashikagas declined; constant civil war. **1543:** Portuguese
sailors were the first Europeans to reach Japan; followed by Spanish, Dutch, and English
traders. **1549:** Spanish missionary St Francis Xavier began to preach Roman Catholic
faith in Japan. **1585–98:** Warlord Hideyoshi took power and attempted to conquer
Korea in 1592 and 1597. **1603:** Ieyasu Tokugawa founded new shogunate at Edo,
reformed administration, and suppressed Christianity. **1630s:** Japan adopted
policy of isolation: all travel forbidden and all foreigners expelled except a
small colony of Dutch traders in Nagasaki harbour. **1853:** USA sent warships
to Edo with demand that Japan open diplomatic and trade relations; Japan
conceded in 1854. **1867:** Revolt by isolationist nobles overthrew the
Tokugawa shogunate. **1868:** Emperor Mutsuhito assumed full powers,
adopted the title *Meiji* ('enlightened rule'), moved imperial
capital from Kyoto to Edo (renamed Tokyo), and launched
policy of swift Westernization. **1894–95:** Sino-Japanese War:
Japan expelled Chinese from Korea. **1902–21:** Japan entered a
defensive alliance with Britain. **1904–05:** Russo-Japanese War:
Japan drove Russians from Manchuria and Korea; Korea annexed
in 1910. **1914:** Japan entered World War I and occupied
German possessions in Far East. **1923:** Earthquake destroyed
much of Tokyo and Yokohama. **1931:** Japan invaded
Chinese province of Manchuria and created
puppet state of Manchukuo; Japanese
government came under control of military and
extreme nationalists. **1937:** Japan resumed
invasion of China. **1940:** After Germany
defeated France, Japan occupied French Indo-
China. **1941:** Japan attacked US fleet at Pearl
Harbor; USA and Britain declared war on Japan.
1942: Japanese conquered Thailand, Burma,
Malaya, Dutch East Indies, Philippines, and
northern New Guinea. **1945:** USA
dropped atomic bombs on Hiroshima and
Nagasaki; Japan surrendered; US general
Douglas MacArthur headed Allied
occupation administration. **1947:**
MacArthur supervised introduction of democratic 'Peace Constitution', accompanied by demilitarization and land reform. **1952:**
Occupation ended. **1955:** Liberal Democratic Party (LDP) founded with support of leading business people. **1956:** Japan admitted
to United Nations. **1950s–70s:** Rapid economic development; growth of manufacturing exports led to great prosperity. **1993:** An
economic recession and financial scandals brought about the downfall of the LDP government in a general election. A coalition
government was formed. **1995:** An earthquake devastated Kobe. **1997:** A financial crash occurred after bank failures. **1998:** The
government introduced a new $200 billion economic stimulus package, after the worst recession since World War II.

Practical information

Visa requirements: UK: visa not required. USA: visa
not required for a stay of up to 90 days
Time difference: GMT +9
Chief tourist attractions: ancient capital of Kyoto;
Buddhist and Shinto temples; pagodas; forests and
mountains; classical Kabuki theatre; traditional festivals
Major holidays: 1–3, 15 January, 11 February, 21
March, 29 April, 3, 5 May, 15, 23 September, 10
October, 3, 23 November

JORDAN

Located in southwest Asia, bounded north by Syria, northeast by Iraq, east, southeast, and south by Saudi Arabia, south by the Gulf of Aqaba, and west by Israel.

National name: *Al-Mamlaka al-Urduniyya al-Hashemiyyah/Hashemite Kingdom of Jordan*
Area: 89,206 sq km/34,442 sq mi (excluding the West Bank 5,879 sq km/2,269 sq mi)
Capital: Amman
Major towns/cities: Zarqa, Irbid, Saet, Ma'an
Major ports: Aqaba
Physical features: desert plateau in east; Rift Valley separates east and west banks of River Jordan

The points of the star represent the first seven verses of the Koran. Red, black, white, and green became the pan-Arab colours. Effective date: 16 April 1928.

Government

Head of state: King Abdullah ibn Hussein from 1999
Head of government: Ali Abu al-Ragheb from 2000
Political system: emergent democracy
Political executive: parliamentary
Administrative divisions: 12 governates
Political parties: independent groups loyal to the king predominate; of the 21 parties registered since 1992, the most significant is the Islamic Action Front (IAF), Islamic fundamentalist
Armed forces: 104,100; plus paramilitary forces of approximately 30,000 (1998)
Conscription: selective
Death penalty: retained and used for ordinary crimes
Defence spend: (% GDP) 7.7 (1998)
Education spend: (% GNP) 6.8 (1997)
Health spend: (% GDP) 5.2 (1997)

Economy and resources

Currency: Jordanian dinar
GDP: (US$) 7.62 billion (1999)
Real GDP growth: (% change on previous year) 1.6 (1999)
GNP: (US$) 7 billion (1999)
GNP per capita (PPP): (US$) 3,542 (1999)
Consumer price inflation: 0.6% (1999)
Unemployment: 20% (1997 est)
Foreign debt: (US$) 8.1 billion (1999 est)
Major trading partners: Saudi Arabia, Germany, India, Iraq, Italy, UK, USA, Japan
Resources: phosphates, potash, shale
Industries: mining and quarrying, petroleum refining, chemical products, alcoholic drinks, food products, phosphate, cement, potash, tourism
Exports: phosphate, potash, fertilizers, foodstuffs, pharmaceuticals, fruit and vegetables, cement. Principal market: Saudi Arabia 15.9% (1999)
Imports: food and live animals, basic manufactures, mineral fuels, machinery and transport equipment. Principal source: Germany 9.9% (1999)
Arable land: 3.6% (1996)

Agricultural products: wheat, barley, maize, tobacco, vegetables, fruits, nuts; livestock rearing (sheep and goats)

Population and society

Population: 6,669,000 (2000 est)
Population growth rate: 3% (1995–2000)
Population density: (per sq km) 73 (1999 est)
Urban population: (% of total) 74 (2000 est)
Age distribution: (% of total population) 0–14 42%, 15–59 53%, 60+ 5% (2000 est)
Ethnic groups: majority of Arab descent (98%); small Circassian, Armenian, and Kurdish minorities
Language: Arabic (official), English
Religion: over 90% Sunni Muslim (official religion), small communities of Christians and Shiite Muslims
Education: (compulsory years) 10
Literacy rate: 95% (men); 84% (women) (2000 est)
Labour force: 27% of population: 15% agriculture, 23% industry, 61% services (1990)
Life expectancy: 69 (men); 72 (women) (1995–2000)
Child mortality rate: (under 5, per 1,000 live births) 31 (1995–2000)
Physicians: 1 per 825 people (1994)
Hospital beds: 1 per 749 people (1994)
TV sets: (per 1,000 people) 52 (1998)
Radios: (per 1,000 people) 287 (1997)
Internet users: (per 10,000 people) 123.4 (1999)
Personal computer users: (per 100 people) 1.4 (1999)

Transport

Airports: international airports: Amman (charter flights only), Zizya (Queen Alia, 30 km south of Amman), Aqaba; internal flights operate between Amman and Aqaba; total passenger km: 4,900 million (1997 est)
Railways: total length: 788 km/490 mi; total passenger km: 1 million (1994)
Roads: total road network: 6,640 km/4,126 mi, of which 100% paved (1996 est); passenger cars: 48.2 per 1,000 people (1996)

Chronology

13th century BC: Oldest known 'states' of Jordan, including Gideon, Ammon, Moab, and Edom, established. **c. 1000 BC:** East Jordan was part of kingdom of Israel, under David and Solomon. **4th century BC:** Southeast Jordan occupied by the independent Arabic-speaking Nabataeans. **64 BC:** Conquered by the Romans and became part of the province of Arabia. **AD 636:** Became largely Muslim after the Byzantine forces of Emperor Heraclius were defeated by Arab armies at battle of Yarmuk, in northern Jordan. **1099–1187:** Part of Latin Kingdom established by Crusaders in Jerusalem. **from early 16th century:** Part of Turkish Ottoman Empire, administered from Damascus. **1920:** Trans-Jordan (the area east of the River Jordan) and Palestine (which includes the West Bank) placed under British administration by League of Nations mandate. **1923:** Trans-Jordan separated from Palestine and recognized by Britain as a substantially independent state under the rule of Emir Abdullah ibn Hussein, a member of the Hashemite dynasty of Arabia. **1946:** Trans-Jordan achieved independence from Britain, with Abd Allah as king; name changed to Jordan. **1948:** British mandate for Palestine expired, leading to fighting between Arabs and Jews, who each claimed the area. **1950:** Jordan annexed West Bank; 400,000 Palestinian refugees flooded into Jordan, putting pressure on the economy. **1952:** Partially democratic constitution introduced. **1958:** Jordan and Iraq formed Arab Federation that ended when Iraqi monarchy was deposed. **1967:** Israel defeated Egypt, Syria, and Jordan in Arab–Israeli Six-Day War, and captured and occupied the West Bank, including Arab Jerusalem. Martial law imposed. **1970–71:** Jordanians moved against increasingly radicalized Palestine Liberation Organization (PLO), which had launched guerrilla raids on Israel from Jordanian territory, resulting in bloody civil war, before the PLO leadership fled abroad. **1976:** Political parties were banned and elections postponed until further notice. **1980:** Jordan emerged as an important ally of Iraq in its war against Iran, an ally of Syria, with whom Jordan's relations were tense. **1984:** Women voted for the first time; the parliament was recalled. **1985:** King Hussein ibn Tal Abdulla el Hashim and PLO leader Yassir Arafat put forward a framework for a Middle East peace settlement. There was a secret meeting between Hussein and the Israeli prime minister. **1988:** Hussein announced his willingness to cease administering the West Bank as part of Jordan, passing responsibility to the PLO; parliament was suspended. **1989:** There were riots over price increases of up to 50% following a fall in oil revenues. In the first parliamentary elections for 23 years the Muslim Brotherhood won 25 of 80 seats but were exiled from government. **1990:** Hussein unsuccessfully tried to mediate after Iraq's invasion of Kuwait. There were huge refugee problems as thousands fled to Jordan from Kuwait and Iraq. **1991:** 24 years of martial law ended, the ban on political parties was lifted, and Jordan remained neutral during the Gulf War involving Iraq. **1993:** Candidates loyal to Hussein won a majority in the parliamentary elections; several leading Islamic fundamentalists lost their seats. **1994:** An economic cooperation pact was signed with the PLO. A peace treaty was signed with Israel, ending the 46-year-old state of war. **1999:** King Hussein died and his eldest son, Abdullah, succeeded him. Ali Abu al-Ragheb was appointed prime minister. In May, Abdullah held talks with Yassir Arafat prior to Israeli peace negotiations.

Practical information

Visa requirements: UK: visa required. USA: visa required
Time difference: GMT +2
Chief tourist attractions: ancient cities of Petra and Jerash

Major holidays: 1 January, 1, 25 May, 10 June, 11 August, 14 November, 25 December; variable: Eid-ul-Adha (4 days), first day of Ramadan, end of Ramadan (4 days), New Year (Muslim), Prophet's Birthday

KAZAKHSTAN

Located in central Asia, bounded north by Russia, west by the Caspian Sea, east by China, and south by Turkmenistan, Uzbekistan, and Kyrgyzstan.

National name: *Kazak Respublikasy/Republic of Kazakhstan*
Area: 2,717,300 sq km/1,049,150 sq mi
Capital: Astana (formerly Akmola)
Major towns/cities: Karagauda, Pavlodar, Semipalatinsk, Petropavlovsk, Shymkent
Physical features: Caspian and Aral seas, Lake Balkhash; Steppe region; natural gas and oil deposits in the Caspian Sea

Blue represents the sky. The golden sun symbolizes the country's hopes for the future.
Effective date: 4 June 1992.

Government

Head of state: Nursultan Nazarbayev from 1991
Head of government: Kasymzhomart Tokaev from 1999
Political system: authoritarian nationalist
Political executive: unlimited presidency
Administrative divisions: 14 regions and one city (Astana)
Political parties: Congress of People's Unity of Kazakhstan, moderate, centrist; People's Congress of Kazakhstan, moderate, ethnic; Socialist Party of Kazakhstan (SPK), left wing; Republican Party, right-of-centre coalition
Armed forces: 55,100 (1998)
Death penalty: retained and used for ordinary crimes
Defence spend: (% GDP) 2.2 (1998)
Education spend: (% GNP) 4.4 (1997)
Health spend: (% GDP) 3.9 (1997)

Economy and resources

Currency: tenge
GDP: (US$) 15.6 billion (1999)
Real GDP growth: (% change on previous year) 1.7 (1999 est)
GNP: (US$) 18.9 billion (1999)
GNP per capita (PPP): (US$) 4,408 (1999)
Consumer price inflation: 8.4% (1999)
Unemployment: 8% (1998)
Foreign debt: (US$) 7.9 billion (1999)
Major trading partners: Russia, China, USA, UK, Germany, EU
Resources: petroleum, natural gas, coal, bauxite, chromium, copper, iron ore, lead, titanium, magnesium, tungsten, molybdenum, gold, silver, manganese
Industries: metal processing, heavy engineering, mining and quarrying, chemicals, fuel, power, machine-building, textiles, food processing, household appliances
Exports: ferrous and non-ferrous metals, mineral products (including petroleum and petroleum products), chemicals. Principal market: Russia 19.8% (1999)

Imports: energy products and electricity, machinery and transport equipment, chemicals. Principal source: Russia 36.7% (1999)
Arable land: 11.9% (1996)
Agricultural products: fruits, sugar beet, vegetables, potatoes, cotton, cereals; livestock rearing (particularly sheep); karakul and astrakhan wool

Population and society

Population: 16,223,000 (2000 est)
Population growth rate: –0.4% (1995–2000)
Population density: (per sq km) 6 (1999 est)
Urban population: (% of total) 56 (2000 est)
Age distribution: (% of total population) 0–14 28%, 15–59 60%, 60+ 12% (2000 est)
Ethnic groups: 50% of Kazakh descent, 32% ethnic Russian, 4.5% Ukrainian, 2% German, 2% Uzbek, and 2% Tatar
Language: Kazakh (related to Turkish; official), Russian
Religion: Sunni Muslim 50–60%, Russian Orthodox 30–35%
Education: (compulsory years) 11
Literacy rate: 99% (men); 99% (women) (2000 est)
Labour force: 22% agriculture, 22.2% industry, 55.8% services (1995)
Life expectancy: 63 (men); 73 (women) (1995–2000)
Child mortality rate: (under 5, per 1,000 live births) 41 (1995–2000)
Physicians: 1 per 265 people (1996)
Hospital beds: 1 per 86 people (1996)
TV sets: (per 1,000 people) 231 (1997)
Radios: (per 1,000 people) 384 (1997)
Internet users: (per 10,000 people) 43.0 (1999)

Transport

Airports: international airports: Almaty, Aktau, Atyrau; 18 domestic airports; total passenger km: 2,429 million (1995)
Railways: total length: 13,826 km/8,591 mi; total passenger km: 15,204 million (1995)
Roads: total road network: 125,796 km/78,170 mi, of which 80.5% paved (1997); passenger cars: 61 per 1,000 people (1997)

Chronology

early Christian era: Settled by Mongol and Turkic tribes. **8th century:** Spread of Islam. **10th century:** Southward migration into east Kazakhstan of Kazakh tribes, displaced from Mongolia by the Mongols. **13th–14th centuries:** Part of Mongol Empire. **late 15th century:** Kazakhs emerged as distinct ethnic group from Kazakh Orda tribal confederation. **early 17th century:** The nomadic, cattle-breeding Kazakhs split into smaller groups, united in the three Large, Middle, and Lesser Hordes (federations), led by khans (chiefs). **1731–42:** Faced by attacks from the east by Oirot Mongols, protection was sought from the Russian tsars, and Russian control was gradually established. **1822–48:** Conquest by tsarist Russia completed; khans deposed. Large-scale Russian and Ukrainian peasant settlement of the steppes after the abolition of serfdom in Russia in 1861. **1887:** Alma-Alta (now Almaty), established in 1854 as a fortified trading centre and captured by the Russians in 1865, destroyed by an earthquake. **1916:** 150,000 killed as anti-Russian rebellion brutally repressed. **1917:** Bolshevik coup in Russia followed by outbreak of civil war in Kazakhstan. **1920:** Autonomous republic in USSR. **early 1930s:** More than 1 million died of starvation during the campaign to collectivize agriculture. **1936:** Joined USSR and became a full union republic. **early 1940s:** Volga Germans deported to the republic by Soviet dictator Joseph Stalin. **1954–56:** Part of Soviet leader Nikita Khrushchev's ambitious 'Virgin Lands' agricultural extension programme; large influx of Russian settlers made Kazakhs a minority in their own republic. **1986:** There were nationalist riots in Alma-Alta (now Almaty) after the reformist Soviet leader Mikhail Gorbachev ousted the local communist leader and installed an ethnic Russian. **1989:** Nursultan Nazarbayev, a reformist and mild nationalist, became leader of the Kazakh Communist Party (KCP) and instituted economic and cultural reform programmes, encouraging foreign inward investment. **1990:** Nazarbayev became head of state; economic sovereignty was declared. **1991:** Nazarbayev became head of state and condemned the attempted anti-Gorbachev coup in Moscow; the KCP was abolished. The country joined the new Commonwealth of Independent States (CIS); and independence was recognized by the USA. **1992:** Kazakhstan was admitted into the United Nations (UN) and the Conference on Security and Cooperation in Europe (CSCE; now the Organization on Security and Cooperation in Europe, OSCE). **1993:** Presidential power was increased by a new constitution. A privatization programme was launched. START-1 (disarmament treaty) and Nuclear Non-Proliferation Treaty were both ratified by Kazakhstan. **1994:** There was economic, social, and military union with Kyrgyzstan and Uzbekistan. **1995:** An economic and military cooperation pact was signed with Russia. Kazakhstan achieved nuclear-free status. **1997:** Astana (formerly known as Akmola) was designated as the new capital. President Nazarbayev appointed Nurlan Balgymbayev, head of the Kazakh Oil state petroleum company, prime minister. **1998:** A treaty of 'eternal friendship' and a treaty of deepening economic cooperation was signed with Uzbekistan. **1999:** Nazarbayev was re-elected president, though international observers claimed the election was flawed. Kasymzhomart Tokaev was appointed prime minister. **2000:** A huge offshore oil field was discovered in the Caspian Sea.

Practical information

Visa requirements: UK: visa required. USA: visa required

Time difference: GMT +6

Major holidays: 1, 28 January, 8, 22 March, 1, 9 May, 25 October, 31 December

KENYA

Located in east Africa, bounded to the north by Sudan and Ethiopia, to the east by Somalia, to the southeast by the Indian Ocean, to the southwest by Tanzania, and to the west by Uganda.

National name: *Jamhuri ya Kenya/Republic of Kenya*
Area: 582,600 sq km/224,941 sq mi
Capital: Nairobi
Major towns/cities: Mombasa, Kisumu, Nakuru, Eldoret, Nyeri
Major ports: Mombasa
Physical features: mountains and highlands in west and centre; coastal plain in south; arid interior and tropical coast; semi-desert in north; Great Rift Valley, Mount Kenya, Lake Nakuru (salt lake with world's largest colony of flamingos), Lake Turkana (Rudolf)

Black stands for the African people. White symbolizes peace. Black, red, and green, the 'black liberation' colours, denote Africa's rebirth. Red represents the blood common to all people. Green recalls the fertile land. Effective date: 12 December 1963.

Government

Head of state and government: Daniel arap Moi from 1978
Political system: authoritarian nationalist
Political executive: unlimited presidency
Administrative divisions: seven provinces and the Nairobi municipality
Political parties: Kenya African National Union (KANU), nationalist, centrist; Forum for the Restoration of Democracy–Kenya (FORD–Kenya), left of centre; Forum for the Restoration of Democracy–Asili (FORD–Asili), left of centre; Democratic Party (DP), centrist; Safina, centrist
Armed forces: 24,200; paramilitary force 5,000 (1998)
Conscription: military service is voluntary
Death penalty: retained and used for ordinary crimes
Defence spend: (% GDP) 3.1 (1998)
Education spend: (% GNP) 6.6 (1996)
Health spend: (% GDP) 4.6 (1997)

Economy and resources

Currency: Kenyan shilling
GDP: (US$) 10.6 billion (1999)
Real GDP growth: (% change on previous year) 1.4 (1999)
GNP: (US$) 10.6 billion (1999)
GNP per capita (PPP): (US$) 975 (1999)
Consumer price inflation: 2.6% (1999)
Unemployment: 30% (urban, 1995 est)
Foreign debt: (US$) 6.2 billion (1999)
Major trading partners: Uganda, UK, Tanzania, Germany, Japan, United Arab Emirates, Pakistan
Resources: soda ash, fluorspar, salt, limestone, rubies, gold, vermiculite, diatonite, garnets
Industries: food processing, petroleum refining and petroleum products, textiles and clothing, leather products, chemicals, cement, paper and paper products, beverages, tobacco, ceramics, rubber and metal products, vehicle assembly, tourism
Exports: coffee, tea, petroleum products, soda ash, horticultural products. Principal market: Uganda 18.3% (1999 est)

Imports: crude petroleum, motor vehicles, industrial machinery, iron and steel, chemicals, basic manufactures. Principal source: UK 11.9% (1999 est)
Arable land: 7% (1996)
Agricultural products: coffee, tea, maize, wheat, sisal, sugar cane, pineapples, cotton, horticulture; dairy products

Population and society

Population: 30,080,000 (2000 est)
Population growth rate: 2% (1995–2000)
Population density: (per sq km) 51 (1999 est)
Urban population: (% of total) 33 (2000 est)
Age distribution: (% of total population) 0–14 43%, 15–59 52%, 60+ 5% (2000 est)
Ethnic groups: main ethnic groups are the Kikuyu (about 21%), the Luhya (14%), the Luo (13%), the Kalenjin (11%), the Kamba (11%), the Kisii (6%), and the Meru (5%); there are also Asian, Arab, and European minorities
Language: English, Kiswahili (both official), many local dialects
Religion: Roman Catholic 28%, Protestant 8%, Muslim 6%, traditional tribal religions
Education: (years) 8 (not compulsory, but free)
Literacy rate: 89% (men); 76% (women) (2000 est)
Labour force: 76.7% agriculture, 8% industry, 15.3% services (1996)
Life expectancy: 51 (men); 53 (women) (1995–2000)
Child mortality rate: (under 5, per 1,000 live births) 104 (1995–2000)
Physicians: 1 per 6,430 people (1994)
Hospital beds: 1 per 786 people (1994)
TV sets: (per 1,000 people) 21 (1997)
Radios: (per 1,000 people) 104 (1997)
Internet users: (per 10,000 people) 11.8 (1999)
Personal computer users: (per 100 people) 0.4 (1999)

Transport

Airports: international airports: Mombasa (Moi), Nairobi (Jomo Kenyatta), Eldoret (opening date of 1997 delayed); three domestic airports; total passenger km: 1,802 million (1997 est)

Railways: total length: 2,740 km/1,702 mi; total passenger km: 464 million (1994)
Roads: total road network: 63,800 km/39,645 mi, of which 13.9% paved (1996 est); passenger cars: 10.9 per 1,000 people (1996 est)

Chronology

8th century: Arab traders began to settle along coast of East Africa.
16th century: Portuguese defeated coastal states and exerted spasmodic control over them. **18th century:** Sultan of Oman reasserted Arab overlordship of East African coast, making it subordinate to Zanzibar.
19th century: Europeans, closely followed by Christian missionaries, began to explore inland.
1887: British East African Company leased area of coastal territory from sultan of Zanzibar. **1895:** Britain claimed large inland region as East African Protectorate. **1903:** Railway from Mombasa to Uganda built using Indian labourers, many of whom settled in the area; British and South African settlers began to farm highlands. **1920:** East African Protectorate became crown colony of Kenya, with legislative council elected by white settlers (and by Indians and Arabs soon afterwards). **1923:** Britain rejected demand for internal self-government by white settlers. **1944:** First African appointment to legislative council; Kenyan African Union (KAU) founded to campaign for African rights.
1947: Jomo Kenyatta became leader of KAU, which was dominated by Kikuyu tribe.
1952: Mau Mau (Kikuyu secret society) began terrorist campaign to drive white settlers from tribal lands; Mau Mau largely suppressed by 1954 but state of emergency lasted for eight years. **1953:** Kenyatta charged with management of Mau Mau activities and imprisoned by the British. He was released in 1959, but exiled to northern Kenya. **1956:** Africans allowed to elect members of legislative council on a restricted franchise. **1960:** Britain announced plans to prepare Kenya for majority African rule. **1961:** Kenyatta allowed to return to help negotiate Kenya's independence. **1963:** Kenya achieved independence with Kenyatta as prime minister.
1964: Kenya became a republic with Kenyatta as president. **1969:** Kenya became one-party state under Kenyan African National Union (KANU). **1978:** President Kenyatta died and was succeeded by Daniel arap Moi. **1984:** There were violent clashes between government troops and the ethnic Somali population at Wajir. **1989:** Moi announced the release of political prisoners. **1991:** A multiparty system was conceded after an opposition group was launched. **1997:** There were demonstrations calling for democratic reform. Constitutional reforms were adopted.
1998: A bomb exploded at the US embassy in Nairobi, killing over 230 people and injuring 5,000; an anti-American Islamic group claimed responsibility. **1999:** A framework agreement was signed with the leaders of Uganda and Tanzania, intending to reestablish the East African Community (EAC) which had collapsed in 1977, hoping to lead to a Common market and political federation similar to that of the European Union (EU).

Practical information

Visa requirements: UK: visa not required. USA: visa required
Time difference: GMT +3
Chief tourist attractions: wildlife – 25 national parks and 23 game reserves, notably the Rift Valley, containing Aberdare National Park, overlooked by Mount Kenya, which also has a national park; Indian Ocean coast
Major holidays: 1 January, 1 May, 1 June, 10, 20 October, 12, 25–26 December; variable: Good Friday, Easter Monday, end of Ramadan

KIRIBATI FORMERLY PART OF THE GILBERT AND ELLICE ISLANDS (UNTIL 1979)

Republic in the west central Pacific Ocean, comprising three groups of coral atolls.

National name: *Ribaberikan Kiribati/Republic of Kiribati*
Area: 717 sq km/277 sq mi
Capital: Bairiki (on Tarawa atoll)
Major towns/cities: principal islands are the Gilbert Islands, the Phoenix Islands, the Line Islands, Banaba
Major ports: Bairiki, Betio (on Tarawa)
Physical features: comprises 33 Pacific coral islands

The flag was selected following a design competition. The waves represent the Pacific Ocean. Effective date: 12 July 1979.

Government

Head of state and government: Teburoro Tito from 1994
Political system: liberal democracy
Political executive: limited presidency
Political parties: Maneaban Te Mauri (MTM), dominant faction; National Progressive Party (NPP), former governing faction 1979–94
Armed forces: no standing army
Death penalty: abolished in 1979

Economy and resources

Currency: Australian dollar
GDP: (US$) 75 million (1999)
Real GDP growth: (% change on previous year) 2.5 (1999 est)
GNP: (US$) 81 million (1999)
GNP per capita (PPP): (US$) 3,186 (1999)
Consumer price inflation: 2% (1999 est)
Unemployment: 2.2% (1992 est)
Foreign debt: (US$) 11 million (1997)
Major trading partners: Australia, Bangladesh, Japan, Fiji Islands, USA, France, New Zealand, China
Resources: phosphate, salt

Industries: handicrafts, coconut-based products, soap, foods, furniture, leather goods, garments, tourism
Exports: copra, fish, seaweed, bananas, breadfruit, taro.
Imports: foodstuffs, machinery and transport equipment, mineral fuels, basic manufactures.
Arable land: 50.7% (1995)
Agricultural products: copra, coconuts, bananas, screw-pine, papaya, breadfruit; livestock rearing

Population and society

Population: 83,000 (2000 est)
Population growth rate: 1.4% (1995–2000)
Population density: (per sq km) 107 (1999 est)
Urban population: (% of total) 39 (2000 est)
Age distribution: (% of total population) 0–14 40%, 15–59 54%, 60+ 6% (2000 est)
Ethnic groups: predominantly Micronesian, with a Polynesian minority; also European and Chinese minorities
Language: English (official), Gilbertese
Religion: Roman Catholic, Protestant (Congregationalist)
Education: (compulsory years) 9
Literacy rate: 90% (men); 90%(women) (1993 est)
Life expectancy: 61 (men); 65 (women) (1998 est)

Chronology

1st millennium BC: Settled by Austronesian-speaking peoples. **1606:** Visited by Spanish explorers. **late 18th century:** Visited by British naval officers. **1857:** Christian mission established. **1892:** Gilbert (Kiribati) and Ellice (Tuvalu) Islands proclaimed a British protectorate.
1916–39: Uninhabited Phoenix Islands, Christmas Island, Ocean Island, and Line Island (Banaba) added to colony. **1942–43:** While occupied by Japanese it was the scene of fierce fighting with US troops. **late 1950s:** UK tested nuclear weapons on Christmas Island (Kiritimati). **1963:** Legislative council established. **1974:** Legislative council replaced by an elected House of Assembly.
1975: The mainly Melanesian-populated Ellice Islands separated to become Tuvalu. **1977:** The predominantly Micronesian-populated Gilbert Islands was granted internal self-government. **1979:** The Gilbert Islands achieved independence within the Commonwealth, as the Republic of Kiribati, with Ieremia Tabai as president. **1985:** Kiribati's first political party, the opposition Christian Democrats, was formed. **1994:** The government resigned after losing a vote of confidence.

Practical information

Visa requirements: UK: visa not required for a stay of up to 28 days. USA: visa required
Time difference: GMT –10/–11
Chief tourist attractions: remoteness; game-fishing; ecotourism, particularly birdwatching

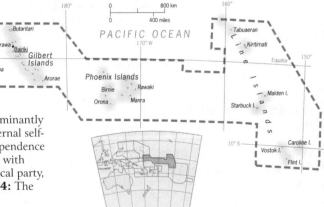

KUWAIT

Located in southwest Asia, bounded north and northwest by Iraq, east by the Persian Gulf, and south and southwest by Saudi Arabia.

National name: *Dowlat al-Kuwayt/State of Kuwait*
Area: 17,819 sq km/6,879 sq mi
Capital: Kuwait (and chief port)
Major towns/cities: as-Salimiya, Hawalli, Al Farwaniyah, Abraq Kheetan, Al Jahra, Al Ahmadi, Al Fuhayhil
Physical features: hot desert; islands of Faylakah, Bubiyan, and Warbah at northeast corner of Arabian Peninsula

The flag uses the pan-Arab colours. The design may have been inspired by the pre-1958 Iraqi flag.
Effective date: 24 November 1961.

Government

Head of state: Sheikh Jabir al-Ahmad al-Jabir al-Sabah from 1977
Head of government: Crown Prince Sheikh Saad al-Abdullah al-Salinas al-Sabah from 1978
Political system: absolutist
Political executive: absolute
Administrative divisions: five governates
Political parties: none
Armed forces: 15,300 (1998)
Conscription: compulsory for two years
Death penalty: retained and used for ordinary crimes
Defence spend: (% GDP) 12.9 (1998)
Education spend: (% GNP) 5.0 (1997)
Health spend: (% GDP) 3.3 (1997)

Economy and resources

Currency: Kuwaiti dinar
GDP: (US$) 29.6 billion (1999)
Real GDP growth: (% change on previous year) –2.4 (1999 est)
GNP: (US$) 35.1 billion (1997)
GNP per capita (PPP): (US$) 24,270 (1997)
Consumer price inflation: 3% (1999)
Unemployment: 1.8% (1996)
Foreign debt: (US$) 8.85 billion (1999 est)
Major trading partners: USA, Japan, Germany, France, Saudi Arabia, United Arab Emirates, India, UK, Italy
Resources: petroleum, natural gas, mineral water
Industries: petroleum refining, petrochemicals, food processing, gases, construction
Exports: petroleum and petroleum products (accounted for more than 93% of export revenue in 1994), chemical fertilizer, gas (natural and manufactured), basic manufactures. Principal market: Japan 22.8% (1999)
Imports: machinery and transport equipment, basic manufactures (especially iron, steel, and textiles) and other manufactured goods, live animals and food. Principal source: USA 15.4% (1999)
Arable land: 0.3% (1996)

Agricultural products: melons, tomatoes, cucumbers, onions; livestock rearing (poultry); fishing

Population and society

Population: 1,972,000 (2000 est)
Population growth rate: 3.1% (1995–2000)
Population density: (per sq km) 106 (1999 est)
Urban population: (% of total) 98 (2000 est)
Age distribution: (% of total population) 0–14 34%, 15–59 62%, 60+ 4% (2000 est)
Ethnic groups: about 45% Kuwaiti, 35% non-Kuwaiti Arab, 9% Indian and Pakistani, 4% Iranian
Language: Arabic (78%) (official), English, Kurdish (10%), Farsi (4%)
Religion: Sunni Muslim 45%, Shiite Muslim 40%; Christian, Hindu, and Parsi about 5%
Education: (compulsory years) 8
Literacy rate: 85% (men); 80% (women) (2000 est)
Labour force: 1% agriculture, 15% industry, 84% services (1996 est)
Life expectancy: 74 (men); 78 (women) (1995–2000)
Child mortality rate: (under 5, per 1,000 live births) 15 (1995–2000)
Physicians: 1 per 540 people (1995)
Hospital beds: 1 per 365 people (1996)
TV sets: (per 1,000 people) 491 (1997)
Radios: (per 1,000 people) 660 (1997)
Internet users: (per 10,000 people) 527.2 (1999)
Personal computer users: (per 100 people) 12.1 (1999)

Transport

Airports: international airports: Kuwait City; total passenger km: 5,997 million (1997 est)
Railways: none
Roads: total road network: 4,450 km/2,765 mi, of which 80.6% paved (1996 est); passenger cars: 358 per 1,000 people (1996 est)

Chronology

c. **3000 BC:** Archaeological evidence suggests that coastal parts of Kuwait may have been part of a commercial civilization contemporary with the Sumerian, based in Mesopotamia (the Tigris and Euphrates valley area of Iraq). *c.* **323 BC:** Visited by Greek colonists at the time of Alexander the Great. **7th century AD:** Islam introduced. **late 16th century:** Fell under nominal control of Turkish Ottoman Empire. **1710:** Control was assumed by the Utab, a member of the Anaza tribal confederation in northern Arabia, and Kuwait city was founded, soon developing from a fishing village into an important port. **1756:** Autonomous Sheikhdom of Kuwait founded by Abd Rahman of the al-Sabah family, a branch of the Utab. **1776:** British East India Company set up a base in the Gulf. **1899:** Concerned at the potential threat of growing Ottoman and German influence, Britain signed a treaty with Kuwait, establishing a self-governing protectorate in which the Emir received an annual subsidy from Britain in return for agreeing not to alienate any territory to a foreign power. **1914:** Britain recognized Kuwait as an 'independent government under British protection'. **1922–33:** Agreement on frontiers with Iraq, to the north, and Nejd (later Saudi Arabia) to the southwest. **1938:** Oil discovered; large-scale exploitation after World War II transformed the economy. **1961:** Full independence achieved from Britain, with Sheikh Abdullah al-Salem al-Sabah as emir. Attempted Iraqi invasion discouraged by dispatch of British troops to the Gulf. **1962:** Constitution introduced, with franchise restricted to 10% of the population. **1977:** Crown Prince Jabir Al Ahmad Al Jabir Al Sabah became Emir. The National Assembly was dissolved. **1978:** Sheikh Saad al-Abdullah al-Salem al-Sabah was appointed prime minister by the emir. **1981:** The National Assembly was reconstituted. **1983:** Shiite guerrillas bombed targets in Kuwait; 17 were arrested. **1986:** The National Assembly was dissolved. **1987:** Kuwaiti oil tankers were reflagged and received US Navy protection; there were missile attacks by Iran. **1988:** Aircraft hijacked by pro-Iranian Shiites demanding the release of convicted guerrillas; Kuwait refused. **1989:** Two of the convicted guerrillas were released. **1990:** Prodemocracy demonstrations were suppressed. Kuwait was annexed by Iraq in August, causing extensive damage to property and environment. The emir set up a government in exile in Saudi Arabia. **1991:** US-led coalition forces defeated Iraqi forces in Kuwait in the Gulf War. The new government omitted any opposition representatives. **1992:** The reconstituted national assembly was elected, with opposition nominees, including Islamic candidates, winning the majority of seats. **1993:** Incursions by Iraq into Kuwait were repelled by US-led air strikes on Iraqi military sites. **1994:** The massing of Iraqi troops on the Kuwaiti border prompted a US-led response. Iraqi president Saddam Hussein publicly renounced any claim to Kuwait. **1999:** A decree to secure a political voice for women in Kuwait was defeated in parliament, in the belief that female participation in politics would violate the principles of Islam and Kuwaiti traditions. **2000:** The high court, and later the constitutional court, upheld parliament's refusal to allow women the vote.

Practical information

Visa requirements: UK: visa required. USA: visa required
Time difference: GMT +3
Major holidays: 1 January, 25 February (3 days); variable: Eid-ul-Adha (3 days), end of Ramadan (3 days), New Year (Muslim), Prophet's Birthday, Ascension of the Prophet (March/April), Standing on Mount Arafat (August)

KYRGYZSTAN

Located in central Asia, bounded north by Kazakhstan, east by China, west by Uzbekistan, and south by Tajikistan.

National name: *Kyrgyz Respublikasy/Kyrgyz Republic*
Area: 198,500 sq km/76,640 sq mi
Capital: Bishkek (formerly Frunze)
Major towns/cities: Osh, Przhevalsk, Kyzyl-Kiya, Tokmak, Djalal-Abad
Physical features: mountainous, an extension of the Tien Shan range

Red recalls the banner of Manas who united the Kyrgyz tribes. The emblem shows a bird's-eye view of a yurt, secured by a lattice of ropes. Effective date: 3 March 1992.

Government

Head of state: Askar Akayev from 1991
Head of government: Kurmanbek Bakiyev from 2000
Political system: emergent democracy
Political executive: limited presidency
Administrative divisions: six regions and the municipality of Bishkek, the capital
Political parties: Party of Communists of Kyrgyzstan (banned 1991–92); Ata Meken, Kyrgyz-nationalist; Erkin Kyrgyzstan, Kyrgyz-nationalist; Social Democratic Party, nationalist, pro-Akayev; Democratic Movement of Kyrgyzstan, nationalist reformist
Armed forces: 12,200 (1998)
Conscription: compulsory for 12–18 months
Death penalty: retained and used for ordinary crimes
Defence spend: (% GDP) 3.6 (1998)
Education spend: (% GNP) 5.7 (1996)
Health spend: (% GDP) 4 (1997)

Economy and resources

Currency: som
GDP: (US$) 1.2 billion (1999)
Real GDP growth: (% change on previous year) 7.8 (1999)
GNP: (US$) 1.4 billion (1999)
GNP per capita (PPP): (US$) 2,223 (1999)
Consumer price inflation: 35.9% (1999)
Unemployment: 3.2% (1998)
Foreign debt: (US$) 1.3 billion (1999 est)
Major trading partners: Germany, Russia, Kazakhstan, Uzbekistan, Turkey, China, UK, Cuba, Ukraine
Resources: petroleum, natural gas, coal, gold, tin, mercury, antimony, zinc, tungsten, uranium
Industries: metallurgy, machinery, electronics and instruments, textiles, food processing (particularly sugar refining), mining
Exports: wool, cotton yarn, tobacco, electric power, electronic and engineering products, non-ferrous metallurgy, food and beverages. Principal market: Germany 32.7% (1999)
Imports: petroleum, natural gas, engineering products,

food products. Principal source: Russia 18.2% (1999)
Arable land: 4.7% (1996)
Agricultural products: grain, potatoes, cotton, tobacco, sugar beet, hemp, kenat, kendyr, medicinal plants; livestock rearing (sheep, cattle, goats, yaks, and horses) is the mainstay of agricultural activity

Population and society

Population: 4,699,000 (2000 est)
Population growth rate: 0.6% (1995–2000)
Population density: (per sq km) 24 (1999 est)
Urban population: (% of total) 33 (2000 est)
Age distribution: (% of total population) 0–14 35%, 15–59 56%, 60+ 9% (2000 est)
Ethnic groups: 60% ethnic Kyrgyz, 15% Russian, 14% Uzbek, 2% Ukrainian; Dungan, German, Kazakh, Korean, Tajik, Tartar and Uighar minorities
Language: Kyrgyz (a Turkic language; official), Russian
Religion: Sunni Muslim 70%, Russian Orthodox 20%
Education: (compulsory years) 9
Literacy rate: 98% (men); 98% (women) (2000 est)
Labour force: 48% agriculture, 10% industry, 42% services (1997)
Life expectancy: 63 (men); 72 (women) (1995–2000)
Child mortality rate: (under 5, per 1,000 live births) 50 (1995–2000)
Physicians: 1 per 276 people (1995)
Hospital beds: 1 per 86 people (1995)
TV sets: (per 1,000 people) 44 (1997)
Radios: (per 1,000 people) 112 (1997)
Internet users: (per 10,000 people) 21.4 (1999)

Transport

Airports: international airports: Bishkek (Bishkek Manas), Osh; three domestic airports; total passenger km: 573 million (1995)
Railways: total length: 376 km/234 mi; passengers carried: 2.3 million (1993)
Roads: total road network: 18,500 km/11,496 mi, of which 91.1% paved (1996 est); passenger cars: 31.9 per 1,000 people (1996 est)

Chronology

8th century: Spread of Islam. **10th century onwards:** Southward migration of Kyrgyz people from upper Yenisey River region to Tien Shan region; accelerated following rise of Mongol Empire in 13th century. **13th–14th centuries:** Part of Mongol Empire. **1685:** Came under control of Mongol Oirots following centuries of Turkic rule. **1758:** Kyrgyz people became nominal subjects of Chinese Empire, following Oirots' defeat by Chinese rulers, the Manchus. **early 19th century:** Came under suzerainty of Khanate (chieftaincy) of Kokand, to the west. **1864–76:** Incorporated into tsarist Russian Empire. **1916–17:** Many Kyrgyz migrated to China after Russian suppression of rebellion in Central Asia and outbreak of civil war following 1917 October Revolution in Russia, with local armed guerrillas (*basmachi*) resisting Bolshevik Red Army. **1917–1924:** Part of independent Turkestan republic. **1920s:** Land reforms resulted in settlement of many formerly nomadic Kyrgyz; literacy and education improved. **1924:** Became autonomous republic within USSR. **1930s:** Agricultural collectivization programme provoked *basmachi* resistance and local 'nationalist communists' were purged from Kyrgyz Communist Party (KCP). **1936:** Became full union republic within USSR. **1990:** A state of emergency was imposed in Bishkek after ethnic clashes. **1991:** Askar Akayev, a reform communist, was chosen as president, and condemned the attempted coup in Moscow against the reformist Mikhail Gorbachev; Kyrgyzstan joined the new Commonwealth of Independent States (CIS) and its independence was recognized by the USA. **1992:** Kyrgyzstan joined the United Nations and Conference on Security and Cooperation in Europe (CSCE; now the Organization on Security and Cooperation in Europe, OSCE). A market-centred economic reform programme was instituted. **1994:** The country joined the Central Asian Union, with Kazakhstan and Uzbekistan. **1996:** A constitutional amendment increased the powers of the president. An agreement was made with Kazakhstan and Uzbekistan to create a single economic market. **1997:** Private ownership of land was legalized but the privatization programme was suspended. An agreement was made on border controls with Russia. **1998:** A referendum approved the private ownership of land. **1999:** Amengeldy Muraliyev was appointed prime minister. **2000:** Islamist rebels crossed into the country from Afghanistan via Tajikistan, seeking to create an Islamic state in east Uzbekistan. Akayev was re-elected president, despite an economy in crisis, though international observers criticized the conduct of the election. Kurmanbek Bakiyev was appointed prime minister.

Practical information

Visa requirements: UK: visa not required. USA: visa not required
Time difference: GMT +5
Chief tourist attractions: tourist facilities are limited – the country is visited mostly by mountaineers; spectacular and largely unspoilt mountain scenery; great crater lake of Issyk-Kul; several historical and cultural sites
Major holidays: 1, 7 January, 8, 21 March, 1, 9 May, 31 August

LAOS

Landlocked country in southeast Asia, bounded north by China, east by Vietnam, south by Cambodia, west by Thailand, and northwest by Myanmar.

National name: *Sathalanalat Praxathipatai Paxaxôn Lao/Democratic People's Republic of Laos*
Area: 236,790 sq km/91,424 sq mi
Capital: Vientiane
Major towns/cities: Louangphrabang (the former royal capital), Pakxé, Savannakhet
Physical features: landlocked state with high mountains in east; Mekong River in west; rainforest covers nearly 60% of land

White symbolizes justice and the promise of the future. Blue stands for prosperity. Red represents unity and purpose and the blood shed during the struggle for freedom. Effective date: 4 December 1975.

Government

Head of state: Gen Khamtay Siphandon from 1998
Head of government: Gen Sisavath Keobounphanh from 1998
Political system: communist
Political executive: communist
Administrative divisions: 16 provinces, one municipality (Vientiane), and one special region
Political party: Lao People's Revolutionary Party (LPRP, the only legal party), socialist
Armed forces: 29,100 (1998)
Conscription: military service is compulsory for a minimum of 18 months
Death penalty: retained and used for ordinary crimes
Defence spend: (% GDP) 3.7 (1998)
Education spend: (% GNP) 2.1 (1997)
Health spend: (% GDP) 3.6 (1997)

Economy and resources

Currency: new kip
GDP: (US$) 1.37 billion (1999)
Real GDP growth: (% change on previous year) 4 (1999 est)
GNP: (US$) 1.4 billion (1999)
GNP per capita (PPP): (US$) 1,726 (1999)
Consumer price inflation: 128.5% (1999)
Unemployment: 2.4% (1995)
Foreign debt: (US$) 2.44 billion (1998)
Major trading partners: Thailand, Japan, Germany, France, China, Italy
Resources: coal, tin, gypsum, baryte, lead, zinc, nickel, potash, iron ore; small quantities of gold, silver, precious stones
Industries: processing of agricultural produce, sawmilling, textiles and garments, handicrafts, basic consumer goods
Exports: timber, textiles and garments, motorcycles, electricity, coffee, tin, gypsum. Principal market: Vietnam 42.7% (1997)
Imports: food (particularly rice and sugar), mineral fuels, machinery and transport equipment, cement, cotton yarn. Principal source: Thailand 56.2% (1997)
Arable land: 3.5% (1996)

Agricultural products: rice, maize, tobacco, cotton, coffee, sugar cane, cassava, potatoes, sweet potatoes; livestock rearing (pigs, poultry, and cattle); fishing; forest resources including valuable wood such as Teruk (logging suspended in 1991 to preserve the forest area); opium is produced but its manufacture is controlled by the state

Population and society

Population: 5,433,000 (2000 est)
Population growth rate: 2.6% (1995–2000)
Population density: (per sq km) 22 (1999 est)
Urban population: (% of total) 24 (2000 est)
Age distribution: (% of total population) 0–14 44%, 15–59 50%, 60+ 6% (2000 est)
Ethnic groups: Laotian, predominantly the lowland Lao Lum (over 60%), the upland and mountain-dwelling Lao Theung (22%), and the tribal Laotai and Lao Soung (9%); Vietnamese Chinese (1%)
Language: Lao (official), French, English, ethnic languages
Religion: Theravada Buddhist 85%, animist beliefs among mountain dwellers
Education: (compulsory years) 5
Literacy rate: 64% (men); 33% (women) (2000 est)
Labour force: 77% agriculture, 6% industry, 17% services (1997 est)
Life expectancy: 52 (men); 55 (women) (1995–2000)
Child mortality rate: (under 5, per 1,000 live births) 150 (1995–2000)
Physicians: 1 per 5,000 people (1994)
Hospital beds: 1 per 385 people (1994)
TV sets: (per 1,000 people) 4 (1997)
Radios: (per 1,000 people) 143 (1997)
Internet users: (per 10,000 people) 3.8 (1999)
Personal computer users: (per 100 people) 0.2 (1999)

Transport

Airports: international airports: Vientiane (Wattai); three domestic airports; total passenger km: 48 million (1995)
Railways: none
Roads: total road network: 22,321 km/13,870 mi (1996); passenger cars: 3.4 per 1,000 people (1996 est)

Chronology

c. **2000–500 BC:** Early Bronze Age civilizations in central Mekong River and Plain of Jars regions. **5th–8th centuries:** Occupied by immigrants from southern China. **8th century onwards:** Theravada Buddhism spread by Mon monks. **9th–13th centuries:** Part of the sophisticated Khmer Empire, centred on Angkor in Cambodia. **12th century:** Small independent principalities, notably Louangphrabang, established by Lao invaders from Thailand and Yunnan, southern China; they adopted Buddhism. **14th century:** United by King Fa Ngum; the first independent Laotian state, Lan Xang, formed. It was to dominate for four centuries, broken only by a period of Burmese rule 1574–1637. **17th century:** First visited by Europeans. **1713:** The Lan Xang kingdom split into three separate kingdoms, Louangphrabang, Vientiane, and Champassac, which became tributaries of Siam (Thailand) from the late 18th century. **1893–1945:** Laos was a French protectorate, comprising the three principalities of Louangphrabang, Vientiane, and Champassac. **1945:** Temporarily occupied by Japan. **1946:** Retaken by France, despite opposition by the Chinese-backed Lao Issara (Free Laos) nationalist movement. **1950:** Granted semi-autonomy in French Union, as an associated state under the constitutional monarchy of the king of Louangphrabang. **1954:** Independence achieved from France under the Geneva Agreements, but civil war broke out between a moderate royalist faction of the Lao Issara, led by Prince Souvanna Phouma, and the communist Chinese-backed Pathet Lao (Land of the Lao) led by Prince Souphanouvong. **1957:** A coalition government, headed by Souvanna Phouma, was established by the Vientiane Agreement. **1959:** Savang Vatthana became king. **1960:** Right-wing pro-Western government seized power, headed by Prince Boun Gum. **1962:** Geneva Agreement established a new coalition government, led by Souvanna Phouma, but civil war continued, the Pathet Lao receiving backing from the North Vietnamese, and Souvanna Phouma from the USA. **1973:** Vientiane ceasefire agreement divided the country between the communists and the Souvanna Phouma regime and brought the withdrawal of US, Thai, and North Vietnamese forces. **1975:** Communists seized power; a republic was proclaimed, with Prince Souphanouvong as head of state and the Communist Party leader Kaysone Phomvihane as the controlling prime minister. **1979:** Food shortages and the flight of 250,000 refugees to Thailand led to an easing of the drive towards nationalization and agricultural collectivization. **1985:** Greater economic liberalization received encouragement from the Soviet Union's reformist leader Mikhail Gorbachev. **1989:** The first assembly elections since communist takeover were held; Vietnamese troops were withdrawn from the country. **1991:** A security and cooperation pact was signed with Thailand, and an agreement reached on the phased repatriation of Laotian refugees. **1995:** The US lifted its 20-year aid embargo. **1996:** The military tightened its grip on political affairs, but inward investment and private enterprise continued to be encouraged, fuelling economic expansion. **1997:** Membership of the Association of South East Asian Nations (ASEAN) was announced. **1998:** Khamtay Siphandon became president and was replaced as prime minister by Sisavath Keobounphanh.

Practical information

Visa requirements: UK: visa required. USA: visa required
Time difference: GMT +7

Chief tourist attractions: spectacular scenery; ancient pagodas and temples; wildlife
Major holidays: 24 January, 13–15 April, 1 May, 2 December

LATVIA

Located in northern Europe, bounded east by Russia, north by Estonia, north and northwest by the Baltic Sea, south by Lithuania, and southeast by Belarus.

It is said that berries were used to dye the flag. Red represents the blood shed in the past and the willingness to offer it again. White stands for right, truth, the honour of free citizens, and trustworthiness. Effective date: 27 February 1990.

National name: *Latvijas Republika/Republic of Latvia*
Area: 63,700 sq km/24,594 sq mi
Capital: Riga
Major towns/cities: Daugavpils, Leipaja, Jurmala, Jelgava, Ventspils
Major ports: Ventspils, Leipaja
Physical features: wooded lowland (highest point 312 m/1,024 ft), marshes, lakes; 472 km/293 mi of coastline; mild climate

Government

Head of state: Vaira Vike-Freiberga from 1999
Head of government: Andris Berzins from 2000
Political system: emergent democracy
Political executive: parliamentary
Administrative divisions: 26 districts and seven municipalities
Political parties: Latvian Way (LW), right of centre; Latvian National and Conservative Party (LNNK), right wing, nationalist; Economic-Political Union (formerly known as Harmony for Latvia and Rebirth of the National Economy), centrist; Ravnopravie (Equal Rights), centrist; For the Fatherland and Freedom (FFF), extreme nationalist; Latvian Peasants' Union (LZS), rural based, left of centre; Union of Christian Democrats, right of centre; Democratic Centre Party, centrist; Movement for Latvia, pro-Russian, populist; Master in Your Own Home (Saimnieks), ex-communist, populist; Latvian National Party of Reforms, right-of-centre nationalist coalition
Armed forces: 5,000 (1998)
Conscription: compulsory for 18 months
Death penalty: abolished for ordinary crimes 1999; laws provide for the death penalty for exceptional crimes only
Defence spend: (% GDP) 2.5 (1998)
Education spend: (% GNP) 6.5 (1996)
Health spend: (% GDP) 6.1 (1997)

Economy and resources

Currency: lat
GDP: (US$) 6.7 billion (1999)
Real GDP growth: (% change on previous year) 0.1 (1999)
GNP: (US$) 6 billion (1999)
GNP per capita (PPP): (US$) 5,938 (1999)
Consumer price inflation: 2.4% (1999)
Unemployment: 14.4% (1997)
Foreign debt: (US$) 900 million (1999)
Major trading partners: Germany, Russia, UK, Lithuania, Finland, Sweden, Estonia
Resources: peat, gypsum, dolomite, limestone, amber, gravel, sand
Industries: food processing, machinery and equipment (major producer of electric railway passenger cars and long-distance telephone exchanges), chemicals and chemical products, sawn timber, paper and woollen goods
Exports: timber and timber products, textiles, food and agricultural products, machinery and electrical equipment, metal industry products. Principal market: Germany 16.4% (1999)
Imports: mineral fuels and products, machinery and electrical equipment, chemical industry products. Principal source: Russia 15.2% (1999)
Arable land: 27.3% (1996)
Agricultural products: oats, barley, rye, potatoes, flax; cattle and dairy farming and pig breeding are the chief agricultural occupations

Population and society

Population: 2,357,000 (2000 est)
Population growth rate: –1.5% (1995–2000)
Population density: (per sq km) 38 (1999 est)
Urban population: (% of total) 69 (2000 est)
Age distribution: (% of total population) 0–14 18%, 15–59 62%, 60+ 20% (2000 est)
Ethnic groups: 56% of Latvian ethnic descent, 32% ethnic Russian, 4% Belorussian, 3% Ukrainian, 2% Polish, 1% Lithuanian
Language: Latvian (official)
Religion: Lutheran, Roman Catholic, Russian Orthodox
Education: (compulsory years) 9
Literacy rate: 99% (men); 99% (women) (2000 est)
Labour force: 18.3% agriculture, 25.5% industry, 56.2% services (1996)
Life expectancy: 63 (men); 74 (women) (1995–2000)
Child mortality rate: (under 5, per 1,000 live births) 25 (1995–2000)
Physicians: 1 per 8,541 people (34 per 10,000 people) (1996)
Hospital beds: 1 per 23,840 people (96.6 per 10,000 people) (1997)
TV sets: (per 1,000 people) 492 (1997)
Radios: (per 1,000 people) 710 (1997)
Internet users: (per 10,000 people) 430.4 (1999)
Personal computer users: (per 100 people) 8.2 (1999)

Transport

Airports: international airports: Riga (Spilva), Jelgava; total passenger km: 282.7 million (1996)
Railways: total length: 2,703 km/1,680 mi; total passenger km: 1,200 million (1996)
Roads: total road network: 55,942 km/34,762 mi, of which 38.3% paved (1997); passenger cars: 174 per 1,000 people (1997)

Chronology

9th–10th centuries: Invaded by Vikings and Russians. **13th century:** Conquered by crusading German Teutonic Knights, who named the area Livonia and converted population to Christianity; Riga joined the Hanseatic League, a northern European union of commercial towns. **1520s:** Lutheranism established as a result of the Reformation. **16th–17th centuries:** Successively under Polish, Lithuanian, and Swedish rule. **1721:** Tsarist Russia took control. **1819:** Serfdom abolished. **1900s:** Emergence of an independence movement. **1914–18:** Under partial German occupation during World War I. **1918–19:** Independence proclaimed and achieved after Russian Red Army troops expelled by German, Polish, and Latvian forces. **1920s:** Land reforms introduced by Farmers' Union government. **1934:** Democracy overthrown and, at time of economic depression, an autocratic regime was established; Baltic Entente mutual defence pact made with Estonia and Lithuania. **1940:** Incorporated into Soviet Union (USSR) as constituent republic, following secret German–Soviet agreement. **1941–44:** Occupied by Germany. **1944:** USSR regained control; mass deportations of Latvians to Central Asia, followed by immigration of ethnic Russians; agricultural collectivization. **1960s and 1970s:** Extreme repression of Latvian cultural and literary life. **1980s:** Nationalist dissent began to grow, influenced by the Polish Solidarity movement and Mikhail Gorbachev's *glasnost* ('openness') initiative in the USSR. **1988:** The Latvian Popular Front was established to campaign for independence. The prewar flag was readopted and official status was given to the Latvian language. **1989:** The Latvian parliament passed a sovereignty declaration. **1990:** The Popular Front secured a majority in local elections and its leader, Ivan Godmanir, became the prime minister. The Latvian Communist Party split into pro-independence and pro-Moscow wings. The country entered a 'transitional period of independence' and the Baltic Council was reformed. **1991:** Soviet troops briefly seized key installations in Riga. There was an overwhelming vote for independence in a referendum. Full independence was achieved following the failure of the anti-Gorbachev coup attempt in Moscow; the Communist Party was outlawed. Joined United Nations (UN); a market-centred economic reform programme was instituted. **1992:** The curbing of rights of non-citizens prompted Russia to request minority protection by the UN. **1993:** The right-of-centre Latvian Way won the general election; a free-trade agreement was reached with Estonia and Lithuania. **1994:** The last Russian troops departed. **1995:** A trade and cooperation agreement was signed with the European Union (EU). A general election produced a hung parliament in which extremist parties received most support. Applied for EU membership. **1996:** Guntis Ulmanis was re-elected president. The finance minister and deputy prime minister resigned from the eight-party coalition. **1997:** A new political party was formed, the Latvian National Party of Reforms. Former Communist leader Alfreds Rubiks was released from prison. **1998:** The DPS withdrew from the government, leaving the coalition as a minority. Citizenship laws were relaxed to make it easier for ethnic Russians to acquire citizenship. **1999:** Andris Skele became prime minister. Vaira Vike-Freiberga was sworn in as president. **2000:** Andris Skele resigned as prime minister after a disagreement within his coalition. He was replaced by Andris Berzins, who headed a coalition of the same parties as before, the Union for the Fatherland and Freedom (FF/LNNK), LW, and Skele's People's Party, as well as the additional New Party.

Practical information

Visa requirements: UK: visa not required. USA: visa required
Time difference: GMT +2
Chief tourist attractions: historic centre of Riga, with medieval and art nouveau architecture; extensive beaches on Baltic coast; Gauja National Park; winter-sports facilities at Sigulda
Major holidays: 1 January, 1 May, 23–24 June, 18 November, 25–26 December; variable: Good Friday

LEBANON

Located in western Asia, bounded north and east by Syria, south by Israel, and west by the Mediterranean Sea.

National name: *Jumhouria al-Lubnaniya/Republic of Lebanon*
Area: 10,452 sq km/4,035 sq mi
Capital: Beirut (and chief port)
Major towns/cities: Tripoli, Zahlé, Baabda, Baalbek, Jezzine
Major ports: Tripoli, Tyre, Sidon, Joûnié
Physical features: narrow coastal plain; fertile Bekka valley running north–south between Lebanon and Anti-Lebanon mountain ranges

Red is said to stand for bloodshed. White represents peace, holiness, and eternity. Effective date: 9 December 1943.

Government

Head of state: Emile Lahoud from 1998
Head of government: Rafik Hariri from 2000
Political system: emergent democracy
Political executive: dual executive
Administrative divisions: five governates
Political parties: Phalangist Party, Christian, radical, nationalist; Progressive Socialist Party (PSP), Druze, moderate, socialist; National Liberal Party (NLP), Maronite, left of centre; National Bloc, Maronite, moderate; Lebanese Communist Party (PCL), nationalist, communist; Parliamentary Democratic Front, Sunni Muslim, centrist
Armed forces: 55,100 (1998); in 1995 there were 30,000 Syrian troops and the pro-Israeli South Lebanese army numbered 2,500
Conscription: compulsory for 12 months
Death penalty: retained and used for ordinary crimes
Defence spend: (% GDP) 3.6 (1998)
Education spend: (% GNP) 2.5 (1996)
Health spend: (% GDP) 10.1 (1997)

Economy and resources

Currency: Lebanese pound
GDP: (US$) 16.6 billion (1999 est)
Real GDP growth: (% change on previous year) –1 (1999)
GNP: (US$) 15.8 billion (1999)
GNP per capita (PPP): (US$) 4,129 (1999)
Consumer price inflation: 0.5% (1999)
Unemployment: 18% (1997 est)
Foreign debt: (US$) 8.4 billion (1999 est)
Major trading partners: Italy, Saudi Arabia, United Arab Emirates, Syria, Germany, USA, France, Kuwait, Switzerland
Resources: there are no commercially viable mineral deposits; small reserves of lignite and iron ore
Industries: food processing, petroleum refining, textiles, furniture and woodworking, paper and paper products
Exports: paper products, textiles, fruit and vegetables, jewellery. Principal market: Saudi Arabia 11% (1999)
Imports: electrical equipment, vehicles, petroleum, metals, machinery, consumer goods. Principal source:

Italy 12.9% (1999)
Arable land: 18.2% (1996)
Agricultural products: citrus fruits, potatoes, melons, apples, grapes (viticulture is significant), wheat, sugar beet, olives, bananas; livestock rearing (goats and sheep); although illegal, hashish is an important export crop

Population and society

Population: 3,282,000 (2000 est)
Population growth rate: 1.7% (1995–2000)
Population density: (per sq km) 310 (1999 est)
Urban population: (% of total) 90 (2000 est)
Age distribution: (% of total population) 0–14 33%, 15–59 58%, 60+ 9% (2000 est)
Ethnic groups: about 95% Arab, with Armenian, Assyrian, Jewish, Turkish, and Greek minorities
Language: Arabic (official), French, Armenian, English
Religion: Muslim 70% (Shiite 35%, Sunni 23%, Druze 7%, other 5%); Christian 30% (mainly Maronite 19%), Druze 3%; other Christian denominations including Greek Orthodox, Armenian, and Roman Catholic
Education: not compulsory
Literacy rate: 92% (men); 80% (women) (2000 est)
Labour force: 3.7% agriculture, 30.3% industry, 66% services (1997 est)
Life expectancy: 68 (men); 72 (women) (1995–2000)
Child mortality rate: (under 5, per 1,000 live births) 35 (1995–2000)
Physicians: 1 per 568 people (1995 est)
Hospital beds: 1 per 606 people (1993 est)
TV sets: (per 1,000 people) 354 (1997)
Radios: (per 1,000 people) 906 (1997)
Internet users: (per 10,000 people) 618.1 (1999)
Personal computer users: (per 100 people) 4.6 (1999)

Transport

Airports: international airports: Beirut (Khaldeh); most major cities have domestic airports; total passenger km: 2,173 million (1997 est)
Railways: total length: 222 km/138 mi
Roads: total road network: 6,350 km/3,946 mi, of which 95% paved (1996 est); passenger cars: 731 per 1,000 people (1996 est)

Chronology

5th century BC–1st century AD: Part of the eastern Mediterranean Phoenician Empire. **1st century:** Came under Roman rule; Christianity introduced. **635:** Islam introduced by Arab tribes, who settled in southern Lebanon. **11th century:** Druze faith developed by local Muslims. **1516:** Became part of the Turkish Ottoman Empire. **1860:** Massacre of thousands of Christian Maronites by the Muslim Druze led to French intervention. **1920–41:** Administered by French under League of Nations mandate. **1943:** Independence achieved as a republic, with a constitution that enshrined Christian and Muslim power-sharing. **1945:** Joined the Arab League. **1948–49:** Lebanon joined the first Arab war against Israel; Palestinian refugees settled in the south. **1958:** Revolt by radical Muslims opposed to pro-Western policies of the Christian president, Camille Chamoun. **1964:** Palestine Liberation Organization (PLO) founded in Beirut. **1967:** More Palestinian refugees settled in Lebanon following the Arab–Israeli war. **1971:** PLO expelled from Jordan; established headquarters in Lebanon. **1975:** Outbreak of civil war between conservative Christians and leftist Muslims backed by PLO. **1976:** Ceasefire agreed; Syrian-dominated Arab deterrent force formed to keep the peace, but considered by Christians as an occupying force. **1978:** Israel launched a limited invasion of southern Lebanon in search of PLO guerrillas. An international United Nations peacekeeping force was unable to prevent further fighting. **1979:** Part of southern Lebanon declared an 'independent free Lebanon' by a right-wing army officer. **1982:** Israel again invaded Lebanon. Palestinians withdrew from Beirut under the supervision of an international peacekeeping force; the PLO moved its headquarters to Tunis. **1983:** An agreement was reached for withdrawal of Syrian and Israeli troops but abrogated under Syrian pressure; intense fighting was seen between Christian Phalangists and Muslim Druze militias. **1984:** Most of the international peacekeeping force were withdrawn. Radical Muslim militia took control of west Beirut. **1985:** Lebanon was in chaos; many foreigners were taken hostage and Israeli troops withdrawn. **1987:** Syrian troops were sent into Beirut. **1988:** Gen Michel Aoun was appointed to head the caretaker military government; Premier Selim el-Hoss set up a rival government; the threat of partition hung over country. **1989:** Gen Aoun declared a 'war of liberation' against Syrian occupation; Arab League-sponsored talks resulted in a ceasefire and a revised constitution recognizing Muslim majority; René Mouhawad was assassinated after 17 days as president; Maronite Christian Elias Hrawi was named as his successor; Aoun occupied the presidential palace, rejecting the constitution. **1990:** The release of Western hostages began. Gen Aoun, crushed by Syrians, surrendered and legitimate government was restored. **1991:** The government extended its control to the whole country. A treaty of cooperation with Syria was signed. **1992:** The remaining Western hostages were released. A pro-Syrian administration led by businessman Rafik al-Hariri was re-elected after many Christians boycotted general election. **1993:** Israel launched attacks against Shia fundamentalist Hezbollah strongholds in southern Lebanon before the USA and Syria brokered an agreement to avoid the use of force. **1996:** Israel launched a rocket attack on southern Lebanon in response to Hezbollah activity. USA, Israel, Syria, and Lebanon attempted to broker a new ceasefire. **1998:** Army chief General Emile Lahoud was elected president. Salim al-Hoss became prime minister. **2000:** Israel withdrew its troops from southern Lebanon, and Lebanese troops assumed control in the region from Hezbollah guerillas. Rafik al-Hariri became prime minister for a second time.

Practical information

Visa requirements: UK: visa required. USA: visa required

Time difference: GMT +2

Chief tourist attractions: sunny climate; scenery; historic sites

Major holidays: 1 January, 9 February, 1 May, 15 August, 1, 22 November, 25 December; variable: Eid-ul-Adha (3 days), Ashora, Good Friday, Easter Monday, end of Ramadan (3 days), New Year (Muslim), Prophet's Birthday

LESOTHO

Landlocked country in southern Africa, an enclave within South Africa.

National name: *Mmuso oa Lesotho/Kingdom of Lesotho*
Area: 30,355 sq km/11,720 sq mi
Capital: Maseru
Major towns/cities: Qacha's Nek, Teyateyaneng, Mafeteng, Hlotse, Roma, Quthing
Physical features: mountainous with plateaux, forming part of South Africa's chief watershed

White stands for peace. The shield and weapons express a willingness to defend the country. Green symbolizes plenty. Blue represents rain. Effective date: 20 January 1987.

Government

Head of state: King Letsie III from 1996
Head of government: Bethuel Pakulitha Mosisili from 1998
Political system: emergent democracy
Political executive: parliamentary
Administrative divisions: ten districts
Political parties: Basotho National Party (BNP), traditionalist, nationalist, right of centre; Basutoland Congress Party (BCP), left of centre
Armed forces: 2,000 (1998)
Conscription: military service is voluntary
Death penalty: retained and used for ordinary crimes
Defence spend: (% GDP) 3.5 (1998)
Education spend: (% GNP) 7 (1996)
Health spend: (% GDP) 5.6 (1997)

Economy and resources

Currency: loti
GDP: (US$) 874 million (1999)
Real GDP growth: (% change on previous year) 2 (1999 est)
GNP: (US$) 1.2 billion (1999)
GNP per capita (PPP): (US$) 2,058 (1999)
Consumer price inflation: 8.7% (1999)
Unemployment: 60% (1998 est)
Foreign debt: (US$) 705 million (1999 est)
Major trading partners: SACU (South African Customs Union) members: Lesotho, Botswana, Swaziland, Namibia, and South Africa; Taiwan, Hong Kong, USA, Canada, Italy, and other EU countries
Resources: diamonds, uranium, lead, iron ore; believed to have petroleum deposits
Industries: food products and beverages, textiles and clothing, mining, baskets, furniture; approximately 35% of Lesotho's adult male labour force was employed in South African mines 1995
Exports: clothing, footwear, furniture, food and live animals (cattle), hides, wool and mohair, baskets. Principal market: SACU 65.1% (1998)
Imports: food and live animals, machinery and transport equipment, electricity, petroleum products. Principal source: SACU 89.6% (1998)

Arable land: 10.5% (1996)
Agricultural products: maize, wheat, sorghum, asparagus, peas, and other vegetables; livestock rearing (sheep, goats, and cattle)

Population and society

Population: 2,153,000 (2000 est)
Population growth rate: 2.2% (1995–2000)
Population density: (per sq km) 69 (1999 est)
Urban population: (% of total) 28 (2000 est)
Age distribution: (% of total population) 0–14 40%, 15–59 54%, 60+ 6% (2000 est)
Ethnic groups: almost entirely Bantus (of Southern Sotho) or Basotho
Language: English (official), Sesotho, Zulu, Xhosa
Religion: Protestant 42%, Roman Catholic 38%, indigenous beliefs
Education: (compulsory years) 7
Literacy rate: 72% (men); 93% (women) (2000 est)
Labour force: 76.3% agriculture, 11.1% industry, 12.6% services (1994 est)
Life expectancy: 55 (men); 57 (women) (1995–2000)
Child mortality rate: (under 5, per 1,000 live births) 130 (1995–2000)
Physicians: 1 per 14,306 people (1993)
Hospital beds: 1 per 789 people (1992)
TV sets: (per 1,000 people) 24 (1997)
Radios: (per 1,000 people) 49 (1997)
Internet users: (per 10,000 people) 4.7 (1999)

Transport

Airports: international airports: Maseru (Moshoeshoe I); 40 airstrips, of which 14 receive charter and regular scheduled air services; total passenger km: 8 million (1995)
Railways: total length: 2.6 km/1.6 mi (a branch line connecting Maseru with the Bloemfontein–Natal line at Marseilles)
Roads: total road network: 4,955 km/3,079 mi, of which 17.9% paved (1996); passenger cars: 5.7 per 1,000 people (1996 est)

Chronology

18th century: Formerly inhabited by nomadic hunter-gatherer San, Zulu-speaking Ngunis, and Sotho-speaking peoples settled in the region. **1820s:** Under the name of Basutoland, Sotho nation founded by Moshoeshoe I, who united the people to repulse Zulu attacks from south. **1843:** Moshoeshoe I negotiated British protection as tension with South African Boers increased. **1868:** Became British territory, administered by Cape Colony (in South Africa) from 1871. **1884:** Became British crown colony, after revolt against Cape Colony control; Basuto chiefs allowed to govern according to custom and tradition, but rich agricultural land west of the Caledon River was lost to South Africa. **1900s:** Served as a migrant labour reserve for South Africa's mines and farms. **1952:** Left-of-centre Basutoland African Congress, later Congress Party (BCP), founded by Ntsu Mokhehle to campaign for self rule. **1966:** Independence achieved within Commonwealth, as Kingdom of Lesotho, with Moshoeshoe II as king and Chief Leabua Jonathan of the conservative Basotho National Party (BNP) as prime minister. **1970:** State of emergency declared; king briefly forced into exile after attempting to increase his authority. **1973:** State of emergency lifted; BNP won majority of seats in general election. **1975:** Members of ruling party attacked by South African-backed guerrillas, who opposed African National Congress (ANC) guerrillas using Lesotho as a base. **1986:** South Africa imposed a border blockade, forcing the deportation of 60 ANC members. **1990:** Moshoeshoe II was dethroned and replaced by his son, as King Letsie III. **1993:** Free multiparty elections ended the military rule. **1994:** Fighting between rival army factions was ended by a peace deal, brokered by the Organization of African Unity. **1995:** King Letsie III abdicated to restore King Moshoeshoe II to the throne. **1996:** King Moshoeshoe II was killed in car accident; King Letsie III was restored to the throne. **1998:** The LCD attained general election victory amidst claims of rigged polls; public demonstrations followed. South Africa sent troops to support the government. An interim political authority was appointed prior to new elections. Bethuel Mosisili became the new prime minister.

Practical information

Visa requirements: UK: visa not required for visits of up to 30 days. USA: visa required
Time difference: GMT +2
Chief tourist attractions: mountain scenery

Major holidays: 1 January, 12, 21 March, 2 May, 4 October, 25–26 December; variable: Ascension Thursday, Good Friday, Easter Monday, Family (July), National Sports (October)

LIBERIA

Located in West Africa, bounded north by Guinea, east by Côte
d'Ivoire, south and southwest by the Atlantic Ocean, and northwest by
Sierra Leone.

National name: *Republic of Liberia*
Area: 111,370 sq km/42,999 sq mi
Capital: Monrovia (and chief port)
Major towns/cities: Bensonville, Saniquillie, Gbarnga, Voinjama,
Buchanan
Major ports: Buchanan, Greenville
Physical features: forested highlands; swampy tropical coast where six
rivers enter the sea

The star depicts Liberia as a shining light. The blue
canton symbolizes the dark continent of Africa. The
stripes represent the 11 signatories to the Declaration of
Independence. Effective date: 26 July 1847.

Government

Head of state and government: Charles Ghankay
Taylor from 1997
Political system: emergent democracy
Political executive: limited presidency
Administrative divisions: 13 counties
Political parties: National Democratic Party of Liberia
(NDPL), nationalist, left of centre; National Patriotic
Front of Liberia (NPFL), left of centre; United Democratic
Movement of Liberia for Democracy (Ulimo), left of
centre; National Patriotic Party (NPP), antidemocratic
Armed forces: 14,000 (1998)
Conscription: military service is voluntary
Death penalty: retained and used for ordinary crimes
Defence spend: (% GDP) 3.9 (1998)
Education spend: (% GNP) N/A
Health spend: (% GDP) 8 (1997 est)

Economy and resources

Currency: Liberian dollar
GDP: (US$) 2.8 billion (1999 est)
Real GDP growth: (% change on previous year) 23
(1999 est)
GNP: (US$) N/A
GNP per capita (PPP): (US$) N/A
Consumer price inflation: 3% (1998 est)
Unemployment: 80% (1995 est)
Foreign debt: (US$) 2.26 billion (1999 est)
Major trading partners: Belgium–Luxembourg, South
Korea, Switzerland, Japan, USA, Germany, Italy, France
Resources: iron ore, diamonds, gold, barytes, kyanite
Industries: beverages (soft drinks and beer), mineral
products, chemicals, tobacco and other agricultural
products, cement, mining, rubber, furniture, bricks,
plastics
Exports: iron ore, rubber, timber, coffee, cocoa, palm-
kernel oil, diamonds, gold. Principal market: Belgium
53.3% (1999 est)
Imports: machinery and transport equipment, mineral
fuels, rice, basic manufactures, food and live animals.
Principal source: South Korea 29.5% (1999 est)
Arable land: 1.3% (1995)

Agricultural products: rice, cassava, coffee, citrus fruits,
cocoa, palm kernels, sugar cane; timber production;
rubber plantation

Population and society

Population: 3,154,000 (2000 est)
Population growth rate: 8.2% (1995–2000)
Population density: (per sq km) 26 (1999 est)
Urban population: (% of total) 45 (2000 est)
Age distribution: (% of total population) 0–14 42%,
15–59 53%, 60+ 5% (2000 est)
Ethnic groups: 95% indigenous peoples, including the
Kpelle, Bassa, Gio, Kru, Grebo, Mano, Krahn, Gola,
Ghandi, Loma, Kissi, Vai, and Bella; 5% descended from
repatriated US slaves
Language: English (official), over 20 Niger-Congo
languages
Religion: animist 70%, Sunni Muslim 20%, Christian
10%
Education: (compulsory years) 9
Literacy rate: 70% (men); 38% (women) (2000 est)
Labour force: 69% agriculture, 8% industry, 23%
services (1997 est)
Life expectancy: 46 (men); 49 (women) (1995–2000)
Child mortality rate: (under 5, per 1,000 live births)
174 (1995–2000)
TV sets: (per 1,000 people) 29 (1997)
Radios: (per 1,000 people) 329 (1997)
Internet users: (per 10,000 people) 1.0 (1999)

Transport

Airports: international airports: Monrovia (Robertsfield
and Spriggs Payne); regular services operate from Monrovia
to major towns (most air services have been suspended
since 1992); total passenger km: 7 million (1992)
Railways: total length: 490 km/304 mi. The railways
were originally constructed for iron-ore transport; after
several years of civil war, there is no report that any of
them are still in operation
Roads: total road network: 10,600 km/6,587 mi, of
which 6.2% paved (1996 est); passenger cars: 2.6 per
1,000 people (1996 est)

Chronology

1821: Purchased by the philanthropic American Colonization Society and turned into settlement for liberated black slaves from southern USA. **1847:** Recognized as an independent republic. **1869:** The True Whig Party founded, which was to dominate politics for more than a century, providing all presidents. **1926:** Large concession sold to Firestone Rubber Company as foreign indebtedness increased. **1980:** President Tolbert was assassinated in military coup led by Sgt Samuel Doe, who banned political parties and launched an anticorruption drive. **1984:** A new constitution was approved in a referendum. The National Democratic Party (NDPL) was founded by Doe as political parties were relegalized. **1985:** Doe and the NDPL won decisive victories in the allegedly rigged elections. **1990:** Doe was killed as bloody civil war broke out, involving Charles Taylor and Gen Hezekiah Bowen, who led rival rebel armies, the National Patriotic Front (NPFL) and the Armed Forces of Liberia (AFL). The war left 150,000 dead and 2 million homeless. A West African peacekeeping force was drafted in. **1992:** Monrovia was under siege by Taylor's rebel forces. **1995:** Ghanaian-backed peace proposals were accepted by rebel factions; an interim Council of State was established, comprising leaders of three main rebel factions. **1996:** There was renewed fighting in the capital. A peace plan was reached in talks convened by the Economic Community of West African States (ECOWAS); Ruth Perry became Liberia's first female head of state. **1997:** The National Patriotic Party (NPP) won a majority in assembly elections and its leader, Charles Taylor, became head of state. **1998:** There was fighting in Monrovia between President Taylor's forces and opposition militias. **1999:** Liberia was accused of supporting rebels in Sierra Leone. There were clashes between Liberian troops and rebels on the border with Guinea. **2000:** Liberia was accused of buying diamonds from Sierra Leonean rebels, in contravention of an international ban. A massive offensive against rebels in northern Liberia was launched in September.

Practical information

Visa requirements: UK: visa required. USA: visa required
Time difference: GMT +/–0
Chief tourist attractions: sandy beaches along Atlantic coast; mountain scenery
Major holidays: 1 January, 11 February, 15 March, 12 April, 14 May, 26 July, 24 August, 29 November, 25 December; variable: Decoration (March), National Fast and Prayer (April), Thanksgiving (November)

LIBYA

Located in North Africa, bounded north by the Mediterranean Sea, east by Egypt, southeast by Sudan, south by Chad and Niger, and west by Algeria and Tunisia.

National name: *Al-Jamahiriyya al-'Arabiyya al-Libiyya ash-Sha'biyya al-Ishtirakiyya al-'Uzma/Great Libyan Arab Socialist People's State of the Masses*
Area: 1,759,540 sq km/679,358 sq mi
Capital: Tripoli
Major towns/cities: Benghazi, Misratah, Az Zawiyah, Tobruk, Ajdabiya, Darnah
Major ports: Benghazi, Misratah, Az Zawiyah, Tobruk, Ajdabiya, Darnah
Physical features: flat to undulating plains with plateaux and depressions stretch southwards from the Mediterranean coast to an extremely dry desert interior

The flag was said to represent the nation's hope for a green revolution in agriculture. Green expresses the people's Muslim faith. Effective date: 20 November 1977.

Government

Head of state: Moamer al-Khaddhafi from 1969
Head of government: Mubarak al-Shamikh from 2000
Political system: nationalistic socialist
Political executive: unlimited presidency
Administrative divisions: 25 municipalities
Political party: Arab Socialist Union (ASU), radical, left wing
Armed forces: 65,000 (1998)
Conscription: conscription is selective for two years
Death penalty: retained and used for ordinary crimes
Defence spend: (% GDP) 5.3 (1998)
Education spend: (% GNP) N/A
Health spend: (% GDP) 3.4 (1997 est)

Economy and resources

Currency: Libyan dinar
GDP: (US$) 42.7 billion (1999 est)
Real GDP growth: (% change on previous year) 5.4 (1999)
GNP: (US$) N/A
GNP per capita (PPP): (US$) N/A
Consumer price inflation: 18% (1999)
Unemployment: 20% (1997 est)
Foreign debt: (US$) 3.9 billion (1999)
Major trading partners: Italy, Germany, Spain, UK, France, Turkey, Tunisia, South Korea
Resources: petroleum, natural gas, iron ore, potassium, magnesium, sulphur, gypsum
Industries: petroleum refining, processing of agricultural products, cement and other building materials, fish processing and canning, textiles, clothing and footwear
Exports: crude petroleum (accounted for 94% of 1991 export earnings), chemicals and related products. Principal market: Italy 33.2% (1999)
Imports: machinery and transport equipment, basic manufactures, food and live animals, miscellaneous manufactured articles. Principal source: Italy 23.7% (1999)

Arable land: 1% (1996)
Agricultural products: barley, wheat, grapes, olives, dates; livestock rearing (sheep, goats, and camels); fishing

Population and society

Population: 5,605,000 (2000 est)
Population growth rate: 2.4% (1995–2000)
Population density: (per sq km) 3 (1999 est)
Urban population: (% of total) 88 (2000 est)
Age distribution: (% of total population) 0–14 38%, 15–59 57%, 60+ 5% (2000 est)
Ethnic groups: majority are of Berber and Arab origin (97%), with a small number of Tebou and Touareg nomads and semi-nomads, mainly in south
Language: Arabic (official), Italian, English
Religion: Sunni Muslim 97%
Education: (compulsory years) 9
Literacy rate: 91% (men); 68% (women) (2000 est)
Labour force: 17.9% agriculture, 29.9% industry, 52.2% services (1996)
Life expectancy: 68 (men); 72 (women) (1995–2000)
Child mortality rate: (under 5, per 1,000 live births) 32 (1995–2000)
Physicians: 1 per 812 people (1996 est)
Hospital beds: 1 per 314 people (1996 est)
TV sets: (per 1,000 people) 140 (1997)
Radios: (per 1,000 people) 259 (1997)
Internet users: (per 10,000 people) 12.8 (1999)

Transport

Airports: international airports: Tripoli, Benghazi (Benina), Sebhah (international civilian links with Libya have been suspended since April 1992, in accordance with a UN Security Council Resolution of March 1992); seven domestic airports; total passenger km: 398 million (1995)
Railways: none
Roads: total road network: 83,200 km/51,700 mi, of which 57.2% paved (1996 est); passenger cars: 80.9 per 1,000 people (1996 est)

Chronology

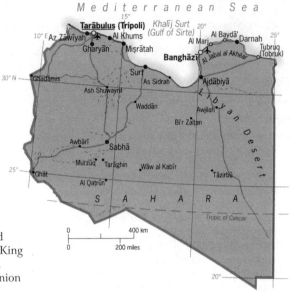

7th century BC: Tripolitania, in western Libya, was settled by Phoenicians, who founded Tripoli; it became an eastern province of Carthaginian kingdom, which was centred on Tunis to the west. **4th century BC:** Cyrenaica, in eastern Libya, colonized by Greeks, who called it Libya. **74 BC:** Became a Roman province, with Tripolitania part of Africa Nova province and Cyrenaica combined with Crete as a province. **19 BC:** The desert region of Fezzan (Phazzania), inhabited by Garmante people, was conquered by Rome. **6th century AD:** Came under control of Byzantine Empire. **7th century:** Conquered by Arabs, who spread Islam: Egypt ruled Cyrenaica and Morrocan Berber Almohads controlled Tripolitania. **mid-16th century:** Became part of Turkish Ottoman Empire, who combined the three ancient regions into one regency in Tripoli. **1711:** Karamanli (Qaramanli) dynasty established virtual independence from Ottomans. **1835:** Ottoman control reasserted. **1911–12:** Conquered by Italy. **1920s:** Resistance to Italian rule by Sanusi order and Umar al-Mukhtar. **1934:** Colony named Libya. **1942:** Italians ousted, and area divided into three provinces: Fezzan (under French control), Cyrenaica, and Tripolitania (under British control). **1951:** Achieved independence as United Kingdom of Libya, under King Idris, former Amir of Cyrenaica and leader of Sanusi order. **1959:** Discovery of oil transformed economy, but also led to unsettling social changes. **1969:** King deposed in military coup led by Col Moamer al Khaddhafi. Revolution Command Council set up and Arab Socialist Union (ASU) proclaimed the only legal party in a new puritanical Islamic-socialist republic which sought Pan-Arab unity. **1970s:** Economic activity collectivized, oil industry nationalized, opposition suppressed by Khaddhafi's revolutionary regime. **1972:** Proposed federation of Libya, Syria, and Egypt abandoned. **1980:** A proposed merger with Syria was abandoned. Libyan troops began fighting in northern Chad. **1986:** The US bombed Khaddhafi's headquarters, following allegations of his complicity in terrorist activities. **1988:** Diplomatic relations with Chad were restored, political prisoners were freed, and the economy was liberalized. **1989:** The US navy shot down two Libyan planes. There was a reconciliation with Egypt. **1992:** Khaddhafi came under international pressure to extradite the alleged terrorists suspected of planting a bomb on a plane that crashed in Lockerbie in Scotland for trial outside Libya. United Nations sanctions were imposed; several countries severed diplomatic and air links with Libya. **1995:** There was an antigovernment campaign of violence by Islamicists. Hundreds of Palestinians and thousands of foreign workers were expelled. **1999:** Lockerbie suspects were handed over for trial in the Netherlands, to be tried by Scottish judges, who ruled that the suspects should be tried on every count. Having handed over the suspects, and after Libya paid compensation to the family of PC Yvonne Fletcher, who was murdered outside the Libyan embassy in London in 1984, full diplomatic relations with the UK were restored and UN sanctions were suspended. **2000:** Khaddhafi installed a new head of government, Prime Minister Mubarak al-Shamikh. He also abolished 12 central government ministries and transferred their powers to provincial bodies.

Practical information

Visa requirements: UK: visa required. USA: visa required
Time difference: GMT +1
Chief tourist attractions: Tripoli, with its beaches and annual International Fair; the ancient Roman towns of Leptis Magna, Sabratha, and Cyrene
Major holidays: 2, 8, 28 March, 11 June, 23 July, 1 September, 7 October; variable: Eid-ul-Adha (4 days), end of Ramadan (3 days), Prophet's Birthday

LIECHTENSTEIN

Landlocked country in west-central Europe, bounded east by Austria and west by Switzerland.

National name: *Fürstentum Liechtenstein/Principality of Liechtenstein*
Area: 160 sq km/62 sq mi
Capital: Vaduz
Major towns/cities: Balzers, Schaan, Ruggell, Triesen, Eschen
Physical features: landlocked Alpine; includes part of Rhine Valley in west

The crown emblem was modernized in 1982. The colours date back to the 18th century. Effective date: 18 September 1982.

Government

Head of state: Prince Hans Adam II from 1989
Head of government: Mario Frick from 1993
Political system: liberal democracy
Political executive: parliamentary
Administrative divisions: 11 communes
Political parties: Patriotic Union (VU), conservative; Progressive Citizens' Party (FBP), conservative; Free Voters' List (FL)
Armed forces: no standing army since 1868; there is a police force of 59 men and 19 auxiliaries
Death penalty: abolished in 1987 (last execution in 1785)

Economy and resources

Currency: Swiss franc
GDP: (US$) 730 million (1998 est)
Real GDP growth: (% change on previous year) N/A
GNP per capita (PPP): (US$) 24,000 (1998 est)
Consumer price inflation: 0.5% (1998 est)
Unemployment: 1.5% (1999)
Major trading partners: Switzerland and other EFTA countries, EU countries
Resources: hydro power
Industries: small machinery, textiles, ceramics, chemicals, furniture, precision instruments, pharmaceutical products, financial services, tourism

Chronology

c. AD 500: Settled by Germanic-speaking Alemanni tribe. 1342: Became a sovereign state. 1434: Present boundaries established. 1719: Former independent lordships of Schellenberg and Vaduz were united by Princes of Liechtenstein to form the present state. 1815–66: A member of the German Confederation. 1868: Abolished standing armed forces. 1871: Liechtenstein was the only German principality to stay outside the newly formed German Empire. 1918: Patriotic Union (VU) party founded, drawing most support from the mountainous south. 1919: Switzerland replaced Austria as the foreign representative of Liechtenstein. 1921: Adopted Swiss currency; constitution created a parliament. 1923: United with Switzerland in customs and monetary union. 1938: Prince Franz Josef II came to power. 1970: After 42 years as the main governing party, the northern-based Progressive Citizens' Party (FBP), was defeated by VU, which became a dominant force in politics. 1978: Joined the Council of Europe. 1984: The franchise was extended to women in the national elections. 1989: Prince Franz Josef II died; succeeded by Hans Adam II. 1990: Joined the United Nations. 1991: Became the seventh member of the European Free Trade Association. 1993: Mario Frick of VU became Europe's youngest head of government, aged 28.

Exports: small machinery, artificial teeth and other material for dentistry, stamps, precision instruments
Imports: machinery and transport equipment, foodstuffs, textiles, metal goods
Arable land: 25% (1995)
Agricultural products: maize, potatoes; cattle rearing

Population and society

Population: 33,000 (2000 est)
Population growth rate: 1.3% (1995–2000)
Population density: (per sq km) 199 (1999 est)
Urban population: (% of total) 23 (2000 est)
Age distribution: (% of total population) 0–14 19%, 15–59 68%, 60+ 13% (2000 est)
Ethnic groups: indigenous population of Alemannic origin; one-third of the population are foreign-born resident workers (mainly Italian and Turkish)
Language: German (official), an Alemannic dialect
Religion: Roman Catholic 80%, Protestant 7%
Education: (compulsory years) 8
Literacy rate: 99% (men); 99% (women) (2000 est)
Labour force: 1.5% agriculture, 46% industry, 52.5% services (1996)
Life expectancy: 78 (men); 83 (women) (1995–2000)

Practical information

Visa requirements: UK/USA: visa not required
Time difference: GMT +1
Chief tourist attractions: Alpine setting; postal museum; Prince's castle; National Museum and State Art Collection at Vaduz

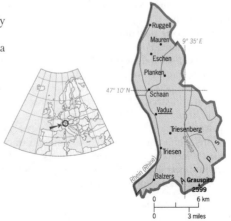

LITHUANIA

Located in northern Europe, bounded north by Latvia, east by Belarus, south by Poland and the Kaliningrad area of Russia, and west by the Baltic Sea.

National name: *Lietuvos Respublika/Republic of Lithuania*
Area: 65,200 sq km/25,173 sq mi
Capital: Vilnius
Major towns/cities: Kaunas, Klaipeda, Siauliai, Panevezys
Physical features: central lowlands with gentle hills in west and higher terrain in southeast; 25% forested; some 3,000 small lakes, marshes, and complex sandy coastline; River Nemunas

Yellow stands for grain and freedom from need. Green symbolizes the forests and hope. Red represents bloodshed and courage. Effective date: 20 March 1989.

Government

Head of state: Valdas Adamkus from 1998
Head of government: Rolandas Paksas from 2000
Political system: emergent democracy
Political executive: dual executive
Administrative divisions: 10 districts subdivided into 56 municipalities
Political parties: Lithuanian Democratic Labour Party (LDLP), reform-socialist (ex-communist); Homeland Union–Lithuanian Conservatives (Tevynes Santara), right of centre, nationalist; Christian Democratic Party of Lithuania, right of centre; Lithuanian Social Democratic Party, left of centre
Armed forces: 11,100 (1998)
Conscription: military service is compulsory for 12 months
Death penalty: abolished in 1998
Defence spend: (% GDP) 1.3 (1998)
Education spend: (% GNP) 5.6 (1996)
Health spend: (% GDP) 6.4 (1997)

Economy and resources

Currency: litas
GDP: (US$) 10.6 billion (1999)
Real GDP growth: (% change on previous year) –4.1 (1999)
GNP: (US$) 9.7 billion (1999)
GNP per capita (PPP): (US$) 6,093 (1999)
Consumer price inflation: 0.8% (1999)
Unemployment: 14.1% (1997)
Foreign debt: (US$) 2.4 billion (1999 est)
Major trading partners: Russia, Germany, Belarus, Latvia, Ukraine, Poland, Italy, the Netherlands
Resources: small deposits of petroleum, natural gas, peat, limestone, gravel, clay, sand
Industries: petroleum refining and petroleum products, cast iron and steel, textiles, mineral fertilizers, fur coats, refrigerators, TV sets, bicycles, paper
Exports: textiles, machinery and equipment, non-precious metals, animal products, timber. Principal market: Germany 15.8% (1999)
Imports: petroleum and natural gas products, machinery and transport equipment, chemicals, fertilizers, consumer goods. Principal source: Russia 19.8% (1999)
Arable land: 45.5% (1996)
Agricultural products: cereals, sugar beet, potatoes, vegetables; livestock rearing and dairy farming (animal husbandry accounted for more than 50% of the value of total agricultural production in 1992)

Population and society

Population: 3,670,000 (2000 est)
Population growth rate: –0.3% (1995–2000)
Population density: (per sq km) 56 (1999 est)
Urban population: (% of total) 68 (2000 est)
Age distribution: (% of total population) 0–14 19%, 15–59 63%, 60+ 18% (2000 est)
Ethnic groups: 80% Lithuanian ethnic descent, 9% ethnic Russian, 8% Polish, 2% Belarussian, 1% Ukrainian
Language: Lithuanian (official)
Religion: predominantly Roman Catholic; Evangelical Lutheran, also Russian Orthodox, Evangelical Reformist, and Baptist
Education: (compulsory years) 9
Literacy rate: 99% (men); 99% (women) (2000 est)
Labour force: 13.9% agriculture, 37.9% industry, 48.2% services (1997 est)
Life expectancy: 64 (men); 76 (women) (1995–2000)
Child mortality rate: (under 5, per 1,000 live births) 24 (1995–2000)
Physicians: 1 per 250 people (1997 est)
Hospital beds: 1 per 90 people (1997 est)
TV sets: (per 1,000 people) 459 (1997)
Radios: (per 1,000 people) 513 (1997)
Internet users: (per 10,000 people) 278.3 (1999)
Personal computer users: (per 100 people) 5.9 (1999)

Transport

Airports: international airports: Vilnius, Kaunas, Siauliai; few domestic flights; total passenger km: 308 million (1995)
Railways: total length: 2,007 km/1,247 mi; total passenger km: 1,574 million (1994)
Roads: total road network: 68,161 km/42,355 mi, of which 88.8% paved (1997); passenger cars: 238 per 1,000 people (1997)

Chronology

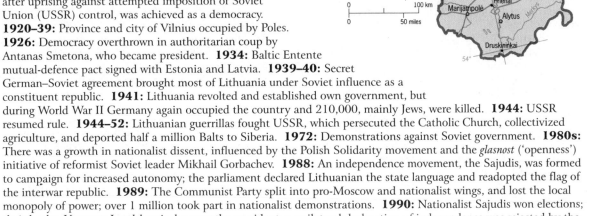

late 12th century: Became a separate nation. **1230:** Mindaugas united Lithuanian tribes to resist attempted invasions by German and Livonian Teutonic Knights, and adopted Christianity. **14th century:** Strong Grand Duchy formed by Gediminas, founder of Vilnius and Jogaila dynasty, and his son, Algirdas; absorbing Ruthenian territories to east and south, it stretched from the Baltic to the Black Sea and east, nearly reaching Moscow. **1410:** Led by Duke Vytautas, and in alliance with Poland, the Teutonic Knights were defeated decisively at the Battle of Tannenberg. **1569:** Joined Poland in a confederation, under the Union of Lublin, in which Poland had the upper hand and Lithuanian upper classes were Polonized. **1795:** Came under control of Tsarist Russia, following partition of Poland; 'Lithuania Minor' (Kaliningrad) fell to Germany. **1831 and 1863:** Failed revolts for independence. **1880s:** Development of organized nationalist movement. **1914–18:** Occupied by German troops during World War I. **1918–19:** Independence declared and, after uprising against attempted imposition of Soviet Union (USSR) control, was achieved as a democracy. **1920–39:** Province and city of Vilnius occupied by Poles. **1926:** Democracy overthrown in authoritarian coup by Antanas Smetona, who became president. **1934:** Baltic Entente mutual-defence pact signed with Estonia and Latvia. **1939–40:** Secret German–Soviet agreement brought most of Lithuania under Soviet influence as a constituent republic. **1941:** Lithuania revolted and established own government, but during World War II Germany again occupied the country and 210,000, mainly Jews, were killed. **1944:** USSR resumed rule. **1944–52:** Lithuanian guerrillas fought USSR, which persecuted the Catholic Church, collectivized agriculture, and deported half a million Balts to Siberia. **1972:** Demonstrations against Soviet government. **1980s:** There was a growth in nationalist dissent, influenced by the Polish Solidarity movement and the *glasnost* ('openness') initiative of reformist Soviet leader Mikhail Gorbachev. **1988:** An independence movement, the Sajudis, was formed to campaign for increased autonomy; the parliament declared Lithuanian the state language and readopted the flag of the interwar republic. **1989:** The Communist Party split into pro-Moscow and nationalist wings, and lost the local monopoly of power; over 1 million took part in nationalist demonstrations. **1990:** Nationalist Sajudis won elections; their leader, Vytautas Landsbergis, became the president; a unilateral declaration of independence was rejected by the USSR, who imposed an economic blockade. **1991:** Soviet paratroopers briefly occupied key buildings in Vilnius, killing 13; the Communist Party was outlawed; Lithuanian independence was recognized by the USSR and Western nations; the country was admitted into the United Nations. **1992:** Economic restructuring caused a contraction in GDP. **1993:** A free-trade agreement was reached with other Baltic States. The last Russian troops departed. **1994:** A friendship and cooperation treaty was signed with Poland. **1994:** A trade and cooperation agreement was reached with the European Union. **1997:** A border treaty was signed with Russia. **1998:** Valdas Adamkus became president. **1999:** Andrius Kubelius became prime minister following the resignation of Rolandas Paksas. **2000:** Paksas returned to power, leading a centre-left coalition.

Practical information

Visa requirements: UK: visa not required for a stay of up to 90 days. USA: visa not required for a stay of up to 90 days
Time difference: GMT +2
Chief tourist attractions: historic cities of Vilnius, Klaipeda, Kaunas, and Trakai; coastal resorts such as Palanga and Kursiu Nerija; picturesque countryside
Major holidays: 1 January, 16 February, 5 May, 6 July, 1 November, 25–26 December; variable: Easter Monday

LUXEMBOURG

Landlocked country in Western Europe, bounded north and west by Belgium, east by Germany, and south by France.

National name: *Grand-Duché de Luxembourg/Grand Duchy of Luxembourg*
Area: 2,586 sq km/998 sq mi
Capital: Luxembourg
Major towns/cities: Esch-sur-Alzette, Differdange, Dudelange, Pétange
Physical features: on the River Moselle; part of the Ardennes (Oesling) forest in north

The blue band is lighter than that of the Dutch tricolour. Effective date: 16 August 1972.

Government

Head of state: Grand Duke Henri from 2000
Head of government: Jean-Claude Juncker from 1995
Political system: liberal democracy
Political executive: parliamentary
Administrative divisions: 12 cantons within three districts
Political parties: Christian Social Party (PCS), moderate, left of centre; Luxembourg Socialist Workers' Party (POSL), moderate, socialist; Democratic Party (PD), left of centre; Communist Party of Luxembourg, pro-European left wing
Armed forces: 800; gendarmerie 600 (1998)
Conscription: military service is voluntary
Death penalty: abolished in 1979
Defence spend: (% GDP) 0.9 (1998)
Education spend: (% GNP) 4.1 (1996)
Health spend: (% GDP) 7 (1997)

Economy and resources

Currency: Luxembourg franc
GDP: (US$) 19.3 billion (1999)
Real GDP growth: (% change on previous year) 7.5 (1999)
GNP: (US$) 19.3 billion (1999)
GNP per capita (PPP): (US$) 38,247 (1999)
Consumer price inflation: 1.0% (1999)
Unemployment: 2.8% (1998)
Major trading partners: EU (principally Belgium, Germany, and France), USA
Resources: iron ore
Industries: steel and rolled steel products, chemicals, rubber and plastic products, metal and machinery products, paper and printing products, food products, financial services
Exports: base metals and manufactures, mechanical and electrical equipment, rubber and related products, plastics, textiles and clothing. Principal market: Germany 24.7% (1999)
Imports: machinery and electrical apparatus, transport equipment, mineral products. Principal source: Belgium 34.6% (1999)
Arable land: 22% (1995)
Agricultural products: maize, roots and tubers,

wheat, forage crops, grapes; livestock rearing and dairy farming

Population and society

Population: 431,000 (2000 est)
Population growth rate: 1.1% (1995–2025)
Population density: (per sq km) 165 (1999 est)
Urban population: (% of total) 92 (2000 est)
Age distribution: (% of total population) 0–14 18%, 15–59 62%, 60+ 20% (2000 est)
Ethnic groups: majority descended from the Moselle Franks (French and German blend), Portuguese, Italian, and other European guest and resident workers
Language: Letzeburgisch (a German-Moselle-Frankish dialect; official), English
Religion: Roman Catholic about 95%, Protestant and Jewish 4%
Education: (compulsory years) 9
Literacy rate: 99% (men); 99% (women) (2000 est)
Labour force: 51.2% of population: 2.6% agriculture, 27.7% industry, 69.7% services (1992)
Life expectancy: 73 (men); 80 (women) (1995–2000)
Child mortality rate: (under 5, per 1,000 live births) 8 (1995–2000)
Physicians: 1 per 455 people (1995)
Hospital beds: 1 per 80 people (1996)
TV sets: (per 1,000 people) 391 (1997)
Radios: (per 1,000 people) 683 (1997)
Internet users: (per 10,000 people) 1,747.4 (1999)
Personal computer users: (per 100 people) 39.6 (1999)

Transport

Airports: international airports: Luxembourg (Findel); no domestic airports; total passenger km: 380 million (1995)
Railways: total length: 275 km/171 mi; total passenger km: 295 million (1997)
Roads: total road network: 5,171 km/3,213 mi, of which 100% paved (1997); passenger cars: 566 per 1,000 people (1997)

Chronology

963: Luxembourg became autonomous within Holy Roman Empire under Siegfried, Count of Ardennes. **1060:** Conrad, descendent of Siegfried, took the title Count of Luxembourg. **1354:** Emperor Charles IV promoted Luxembourg to the status of duchy. **1441:** Luxembourg ceded to dukes of Burgundy. **1482:** Luxembourg came under Habsburg control. **1555:** Luxembourg became part of Spanish Netherlands on division of Habsburg domains. **1684–97:** Much of Luxembourg occupied by France. **1713:** Treaty of Utrecht transferred Spanish Netherlands to Austria. **1797:** Conquered by revolutionary France. **1815:** Congress of Vienna made Luxembourg a grand duchy, under King William of the Netherlands. **1830:** Most of Luxembourg supported Belgian revolt against the Netherlands. **1839:** Western part of Luxembourg assigned to Belgium. **1842:** Luxembourg entered the Zollverein (German customs union). **1867:** Treaty of London confirmed independence and neutrality of Luxembourg to allay French fears about possible inclusion in a unified Germany. **1870s:** Development of iron and steel industry. **1890:** Link with Dutch crown ended on accession of Queen Wilhelmina, since Luxembourg's law of succession did not permit a woman to rule. **1912:** Revised law of succession allowed Marie-Adelaide to become grand duchess. **1914–18:** Occupied by Germany. **1919:** Plebiscite overwhelmingly favoured continued independence; Marie-Adelaide abdicated after allegations of collaboration with Germany; Charlotte became grand duchess. **1921:** Entered into close economic links with Belgium. **1940:** Invaded by Germany. **1942–44:** Annexed by Germany. **1948:** Luxembourg formed Benelux customs union with Belgium and the Netherlands. **1949:** Luxembourg became founding member of North Atlantic Treaty Organization (NATO). **1958:** Luxembourg became founding member of European Economic Community (EEC). **1964:** Grand Duchess Charlotte abdicated in favour of her son Jean. **1994:** Former premier Jacques Santer became the president of the European Commission (EC). **1995:** Jean-Claude Juncker became prime minister. **2000:** Grand Duke Jean abdicated in favour of his son Henri.

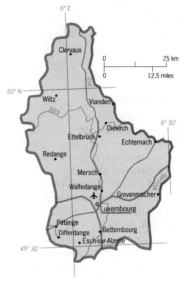

Practical information

Visa requirements: UK: visa not required. USA: visa not required for a stay of up to 90 days
Time difference: GMT +1
Chief tourist attractions: Luxembourg-Ville, with its historic monuments and many cultural events; medieval castles (Clerf, Esch/Sauer, Vianden, Wietz);

Benedictine abbey at Echternach; thermal centre at Mandorf-les-Bains; footpaths and hiking trails
Major holidays: 1 January, 1 May, 23 June, 15 August, 1–2 November, 25–26, 31 December; variable: Ascension Thursday, Easter Monday, Whit Monday, Shrove Monday

MACEDONIA

Landlocked country in southeast Europe, bounded north by Serbia, west by Albania, south by Greece, and east by Bulgaria.

National name: *Republika Makedonija/Republic of Macedonia* (official internal name); *Poranesna Jugoslovenska Republika Makedonija/Former Yugoslav Republic of Macedonia* (official international name)
Area: 25,700 sq km/9,922 sq mi
Capital: Skopje
Major towns/cities: Bitola, Prilep, Kumanovo, Tetovo
Physical features: mountainous; rivers: Struma, Vardar; lakes: Ohrid, Prespa, Scutari; partly Mediterranean climate with hot summers

Red and yellow were the colours of Macedonia's flag when the republic was part of Yugoslavia. Effective date: 6 October 1995.

Government

Head of state: Boris Trajkovski from 1999
Head of government: Ljubco Georgievski from 1998
Political system: emergent democracy
Political executive: limited presidency
Administrative divisions: 123 municipalities
Political parties: Socialist Party (SP); Social Democratic Alliance of Macedonia (SDSM) bloc, former Communist Party; Party for Democratic Prosperity (PDP), ethnic Albanian, left of centre; Internal Macedonian Revolutionary Organization–Democratic Party for Macedonian National Unity (VMRO–DPMNE), radical nationalist; Democratic Party of Macedonia (DPM), nationalist, free market
Armed forces: 20,000; plus paramilitary force of 7,500 (1998)
Conscription: military service is compulsory for nine months
Death penalty: laws do not provide for the death penalty for any crime
Defence spend: (% GDP) 9.9 (1998)
Education spend: (% GNP) 5.1 (1997 est)
Health spend: (% GDP) 6.1 (1997 est)

Economy and resources

Currency: Macedonian denar
GDP: (US$) 3.4 billion (1999)
Real GDP growth: (% change on previous year) 2.7 (1999)
GNP: (US$) 3.4 billion (1999)
GNP per capita (PPP): (US$) 4,339 (1999)
Consumer price inflation: –1.1% (1999)
Unemployment: 38.8% (1997)
Foreign debt: (US$) 1.7 billion (1999)
Major trading partners: Germany, Bulgaria, Yugoslavia, Slovenia, Ukraine, USA, Greece, Italy
Resources: coal, iron, zinc, chromium, manganese, lead, copper, nickel, silver, gold
Industries: metallurgy, chemicals, textiles, buses, refrigerators, detergents, medicines, wood pulp, wine
Exports: manufactured goods, machinery and transport equipment, miscellaneous manufactured articles, sugar beet, vegetables, cheese, lamb, tobacco. Principal market: Germany 21.4% (1999)
Imports: mineral fuels and lubricants, manufactured goods, machinery and transport equipment, food and live animals, chemicals. Principal source: Germany 13.7% (1999)
Arable land: 23.9% (1996)
Agricultural products: rice, wheat, barley, sugar beet, fruit and vegetables, tobacco, sunflowers, potatoes, grapes (wine industry is important); livestock rearing and dairy farming

Population and society

Population: 2,024,000 (2000 est)
Population growth rate: 0.6% (1995–2000)
Population density: (per sq km) 78 (1999 est)
Urban population: (% of total) 62 (2000 est)
Age distribution: (% of total population) 0–14 23%, 15–59 62%, 60+ 15% (2000 est)
Ethnic groups: 67% Macedonian ethnic descent, 23% ethnic Albanian, 4% Turkish, 2% Romanian, 2% Serb, and 2% Muslim, comprising Macedonian Slavs who converted to Islam during the Ottoman era, and are known as Pomaks. This ethnic breakdown is disputed by Macedonia's ethnic Albanian population, who claim that they form 40% of the population, and seek autonomy, and by ethnic Serbs, who claim that they form 11.5%.
Language: Macedonian (related to Bulgarian; official), Albanian
Religion: Christian, mainly Orthodox 67%; Muslim 30%
Education: (compulsory years) 8
Literacy rate: 98% (men); 98% (women) (2000 est)
Labour force: 13.4% agriculture, 33.7% industry, 52.9% services (1997 est)
Life expectancy: 71 (men); 75 (women) (1995–2000)
Child mortality rate: (under 5, per 1,000 live births) 26 (1995–2000)
Physicians: 1 per 445 people (1997)
Hospital beds: 1 per 194 people (1997)
TV sets: (per 1,000 people) 250 (1998)
Radios: (per 1,000 people) 200 (1997)
Internet users: (per 10,000 people) 149.2 (1999)

Transport

Airports: international airports: Skopje, Ohrid; domestic services between Skopje and Ohrid; total passenger km: 340 million (1995)
Railways: total length: 699 km/434 mi; passengers carried: 1.25 million (1994)
Roads: total road network: 8,684 km/5,396 mi, of which 63.8% paved (1997); passenger cars: 141 per 1,000 people (1996)

Chronology

4th century BC: Part of ancient great kingdom of Macedonia, which included northern Greece and southwest Bulgaria and, under Alexander the Great, conquered a vast empire; Thessaloniki founded. **146 BC:** Macedonia became a province of the Roman Empire. **395 AD:** On the division of the Roman Empire, came under the control of the Byzantine Empire, with its capital at Constantinople. **6th century:** Settled by Slavs, who later converted to Christianity. **9th–14th centuries:** Under successive rule by Bulgars, Byzantium, and Serbia. **1371:** Became part of Islamic Ottoman Empire. **late 19th century:** The 'Internal Macedonian Revolutionary Organization', through terrorism, sought to provoke Great Power intervention against Turks. **1912–13:** After First Balkan War, partitioned between Bulgaria, Greece, and the area that constitutes the current republic of Serbia. **1918:** Serbian part included in what was to become Yugoslavia; Serbian imposed as official language. **1941–44:** Occupied by Bulgaria. **1945:** Created a republic within Yugoslav Socialist Federation. **1967:** The Orthodox Macedonian archbishopric of Skopje, forcibly abolished 200 years earlier by the Turks, was restored. **1980:** The rise of nationalism was seen after the death of Yugoslav leader Tito. **1990:** Multiparty elections produced an inconclusive result. **1991:** Kiro Gligorov, a pragmatic former communist, became president. A referendum supported independence. **1992:** Independence was declared, and accepted by Serbia/Yugoslavia, but international recognition was withheld because of Greece's objections to the name. **1993:** Sovereignty was recognized by the UK and Albania; United Nations membership was won under the provisional name of the Former Yugoslav Republic of Macedonia; Greece blocked full European Union (EU) recognition. **1994:** Independence was recognized by the USA; a trade embargo was imposed by Greece, causing severe economic damage. **1995:** Independence was recognized by Greece and the trade embargo lifted. President Gligorov survived a car bomb assassination attempt. **1997:** Plans to reduce the strength of the UN Preventive Deployment Force (UNPREDEP) were abandoned. The government announced compensation for the public's losses in failed investment schemes. **1998:** The UN extended the mandate of UNPREDEP. A general election resulted in Ljubco Georgievski, the VRMO-DPMNE leader, becoming prime minister. A 1,700-strong NATO force was deployed in Macedonia to safeguard the 2,000 ceasefire verification monitors in neighbouring Kosovo, Yugoslavia. **1999:** Boris Trajkovski was elected president.

Practical information

Visa requirements: UK: visa not required. USA: visa required
Time difference: GMT +1

Chief tourist attractions: mountain scenery
Major holidays: 1–2 January, 1–2 May, 2 August, 11 October

MADAGASCAR

Island in the Indian Ocean, off the coast of East Africa, about 400 km/280 mi from Mozambique.

National name: *Repoblikan'i Madagasikara/République de Madagascar/ Republic of Madagascar*
Area: 587,041 sq km/226,656 sq mi
Capital: Antananarivo
Major towns/cities: Antsirabe, Mahajanga, Fianarantsoa, Toamasina, Ambatondrazaka
Major ports: Toamasina, Antsiranana, Toliary, Mahajanga
Physical features: temperate central highlands; humid valleys and tropical coastal plains; arid in south

Red stands for sovereignty. White represents purity. Green recalls the coastal inhabitants and is a symbol of hope. Effective date: 14 October 1958.

Government

Head of state: Didier Ratsiraka from 1996
Head of government: René Tantely Gabrio Andrianarivo from 1998
Political system: emergent democracy
Political executive: limited presidency
Administrative divisions: six provinces
Political parties: Association for the Rebirth of Madagascar (AREMA), left of centre; One Should Not Be Judged By One's Works (AVI), left of centre; Rally for Socialism and Democracy (RPSD), left of centre
Armed forces: 21,000; plus paramilitary gendarmerie of 7,500 (1998)
Conscription: military service is compulsory for 18 months
Death penalty: retains the death penalty for ordinary crimes but can be considered abolitionist in practice (last known execution in 1958)
Defence spend: (% GDP) 0.9 (1998)
Education spend: (% GDP) 1.9 (1997)
Health spend: (% GDP) 1.1 (1990–95)

Economy and resources

Currency: Malagasy franc
GDP: (US$) 3.73 billion (1999)
Real GDP growth: (% change on previous year) 4.7 (1999)
GNP: (US$) 3.7 billion (1999)
GNP per capita (PPP): (US$) 766 (1999)
Consumer price inflation: 9.9% (1999)
Unemployment: 6% (1995 est)
Foreign debt: (US$) 4.37 billion (1999)
Major trading partners: France, Japan, Germany, USA, UK, China, Singapore
Resources: graphite, chromite, mica, titanium ore, small quantities of precious stones, bauxite and coal deposits, petroleum reserves
Industries: food products, textiles and clothing, beverages, chemical products, cement, fertilizers, pharmaceuticals
Exports: coffee, shrimps, cloves, vanilla, petroleum products, chromium, cotton fabrics. Principal market: France 41.4% (1999 est)
Imports: minerals (crude petroleum), chemicals, machinery, vehicles and parts, metal products, electrical equipment. Principal source: France 34.2% (1999 est)
Arable land: 4.4% (1996)
Agricultural products: rice, cassava, mangoes, bananas, potatoes, sugar cane, seed cotton, sisal, vanilla, cloves, coconuts, tropical fruits; cattle-farming; sea-fishing

Population and society

Population: 15,942,000 (2000 est)
Population growth rate: 2.9% (1995–2000)
Population density: (per sq km) 26 (1999 est)
Urban population: (% of total) 30 (2000 est)
Age distribution: (% of total population) 0–14 44%, 15–59 51%, 60+ 5% (2000 est)
Ethnic groups: 18 main Malagasy tribes of Malaysian–Polynesian origin; also French, Chinese, Indians, Pakistanis, and Comorans
Language: Malagasy, French (both official), local dialects
Religion: over 50% traditional beliefs, Roman Catholic, Protestant about 40%, Muslim 7%
Education: (compulsory years) 5
Literacy rate: 74% (men); 60% (women) (2000 est)
Labour force: 75.5% agriculture, 7% industry, 17.5% services (1997 est)
Life expectancy: 56 (men); 59 (women) (1995–2000)
Child mortality rate: (under 5, per 1,000 live births) 116 (1995–2000)
Physicians: 1 per 8,385 people (1993 est)
Hospital beds: 1 per 1,072 people (1993 est)
TV sets: (per 1,000 people) 45 (1997)
Radios: (per 1,000 people) 192 (1997)
Internet users: (per 10,000 people) 5.2 (1999)
Personal computer users: (per 100 people) 0.2 (1999)

Transport

Airports: international airports: Antananarivo (Ivato), Mahajunga (Amborovi); two domestic airports and 57 airfields open to public air traffic; total passenger km: 631 million (1995)
Railways: total length: 1,054 km/655 mi; total passenger km: 60 million (1994)
Roads: total road network: 49,837 km/30,969 mi, of which 11.6% paved (1996); passenger cars: 4.1 per 1,000 people (1996 est)

Chronology

c. **6th–10th centuries** AD: Settled by migrant Indonesians. **1500:** First visited by European navigators. **17th century:** Development of Merina and Sakalava kingdoms in the central highlands and west coast. **1642–74:** France established a coastal settlement at Fort-Dauphin, which they abandoned after a massacre by local inhabitants. **late 18th–early 19th century:** Merinas, united by their ruler Andrianampoinimerina, became dominant kingdom; court converted to Christianity. **1861:** Ban on Christianity (imposed in 1828) and entry of Europeans lifted by Merina king, Radama II. **1885:** Became French protectorate. **1895:** Merina army defeated by French and became a colony; slavery abolished. **1942–43:** British troops invaded to overthrow French administration allied to the pro-Nazi Germany Vichy regime and install anti-Nazi Free French government. **1947–48:** Nationalist uprising brutally suppressed by French. **1960:** Independence achieved from France, with Philibert Tsiranana, the leader of the Social Democratic Party (PSD), as president. **1972:** Merina-dominated army overthrew Tsiranana's government, dominated by the cotier (coastal tribes), as the economy deteriorated. **1975:** Martial law imposed; new one-party state Marxist constitution adopted, with Lt-Commander Didier Ratsiraka as president. **1978:** More than 1,000 people were killed in race riots in Majunga city in the northwest. **1980:** Ratsiraka abandoned the Marxist experiment, which had involved nationalization and the severing of ties with France. **1990:** Political opposition was legalized and 36 new parties were created. **1991:** Antigovernment demonstrations were held. Ratsiraka formed a new unity government, which included opposition members. **1992:** Constitutional reform was approved by a referendum. **1995:** A referendum backed the appointment of a prime minister by the president, rather than the assembly. **1996:** Didier Ratsiraka was elected president again. **1998:** ARES largest party following election. Tantely Andrianarivo appointed prime minister. **2000:** Around 600,000 people were made homeless when cyclones which had been striking southern Africa swept through the island.

Practical information

Visa requirements: UK: visa required. USA: visa required
Time difference: GMT +3
Chief tourist attractions: unspoilt scenery; unusual wildlife; much of Madagascar's flora and fauna is unique to the island – there are 3,000 endemic species of butterfly
Major holidays: 1 January, 29 March, 1 May, 26 June, 15 August, 1 November, 25, 30 December; variable: Ascension Thursday, Good Friday, Easter Monday, Whit Monday

MALAWI

Located in southeast Africa, bounded north and northeast by Tanzania; east, south, and west by Mozambique; and west by Zambia.

National name: *Republic of Malawi*
Area: 118,484 sq km/45,735 sq mi
Capital: Lilongwe
Major towns/cities: Blantyre, Mzuzu, Zomba
Physical features: landlocked narrow plateau with rolling plains; mountainous west of Lake Nyasa

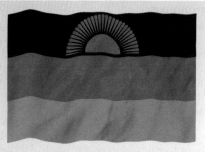

Black, red, and green are known as the 'black liberation' colours, recalling Jamaican black activist Marcus Garvey. Taken from the arms of Nyasaland, the sun indicates the dawning of a new era. Effective date: 6 July 1964.

Government

Head of state and government: Bakili Muluzi from 1994
Political system: emergent democracy
Political executive: limited presidency
Administrative divisions: three regions, subdivided into 24 districts
Political parties: Malawi Congress Party (MCP), multiracial, right wing; United Democratic Front (UDF), left of centre; Alliance for Democracy (AFORD), left of centre
Armed forces: 5,000 (1998)
Conscription: military service is voluntary
Death penalty: retained and used for ordinary crimes
Defence spend: (% GDP) 1.2 (1998)
Education spend: (% GNP) 5.4 (1997 est)
Health spend: (% GDP) 2.3 (1990–95)

Economy and resources

Currency: Malawi kwacha
GDP: (US$) 1.8 billion (1999)
Real GDP growth: (% change on previous year) 4.7 (1999)
GNP: (US$) 2 billion (1999)
GNP per capita (PPP): (US$) 581 (1999)
Consumer price inflation: 44.8% (1999)
Unemployment: N/A
Foreign debt: (US$) 2.7 billion (1999)
Major trading partners: South Africa, USA, Germany, the Netherlands, Zimbabwe, Zambia, Japan
Resources: marble, coal, gemstones, bauxite and graphite deposits, reserves of phosphates, uranium, glass sands, asbestos, vermiculite
Industries: food products, chemical products, textiles, beverages, cement
Exports: tobacco, tea, sugar, cotton, groundnuts. Principal market: South Africa 14.8% (1998 est)
Imports: petroleum products, fertilizers, coal, machinery and transport equipment, miscellaneous manufactured articles. Principal source: South Africa 36.1% (1998 est)
Arable land: 17% (1996)

Agricultural products: maize, cassava, groundnuts, pulses, tobacco, tea, sugar cane

Population and society

Population: 10,925,000 (2000 est)
Population growth rate: 2.4% (1995–2000)
Population density: (per sq km) 90 (1999 est)
Urban population: (% of total) 25 (2000 est)
Age distribution: (% of total population) 0–14 47%, 15–59 48%, 60+ 5% (2000 est)
Ethnic groups: almost all indigenous Africans, divided into numerous ethnic groups, such as the Chewa, Nyanja, Tumbuka, Yao, Lomwe, Sena, Tonga, and Ngoni. There are also Asian and European minorities
Language: English, Chichewa (both official), other Bantu languages
Religion: Protestant 50%, Roman Catholic 20%, Muslim 2%, animist
Education: (compulsory years) 8
Literacy rate: 74% (men); 46% (women) (2000 est)
Labour force: 84.1% agriculture, 5.9% industry, 10% services (1997 est)
Life expectancy: 39 (men); 40 (women) (1995–2000)
Child mortality rate: (under 5, per 1,000 live births) 220 (1995–2000)
Physicians: 1 per 44,205 people (1993 est)
Hospital beds: 1 per 1,184 people (1993 est)
TV sets: (per 1,000 people) 2 (1997)
Radios: (per 1,000 people) 249 (1997)
Internet users: (per 10,000 people) 9.4 (1999)
Personal computer users: (per 100 people) 0.1 (1999)

Transport

Airports: international airports: Lilongwe (Kamuzu), Blantyre (Chileka); three domestic airports; total passenger km: 110 million (1995)
Railways: total length: 797 km/495 mi; total passenger km: 19 million (1994)
Roads: total road network: 16,451 km/10,223 mi, of which 19% paved (1997); passenger cars: 2.3 per 1,000 people (1996 est)

Chronology

1st–4th centuries AD: Immigration by Bantu-speaking peoples. **1480:** Foundation of Maravi (Malawi) Confederacy, which covered much of central and southern Malawi and lasted into the 17th century. **1530:** First visited by the Portuguese. **1600:** Ngonde kingdom founded in northern Malawi by immigrants from Tanzania. **18th century:** Chikulamayembe state founded by immigrants from east of Lake Nyasa; slave trade flourished and Islam introduced in some areas. **mid-19th century:** Swahili-speaking Ngoni peoples, from South Africa, and Yao entered the region, dominating settled agriculturists; Christianity introduced by missionaries, such as David Livingstone. **1891:** Became British protectorate of Nyasaland; cash crops, particularly coffee, introduced. **1915:** Violent uprising, led by Rev John Chilembwe, against white settlers who had moved into the fertile south, taking land from local population. **1953:** Became part of white-dominated Central African Federation, which included South Rhodesia (Zimbabwe) and North Rhodesia (Zambia). **1958:** Dr Hastings Kamuzu Banda returned to the country after working abroad and became head of the conservative-nationalist Nyasaland/Malawi Congress Party (MCP), which spearheaded the campaign for independence. **1963:** Central African Federation dissolved. **1964:** Independence achieved within Commonwealth as Malawi, with Banda as prime minister. **1966:** Became one-party republic, with Banda as president. **1967:** Banda became pariah of Black Africa by recognizing racist, white-only republic of South Africa. **1971:** Banda was made president for life. **1970s:** There were reports of human-rights violations and the murder of Banda's opponents. **1980s:** The economy began to deteriorate after nearly two decades of expansion. **1986–89:** There was an influx of nearly a million refugees from Mozambique. **1992:** There were calls for a multiparty political system. Countrywide industrial riots caused many fatalities. Western aid was suspended over human-rights violations. **1993:** A referendum overwhelmingly supported the ending of one-party rule. **1994:** A new multiparty constitution was adopted. Bakili Muluzi, of the United Democratic Front (UDF), was elected president in the first free elections for 30 years. **1995:** Banda and the former minister of state John Tembo were charged with conspiring to murder four political opponents in 1983, but were cleared. **1999:** Violent protests followed the announcement that Muluzi had been re-elected as president. **2000:** Muluzi sacked his entire cabinet after high-ranking officials were accused of corruption, in a move aimed at placating foreign donors. However, his new government included many of the same people.

Practical information

Visa requirements: UK: visa not required. USA: visa not required
Time difference: GMT +2
Chief tourist attractions: beaches on Lake Nyasa; varied scenery; big game; excellent climate
Major holidays: 1 January, 3 March, 14 May, 6 July, 17 October, 22, 25–26 December; variable: Good Friday, Easter Monday, Holy Saturday

MALAYSIA

Located in southeast Asia, comprising the Malay Peninsula, bounded north by Thailand, and surrounded east and south by the South China Sea and west by the Strait of Malacca; and the states of Sabah and Sarawak in the northern part of the island of Borneo (southern Borneo is part of Indonesia).

Red and white are the traditional colours of South East Asia. The blue canton recalls the British Empire and represents unity. Yellow is the colour of the Sultans of Malaysia. Effective date: 16 September 1963.

National name: *Persekutuan Tanah Malaysia/Federation of Malaysia*
Area: 329,759 sq km/127,319 sq mi
Capital: Kuala Lumpur
Major towns/cities: Johor Bahru, Ipoh, George Town (on Penang island), Kuala Terengganu, Kuala Bahru, Petalong Jaya, Kelang, Kuching (on Sarawak), Kota Kinabalu (on Sabah)
Major ports: Kelang
Physical features: comprises peninsular Malaysia (the nine Malay states – Johore, Kedah, Kelantan, Negri Sembilan, Pahang, Perak, Perlis, Selangor, Terengganu – plus Malacca and Penang); states of Sabah and Sarawak on the island of Borneo; and the federal territory of Kuala Lumpur; 75% tropical rainforest; central mountain range; Mount Kinabalu, the highest peak in southeast Asia, is in Sabah; swamps in east; Niah caves (Sarawak)

Government

Head of state: Tuanku Salehuddin Abdul Aziz Shan bin al-Marhum Hisamuddin Alam Shah from 1999
Head of government: Mahathir bin Muhammad from 1981
Political system: liberal democracy
Political executive: parliamentary
Administrative divisions: 13 states
Political parties: New United Malays' National Organization (UMNO Baru), Malay-oriented nationalist; Malaysian Chinese Association (MCA), Chinese-oriented, conservative; Gerakan Party, Chinese-oriented, socialist; Malaysian Indian Congress (MIC), Indian-oriented; Democratic Action Party (DAP), multiracial but Chinese-dominated, left of centre; Pan-Malayan Islamic Party (PAS), Islamic; Semangat '46 (Spirit of 1946), moderate, multiracial
Armed forces: 110,000; reserve force 40,600; paramilitary force 20,100 (1998)
Conscription: military service is voluntary
Death penalty: retained and used for ordinary crimes
Defence spend: (% GDP) 3.7 (1998)
Education spend: (% GNP) 4.9 (1997)
Health spend: (% GDP) 2.4 (1997)

Economy and resources

Currency: ringgit
GDP: (US$) 74.6 billion (1999)
Real GDP growth: (% change on previous year) 5.6 (1999 est)
GNP: (US$) 77.3 billion (1999)
GNP per capita (PPP): (US$) 7,963 (1999)
Consumer price inflation: 2.8% (1999 est)
Unemployment: 2.5% (1997)
Foreign debt: (US$) 42 billion (1999)
Major trading partners: Japan, USA, Singapore, Taiwan, UK and other EU countries, South Korea
Resources: tin, bauxite, copper, iron ore, petroleum, natural gas, forests
Industries: electrical and electronic appliances (particularly radio and TV receivers), food processing, rubber products, industrial chemicals, wood products, petroleum refinery, motor vehicles, tourism
Exports: palm oil, rubber, crude petroleum, machinery and transport equipment, timber, tin, textiles, electronic goods. Principal market: USA 21.9% (1999)
Imports: machinery and transport equipment, chemicals, foodstuffs, crude petroleum, consumer goods. Principal source: Japan 20.8% (1999)
Arable land: 5.5% (1996)
Agricultural products: rice, cocoa, palm, rubber, pepper, coconuts, tea, pineapples

Population and society

Population: 22,244,000 (2000 est)
Population growth rate: 2.0% (1995–2000); 1.7% (2000–05)
Population density: (per sq km) 66 (1999 est)
Urban population: (% of total) 57 (2000 est)
Age distribution: (% of total population) 0–14 34%, 15–59 59%, 60+ 7% (2000 est)
Ethnic groups: 56% of the population is Malay, four-fifths of whom live in rural areas; 34% is Chinese, four-fifths of whom are in towns; 9% is Indian, mainly Tamil
Language: Bahasa Malaysia (Malay; official), English, Chinese, Tamil, Iban, many local dialects
Religion: Muslim (official) about 53%, Buddhist 19%, Hindu, Christian, local beliefs
Education: (compulsory years) 11
Literacy rate: 91% (men); 83% (women) (2000 est)
Labour force: 15.2% agriculture, 38.2% industry, 46.6% services (1997)
Life expectancy: 70 (men); 74 (women) (1995–2000)
Child mortality rate: (under 5, per 1,000 live births) 15 (1995–2000)
Physicians: 1 per 1,935 people (1996 est)
Hospital beds: 1 per 495 people (1995)
TV sets: (per 1,000 people) 166 (1998)
Radios: (per 1,000 people) 420 (1997)
Internet users: (per 10,000 people) 687.1 (1999)
Personal computer users: (per 100 people) 6.9 (1999)

Transport

Airports: international airports: Kuala Lumpur, Penang (Bayan Lepas), Kota Kinabalu, Kuching; 15 domestic airports; total passenger km: 28,762 million (1997 est)
Railways: total length: 1,882 km/1,169 mi; total

passenger km: 1,367 million (1994)
Roads: total road network: 94,500 km/58,722 mi, of which 75.1% paved (1996 est); passenger cars: 145 per 1,000 people (1996)

Chronology

1st century AD: Peoples of Malay peninsula influenced by Indian culture and Buddhism.
8th–13th centuries: Malay peninsula formed part of Buddhist Srivijaya Empire based in Sumatra. **14th century:** Siam (Thailand) expanded to include most of Malay peninsula. **1403:** Muslim traders founded port of Malacca, which became a great commercial centre, encouraging the spread of Islam. **1511:** The Portuguese attacked and captured Malacca. **1641:** The Portuguese were ousted from Malacca by the Dutch after a seven-year blockade. **1786:** The British East India Company established a trading post on island of Penang. **1795–1815:** Britain occupied the Dutch colonies after France conquered the Netherlands. **1819:** Stamford Raffles of East India Company obtained Singapore from Sultan of Johore. **1824:** Anglo-Dutch Treaty ceded Malacca to Britain in return for territory in Sumatra.
1826: British possessions of Singapore, Penang, and Malacca formed the Straits Settlements, ruled by the governor of Bengal; ports prospered and expanded. **1840:** The Sultan of Brunei gave Sarawak to James Brooke, whose family ruled it as an independent state until 1946. **1851:** Responsibility for Straits Settlements assumed by the governor general of India. **1858:** British government, through India Office, took over administration of Straits Settlements. **1867:** Straits Settlements became crown colony of British Empire. **1874:** British protectorates established over four Malay states of Perak, Salangor, Pahang, and Negri Sembilan, which federated in 1896. **1888:** Britain declared protectorate over northern Borneo (Sabah). **late 19th century:** Millions of Chinese and thousands of Indians migrated to Malaya to work in tin mines and on rubber plantations. **1909–14:** Britain assumed indirect rule over five northern Malay states after agreement with Siam (Thailand). **1941–45:** Japanese occupation. **1946:** United Malay National Organization (UMNO) founded to oppose British plans for centralized Union of Malaya. **1948:** Britain federated nine Malay states with Penang and Malacca to form the single colony of the Federation of Malaya. **1948–60:** Malayan emergency: British forces suppressed insurrection by communist guerrillas. **1957:** Federation of Malaya became independent with Prince Abdul Rahman (leader of UMNO) as prime minister. **1963:** Federation of Malaya combined with Singapore, Sarawak, and Sabah to form Federation of Malaysia. **1963–66:** 'The Confrontation' – guerrillas supported by Indonesia opposed federation with intermittent warfare. **1965:** Singapore withdrew from the Federation of Malaysia. **1968:** Philippines claimed sovereignty over Sabah. **1969:** Malay resentment of Chinese economic dominance resulted in race riots in Kuala Lumpur. **1971:** *Bumiputra* policies which favoured ethnic Malays in education and employment introduced by Tun Abul Razak of UMNO. **1981:** Mahathir bin Muhammad (UMNO) became the prime minister; the government became increasingly dominated by Muslim Malays. **1987:** Malay–Chinese relations deteriorated; over 100 opposition activists were arrested. **1991:** An economic development policy was launched which aimed at 7% annual growth. **1997:** The currency was allowed to float. Parts of Borneo and Sumatra were covered by thick smog for several weeks following forest-clearing fires. **1998:** The repatriation of foreign workers commenced. Currency controls were introduced as the GDP contracted sharply. **1999:** Mahathir bin Muhammad's ruling coalition party was elected to retain power. Tuanku Salehuddin Abdul Aziz Shan bin al-Marhum Hisamuddin Alam Shah was appointed president. **2000:** Ex-deputy prime minister Anwar Ibrahim was found guilty of charges of sodomy by the high court and sentenced to nine years' imprisonment, to be served in addition to the six-year sentence he received in April 1999 for corruption. The International Commission of Jurists condemned the verdict as politically motivated.

Practical information

Visa requirements: UK/USA: visa not required
Time difference: GMT +8
Chief tourist attractions: cultures of the country's many ethnic groups; tranquil beaches backed by dense rainforest

Major holidays: 1 January (in some states), 1 May, 3 June, 31 August, 25 December; variable: Eid-ul Adha, Diwali (in most states), end of Ramadan (2 days), New Year (Chinese, January/February, most states), New Year (Muslim), Prophet's Birthday, Wesak (most states), several local festivals

MALDIVES

Group of 1,196 islands in the north Indian Ocean, about 640 km/400
mi southwest of Sri Lanka.
National name: *Divehi Raajjeyge Jumhuriyya/Republic of the Maldives*
Area: 298 sq km/115 sq mi
Capital: Malé
Physical features: comprises 1,196 coral islands, grouped into 12
clusters of atolls, largely flat, none bigger than 13 sq km/5 sq mi,
average elevation 1.8 m/6 ft; 203 are inhabited

The green panel and the crescent represent Islam. Red
recalls the original flag. Effective date: 26 July 1965.

Government

Head of state and government: Maumoon Abd
Gayoom from 1978
Political system: authoritarian nationalist
Political executive: unlimited presidency
Administrative divisions: 20 districts
Political parties: none; candidates elected on
basis of personal influence and clan loyalties
Armed forces: no standing army
Death penalty: retains the death penalty for
ordinary crimes but can be considered abolitionist
in practice (last known execution in 1952)
Education spend: (% GNP) 6.4 (1996)
Health spend: (% GDP) 8.2 (1997 est)

Economy and resources

Currency: rufiya
GDP: (US$) 368 million (1999 est)
Real GDP growth: (% change on previous year)
7 (1999 est)
GNP: (US$) 322 million (1999)
GNP per capita (PPP): (US$) 3,545 (1999)
Consumer price inflation: 3% (1999 est)
Unemployment: 0.9% (1997 est)
Foreign debt: (US$) 203 million (1999)
Major trading partners: UK, Singapore, USA,
India, Sri Lanka, Thailand, Germany, Japan,
Qatar
Resources: coral (mining was banned as a
measure against the encroachment of the sea)
Industries: fish canning, clothing, soft-drink
bottling, shipping, lacquer work, shell craft,
tourism
Exports: marine products (tuna bonito –
'Maldive fish'), clothing. Principal market:
Germany 33% (1996)
Imports: consumer manufactured goods,
petroleum products, food, intermediate and
capital goods. Principal source: Singapore 29.1%
(1996)
Arable land: 10% (1995)
Agricultural products: coconuts, maize, cassava,
sweet potatoes, chillies; fishing (the Maldives'

second-largest source of foreign exchange after
tourism in 1995)

Population and society

Population: 286,000 (2000 est)
Population growth rate: 2.8% (1995–2025)
Population density: (per sq km) 933 (1999 est)
Urban population: (% of total) 26 (2000 est)
Age distribution: (% of total population) 0–14
43%, 15–59 52%, 60+ 5% (2000 est)
Ethnic groups: four main groups: Dravidian in
the northern islands, Arab in the middle islands,
Sinhalese in the southern islands, and African
Language: Divehi (a Sinhalese dialect; official),
English, Arabic
Religion: Sunni Muslim
Education: not compulsory
Literacy rate: 96% (men); 96% (women) (2000
est)
Labour force: 41% of population: 32%
agriculture, 31% industry, 37% services (1990)
Life expectancy: 66 (men); 63 (women)
(1995–2000)
Child mortality rate: (under 5, per 1,000 live
births) 66 (1995–2000)
Physicians: 1 per 1,955 people (1996)
Hospital beds: 1 per 806 people (1996)
TV sets: (per 1,000 people) 28 (1997)
Radios: (per 1,000 people) 129 (1997)
Internet users: (per 10,000 people) 71.8 (1999)
Personal computer users: (per 100 people) 1.8
(1999)

Transport

Airports: international airports: Malé, Gan; total
passenger km: 71 million (1995)
Railways: none
Roads: total road network: 9.6 km/6 mi;
passenger cars: 4.1 per 1,000 people (1996 est)

Chronology

12th century AD: Islam introduced by seafaring Arabs, who displaced the indigenous Dravidian population. **14th century:** Ad-Din sultanate established. **1558–73:** Under Portuguese rule. **1645:** Became a dependency of Ceylon (Sri Lanka), which was ruled by the Dutch until 1796 and then by the British, with Sinhalese and Indian colonies being established. **1887:** Became an internally self-governing British protectorate, which remained a dependency of Sri Lanka until 1948. **1932:** The sultanate became an elected position when the Maldives' first constitution was introduced. **1953:** Maldive Islands became a republic within the Commonwealth, as the ad-Din sultanate was abolished. **1954:** Sultan restored. **1959–60:** Secessionist rebellion in Suvadiva (Huvadu) and Addu southern atolls. **1965:** Achieved full independence outside Commonwealth. **1968:** Sultan deposed after referendum; republic reinstated with Ibrahim Nasir as president. **1975:** The closure of a British airforce staging post on the southern island of Gan led to a substantial loss in income. **1978:** The autocratic Nasir retired and was replaced by the progressive Maumoon Abd Gayoom. **1980s:** Economic growth was boosted by the rapid development of the tourist industry. **1982:** Rejoined the Commonwealth. **1985:** Became a founder member of the South Asian Association for Regional Cooperation. **1988:** A coup attempt by Sri Lankan mercenaries, thought to have the backing of former president Nasir, was foiled by Indian paratroops. **1998:** Gayoom was re-elected for a further presidential term.

Practical information

Visa requirements: UK: visa required. USA: visa required
Time difference: GMT +5
Chief tourist attractions: white sandy beaches; multi-coloured coral formations; water and underwater sports/activities

Major holidays: 1 January, 26 July (2 days), 11 November (2 days); variable: Eid-ul-Adha (4 days), end of Ramadan (3 days), New Year (Muslim), Prophet's Birthday, first day of Ramadan (2 days), Huravee (February), Martyrs (April), National (October/November, 2 days)

MALI

Landlocked country in northwest Africa, bounded to the northeast by Algeria, east by Niger, southeast by Burkina Faso, south by Côte d'Ivoire, southwest by Senegal and Guinea, and west and north by Mauritania.

National name: République du Mali/Republic of Mali
Area: 1,240,142 sq km/478,818 sq mi
Capital: Bamako
Major towns/cities: Mopti, Kayes, Ségou, Tombouctou, Sikasso
Physical features: landlocked state with River Niger and savannah in south; part of the Sahara in north; hills in northeast; Senegal River and its branches irrigate the southwest

Green, yellow, and red are the pan-African colours. The flag is modelled on the French tricolour. The design was identical to the Rwandan tricolour, obliging that country to modify its flag. Effective date: 1 March 1961.

Government

Head of state: Alpha Oumar Konare from 1992
Head of government: Mande Sidibe from 2000
Political system: emergent democracy
Political executive: limited presidency
Administrative divisions: capital district of Bamako and eight regions
Political parties: Alliance for Democracy in Mali (ADEMA), left of centre; National Committee for Democratic Initiative (CNID), left of centre; Assembly for Democracy and Progress (RDP), left of centre; Civic Society and the Democracy and Progress Party (PDP), left of centre; Malian People's Democratic Union (UDPM), nationalist socialist
Armed forces: 7,400; plus paramilitary forces of 7,800 (1998)
Conscription: selective conscription for two years
Death penalty: retains the death penalty for ordinary crimes but can be considered abolitionist in practice (last execution 1980)
Defence spend: (% GDP) 2.0 (1998)
Education spend: (% GNP) 2.2 (1996)
Health spend: (% GDP) 4.2 (1997 est)

Economy and resources

Currency: franc CFA
GDP: (US$) 2.7 billion (1999)
Real GDP growth: (% change on previous year) 5.3 (1999)
GNP: (US$) 2.6 billion (1999)
GNP per capita (PPP): (US$) 693 (1999)
Consumer price inflation: 1.3% (1999)
Foreign debt: (US$) 3 billion (1999)
Major trading partners: Côte d'Ivoire, Italy, Thailand, Belgium, France, Portugal, Senegal, Germany
Resources: iron ore, uranium, diamonds, bauxite, manganese, copper, lithium, gold
Industries: food processing, cotton processing, textiles, clothes, cement, pharmaceuticals
Exports: cotton, livestock, gold, miscellaneous manufactured articles. Principal market: Italy 18.2% (1999 est)
Imports: machinery and transport equipment, food products, petroleum products, other raw materials, chemicals, miscellaneous manufactured articles. Principal

source: Côte d'Ivoire 18.7% (1999 est)
Arable land: 3.8% (1996)
Agricultural products: seed cotton, cotton lint, groundnuts, millet, sugar cane, rice, sorghum, sweet potatoes, mangoes, vegetables; livestock rearing (cattle, sheep, and goats); fishing

Population and society

Population: 11,234,000 (2000 est)
Population growth rate: 2.4% (1995–2000)
Population density: (per sq km) 9 (1999 est)
Urban population: (% of total) 30 (2000 est)
Age distribution: (% of total population) 0–14 46%, 15–59 48%, 60+ 6% (2000 est)
Ethnic groups: around 50% belong to the Mande group, including the Bambara, Malinke, and Sarakole; other significant groups include the Fulani, Minianka, Senutu, Songhai, and the nomadic Tuareg in the north
Language: French (official), Bambara, other African languages
Religion: Sunni Muslim 80%, animist, Christian
Literacy rate: 49% (men); 33% (women) (2000 est)
Labour force: 82.6% agriculture, 2.4% industry, 15% services (1997 est)
Life expectancy: 52 (men); 55 (women) (1995–2000)
Child mortality rate: (under 5, per 1,000 live births) 236 (1995–2000)
Physicians: 1 per 18,376 people (1993 est)
Hospital beds: 1 per 2,623 people (1990)
TV sets: (per 1,000 people) 10 (1997)
Radios: (per 1,000 people) 54 (1997)
Internet users: (per 10,000 people) 9.1 (1999)
Personal computer users: (per 100 people) 0.1 (1999)

Transport

Airports: international airports: Bamako (Senou), Mopti; ten domestic airports; total passenger km: 223 million (1995)
Railways: total length: 641 km/398 mi; total passenger km: 930 million (1995)
Roads: total road network: 15,100 km/9,383 mi, of which 12.1% paved (1996 est); passenger cars: 2.9 per 1,000 people (1996 est)

Chronology

5th–13th centuries: Ghana Empire founded by agriculturist Soninke people, based on the Saharan gold trade for which Timbuktu became an important centre. At its height in the 11th century it covered much of the western Sahel, comprising parts of present-day Mali, Senegal, and Mauritania. Wars with Muslim Berber tribes from the north led to its downfall. **13th–15th centuries:** Ghana Empire superseded by Muslim Mali Empire of Malinke (Mandingo) people of southwest, from which Mali derives its name. At its peak, under Mansa Musa in the 14th century, it covered parts of Mali, Senegal, Gambia, and southern Mauritania. **15th–16th centuries:** Muslim Songhai Empire, centred around Timbuktu and Gao, superseded Mali Empire. It covered Mali, Senegal, Gambia, and parts of Mauritania, Niger, and Nigeria, and included a professional army and civil service. **1591:** Songhai Empire destroyed by Moroccan Berbers, under Ahmad al-Mansur, who launched an invasion to take over western Sudanese gold trade and took control over Timbuktu. **18th–19th centuries:** Niger valley region was divided between the nomadic Tuareg, in the area around Gao in the northeast, and the Fulani and Bambara kingdoms, around Macina and Bambara in the centre and southwest. **late 18th century:** Western Mali visited by Scottish explorer Mungo Park. **mid-19th century:** The Islamic Tukolor, as part of a jihad (holy war) conquered much of western Mali, including Fulani and Bambara kingdoms, while in the south, Samori Ture, a Muslim Malinke (Mandingo) warrior, created a small empire. **1880–95:** Region conquered by French, who overcame Tukolor and Samori resistance to establish colony of French Sudan. **1904:** Became part of the Federation of French West Africa. **1946:** French Sudan became an overseas territory within the French Union, with its own territorial assembly and representation in the French parliament; the pro-autonomy Sudanese Union and Sudanese Progressive Parties founded in Bamako. **1959:** With Senegal, formed the Federation of Mali. **1960:** Separated from Senegal and became independent Republic of Mali, with Modibo Keita, an authoritarian socialist of the Sudanese Union party, as president. **1968:** Keita replaced in army coup by Lt Moussa Traoré, as economy deteriorated: constitution suspended and political activity banned. **1974:** A new constitution made Mali a one-party state, dominated by Traoré's nationalistic socialist Malian People's Democratic Union (UDPM), formed in 1976. **1979:** More than a dozen were killed after a student strike was crushed. **1985:** There was a five-day conflict with Burkina Faso over a long-standing border dispute which was mediated by the International Court of Justice. **late 1980s:** Closer ties developed with the West and free-market economic policies were pursued, including privatization, as the Soviet influence waned. **1991:** Violent demonstrations and strikes against one-party rule led to 150 deaths; Traoré was ousted in a coup. **1992:** A referendum endorsed a new democratic constitution. The opposition Alliance for Democracy in Mali (ADEMA) won multiparty elections; Alpha Oumar Konare was elected president. A peace pact was signed with Tuareg rebels fighting in northern Mali for greater autonomy. **1993–94:** Ex-president Traoré was sentenced to death for his role in suppressing the 1991 riots. **1997:** President Konare was re-elected. **2000:** Mande Sidibe became prime minister.

Practical information

Visa requirements: UK: visa required. USA: visa required
Time difference: GMT +/–0
Chief tourist attractions: cultural heritage, including the historic town of Timbuktu, with its 13th–15th-century mosques; oases along the ancient trans-Sahara camel route
Major holidays: 1, 20 January, 1, 25 May, 22 September, 19 November, 25 December; variable: Eid-ul-Adha, end of Ramadan, Prophet's Birthday, Prophet's Baptism (November)

MALTA

Island in the Mediterranean Sea, south of Sicily, east of Tunisia, and north of Libya.

National name: *Repubblika ta'Malta/Republic of Malta*
Area: 320 sq km/124 sq mi
Capital: Valletta (and chief port)
Major towns/cities: Rabat, Birkirkara, Qormi, Sliema, Zejtun, Zabor
Major ports: Marsaxlokk, Valletta
Physical features: includes islands of Gozo 67 sq km/26 sq mi and Comino 3 sq km/1 sq mi

The George Cross was awarded by King George VI and originally appeared in a small blue canton. The present design dates from 1964 when the islands gained independence. Effective date: 21 September 1964.

Government

Head of state: Guido de Marco from 1999
Head of government: Edward Fenech Adami from 1998
Political system: liberal democracy
Political executive: parliamentary
Administrative divisions: 67 local councils
Political parties: Malta Labour Party (MLP), moderate, left of centre; Nationalist Party (PN), Christian, centrist, pro-European
Armed forces: 1,900 (1998)
Conscription: military service is voluntary
Death penalty: laws provide for the death penalty only for exceptional crimes such as crimes under military law or crimes committed in exceptional circumstances such as wartime (last execution 1943)
Defence spend: (% GDP) 0.9 (1998)
Education spend: (% GNP) 5.2 (1995)
Health spend: (% GDP) 6.3 (1997)

Economy and resources

Currency: Maltese lira
GDP: (US$) 3.47 billion (1999)
Real GDP growth: (% change on previous year) 4.6 (1999)
GNP: (US$) 3.49 billion (1999)
GNP per capita (PPP): (US$) 15,066 (1999 est)
Consumer price inflation: 2.1% (1999)
Unemployment: 5% (1997)
Foreign debt: (US$) 1.03 billion (1997)
Major trading partners: USA, France, Italy, Germany, UK
Resources: stone, sand; offshore petroleum reserves were under exploration 1988–95
Industries: transport equipment and machinery, food and beverages, textiles and clothing, chemicals, ship repair and shipbuilding, tourism
Exports: machinery and transport equipment, manufactured articles (including clothing), beverages, chemicals, tobacco. Principal market: USA 21.4% (1999)
Imports: machinery and transport equipment, basic manufactures (including textile yarn and fabrics), food and live animals, mineral fuels. Principal source: France

19.1% (1999)
Arable land: 31.3% (1995)
Agricultural products: potatoes, tomatoes, peaches, plums, nectarines, apricots, melons, strawberries, wheat, barley; livestock rearing (cattle, pigs, and poultry) and livestock products (chicken eggs, pork, and dairy products)

Population and society

Population: 389,000 (2000 est)
Population growth rate: 0.7% (1995–2000)
Population density: (per sq km) 1,206 (1999 est)
Urban population: (% of total) 91 (2000 est)
Age distribution: (% of total population) 0–14 21%, 15–59 62%, 60+ 17% (2000 est)
Ethnic groups: essentially European, supposedly originated from ancient North African kingdom of Carthage
Language: Maltese, English (both official)
Religion: Roman Catholic 98%
Education: (compulsory years) 10
Literacy rate: 91% (men); 93% (women) (2000 est)
Labour force: 2% agriculture, 27% industry, 71% services (1998)
Life expectancy: 75 (men); 79 (women) (1995–2000)
Child mortality rate: (under 5, per 1,000 live births) 9 (1995–2000)
Physicians: 1 per 406 people (1995)
Hospital beds: 1 per 172 people (1995)
TV sets: (per 1,000 people) 735 (1997)
Radios: (per 1,000 people) 669 (1997)
Internet users: (per 10,000 people) 388.5 (1999)
Personal computer users: (per 100 people) 18.1 (1999)

Transport

Airports: international airports: Luga (Malta, 8 km/5 mi from Valetta); helicopter service between Malta and Gozo; total passenger km: 1,681 million (1997 est)
Railways: none
Roads: total road network: 1,604 km/997 mi, of which 94% paved; passenger cars: 442 per 1,000 people (1994)

Chronology

7th century BC: Invaded and subjugated by Carthaginians from North Africa. **218 BC:** Came under Roman control. **AD 60:** Converted to Christianity by the apostle Paul. **395:** On the division of the Roman Empire, became part of Eastern (Byzantine) portion, dominated by Constantinople. **870:** Came under Arab rule. **1091:** Arabs defeated by Norman Count Roger I of Sicily; Roman Catholic Church re-established. **1530:** Handed over by Holy Roman Emperor Charles V to a religious military order, the Hospitallers (Knights of St John of Jerusalem). **1798–1802:** Occupied by French. **1814:** Annexed to Britain by the Treaty of Paris on condition that Roman Catholic Church was maintained and Maltese Declaration of Rights honoured. **later 19th century– early 20th century:** Became vital British naval base, with famous dockyard that developed as the island's economic mainstay. **1942:** Awarded the George Cross for valour in resisting severe Italian aerial attacks during World War II. **1947:** Achieved self-government. **1956:** Referendum approved MLP's proposal for integration with UK. Plebiscite opposed and boycotted by right-of-centre Nationalist Party (PN). **1958:** MLP rejected final British integration proposal. **1964:** Independence achieved from Britain, within Commonwealth. A ten-year defence and economic-aid treaty with the UK was signed. **1971:** Prime Minister Mintoff adopted a policy of nonalignment and declared the 1964 treaty invalid; negotiations began for leasing NATO base in Malta. **1972:** Seven-year NATO agreement signed. **1974:** Became a republic. **1979:** British military base closed; closer links were established with communist and Arab states, including Libya. **1987:** Edward Fenech Adami (PN) was narrowly elected prime minister; he adopted a more pro-European and pro-American policy stance than the preceding administration. **1990:** A formal application was made for European Community membership. **1998:** The PN was returned to power after a snap election, with Edward Fenech Adami returning as prime minister. **1999:** Guido de Marco was elected president.

Practical information

Visa requirements: UK: visa not required. USA: visa not required
Time difference: GMT +1
Chief tourist attractions: fine climate; sandy beaches and rocky coves; Blue Lagoon of Comino; prehistoric temples and Ta' Pinh church of Gozo; Valetta, with its 16th-century churches, palaces, hospitals, and aqueducts, and its yachting centre
Major holidays: 1 January, 31 March, 1 May, 15 August, 13, 25 December; variable: Good Friday

MARSHALL ISLANDS

Located in the west Pacific Ocean, part of Micronesia, occupying 31 atolls.

National name: *Majol/Republic of the Marshall Islands*
Area: 181 sq km/70 sq mi
Capital: Dalap-Uliga-Darrit (on Majuro atoll)
Major towns/cities: Ebeye (the only other town)
Physical features: comprises the Ratak and Ralik island chains in the West Pacific, which form an archipelago of 31 coral atolls, 5 islands, and 1,152 islets

Orange stands for bravery. White symbolizes peace. The blue field represents the Pacific Ocean. Effective date: 1 May 1979.

Government

Head of state and government: Kessai H Note from 2000
Political system: liberal democracy
Political executive: limited presidency
Political parties: no organized party system, but in 1991 an opposition grouping, the Ralik Ratak Democratic Party, was founded to oppose the ruling group
Death penalty: abolished in 1991
Health spend: (% GDP) 9 (1997 est)

Economy and resources

Currency: US dollar
GDP: (US$) 97 million (1999 est)
Real GDP growth: (% change on previous year) –0.4 (1999 est)
GNP: (US$) 100 million (1999)
GNP per capita (PPP): (US$) 1,860 (1999 est)
Consumer price inflation: 4% (1998)
Unemployment: 16% (1991 est)
Foreign debt: (US$) 125 million (1997 est)
Major trading partners: USA, Japan, Australia
Industries: processing of agricultural products, handicrafts, fish products and canning, tourism
Exports: coconut products, trochus shells, copra, handicrafts, fish, live animals. Principal market: USA

Chronology

after *c.* 1000 BC: Micronesians first settled the islands.
1529: Visited by Spanish navigator Miguel de Saavedra and thereafter came under Spanish influence. **1875:** Spanish rule formally declared in face of increasing encroachment by German traders. **1885:** German protectorate established.
1914: Seized by Japan on the outbreak of World War I.
1920–44: Administered under League of Nations mandate by Japan and vigorously colonized. **1944:** Japanese removed after heavy fighting with US troops during World War II.
1946–63: Eniwetok and Bikini atolls used for US atom-bomb tests; islanders later demanded rehabilitation and compensation for the damage. **1947:** Became part of United Nations (UN) Pacific Islands Trust Territory, administered by USA. **1979:** Amata Kabua was elected president as internal self-government was established. **1986:** The Compact of Free Association with the USA granted the islands self-government, with the USA retaining the responsibility for defence and security until 2001.
1990: UN trust status was terminated. **1991:** Independence was agreed with Kabua as president; UN membership was granted. **2000:** Kessai H Note was elected as president.

Imports: foodstuffs, beverages and tobacco, building materials, machinery and transport equipment, mineral fuels, chemicals. Principal source: USA
Agricultural products: coconuts, tomatoes, melons, breadfruit, cassava, sweet potatoes, copra; fishing

Population and society

Population: 64,000 (2000 est)
Population growth rate: 3.2% (1995–2025)
Population density: (per sq km) 343 (1999 est)
Urban population: (% of total) 72 (2000 est)
Age distribution: (% of total population) 0–14 49%, 15–59 47%, 60+ 4% (2000 est)
Ethnic groups: 90% Marshallese, of predominantly Micronesian descent; remainder European origin, Indian, Chinese, Lebanese
Language: Marshallese, English (both official)
Religion: Christian (mainly Protestant) and Baha'i
Education: (compulsory years) 8
Literacy rate: 91% (men); 90% (women) (1994 est)
Labour force: 25.6% agriculture, 9.4% industry, 65% services (1
Life expectancy: 63 (men); 66 (women) (1998 est)
Physicians: 1 per 3,111 people (1995)
Internet users: (per 10,000 people) 80.3 (1999)

Practical information

Visa requirements: UK/USA: visa required
Time difference: GMT +12
Chief tourist attractions: sandy beaches; coral atolls; tropical vegetation

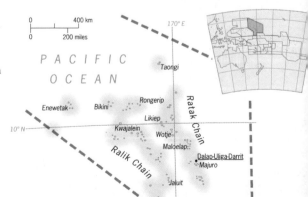

MAURITANIA

Located in northwest Africa, bounded northeast by Algeria, east and south by Mali, southwest by Senegal, west by the Atlantic Ocean, and northwest by Western Sahara.

National name: *Al-Jumhuriyya al-Islamiyya al-Mawritaniyya/République Islamique Arabe et Africaine de Mauritanie/Islamic Republic of Mauritania*
Area: 1,030,700 sq km/397,953 sq mi
Capital: Nouakchott (and chief port)
Major towns/cities: Nouâdhibou, Kaédi, Zouerate, Kiffa, Rosso, Atâr
Major ports: Nouâdhibou
Physical features: valley of River Senegal in south; remainder arid and flat

Green and yellow are Islamic and pan-African colours. The crescent and star represent Islam. Effective date: 1 April 1959.

Government

Head of state: Maaoya Sid'Ahmed Ould Taya from 1984
Head of government: Cheik el Avia Ould Muhammad Khouna from 1998
Political system: emergent democracy
Political executive: limited presidency
Administrative divisions: 12 regions and one capital district
Political parties: Democratic and Social Republican Party (PRDS), left of centre, militarist; Rally for Democracy and National Unity (RDNU), centrist; Mauritanian Renewal Party (MPR), centrist; Umma, Islamic fundamentalist
Armed forces: 15,700; plus paramilitary force of around 5,000 (1998)
Conscription: military service is by authorized conscription for two years
Death penalty: retained and used for ordinary crimes
Defence spend: (% GDP) 2.2 (1998)
Education spend: (% GNP) 5.1 (1997 est)
Health spend: (% GDP) 5.6 (1997 est)

Economy and resources

Currency: ouguiya
GDP: (US$) 959 million (1999)
Real GDP growth: (% change on previous year) 4.3 (1999)
GNP: (US$) 1 billion (1999)
GNP per capita (PPP): (US$) 1,522 (1999 est)
Consumer price inflation: 4.1% (1999)
Unemployment: 23% (1995 est)
Foreign debt: (US$) 2.1 billion (1999 est)
Major trading partners: Japan, France, Spain, Italy, Belgium, Germany
Resources: copper, gold, iron ore, gypsum, phosphates, sulphur, peat
Industries: fish products, cheese and butter, processing of minerals (including imported petroleum), mining
Exports: fish and fish products, iron ore. Principal market: Japan 18.5% (1999 est)
Imports: machinery and transport equipment, foodstuffs, consumer goods, building materials, mineral fuels. Principal source: France 24.9% (1999 est)
Arable land: 0.5% (1996)
Agricultural products: millet, sorghum, dates, maize, rice, pulses, groundnuts, sweet potatoes; livestock rearing (the principal occupation of rural population); fishing (providing 56% of export earnings in 1993). Only 1% of Mauritania receives enough rain to grow crops

Population and society

Population: 2,670,000 (2000 est)
Population growth rate: 2.7% (1995–2000)
Population density: (per sq km) 3 (1999 est)
Urban population: (% of total) 58 (2000 est)
Age distribution: (% of total population) 0–14 43%, 15–59 52%, 60+ 5% (2000 est)
Ethnic groups: over 80% of the population is of Moorish or Moorish-black origin; about 18% is black African (concentrated in the south); there is a small European minority
Language: Hasaniya Arabic (official), Pulaar, Soninke, Wolof (all national languages), French (particularly in the south)
Religion: Sunni Muslim (state religion)
Education: not compulsory
Literacy rate: 53% (men); 32% (women) (2000 est)
Labour force: 53% agriculture, 15% industry, 32% services (1997 est)
Life expectancy: 52 (men); 55 (women) (1995–2000)
Child mortality rate: (under 5, per 1,000 live births) 148 (1995–2000)
Physicians: 1 per 11,316 people (1994)
Hospital beds: 1 per 1,486 people (1993)
TV sets: (per 1,000 people) 91 (1998)
Radios: (per 1,000 people) 151 (1997)
Internet users: (per 10,000 people) 7.7 (1999)
Personal computer users: (per 100 people) 0.6 (1999)

Transport

Airports: international airports: Nouakchott, Nouâdhibou; six domestic airports; total passenger km: 301 million (1995)
Railways: total length: 704 km/437 mi; the principal traffic is iron ore; passenger traffic is negligible
Roads: total road network: 7,660 km/4,760 mi, of which 11.3% paved (1996 est); passenger cars: 7.8 per 1,000 people (1996 est)

Chronology

early Christian era: A Roman province with the name Mauritania, after the Mauri, its Berber inhabitants who became active in the long-distance salt trade. **7th–11th centuries:** Eastern Mauritania was incorporated in the larger Ghana Empire, centred on Mali to the east, but with its capital at Kumbi in southeast Mauritania. The Berbers were reduced to vassals and converted to Islam in the 8th century. **11th–12th centuries:** The area's Sanhadja Berber inhabitants, linked to the Morocco-based Almoravid Empire, destroyed the Ghana Empire and spread Islam among neighbouring peoples. **13th–15th centuries:** Southeast Mauritania formed part of the Muslim Mali Empire, which extended to the east and south. **1441:** Coast visited by Portuguese, who founded port of Arguin and captured Africans to sell as slaves. **15th–16th centuries:** Eastern Mauritania formed part of Muslim Songhai Empire, which spread across western Sahel, and Arab tribes migrated into the area. **1817:** Senegal Treaty recognized coastal region (formerly disputed by European nations) as French sphere of influence. **1903:** Formally became French protectorate. **1920:** Became French colony, within French West Africa. **1960:** Independence achieved, with Moktar Ould Daddah, leader of Mauritanian People's Party (PPM), as president. New capital built at Nouakchott. **1968:** Underlying tensions between agriculturalist black population of south and economically dominant semi-nomadic Arabo-Berber peoples, or Moors, of desert north became more acute after Arabic was made an official language (with French). **1976:** Western Sahara, to the northwest, ceded by Spain to Mauritania and Morocco. Mauritania occupied the southern area and Morocco the mineral-rich north. Polisario Front formed in Sahara to resist this occupation and guerrilla war broke out, with the Polisario receiving backing from Algeria and Libya. **1979:** A peace accord was signed with the Polisario Front in Algiers, in which Mauritania renounced its claims to southern Western Sahara and recognized the Polisario regime; diplomatic relations were restored with Algeria. **1981:** Diplomatic relations with Morocco were broken after it annexed southern Western Sahara. **1984:** Col Maaoya Sid'Ahmed Ould Taya became president. **1985:** Relations with Morocco were restored. **1989:** There were violent clashes in Mauritania and Senegal between Moors and black Africans, chiefly of Senegalese origins; over 50,000 Senegalese were expelled. **1991:** An amnesty was called for political prisoners. Political parties were legalized and a new multiparty constitution was approved in a referendum. **1992:** The first multiparty elections were largely boycotted by the opposition; Taya and his Social Democratic Republican Party (DSRP) were re-elected. Diplomatic relations with Senegal resumed. **1998:** Cheik el Avia Ould Muhammad Khouna was appointed prime minister.

Practical information

Visa requirements: UK: visa required. USA: visa required
Time difference: GMT +/–0
Chief tourist attractions: game reserves; national parks; historic sites, several of which have been listed by UNESCO under its World Heritage programme
Major holidays: 1 January, 1, 25 May, 10 July, 28 November; variable: Eid-ul-Adha, end of Ramadan, New Year (Muslim), Prophet's Birthday

MAURITIUS

Island in the Indian Ocean, east of Madagascar.

National name: *Republic of Mauritius*
Area: 1,865 sq km/720 sq mi
Capital: Port Louis (and chief port)
Major towns/cities: Beau Bassin, Curepipe, Quatre Bornes, Vacoas
Physical features: mountainous, volcanic island surrounded by coral reefs; the island of Rodrigues is part of Mauritius; there are several small island dependencies

Red recalls the struggle for independence. Blue stands for the Indian Ocean. Yellow expresses hope for a bright future. Green represents agriculture and vegetation. Effective date: 12 March 1968.

Government

Head of state: Cassam Uteem from 1992
Head of government: Anerood Jugnauth from 2000
Political system: liberal democracy
Political executive: parliamentary
Administrative divisions: five municipalities and four district councils
Political parties: Mauritius Socialist Movement (MSM), moderate socialist-republican; Mauritius Labour Party (MLP), democratic socialist, Hindu-oriented; Mauritius Social Democratic Party (PMSD), conservative, Francophile; Mauritius Militant Movement (MMM), Marxist-republican; Organization of Rodriguan People (OPR), left of centre
Armed forces: no standing defence forces; 1,800-strong police mobile unit (1998)
Death penalty: abolished in 1985
Defence spend: (% GDP) 2.1 (1998)
Education spend: (% GNP) 4.6 (1997 est)
Health spend: (% GDP) 3.5 (1997 est)

Economy and resources

Currency: Mauritian rupee
GDP: (US$) 4.23 billion (1999)
Real GDP growth: (% change on previous year) 2.7 (1999)
GNP: (US$) 4.2 billion (1999)
GNP per capita (PPP): (US$) 8,652 (1999)
Consumer price inflation: 6.9% (1999)
Unemployment: 9.8% (1995)
Foreign debt: (US$) 2.5 billion (1999)
Major trading partners: UK, France, South Africa, India, Australia, Germany
Industries: textiles and clothing, footwear and other leather products, food products, diamond cutting, jewellery, electrical components, chemical products, furniture, tourism
Exports: raw sugar, clothing, tea, molasses, jewellery. Principal market: UK 32.1% (1999)
Imports: textile yarn and fabrics, petroleum products, industrial machinery, motor vehicles, manufactured goods. Principal source: France 14.9% (1999)
Arable land: 49.3% (1995)

Agricultural products: sugar cane, tea, tobacco, potatoes, maize; poultry farming; fishing; forest resources

Population and society

Population: 1,158,000 (2000 est)
Population growth rate: 0.8% (1995–2000)
Population density: (per sq km) 616 (1999 est)
Urban population: (% of total) 41 (2000 est)
Age distribution: (% of total population) 0–14 25%, 15–59 66%, 60+ 9% (2000 est)
Ethnic groups: five principal ethnic groups: French, black Africans, Indians, Chinese, and Mulattos (or Creoles). Indo-Mauritians predominate, constituting 68% of the population, followed by Creoles (27%), Sino-Mauritians (3%), Franco-Mauritians (2%), and Europeans (0.5%)
Language: English (official), French, Creole (36%), Bhojpuri (32%), other Indian languages
Religion: Hindu over 50%, Christian (mainly Roman Catholic) about 30%, Muslim 17%
Education: (compulsory years) 7
Literacy rate: 88% (men); 81% (women) (2000 est)
Labour force: 13% agriculture, 38% industry, 49% services (1996)
Life expectancy: 68 (men); 75 (women) (1995–2000)
Child mortality rate: (under 5, per 1,000 live births) 18 (1995–2000)
Physicians: 1 per 1,000 people (1993)
Hospital beds: 1 per 350 people (1993)
TV sets: (per 1,000 people) 228 (1997)
Radios: (per 1,000 people) 371 (1997)
Internet users: (per 10,000 people) 478.4 (1999)
Personal computer users: (per 100 people) 9.6 (1999)

Transport

Airports: international airport: Plaisance (Sir Seewoosagur Ramgoolam, 48 km/30 mi southeast of Port Louis); two domestic airports; total passenger km: 3,875 million (1997 est)
Railways: none
Roads: total road network: 1,905 km/1,184 mi, of which 93% paved (1997); passenger cars: 64 per 1,000 people (1997)

Chronology

1598: Previously uninhabited, the island was discovered by the Dutch and named after Prince Morris of Nassau. **1710:** Dutch colonists withdrew. **1721:** Reoccupied by French East India Company, who renamed it Île de France, and established sugar cane and tobacco plantations worked by imported African slaves. **1814:** Ceded to Britain by the Treaty of Paris. **1835:** Slavery abolished; indentured Indian and Chinese labourers imported to work the sugar-cane plantations, which were later hit by competition from beet sugar. **1903:** Formerly administered with Seychelles, it became a single colony. **1936:** Mauritius Labour Party (MLP) founded, drawing strong support from sugar workers. **1957:** Internal self-government granted. **1968:** Independence achieved from Britain within Commonwealth, with Seewoosagur Ramgoolam of centrist Indian-dominated MLP as prime minister. **1971:** A state of emergency was temporarily imposed as a result of industrial unrest. **1982:** Anerood Jugnauth, of the moderate socialist Mauritius Socialist Movement (MSM), became prime minister, pledging a programme of nonalignment, nationalization, and the creation of a republic. **1992:** Became a republic within the Commonwealth, with Cassam Uteem elected as president. **1995:** The MLP and the cross-community Mauritian Militant Movement (MMM) coalition won election victory; Navin Ramgoolam (MLP) became the prime minister. **2000:** General elections in mid-September 2000 were won by an opposition alliance, led by a former prime minister, Anerood Jugnauth.

Practical information

Visa requirements: UK: visa not required. USA: visa not required
Time difference: GMT +4
Chief tourist attractions: fine scenery and beaches; pleasant climate; blend of cultures
Major holidays: 1–2 January, 12 March, 1 May, 1 November, 25 December; variable: Eid-ul-Adha, Diwali, end of Ramadan, Prophet's Birthday, Chinese Spring Festival (February)

MEXICO

Located in the North American continent, bounded north by the USA, east by the Gulf of Mexico, southeast by Belize and Guatemala, and southwest and west by the Pacific Ocean. It is the northernmost country in Latin America.

National name: *Estados Unidos Mexicanos/United States of Mexico*
Area: 1,958,201 sq km/756,061 sq mi
Capital: Mexico City
Major towns/cities: Guadalajara, Monterrey, Puebla, Netzahualcóyotl, Ciudad Juárez, Tijuana
Major ports: 49 ocean ports
Physical features: partly arid central highlands; Sierra Madre mountain ranges east and west; tropical coastal plains; volcanoes, including Popocatepetl; Rio Grande

The emblem was added to distinguish the flag from that of Italy. The design is based on the French tricolour. The colours are those of the Mexican liberation army.
Effective date: 23 November 1968.

Government

Head of state and government: Vicente Fox from 2000
Political system: liberal democracy
Political executive: limited presidency
Administrative divisions: 31 states and a Federal District
Political parties: Institutional Revolutionary Party (PRI), moderate, left wing; National Action Party (PAN), moderate, Christian, right of centre; Party of the Democratic Revolution (PRD), left of centre
Armed forces: 175,000; rural defence militia of 15,000 (1998)
Conscription: one year, part-time (conscripts selected by lottery)
Death penalty: only for exceptional crimes; last execution 1937
Defence spend: (% GDP) 1.0 (1998)
Education spend: (% GNP) 4.9 (1996)
Health spend: (% GDP) 4.7 (1997)

Economy and resources

Currency: Mexican peso
GDP: (US$) 474.9 billion (1999)
Real GDP growth: (% change on previous year) 3.7 (1999)
GNP: (US$) 428.8 billion (1999)
GNP per capita (PPP): (US$) 7,719 (1999)
Consumer price inflation: 16.6% (1999)
Unemployment: 3.2% (1998)
Foreign debt: (US$) 166.6 billion (1999 est)
Major trading partners: USA, Japan, Canada, Spain, France, Germany, Brazil
Resources: petroleum, natural gas, zinc, salt, silver, copper, coal, mercury, manganese, phosphates, uranium, strontium sulphide
Industries: motor vehicles, food processing, iron and steel, chemicals, beverages, electrical machinery, electronic goods, petroleum refining, cement, metals and metal products, tourism
Exports: petroleum and petroleum products, engines and spare parts for motor vehicles, motor vehicles, electrical and electronic goods, fresh and preserved vegetables, coffee, cotton. Principal market: USA 83.2% (1999)
Imports: motor vehicle chassis, industrial machinery and equipment, iron and steel, telecommunications apparatus, organic chemicals, cereals and cereal preparations, petroleum and petroleum products. Principal source: USA 74.8% (1999)
Arable land: 13.2% (1996)
Agricultural products: maize, wheat, sorghum, barley, rice, beans, potatoes, coffee, cotton, sugar cane, fruit and vegetables; livestock raising and fisheries

Population and society

Population: 98,881,000 (2000 est)
Population growth rate: 1.6% (1995–2000); 1.5% (2000–05)
Population density: (per sq km) 50 (1999 est)
Urban population: (% of total) 74 (2000 est)
Age distribution: (% of total population) 0–14 33%, 15–59 60%, 60+ 7% (2000 est)
Ethnic groups: around 60% mestizo (mixed American Indian and Spanish descent), 30% American Indians, remainder mainly of European origin
Language: Spanish (official), Nahuatl, Maya, Zapoteco, Mixteco, Otomi
Religion: Roman Catholic about 90%
Education: (compulsory years) 6
Literacy rate: 93% (men); 89% (women) (2000 est)
Labour force: 23.3% agriculture, 22.7% industry, 54.0% services (1997)
Life expectancy: 70 (men); 76 (women) (1995–2000)
Child mortality rate: (under 5, per 1,000 live births) 38 (1995–2000)
Physicians: 1 per 627 people (1996)
Hospital beds: 1 per 909 people (1996)
TV sets: (per 1,000 people) 261 (1999)
Radios: (per 1,000 people) 325 (1999)
Internet users: (per 10,000 people) 256.8 (1999)
Personal computer users: (per 100 people) 4.4 (1999)

Transport

Airports: international airports: Mexico City (Benito Juárez), Guadalajara (Miguel Hidalgo), Acapulco (General Juan N Alvarez), Monterrey (General Mariano Escobeno), and 40 others; 39 domestic airports; total passenger km: 22,243 million (1997 est)

Railways: total length: 20,596 km/12,798 mi; total passenger km: 1,855 million (1994)
Roads: total road network: 252,000 km/156,593 mi, of which 37.4% paved (1996 est); passenger cars: 94.6 per 1,000 people (1996 est)

Chronology

c. **2600 BC:** Mayan civilization originated in Yucatán peninsula. **1000–500 BC:** Zapotec civilization developed in southern Mexico. **4th–10th centuries AD:** Mayan Empire at its height. **10th–12th centuries:** Toltecs ruled much of Mexico. **12th century:** Aztecs migrated south into the valley of Mexico. *c.* **1325:** Aztecs began building their capital Tenochtitlán on site of present-day Mexico City. **15th century:** Montezuma I built up the Aztec Empire in central Mexico. **1519–21:** Hernán Cortes conquered Aztec Empire and secured Mexico for Spain. **1520:** Montezuma II, last king of the Aztecs, was killed. **1535:** Mexico became Spanish viceroyalty of New Spain; plantations and mining developed with Indian labour. **1519–1607:** Indigenous population reduced from 21 million to 1 million, due mainly to lack of resistance to diseases transported from Old World. **1821:** Independence proclaimed by Augustín de Iturbide with support of Church and landowners. **1822:** Iturbide overthrew provisional government and proclaimed himself Emperor Augustín I. **1824:** Federal republic

established amid continuing public disorder. **1824–55:** Military rule of Antonio López de Santa Anna, who imposed stability (he became president in 1833). **1846–48:** Mexican War: Mexico lost California and New Mexico to USA. **1855:** Benito Juárez aided overthrow of Santa Anna's dictatorship. **1857–60:** Sweeping liberal reforms and anti-clerical legislation introduced by Juárez led to civil war with conservatives. **1861:** Mexico suspended payment on foreign debt leading to French military intervention; Juárez resisted with US support. **1864:** Supported by conservatives, France installed Archduke Maximilian of Austria as emperor of Mexico. **1867:** Maximilian shot by republicans as French troops withdrew; Juárez resumed presidency. **1872:** Death of Juárez. **1876:** Gen Porfirio Diaz established dictatorship; Mexican economy modernized through foreign investment. **1911:** Revolution overthrew Diaz; liberal president Francisco Madero introduced radical land reform and labour legislation but political disorder increased. **1917:** New constitution, designed to ensure permanent democracy, adopted with US encouragement. **1924–35:** Government dominated by anti-clerical Gen Plutarco Calles, who introduced further social reforms. **1929:** Foundation of National Revolutionary Party (PRFN), renamed the Institutional Revolutionary Party (PRI) in 1946. **1938:** President Lázaro Cárdenas nationalized all foreign-owned oil wells in face of US opposition. **1942:** Mexico declared war on Germany and Japan (and so regained US favour). **1946–52:** Miguel Alemán first of succession of authoritarian PRI presidents to seek moderation and stability rather than further radical reform. **1960s:** Rapid industrial growth partly financed by borrowing. **1976:** Huge oil reserves were discovered in the southeastern state of Chiapas; oil production tripled in six years. **1982:** Falling oil prices caused a grave financial crisis; Mexico defaulted on debt. **1985:** An earthquake in Mexico City killed thousands. **1994:** There was an uprising in Chiapas by the Zapatista National Liberation Army (EZLN), seeking rights for the Mayan Indian population; Mexico formed the North American Free Trade Area (NAFTA) with the USA and Canada. **1995:** The government agreed to offer greater autonomy to Mayan Indians in Chiapas. **1996:** There were short-lived peace talks with the EZLN; and violent attacks against the government by the new leftist Popular Revolutionary Army (EPR) increased. **1997:** The PRI lost its assembly majority. A civilian counterpart to the Zapatista rebels, the Zapatista National Liberation Front (EZLN), was formed. **1998:** A lapsed peace accord with Zapatist rebels was reactivated, but talks between the government and the rebels broke down. **2000:** After 71 years, the PRI lost power, and Vicente Fox of the conservative National Action Party was elected president. He promised national unity, job creation, and an attack on government corruption. He signed a bill on indigenous rights, in response to which the leader of the rebels offered to open peace talks.

Practical information

Visa requirements: UK/USA: visa (tourist card) required
Time difference: GMT –6/8
Chief tourist attractions: coastal scenery; volcanoes; Sierra Nevada (Sierra Madre) mountain range; Mayan and Aztec monuments and remains; Spanish colonial churches and other buildings
Major holidays: 1 January, 5 February, 21 March, 1, 5 May, 1, 16 September, 12 October, 2, 20 November, 12, 25, 31 December; variable: Holy Thursday, Good Friday

MICRONESIA, FEDERATED STATES OF

Located in the west Pacific Ocean, forming part of the archipelago of the
Caroline Islands, 800 km/497 mi east of the Philippines.

National name: *Federated States of Micronesia (FSM)*
Area: 700 sq km/270 sq mi
Capital: Palikir (in Pohnpei island state)
Major towns/cities: Kolonia (in Pohnpei), Weno (in Truk), Lelu (in
Kosrae)
Major ports: Teketik, Lepukos, Okak
Physical features: an archipelago of 607 equatorial, volcanic islands in
the West Pacific

The stars represent the states of Pohnpei, Kosrae, Yap,
and Chuuk. The blue field is said to represent the Pacific
Ocean. Effective date: 30 November 1978.

Government

Head of state and government: Jacob Nena from
1997
Political system: liberal democracy
Political executive: limited presidency
Administrative divisions: four states
Political parties: no formally organized political parties
Armed forces: USA is responsible for country's
defence
Death penalty: laws do not provide for the death
penalty for any crime
Health spend: (% GDP) 7.4 (1997 est)

Economy and resources

Currency: US dollar
GDP: (US$) 219 million (1999)
Real GDP growth: (% change on previous year) 0.3
(1999)
GNP: (US$) 210 million (1999)
GNP per capita (PPP): (US$) 3,860 (1999 est)
Consumer price inflation: 4% (1997 est)
Unemployment: 18.2% (1994)
Foreign debt: (US$) 111 million (1997 est)
Major trading partners: USA, Japan, Guam
Industries: food processing (coconut products),
tourism
Exports: copra, pepper, fish. Principal market: Japan
84.8% (1996)
Imports: manufactured goods, machinery and
transport equipment, mineral fuels. Principal source:
USA 73.2% (1996)
Agricultural products: mainly subsistence farming;
coconuts, cassava, sweet potatoes, breadfruit, bananas,
copra, citrus fruits, taro, peppers; fishing

Population and society

Population: 119,000 (2000 est)
Population growth rate: 2% (1995–2000)
Population density: (per sq km) 165 (1999 est)
Urban population: (% of total) 28 (2000 est)
Age distribution: (% of total population) 0–14 44%,
15–59 51%, 60+ 5% (2000 est)
Ethnic groups: main ethnic groups are the Trukese
(41%) and Pohnpeian (26%), both Micronesian
Language: English (official), eight officially recognized
local languages (including Trukese, Pohnpeian, Yapese,
and Kosrean), a number of other dialects
Religion: Christianity (mainly Roman Catholic in Yap
state, Protestant elsewhere)
Education: (compulsory years) 8
Literacy rate: 91% (men); 88% (women) (1980 est)
Labour force: 29.1% agriculture, 3.5% industry, 67.4%
services (1994)
Life expectancy: 67 (men); 71 (women) (1995–2000)
Child mortality rate: (under 5, per 1,000 live births)
37 (1994)
Physicians: 1 per 2,380 people (1993)
Hospital beds: 1 per 365 people (1993)
TV sets: (per 1,000 people) 12 (1993)
Radios: (per 1,000 people) 16 (1993)
Internet users: (per 10,000 people) 172.4 (1999)

Transport

Airports: international airports: Pohnpei, Chuuk, Yap,
Kosrae; domestic services also operate between these
airports
Railways: none
Roads: total road network: 240 km/149 mi, of which
17.7% paved (1996 est)

Chronology

c. **1000 BC:** Micronesians first settled the islands. **1525:** Portuguese navigators first visited Yap and Ulithi islands
in the Carolines (Micronesia). **later 16th century:** Fell under Spanish influence. **1874:** Spanish rule formally
declared in face of increasing encroachment by German traders. **1885:** Yap seized by German naval forces, but was
restored to Spain after arbitration by Pope Leo XIII on the condition that Germany was allowed freedom of trade.
1899: Purchased for $4.5 million by Germany from Spain, after the latter's defeat in the Spanish–American War.
1914: Occupied by Japan at the outbreak of World War I. **1919:** Administered under League of Nations mandate

by Japan, and vigorously colonized. **1944:** Occupied by USA after Japanese forces defeated in World War II. **1947:** Administered by USA as part of the United Nations (UN) Trust Territory of the Pacific Islands, under the name of the Federated States of Micronesia (FSM). **1979:** A constitution was adopted that established a federal system for its four constituent states (Yap, Chuuk, Pohnpei, and Kosrae) and internal self-government. **1986:** The Compact of Free Association was entered into with the USA, granting the islands self-government with the USA retaining responsibility for defence and security until 2001. **1990:** UN trust status was terminated. **1991:** Independence agreed, with Bailey Olter as president. Entered into United Nations (UN) membership. **1997:** Jacob Nena was sworn in as president after the existing president, Bailey Olter, was incapacitated by a stroke.

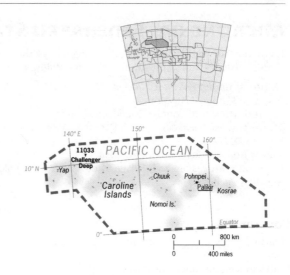

Practical information

Visa requirements: UK: visa not required for a stay of up to 30 days. USA: visa not required
Time difference: GMT +10 (Chuuk and Yap); +11 (Kosrae and Pohnpei)
Chief tourist attractions: excellent conditions for scuba-diving (notably in Chuuk Lagoon); ancient ruined city of Nan Madol on Pohnpei; World War II battle sites and relics (many underwater)
Major holidays: 1 January, 10 May, 24 October, 3 November, 25 December (some variations from island to island)

MOLDOVA

Located in east-central Europe, bounded north, south, and east by Ukraine, and west by Romania.

National name: *Republica Moldova/Republic of Moldova*
Area: 33,700 sq km/13,011 sq mi
Capital: Chisinau (Russian Kishinev)
Major towns/cities: Tiraspol, Balti, Bendery
Physical features: hilly land lying largely between the rivers Prut and Dniester; northern Moldova comprises the level plain of the Balti Steppe and uplands; the climate is warm and moderately continental

The bison's head, star, rose, and crescent are traditional symbols of Moldavia. The colours are based on the Romanian flag. Effective date: 3 November 1990.

Government

Head of state: Petru Lucinschi from 1997
Head of government: Vladimir Voronin from 1999
Political system: emergent democracy
Political executive: limited presidency
Administrative divisions: 38 districts, four municipalities, and two autonomous territorial units – Gauguz (Gagauzi Yeri) and Trans-Dniestr (status of latter was under dispute in 1996)
Political parties: Agrarian Democratic Party (ADP), nationalist, centrist; Socialist Party and Yedinstvo/Unity Movement, reform-socialist; Peasants and Intellectuals, Romanian nationalist; Christian Democratic Popular Front (CDPF), Romanian nationalist; Gagauz-Khalky (GKPM; Gagauz People's Movement), Gagauz separatist; Moldovan Party of Communists (MPC), former Communist Party of Moldova (banned in 1991, revived under new name in 1994)
Armed forces: 11,100 (1998)
Conscription: military service is compulsory for up to 18 months
Death penalty: abolished in 1995
Defence spend: (% GDP) 4.3 (1998)
Education spend: (% GNP) 10.6 (1997 est)
Health spend: (% GDP) 8.3 (1997)

Economy and resources

Currency: leu
GDP: (US$) 1.2 billion (1999)
Real GDP growth: (% change on previous year) –4.4 (1999)
GNP: (US$) 1.6 billion (1999)
GNP per capita (PPP): (US$) 2,358 (1999)
Consumer price inflation: 39.3% (1999)
Unemployment: 10.4% (1998 est)
Foreign debt: (US$) 1 billion (1999)
Major trading partners: Russia, Ukraine, Romania, Germany, Italy
Resources: lignite, phosphorites, gypsum, building materials; petroleum and natural gas deposits discovered in the early 1990s were not yet exploited in 1996
Industries: food processing, wine, tobacco, metalworking, light industry, machine building, cement, textiles, footwear
Exports: food and agricultural products, machinery and equipment, textiles, clothing. Principal market: Russia 40.9% (1999)

Imports: mineral fuels, energy and mineral products, mechanical engineering products, foodstuffs, chemicals, textiles, clothing. Principal source: Russia 20.8% (1999)
Arable land: 53.8% (1996)
Agricultural products: grain, sugar beet, potatoes, vegetables, wine grapes and other fruit, tobacco; livestock products (milk, pork, and beef)

Population and society

Population: 4,380,000 (2000 est)
Population growth rate: 0.02% (1995–2000)
Population density: (per sq km) 130 (1999 est)
Urban population: (% of total) 46 (2000 est)
Age distribution: (% of total population) 0–14 23%, 15–59 62%, 60+ 15% (2000 est)
Ethnic groups: 65% ethnic Moldovan (Romanian), 14% Ukrainian, 13% ethnic Russian, 4% Gagauzi, 2% Bulgarian, 2% Jewish
Language: Moldovan (official), Russian, Gaganz (a Turkish dialect)
Religion: Eastern Orthodox 98.5%; remainder Jewish
Education: (compulsory years) 11
Literacy rate: 99% (men); 98% (women) (2000 est)
Labour force: 45.5% agriculture, 19.2% industry, 35.3% services (1994)
Life expectancy: 64 (men); 72 (women) (1995–2000)
Child mortality rate: (under 5, per 1,000 live births) 32 (1995–2000)
Physicians: 1 per 258 people (1994)
Hospital beds: 1 per 82 people (1994)
TV sets: (per 1,000 people) 297 (1998)
Radios: (per 1,000 people) 740 (1997)
Internet users: (per 10,000 people) 34.3 (1999)
Personal computer users: (per 100 people) 0.8 (1999)

Transport

Airports: international airports: Chisinau; no domestic services; total passenger km: 211 million (1995)
Railways: total length: 1,328km/825 mi; total passenger km: 949.3 million (1996)
Roads: total road network: 12,300 km/7,643 mi, of which 87.3% paved (1996 est); passenger cars: 46.4 per 1,000 people (1996)

Chronology

AD 106: The current area covered by Moldova, which lies chiefly between the Prut River, bordering Romania in the west, and the Dniestr River, with Ukraine in the east, was conquered by the Roman Emperor Trajan and became part of the Roman province of Dacia. It was known in earlier times as Bessarabia. **mid-14th century:** Formed part of an independent Moldovan principality, which included areas, such as Bukovina to the west, that are now part of Romania. **late 15th century:** Under Stephen IV the Great the principality reached the height of its power. **16th century:** Became a tributary of the Ottoman Turks. **1774–75:** Moldovan principality, though continuing to recognize Turkish overlordship, was placed under Russian protectorship; Bukovina was lost to Austria. **1812:** Bessarabia ceded to tsarist Russia. **1856:** Remainder of Moldovan principality became largely independent of Turkish control. **1859:** Moldovan Assembly voted to unite with Wallachia, to the southwest, to form the state of Romania, ruled by Prince Alexandru Ion Cuza. The state became fully independent in 1878. **1918:** Following the Russian Revolution, Bessarabia was seized and incorporated within Romania. **1924:** Moldovan autonomous Soviet Socialist Republic (SSR) created, as part of Soviet Union, comprising territory east of Dniestr River. **1940:** Romania returned Bessarabia, east of Prut River, to Soviet Union, which divided it between Moldovan SSR and Ukraine, with Trans-Dniestr region transferred from Ukraine to Moldova. **1941:** Moldovan SSR occupied by Romania and its wartime ally Germany. **1944:** Red Army reconquered Bessarabia. **1946–47:** Widespread famine as agriculture was collectivized; rich farmers and intellectuals were liquidated. **1950:** Immigration by settlers from Russia and Ukraine as industries were developed. **late 1980s:** There was an upsurge in Moldovan nationalism, encouraged by the *glasnost* initiative of reformist Soviet leader Mikhail Gorbachev. **1988:** The Moldovan Movement in Support of Perestroika (economic restructuring) campaigned for accelerated political reform. **1989:** There were nationalist demonstrations in Kishinev (now Chisinau). The Moldovan Popular Front (MPF) was founded; Moldovan was made the state language. There were campaigns for autonomy among ethnic Russians, strongest in industrialized Trans-Dniestr region, and Turkish-speaking but Orthodox Christian Gagauz minority in southwest. **1990:** The MPF polled strongly in parliamentary elections and Mircea Snegur, a reform-nationalist communist, became president. Economic and political sovereignty was declared. **1991:** Independence was declared and the Communist Party outlawed after a conservative coup in Moscow against Gorbachev; joined Commonwealth of Independent States (CIS). There was insurrection in the Trans-Dniestr region. **1992:** Admitted into United Nations and the Conference on Security and Cooperation in Europe; a peace agreement was signed with Russia to end the civil war in Trans-Dniestr, giving special status to the region. The MPF-dominated government fell; a 'government of national accord' was formed, headed by Andrei Sangheli and dominated by the ADP. **1993:** A new currency, the leu, was introduced. A privatization programme was launched and closer ties were established with Russia. **1994:** Parliamentary elections were won by the ADP. Plebiscite rejected nationalist demands for a merger with Romania. Russia agreed to withdraw Trans-Dniestr troops by 1997. **1995:** Joined Council of Europe; economic growth resumed. **1996:** Petru Lucinschi was elected president. **1997:** A cooperation agreement was signed with the Dniestr region. A law was passed that provided for elections using proportional representation. **1999:** A new coalition government was formed, headed by Ion Sturza. It fell in November, and Vladimir Voronin, a communist, succeeded as prime minister. **2000:** Constitutional changes increased the powers of the Parlamentul (legislature) and the president was now to be elected by the legislature rather than the people. However, the incumbent president Lucinschi refused to stand, and neither of the two presidential candidates in the December contest were able to secure the required majority.

Practical information

Visa requirements: UK: visa required. USA: visa required
Time difference: GMT +2

Major holidays: 1, 7–8 January, 8 March, 9 May, 27, 31 August; variable: Mertsishor (Spring Festival, first week in March), Good Friday, Easter Monday

MONACO

Small sovereign state forming an enclave in southern France, with the Mediterranean Sea to the south.

National name: *Principauté de Monaco/Principality of Monaco*
Area: 1.95 sq km/0.75 sq mi
Physical features: steep and rugged; surrounded landwards by French territory; being expanded by filling in the sea

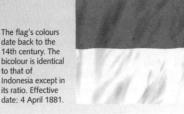

The flag's colours date back to the 14th century. The bicolour is identical to that of Indonesia except in its ratio. Effective date: 4 April 1881.

Government

Head of state: Prince Rainier III from 1949
Head of government: Patrick Leclercq from 2000
Political system: liberal democracy
Political executive: parliamentary
Administrative divisions: four districts: Monaco-Ville, Monte Carlo, La Condamine, Fontvieille
Armed forces: no standing defence forces; defence is the responsibility of France
Education spend: (% GNP) 5.6 (1992)
Health spend: (% GDP) 8 (1997 est)

Economy and resources

Currency: French franc
GDP: (US$) 870 million (1999 est)
GNP per capita (PPP): (US$) 27,000 (1999 est)
Unemployment: 2.8% (1998 est)
Industries: chemicals, pharmaceuticals, plastics, microelectronics, electrical goods, paper, textiles and clothing, gambling, banking and finance, real estate, tourism (which provided an estimated 25% of total government revenue in 1991)
Imports and exports: largely dependent on imports from France; full customs integration with France

Population and society

Population: 34,000 (2000 est)
Population growth rate: 1.1% (1995–2025)
Population density: (per sq km) 16,074 (1999 est)
Urban population: (% of total) 100 (2000 est)
Age distribution: (% of total population) 0–14 12%, 15–59 67%, 60+ 21% (2000 est)
Ethnic groups: 47% French; 10% Monégasque; 20% Italian and other European
Language: French (official), Monégasgne (a mixture of the French Provençal and Italian Ligurian dialects), Italian
Religion: Roman Catholic about 90%
Education: (compulsory years) 10
Literacy rate: 99% (men); 99% (women) (2000 est)
Life expectancy: 75 (men); 82 (women) (1998 est)
Child mortality rate: (under 5, per 1,000 live births) 7 (1998 est)
Physicians: 1 per 254 people (1994)
TV sets: (per 1,000 people) 727 (1996)
Radios: (per 1,000 people) 1,021 (1996)

Chronology

1191: The Genoese took control of Monaco, which had formerly been part of the Holy Roman Empire. **1297:** Came under the rule of the Grimaldi dynasty, the current ruling family, who initially allied themselves to the French. **1524–1641:** Came under Spanish protection. **1793:** Annexed by France during French Revolutionary Wars. **1815:** Placed under protection of Sardinia. **1848:** The towns of Menton and Roquebrune, which had formed the greater part of the principality, seceded and later became part of France. **1861:** Franco-Monegasque treaty restored Monaco's independence under French protection; the first casino was built. **1865:** Customs union established with France. **1918:** France given veto over succession to throne and established that if a reigning prince dies without a male heir, Monaco is to be incorporated into France. **1941–45:** Occupied successively by Italians and Germans during World War II. **1949:** Prince Rainier III ascended the throne. **1956:** Prince Rainier married US actor Grace Kelly. **1958:** Birth of male heir, Prince Albert. **1962:** A new, more liberal constitution was adopted. **1982:** Princess Grace died in a car accident. **1993:** Joined United Nations. **2000:** Patrick Leclercq replaced Michel Leveque as minister of state. France threatened to take punitive measures against Monaco unless it took action against money-laundering and tax-evasion.

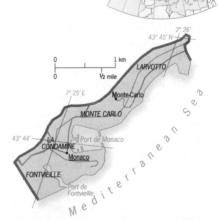

Practical information

Visa requirements: UK/USA: visa not required
Time difference: GMT +1
Chief tourist attractions: dramatic scenery; Mediterranean climate; 13th-century Palais du Prince; numerous entertainment facilities, including casinos, Jardin Exotique, Musée Océanographique, world championship Grand Prix motor race and annual car rally at Monte Carlo

MONGOLIA FORMERLY OUTER MONGOLIA (UNTIL 1924), PEOPLE'S REPUBLIC OF MONGOLIA (1924–91)

Located in east-Central Asia, bounded north by Russia and south by China.

National name: *Mongol Uls/State of Mongolia*
Area: 1,565,000 sq km/604,246 sq mi
Capital: Ulaanbaatar
Major towns/cities: Darhan, Choybalsan, Erdenet
Physical features: high plateau with desert and steppe (grasslands); Altai Mountains in southwest; salt lakes; part of Gobi desert in southeast; contains both the world's southernmost permafrost and northernmost desert

Red represents progress. Blue is Mongolia's national colour. Effective date: 12 February 1992.

Government

Head of state: Natsagiyn Bagabandi from 1997
Head of government: Nambariin Enkhbayar from 2000
Political system: emergent democracy
Political executive: limited presidency
Administrative divisions: 21 provinces and one municipality (Ulaanbaatar)
Political parties: Mongolian People's Revolutionary Party (MPRP), reform-socialist (ex-communist); Mongolian National Democratic Party (MNDP), traditionalist, promarket economy; Union Coalition (UC, comprising the MNPD and the Social Democratic Party (SDP)), democratic, promarket economy
Armed forces: 9,800; plus a paramilitary force of around 7,200 (1998)
Conscription: military service is compulsory for 12 months
Death penalty: retained and used for ordinary crimes
Defence spend: (% GDP) 2.1 (1998)
Education spend: (% GNP) 5.7 (1997)
Health spend: (% GDP) 4.3 (1997)

Economy and resources

Currency: tugrik
GDP: (US$) 905 million (1999)
Real GDP growth: (% change on previous year) 3.5 (1999 est)
GNP: (US$) 900 million (1999)
GNP per capita (PPP): (US$) 1,496 (1999)
Consumer price inflation: 7.6% (1999)
Unemployment: 7.6% (1997 est)
Foreign debt: (US$) 718 million (1997)
Major trading partners: Russia, China, Japan, Switzerland, South Korea, Germany, USA, Italy
Resources: copper, nickel, zinc, molybdenum, phosphorites, tungsten, tin, fluorospar, gold, lead; reserves of petroleum discovered in 1994
Industries: mostly small-scale; food products, copper and molybdenum concentrates, cement, lime, wood and metal-worked products, beverages, leather articles
Exports: minerals and metals (primarily copper concentrate), consumer goods, foodstuffs, agricultural products. Principal market: China 58.2% (1999 est)
Imports: engineering goods, mineral fuels and products, industrial consumer goods, foodstuffs. Principal source: Russia 45.9% (1999 est)

Arable land: 0.8% (1996)
Agricultural products: wheat, oats, barley, potatoes, vegetables; animal herding (particularly cattle rearing) is country's main economic activity (there were 28.6 million cattle, sheep, goats, horses, and camels in 1995)

Population and society

Population: 2,662,000 (2000 est)
Population growth rate: 1.7% (1995–2000)
Population density: (per sq km) 2 (1999 est)
Urban population: (% of total) 64 (2000 est)
Age distribution: (% of total population) 0–14 35%, 15–59 59%, 60+ 6% (2000 est)
Ethnic groups: 91% Mongol, 6% Kazakh; very small groups of Russian (2,000) and Chinese (1,500)
Language: Khalkha Mongolian (official), Kazakh (in the province of Bagan-Ölgiy), Chinese, Russian, Turkic languages
Religion: there is no state religion, but traditional lamaism (Mahayana Buddhism) is gaining new strength; the Sunni Muslim Kazakhs of Western Mongolia have also begun the renewal of their religious life, and Christian missionary activity has increased
Education: (compulsory years) 8
Literacy rate: 99% (men); 99% (women) (2000 est)
Labour force: 43% agriculture, 16% industry, 41% services (1995)
Life expectancy: 64 (men); 67 (women) (1995–2000)
Child mortality rate: (under 5, per 1,000 live births) 73 (1995–2000)
Physicians: 1 per 294 people (1996 est)
Hospital beds: 1 per 87 people (1993 est)
TV sets: (per 1,000 people) 63 (1997)
Radios: (per 1,000 people) 151 (1997)
Internet users: (per 10,000 people) 11.5 (1999)
Personal computer users: (per 100 people) 0.7 (1999)

Transport

Airports: international airports: Ulaanbaatar (Buyant Ukha); six domestic airports; total passenger km: 516 million (1995)
Railways: total track: 2,052 km/1,275 mi; total passenger km: 681 million (1995)
Roads: total road network: 49,250 km/30,604 mi, of which 3.4% paved (1997); passenger cars: 15 per 1,000 people (1997)

Chronology

AD 1206: Nomadic Mongol tribes united by Genghis Khan to form nucleus of vast Mongol Empire which, stretching across central Asia, reached its zenith under Genghis Khan's grandson, Kublai Khan. **late 17th century:** Conquered by China to become province of Outer Mongolia. **1911:** Independence proclaimed by Mongolian nationalists after Chinese 'republican revolution'; tsarist Russia helped Mongolia to secure autonomy, under a traditionalist Buddhist monarchy in the form of a reincarnated lama.
1915: Chinese sovereignty reasserted. **1921:** Chinese rule overthrown with Soviet help. **1924:** People's Republic proclaimed on death of king, when the monarchy was abolished; defeudalization programme launched, entailing collectivization of agriculture and suppression of lama Buddhism. **1932:** Armed antigovernment uprising suppressed with Soviet assistance; 100,000 killed in political purges. **1946:** China recognized Mongolia's independence. **1952:** Death of Marshal Horloogiyn Choybalsan, the dominant force in the ruling communist Mongolian People's Revolutionary Party (MPRP) since 1939.

1958: Yumjaagiyn Tsedenbal became the dominant figure in MPRP and country. **1962:** Joined Comecon. **1966:** 20-year friendship, cooperation, and mutual-assistance pact signed with Soviet Union (USSR). Relations with China deteriorated. **1987:** There was a reduction in the number of Soviet troops; Mongolia's external contacts broadened. The tolerance of traditional social customs encouraged a nationalist revival. **1989:** Further Soviet troop reductions.
1990: A demonstrations and democratization campaign was launched, influenced by events in Eastern Europe. Ex-communist MPRP elected in the first free multiparty elections; Punsalmaagiyn Ochirbat was indirectly elected president. Mongolian script was readopted. **1991:** A privatization programme was launched. GDP declined by 10%.
1992: The MPRP returned to power in assembly elections held under a new, noncommunist constitution. The economic situation worsened; GDP again declined by 10%. **1993:** Ochirbat won the first direct presidential elections. **1996:** The economy showed signs of revival. The Union Coalition won assembly elections, defeating the MPRP and ending 75 years of communist rule. A defence cooperation agreement was signed with the USA. **1997:** The ex-communist Natsagiyn Bagabandi was elected MPRP chairman and then became president. An economic shock therapy programme, supervised by IMF and World Bank, created unemployment and made the government unpopular. All taxes and tariffs on trade were abolished. **1998:** The National Democratic Party (DU) government was toppled after losing a no-confidence vote. Attempts to form a new DU-led government, led by Rinchinnyamiyn Amarjargal, failed, and Janlaviyn Narantsatsralt of the MNDP became prime minister. **1999:** Rinchinnyamiyn Marajargal became prime minister. **2000:** Mongolia's former communists, the MPRP, now branding themselves a centre-left party, won a landslide victory in parliamentary elections, led by Nambariin Enkhbayar.

Practical information

Visa requirements: UK: visa required. USA: visa required
Time difference: GMT +8
Chief tourist attractions: spectacular scenery, including Gobi Desert and Altai chain; wildlife; historical relics; tourism is relatively undeveloped
Major holidays: 1–2 January, 8 March, 1 May, 1 June, 10 July (3 days), 7 November; variable: Tsagaan (Lunar New Year, January/February, 2 days)

MOROCCO

Located in northwest Africa, bounded to the north and northwest by the Mediterranean Sea, to the east and southeast by Algeria, and to the south by Western Sahara.

National name: *Al-Mamlaka al-Maghribyya/Kingdom of Morocco*
Area: 458,730 sq km/177,115 sq mi (excluding Western Sahara)
Capital: Rabat
Major towns/cities: Casablanca, Marrakesh, Fès, Oujda, Kenitra, Tétouan, Meknès
Major ports: Casablanca, Tangier, Agadir
Physical features: mountain ranges, including the Atlas Mountains northeast–southwest; fertile coastal plains in west

The 'Solomon's Seal' pentagram was added to distinguish the flag from other plain red Arab banners. Effective date: 17 November 1915.

Government

Head of state: Sayyid Muhammad VI ibn-Hassan from 1999
Head of government: Abderrahmane Youssoufi from 1998
Political system: emergent democracy
Political executive: dual executive
Administrative divisions: 16 regions subdivided into 65 provinces and prefectures
Political parties: Constitutional Union (UC), right wing; National Rally of Independents (RNI), royalist; Popular Movement (MP), moderate, centrist; Istiqlal, nationalist, centrist; Socialist Union of Popular Forces (USFP), progressive socialist; National Democratic Party (PND), moderate, nationalist
Armed forces: 196,300; paramilitary forces of 42,000 (1998)
Conscription: 18 months
Death penalty: retained and used for ordinary crimes
Defence spend: (% GDP) 4.6 (1998)
Education spend: (% GNP) 5.0 (1997)
Health spend: (% GDP) 1.6 (1990–95)

Economy and resources

Currency: dirham
GDP: (US$) 35.2 billion (1999)
Real GDP growth: (% change on previous year) –0.7 (1999)
GNP: (US$) 33.8 billion (1999)
GNP per capita (PPP): (US$) 3,190 (1999)
Consumer price inflation: 1.6% (1999)
Unemployment: 17.8% (1996)
Foreign debt: (US$) 19.5 billion (1999 est)
Major trading partners: France, Spain, USA, Saudi Arabia, Italy, Germany, India
Resources: phosphate rock and phosphoric acid, coal, iron ore, barytes, lead, copper, manganese, zinc, petroleum, natural gas, fish
Industries: phosphate products (chiefly fertilizers), petroleum refining, food processing, textiles, clothing, leather goods, paper and paper products, tourism
Exports: phosphates and phosphoric acid, mineral products, seafoods and seafood products, citrus fruit, tobacco, clothing, hosiery. Principal market: France 39.2% (1999)
Imports: crude petroleum, raw materials, wheat, chemicals, sawn wood, consumer goods. Principal source:

France 29.1% (1999)
Arable land: 19.7% (1996)
Agricultural products: wheat, barley, sugar beet, citrus fruits, tomatoes, potatoes; fishing (seafoods)

Population and society

Population: 28,351,000 (2000 est)
Population growth rate: 1.8% (1995–2000); 1.6% (2000–05)
Population density: (per sq km) 61 (1999 est)
Urban population: (% of total) 56 (2000 est)
Age distribution: (% of total population) 0–14 33%, 15–59 60%, 60+ 7% (2000 est)
Ethnic groups: majority indigenous Berbers (99%); sizeable Jewish minority
Language: Arabic (75%) (official), Berber dialects (25%), French, Spanish
Religion: Sunni Muslim; Christian and Jewish minorities
Education: (compulsory years) 6
Literacy rate: 62% (men); 36% (women) (2000 est)
Labour force: 38% of population: 40.2% agriculture, 23.3% industry, 36.5% services (1994)
Life expectancy: 65 (men); 69 (women) (1995–2000)
Child mortality rate: (under 5, per 1,000 live births) 68 (1995–2000)
Physicians: 1 per 3,790 people (1994)
Hospital beds: 1 per 1,168 people (1994)
TV sets: (per 1,000 people) 160 (1998)
Radios: (per 1,000 people) 241 (1997)
Internet users: (per 10,000 people) 17.9 (1999)
Personal computer users: (per 100 people) 1.1 (1999)

Transport

Airports: international airports: Casablanca (Mohammed V), Rabat (Salé), Tangier (Boukhalef Sohahel), Agadir (Al Massira), Fès (Sais), Marrakesh, Oujda, Al-Hocina el-Aaiun, Ouarzazate; domestic services operate between these; total passenger km: 5,247 million (1997)
Railways: total length: 1,907 km/1,185 mi; total passenger km: 1,856 million (1997)
Roads: total road network: 57,810 km/35,923 mi, of which 51.8% paved (1997); passenger cars: 39.9 per 1,000 people (1996 est)

Chronology

10th–3rd centuries BC: Phoenicians from Tyre settled along north coast.
1st century AD: Northwest Africa became Roman province of Mauritania.
5th–6th centuries: Invaded by Vandals and Visigoths. **682:** Start of Arab
conquest, followed by spread of Islam. **8th century:** King Idris I
established small Arab kingdom. **1056–1146:** The Almoravids, a Berber
dynasty based at Marrakesh, built an empire embracing Morocco and parts
of Algeria and Spain. **1122–1268:** After a civil war, the Almohads, a rival
Berber dynasty, overthrew the Almoravids; Almohads extended empire but
later lost most of Spain. **1258–1358:** Beni Merin dynasty supplanted Almohads.
14th century: Moroccan Empire fragmented into separate kingdoms, based in
Fès and Marrakesh. **15th century:** Spain and Portugal occupied Moroccan
ports; expulsion of Muslims from Spain in 1492. **16th
century:** Saadian dynasty restored unity of Morocco and
resisted Turkish invasion. **1649:** Foundation of current
Alaouite dynasty of sultans; Morocco remained an independent
and isolated kingdom. **1856:** Under British pressure, the
sultan opened Morocco to European commerce. **1860:** Spain
invaded Morocco, which was forced to cede the southwestern
region of Ifni. **1905:** A major international crisis was caused by
German objections to increasing French influence in Morocco.
1911: Agadir Crisis: further German objections to French
imperialism in Morocco were overcome by territorial compensation in
central Africa. **1912:** Morocco was divided into French and Spanish
protectorates; the sultan was reduced to puppet ruler. **1921:** Moroccan rebels, the Riffs, led by Abd el-Krim, defeated
a large Spanish force at Anual. **1923:** The city of Tangier was separated from Spanish Morocco and made a neutral
international zone. **1926:** French forces crushed Riff revolt. **1944:** A nationalist party, Istiqlal, was founded to
campaign for full independence. **1948:** Consultative assemblies introduced. **1953–55:** Serious anti-French riots.
1956: French and Spanish forces withdrew; Morocco regained effective independence under Sultan Muhammad V,
who took title of king in 1957. **1961:** Muhammad V succeeded by Hassan II. **1962:** First constitution adopted;
replaced in 1970 and 1972. **1965–77:** King Hassan suspended the constitution and ruled by decree. **1969:** Spanish
overseas province of Ifni returned to Morocco. **1975:** Spain withdrew from Western Sahara, leaving Morocco and
Mauritania to divide it between themselves. **1976:** Polisario Front, supported by Algeria, began guerrilla war in
Western Sahara with the aim of securing its independence as the Sahrahwi Arab Democratic Republic. **1979:**
Mauritania withdrew from its portion of Western Sahara, which Morocco annexed after major battles with Polisario.
1984: Morocco signed mutual defence with Libya, which had previously supported Polisario. **1991:** A UN-
sponsored ceasefire came into effect in the Western Sahara. **1992:** The constitution was amended in an attempt to
increase the influence of parliament. **1996:** A new two-chamber assembly was approved. **1998:** Prime Minister
Abderrahmane Youssoufi formed a centre–left coalition. **1999:** King Hassan II died and was succeeded by his son,
Muhammad VI. **2000:** King Muhammad VI embarked on a programme of social and political reform, including
strengthening the rights of women.

Practical information

Visa requirements: UK: visa not required. USA: visa
not required
Time difference: GMT +/–0
Chief tourist attractions: sunny climate; ancient sites
and cities (notably Marrakesh, Fès, Meknès, Rabat);
spectacular scenery; resorts on Atlantic and
Mediterranean coasts
Major holidays: 1 January, 3 March, 1, 23 May, 9 July,
14 August, 6, 18 November; variable: Eid-ul-Adha (2
days), end of Ramadan (2 days), New Year (Muslim),
Prophet's Birthday

MOZAMBIQUE

Located in southeast Africa, bounded north by Zambia, Malawi, and Tanzania; east and south by the Indian Ocean; southwest by South Africa and Swaziland; and west by Zimbabwe.

National name: *República de Moçambique/Republic of Mozambique*
Area: 799,380 sq km/308,640 sq mi
Capital: Maputo (and chief port)
Major towns/cities: Beira, Nampula, Nacala, Chimoio
Major ports: Beira, Nacala, Quelimane
Physical features: mostly flat tropical lowland; mountains in west; rivers Zambezi and Limpopo

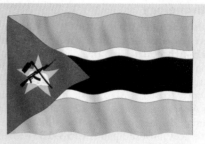

Green stands for agriculture. Red recalls the struggle for independence. White denotes peace. Yellow symbolizes mineral wealth. Effective date: April 1983.

Government

Head of state: Joaquim Alberto Chissano from 1986
Head of government: Pascoal Mocumbi from 1994
Political system: emergent democracy
Political executive: limited presidency
Administrative divisions: 11 provinces
Political parties: National Front for the Liberation of Mozambique (Frelimo), free market; Renamo, or Mozambique National Resistance (MNR), former rebel movement, right of centre
Armed forces: 6,100 (1998)
Conscription: early 1996 government was seeking to reintroduce compulsory military service, which had been suspended under the General Peace Accord
Death penalty: abolished in 1990
Defence spend: (% GDP) 3.9 (1998)
Education spend: (% GNP) 6.2 (1992)
Health spend: (% GDP) 5.8 (1997)

Economy and resources

Currency: metical
GDP: (US$) 4.2 billion (1999)
Real GDP growth: (% change on previous year) 10 (1999)
GNP: (US$) 3.9 billion (1999)
GNP per capita (PPP): (US$) 797 (1999 est)
Consumer price inflation: 4.8% (1999)
Unemployment: N/A
Foreign debt: (US$) 7.7 billion (1999)
Major trading partners: Spain, South Africa, USA, Japan, Zimbabwe, Portugal
Resources: coal, salt, bauxite, graphite; reserves of iron ore, gold, precious and semi-precious stones, marble, natural gas (all largely unexploited in 1996)
Industries: food products, steel, engineering, textiles and clothing, beverages, tobacco, chemical products
Exports: shrimps, lobsters, and other crustaceans, cashew nuts, raw cotton, coal, sugar, sisal, copra. Principal market: Spain 15.9% (1999 est)
Imports: foodstuffs, capital goods, crude petroleum and petroleum products, machinery and spare parts, chemicals. Principal source: South Africa 51.5% (1999 est)
Arable land: 3.8% (1996)
Agricultural products: cassava, maize, bananas, rice,

groundnuts, copra, cashew nuts, cotton, sugar cane; fishing (shrimps, prawns, and lobsters) is principal export activity (42% of export earnings in 1994); forest resources (eucalyptus, pine, and rare hardwoods)

Population and society

Population: 19,680,000 (2000 est)
Population growth rate: 2.5% (1995–2000); 2.8% (2000–05)
Population density: (per sq km) 24 (1999 est)
Urban population: (% of total) 40 (2000 est)
Age distribution: (% of total population) 0–14 45%, 15–59 50%, 60+ 5% (2000 est)
Ethnic groups: the majority belong to local groups, the largest being the Makua-Lomue, who comprise about 38% of the population; the other significant group is the Tsonga (24%)
Language: Portuguese (official), 16 African languages
Religion: animist 48%, Muslim 20%, Roman Catholic 16%, Protestant 16%
Education: (compulsory years) 7
Literacy rate: 60% (men); 29% (women) (2000 est)
Labour force: 53% of population: 83% agriculture, 8% industry, 9% services (1990)
Life expectancy: 44 (men); 47 (women) (1995–2000)
Child mortality rate: (under 5, per 1,000 live births) 183 (1995–2000)
Physicians: 1 per 36,225 people (1993 est)
Hospital beds: 1 per 1,205 people (1993)
TV sets: (per 1,000 people) 4 (1997)
Radios: (per 1,000 people) 40 (1997)
Internet users: (per 10,000 people) 7.8 (1999)
Personal computer users: (per 100 people) 0.3 (1999)

Transport

Airports: international airports: Maputo (Mavalane), Beira, Nampula; five domestic airports; total passenger km: 290 million (1995)
Railways: total length: 3,131 km/1,946 mi; passengers carried: 0.9 million (1994)
Roads: total road network: 30,400 km/18,890 mi, of which 18.7% paved (1996 est); passenger cars: 0.3 per 1,000 people (1996 est)

Chronology

1st–4th centuries AD: Bantu-speaking peoples settled in Mozambique. **8th–15th century:** Arab gold traders established independent city-states on the coast. **1498:** Portuguese navigator Vasco da Gama was the first European visitor; at this time the most important local power was the Maravi kingdom of the Mwene Matapa peoples, who controlled much of the Zambezi basin. **1626:** The Mwene Matapa formally recognized Portuguese sovereignty. Portuguese soldiers set up private agricultural estates and used slave labour to exploit gold and ivory resources. **late 17th century:** Portuguese temporarily pushed south of Zambezi by the ascendant Rozwi kingdom. **1752:** First Portuguese colonial governor appointed; slave trade outlawed. **late 19th century:** Concessions given by Portugal to private companies to develop and administer parts of Mozambique. **1930:** Colonial Act established more centralized Portuguese rule, ending concessions to monopolistic companies and forging closer integration with Lisbon. **1951:** Became an overseas province of Portugal and, economically, a cheap labour reserve for South Africa's mines. **1962:** Frelimo (National Front for the Liberation of Mozambique) established in exile in Tanzania by Marxist guerrillas, including Samora Machel, to fight for independence. **1964:** Fighting broke out between Frelimo forces and Portuguese troops, starting a ten-year liberation war; Portugal despatched 70,000 troops to Mozambique. **1969:** Eduardo Mondlane, leader of Frelimo, was assassinated. **1975:** Following revolution in Portugal, independence was achieved as a socialist republic, with Machel as president, Joaquim Chissano as prime minister, and Frelimo as the sole legal party; Portuguese settlers left the country. Lourenço Marques renamed Maputo. Key enterprises were nationalized. **1977:** Renamo resistance group formed, with covert backing of South Africa. **1979:** Machel encouraged Patriotic Front guerrillas in Rhodesia to accept Lancaster House Agreement, creating Zimbabwe. **1983:** Good relations were restored with Western powers. **1984:** The Nkomati Accord of nonaggression was signed with South Africa. **1986:** Machel was killed in air crash near the South African border and was succeeded by Chissano. **1988:** Tanzanian troops withdrawn from Mozambique. **1989:** Renamo continued attacks on government facilities and civilians. **1990:** One-party rule officially ended, and Frelimo abandoned Marxist–Leninism and embraced market economy. **1992:** A peace accord was signed with Renamo. **1993:** There were price riots in Maputo as a result of the implementation of IMF-promoted reforms to restructure the economy, which was devastated by war and drought. **1994:** The demobilization of contending armies was completed. Chissano and Frelimo were re-elected in the first multiparty elections; Renamo (now a political party) agreed to cooperate with the government. Pascoal Mocumbi was appointed prime minister by President Chissano. **1995:** Mozambique was admitted to the Commonwealth. **2000:** Severe flooding was estimated to involve the loss of 10,000 lives and 1 million homes. The Paris Club of rich countries agreed to suspend Mozambique's repayment of foreign debts.

Practical information

Visa requirements: UK: visa required. USA: visa required
Time difference: GMT +2
Chief tourist attractions: Indian Ocean coastline with beaches bordered by lagoons, coral reefs, and strings of islands (travel within Mozambique can be dangerous due to risk of armed robbery and unexploded landmines)
Major holidays: 1 January, 3 February, 7 April, 1 May, 25 June, 7, 25 September, 25 December

MYANMAR FORMERLY BURMA (UNTIL 1989)

Located in Southeast Asia, bounded northwest by India and Bangladesh, northeast by China, southeast by Laos and Thailand, and southwest by the Bay of Bengal.

National name: *Pyedawngsu Myanma Naingngan/Union of Myanmar*
Area: 676,577 sq km/261,226 sq mi
Capital: Yangon (formerly Rangoon) (and chief port)
Major towns/cities: Mandalay, Mawlamyine, Bago, Bassein, Taung-gyi, Sittwe, Monywa
Physical features: over half is rainforest; rivers Irrawaddy and Chindwin in central lowlands ringed by mountains in north, west, and east

The cog-wheel and rice plant stand for industry and agriculture. The stars represent the 14 states. Blue symbolizes peace. Red denotes courage. Effective date: 4 January 1974.

Government

Head of state and government: Than Shwe from 1992
Political system: military
Political executive: military
Administrative divisions: seven states and seven divisions
Political parties: National Unity Party (NUP), military-socialist ruling party; National League for Democracy (NLD), pluralist opposition grouping
Armed forces: 349,600; plus two paramilitary units totalling 85,300 (1998)
Conscription: military service is voluntary
Death penalty: retained and used for ordinary crimes
Defence spend: (% GDP) 6.8 (1998)
Education spend: (% GNP) 1.2 (1997 est)
Health spend: (% GDP) 2.6 (1997 est)

Economy and resources

Currency: kyat
GDP: (US$) 5.7 billion (1999 est)
Real GDP growth: (% change on previous year) 5.5 (1999 est)
GNP: (US$) N/A
GNP per capita (PPP): (US$) 1,200 (1999 est)
Consumer price inflation: 18.4% (1999)
Unemployment: 4.1% (1997)
Foreign debt: (US$) 6.3 billion (1999)
Major trading partners: India, Singapore, China, Thailand, Japan, Malaysia
Resources: natural gas, petroleum, zinc, tin, copper, tungsten, coal, lead, gems, silver, gold
Industries: food processing, beverages, cement, fertilizers, plywood, petroleum refining, textiles, paper, motor cars, tractors, bicycles
Exports: teak, rice, pulses and beans, rubber, hardwood, base metals, gems, cement. Principal market: India 12.5% (1999)
Imports: raw materials, machinery and transport equipment, tools and spares, construction materials, chemicals, consumer goods. Principal source: Singapore 27.7% (1999)
Arable land: 14.5 (1996)
Agricultural products: rice, sugar cane, maize, groundnuts, pulses, rubber, tobacco; fishing; forest resources (teak and hardwood) – teak is frequently felled illegally and smuggled into Thailand; cultured pearls and oyster shells are part of aquacultural fish production

Population and society

Population: 45,611,000 (2000 est)
Population growth rate: 1.2% (1995–2000)
Population density: (per sq km) 70 (1999 est)
Urban population: (% of total) 28 (2000 est)
Age distribution: (% of total population) 0–14 28%, 15–59 64%, 60+ 8% (2000 est)
Ethnic groups: Burmans, who predominate in the fertile central river valley and southern coastal and delta regions, constitute the ethnic majority, comprising 72% of the total population. Out of more than 100 minority communities, the most important are the Karen (7%), Shan (6%), Indians (6%), Chinese (3%), Kachin (2%), and Chin (2%). The indigenous minority communities, who predominate in mountainous border regions, show considerable hostility towards the culturally and politically dominant Burmans, undermining national unity. There are also minority groups of Indians, Tamils, and Chinese
Language: Burmese (official), English, tribal dialects
Religion: Hinayana Buddhist 89%, Christian 5%, Muslim 4%, animist 1.5%
Education: (compulsory years) 5
Literacy rate: 89% (men); 81% (women) (2000 est)
Labour force: 68.7% agriculture, 9.8% industry, 21.5% services (1994)
Life expectancy: 59 (men); 62 (women) (1995–2000)
Child mortality rate: (under 5, per 1,000 live births) 113 (1995–2000)
Physicians: 1 per 3,359 people (1996)
Hospital beds: 1 per 1,588 people (1996)
TV sets: (per 1,000 people) 7 (1998)
Radios: (per 1,000 people) 95 (1997)
Internet users: (per 10,000 people) 0.1 (1999)
Personal computer users: (per 100 people) 0.1 (1999)

Transport

Airports: international airports: Yangon (Mingaladon); 21 domestic airports; total passenger km: 147 million (1995)
Railways: total length: 4,621 km/2,871 mi; total passenger km: 4,296 million (1995)
Roads: total road network: 28,200 km/17,523 mi, of which 12.2% paved (1996 est); passenger cars: 0.6 per 1,000 people (1996 est)

Chronology

3rd century BC: Sittoung valley settled by Mons; Buddhism introduced by missionaries from India. **3rd century AD:** Arrival of Burmans from Tibet. **1057:** First Burmese Empire established by King Anawrahta, who conquered Thaton, established capital inland at Pagan, and adopted Theravada Buddhism. **1287:** Pagan sacked by Mongols. **1531:** Founding of Toungoo dynasty, which survived until mid-18th century. **1755:** Nation reunited by Alaungpaya, with port of Rangoon as capital. **1824–26:** First Anglo-Burmese war resulted in Arakan coastal strip, between Chittagong and Cape Negrais, being ceded to British India. **1852:** Following defeat in second Anglo-Burmese war, Lower Burma, including Rangoon, was annexed by British. **1886:** Upper Burma ceded to British after defeat of Thibaw in third Anglo-Burmese war; British united Burma, which was administered as a province of British India. **1886–96:** Guerrilla warfare waged against British in northern Burma. **early 20th century:** Burma developed as a major rice, teak and, later, oil exporter, drawing in immigrant labourers and traders from India and China. **1937:** Became British crown colony in Commonwealth, with a degree of internal self-government. **1942:** Invaded and occupied by Japan, who installed anti-British nationalist puppet government headed by Ba Maw. **1945:** Liberated from Japanese control by British, assisted by nationalists Aung San and U Nu, formerly ministers in puppet government, who had formed the socialist Anti Fascist People's Freedom League (AFPFL). **1947:** Assassination of Aung San and six members of interim government by political opponents. **1948:** Independence achieved from Britain as Burma, with U Nu as prime minister. Left Commonwealth. Quasi-federal state established. **1958–60:** Administered by emergency government, formed by army chief of staff Gen Ne Win. **1962:** Gen Ne Win reassumed power in left-wing army coup; he proceeded to abolish federal system and follow the 'Burmese way to socialism', involving sweeping nationalization and international isolation, which crippled the economy. **1973–74:** Adopted presidential-style 'civilian' constitution. **1975:** The opposition National Democratic Front was formed by regionally-based minority groups, who mounted guerrilla insurgencies. **1987:** There were student demonstrations in Rangoon as food shortages worsened. **1988:** The government resigned after violent student demonstrations and workers' riots. Gen Saw Maung seized power in a military coup; over 2,000 were killed. **1989:** Martial law was declared; thousands were arrested including advocates of democracy and human rights. The country was renamed Myanmar, and its capital Yangon. **1990:** The landslide general election victory for opposition National League for Democracy (NLD) was ignored by the military junta; NLD leaders U Nu and Suu Kyi, the daughter of Aung San, were placed under house arrest. **1991:** Martial law and human-rights abuses continued. Suu Kyi, still imprisoned, was awarded the Nobel Peace Prize. There was a pogrom against the Muslim community in the Arakan province in southwest Myanmar. Western countries imposed sanctions. **1992:** Saw Maung was replaced as head of state by Than Shwe. Several political prisoners were liberated. Martial law was lifted, but restrictions on political freedom remained. **1993:** A ceasefire was agreed with Kachin rebels in the northeast. **1995:** Suu Kyi was released from house arrest, but her appointment as NLD leader was declared illegal. NLD boycotted the constitutional convention. **1996:** Suu Kyi held the first party congress since her release; 200 supporters were detained by the government. There were major demonstrations in support of Suu Kyi. **1997:** Admission to Association of South East Asian Nations (ASEAN) granted, despite US sanctions for human-rights abuses. **1998:** Japan resumed a flow of aid, which had been stopped in 1988. The military junta ignored pro-democracy roadside protests by Aung San Suu Kyi and broke up student demonstrations. 300 members of the opposition NLD were released from detention. **2000:** Aung San Suu Kyi was forced to give up a pro-democracy roadside protest after nine days and placed under house arrest for two weeks. **2001:** The government began talks with Suu Kyi and released 84 members of the NLD from prison.

Practical information

Visa requirements: UK: visa required. USA: visa required
Time difference: GMT +6.5
Chief tourist attractions: palaces; Buddhist temples and shrines in Yangon, Mandalay, Taung-gyi, and Pagan

(notably Yangon's ancient Sule, Botataung, and Shwedagon pagodas); Indian Ocean coast; mountainous interior
Major holidays: 4 January, 12 February, 2, 27 March, 1 April, 1 May, 19 July, 1 October, 25 December; variable: New Year (Burmese), Thingyan (April, 4 days), end of Buddhist Lent (October), Full Moon days

NAMIBIA FORMERLY SOUTH WEST AFRICA (UNTIL 1968)

Located in southwest Africa, bounded north by Angola and Zambia, east by Botswana and South Africa, and west by the Atlantic Ocean.

National name: *Republic of Namibia*
Area: 824,300 sq km/318,262 sq mi
Capital: Windhoek
Major towns/cities: Swakopmund, Rehoboth, Rundu
Major ports: Walvis Bay
Physical features: mainly desert (Namib and Kalahari); Orange River; Caprivi Strip links Namibia to Zambezi River; includes the enclave of Walvis Bay (area 1,120 sq km/432 sq mi)

Blue recalls the clear sky, the Atlantic Ocean, water, and rain. Red represents Namibia's people reflecting their heroism and desire for equal opportunity. White stands for peace and unity. Green symbolizes vegetation and agriculture. Effective date: 21 March 1990.

Government

Head of state: Sam Nujoma from 1990
Head of government: Hage Geingob from 1990
Political system: emergent democracy
Political executive: limited presidency
Administrative divisions: 13 regions
Political parties: South West Africa People's Organization (SWAPO), socialist Ovambo-oriented; Democratic Turnhalle Alliance (DTA), moderate, multiracial coalition; United Democratic Front (UDF), disaffected ex-SWAPO members; National Christian Action (ACN), white conservative
Armed forces: 9,000 (1998)
Conscription: military service is voluntary
Death penalty: abolished in 1990
Defence spend: (% GDP) 3.6 (1998)
Education spend: (% GNP) 9.1 (1997)
Health spend: (% GDP) 7.5 (1997 est)

Economy and resources

Currency: Namibian dollar
GDP: (US$) 3.1 billion (1999)
Real GDP growth: (% change on previous year) 3.7 (1999)
GNP: (US$) 3.2 billion (1999)
GNP per capita (PPP): (US$) 5,369 (1999 est)
Consumer price inflation: 8.6% (1999)
Unemployment: 38% (1997 est)
Foreign debt: (US$) 208 million (1999 est)
Major trading partners: South Africa, UK, Germany, France, Spain, Russia
Resources: uranium, copper, lead, zinc, silver, tin, gold, salt, semi-precious stones, diamonds (one of the world's leading producers of gem diamonds), hydrocarbons, lithium, manganese, tungsten, cadmium, vanadium
Industries: food processing (fish), mining and quarrying, metal and wooden products, brewing, meat processing, chemicals, textiles, cement, leather shoes
Exports: diamonds, fish and fish products, live animals and meat, uranium, karakul pelts. Principal market: UK 33% (1998)

Imports: food and live animals, beverages, tobacco, transport equipment, mineral fuels, chemicals, electrical and other machinery. Principal source: South Africa 87% (1998)
Arable land: 1% (1996)
Agricultural products: wheat, maize, sunflower seed, sorghum, vegetables (crop farming is greatly limited by scarcity of water and poor rainfall); fishing; principal agricultural activity is livestock rearing (cattle, sheep, and goats); beef and karakul sheepskin are also produced

Population and society

Population: 1,726,000 (2000 est)
Population growth rate: 2.2% (1995–2000)
Population density: (per sq km) 2 (1999 est)
Urban population: (% of total) 31 (2000 est)
Age distribution: (% of total population) 0–14 42%, 15–59 53%, 60+ 5% (2000 est)
Ethnic groups: 85% black African, of which 51% belong to the Ovambo tribe; the remainder includes the pastoral Nama and hunter-gatherer groups. There is a 6% white minority
Language: English (official), Afrikaans, German, Ovambo (51%), Nama (12%), Kavango (10%), other indigenous languages
Religion: about 90% Christian (Lutheran, Roman Catholic, Dutch Reformed Church, Anglican)
Education: (compulsory years) 7
Literacy rate: 83% (men); 81% (women) (2000 est)
Labour force: 42% of population: 49% agriculture, 15% industry, 36% services (1990)
Life expectancy: 52 (men); 53 (women) (1995–2000)
Child mortality rate: (under 5, per 1,000 live births) 122 (1995–2000)
Physicians: 1 per 4,328 people (1993 est)
Hospital beds: 1 per 207 people (1993 est)
TV sets: (per 1,000 people) 32 (1997)
Radios: (per 1,000 people) 144 (1997)
Internet users: (per 10,000 people) 35.4 (1999)
Personal computer users: (per 100 people) 3 (1999)

Transport

Airports: international airports: Windhoek; all major towns have domestic airports or landing strips; total passenger km: 906 million (1997 est)
Railways: total length: 2,382 km/1,480 mi; total passenger km: 35 million (1994)
Roads: total road network: 63,258 km/39,309 mi, of which 8.3% paved (1997); passenger cars: 45.6 per 1,000 people (1996)

Chronology

1480s: Coast visited by European explorers. **16th century:** Bantu-speaking Herero migrated into northwest and Ovambo settled in northernmost areas. **1840s:** Rhenish Missionary Society began to spread German influence; Jonkar Afrikaner conquest state dominant in southern Namibia. **1884:** Germany annexed most of the area, calling it South West Africa, with Britain incorporating a small enclave around Walvis Bay in the Cape Colony of South Africa. **1892:** German farmers arrived to settle in the region. **1903–04:** Uprisings by the long-settled Nama (Khoikhoi) and Herero peoples brutally repressed by Germans, with over half the local communities slaughtered. **1908:** Discovery of diamonds led to a larger influx of Europeans. **1915:** German colony invaded and seized by South Africa during World War I and the Ovambo, in the north, were conquered. **1920:** Administered by South Africa, under League of Nations mandate. **1946:** Full incorporation in South Africa refused by United Nations (UN). **1949:** White voters in South West Africa given representation in the South African parliament. **1958:** South West Africa People's Organization (SWAPO) formed to campaign for racial equality and full independence. **1960:** Radical wing of SWAPO, led by Sam Nujoma, forced into exile. **1964:** UN voted to end South Africa's mandate, but South Africa refused to relinquish control or soften its policies towards the economically disenfranchised black majority. **1966:** South Africa's apartheid laws extended to the country; 60% of land was allocated to whites, who formed 10% of the population. **1968:** South West Africa redesignated Namibia by UN; SWAPO, drawing strong support from the Ovambo people of the north, began armed guerrilla struggle against South African rule, establishing People's Liberation Army of Namibia (PLAN). **1971:** Prolonged general strike by black Namibian contract workers. **1973:** The UN recognized SWAPO as the 'authentic representative of the Namibian people'. **1975–76:** The establishment of a new Marxist regime in independent Angola strengthened the position of SWAPO guerrilla movement, but also led to the increased military involvement of South Africa in the region. **1978:** UN Security Council Resolution 435 for the granting of full independence was accepted by South Africa, and then rescinded. **1983:** Direct rule was reimposed by Pretoria after the resignation of the Democratic Turnhalle Alliance (DTA), a conservative administration dominated by whites. **1985:** South Africa installed a new puppet administration, the Transitional Government of National Unity (TGNU), which tried to reform the apartheid system, but was not recognized by the UN. **1988:** Peace talks between South Africa, Angola, and Cuba led to an agreement on troop withdrawals and full independence for Namibia. **1989:** UN peacekeeping force were stationed to oversee free elections to the assembly to draft a new constitution; SWAPO won the elections. **1990:** A liberal multiparty constitution was adopted and independence was achieved. Sam Nujoma, SWAPO's former guerrilla leader, was elected president. Joined the Commonwealth. Hage Geingob was appointed prime minister. **1993:** South Africa, with its new multiracial government, relinquished its claim to Walvis Bay sovereignty. Namibia dollar was launched with South African rand parity. **1994:** SWAPO won assembly elections; Nujoma was re-elected president.

Practical information

Visa requirements: UK: visa not required. USA: visa not required
Time difference: GMT +1
Chief tourist attractions: game parks; nature reserves (notably the Etosha National Park and Game Reserve); the government is promoting the development of ecotourism
Major holidays: 1 January, 21 March, 1, 4, 16, 25 May, 26 August, 10, 25–26 December; variable: Good Friday, Easter Monday

NAURU

Island in Polynesia, southwest Pacific, west of Kiribati.

National name: *Republic of Nauru*
Area: 21 sq km/8.1 sq mi
Capital: Yaren (seat of government)
Physical features: tropical coral island in southwest Pacific; plateau encircled by coral cliffs and sandy beaches

Blue stands for the Pacific Ocean. The yellow stripe represents the Equator. The points of the star symbolize the island's 12 original tribes. Effective date: 31 January 1968.

Government

Head of state and government: Bernard Dowiyogo from 2000
Political system: liberal democracy
Political executive: limited presidency
Political parties: candidates are traditionally elected as independents, grouped into pro- and antigovernment factions; Democratic Party of Nauru (DPN), only formal political party, antigovernment
Armed forces: no standing army; Australia is responsible for Nauru's defence
Death penalty: retains the death penalty for ordinary crimes but can be considered abolitionist in practice (no executions since independence)
Health spend: (% GDP) 5 (1997 est)

Economy and resources

Currency: Australian dollar
GDP: (US$) 368 million (1995)
GNP: (US$) 304 million (1994 est)
GNP per capita (PPP): (US$) 11,800 (1994 est)
Consumer price inflation: 10.1% (1996)
Unemployment: 0% (1996)
Major trading partners: Australia, New Zealand, Philippines, Japan

Industries: phosphate mining, financial services
Exports: phosphates. Principal market: Australia
Imports: food and live animals, building construction materials, petroleum, machinery, medical supplies.
Agricultural products: small-scale production; coconuts, bananas, pineapples, screw-pines, livestock rearing (pigs and chickens); almost all the country's requirements (including most of its drinking water) are imported

Population and society

Population: 12,000 (2000 est)
Population growth rate: 1.8% (1995–2025)
Population density: (per sq km) 524 (1999 est)
Urban population: (% of total) 100 (2000 est)
Age distribution: (% of total population) 0–14 43%, 15–59 53%, 60+ 4% (2000 est)
Ethnic groups: about 68% indigenous Nauruan (mixture of Micronesian, Polynesian, and Melanesian descent), about 18% Pacific Islander, 8% European, 6% Chinese
Language: Nauruan, English (both official)
Religion: majority Protestant, Roman Catholic
Education: (compulsory years) 10
Literacy rate: 99% (men); 99% (women) (1997 est)
Life expectancy: 64 (men); 69 (women) (1998 est)

Chronology

1798: British whaler Capt John Fearn first visited Nauru and named it Pleasant Island. **1830s–80s:** The island was a haven for white runaway convicts and deserters. **1888:** Annexed by Germany at the request of German settlers who sought protection from local clan unrest. **1899:** Phosphate deposits discovered; mining began eight years later, with indentured Chinese labourers brought in to work British Australian-owned mines. **1914:** Occupied by Australia on the outbreak of World War I. **1920:** Administered by Australia on behalf of itself, New Zealand, and the UK until independence, except 1942–43, when it was occupied by Japan, and two-thirds of the population were deported briefly to Micronesia. **1968:** Independence achieved, with 'special member' British Commonwealth status. **1987:** Kennan Adeang established the Democratic Party of Nauru. **1994:** Australia agreed to an out-of-court settlement of A$107 million, payable over 20 years, for environmental damage caused by phosphate mining which had left 80% of land agriculturally barren. **2000:** Bernard Dowiyogo was elected president for the sixth time. General elections saw Rene Harris win the popular vote, but he resigned and Dowiyogo was installed.

Practical information

Visa requirements: UK/USA: visa required
Time difference: GMT +12
Chief tourist attractions: beautiful beaches interspersed by coral pinnacles

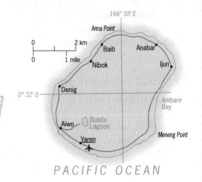

NEPAL

Located in the Himalayan mountain range in Central Asia, bounded north by Tibet (an autonomous region of China), east, south, and west by India.

National name: *Nepál Adhirajya/Kingdom of Nepal*
Area: 147,181 sq km/56,826 sq mi
Capital: Kathmandu
Major towns/cities: Moráng, Biratnagar, Lalitpur, Bhaktapur, Pokhara, Birganj, Dharan
Physical features: descends from the Himalayas in the north through foothills to the River Ganges plain in the south; Mount Everest, Mount Kangchenjunga

Initially the sun and moon had human faces, but they were removed when the flag was updated in 1962. The flag is said to express the hope that Nepal will endure as long as the sun and the moon. The blue border symbolizes peace. Effective date: 16 December 1962.

Government

Head of state: King Birendra Bir Bikram Shah Dev from 1972
Head of government: Girja Prasad Koirala from 2000
Political system: emergent democracy
Political executive: parliamentary
Administrative divisions: 14 zones and 75 districts
Political parties: Nepali Congress Party (NCP), left of centre; United Nepal Communist Party (UNCP; Unified Marxist–Leninist), left wing; Rashtriya Prajatantra Party (RPP), monarchist
Armed forces: 50,000 (1998)
Conscription: military service is voluntary
Death penalty: abolished in 1997
Defence spend: (% GDP) 0.7 (1998)
Education spend: (% GNP) 3.2 (1997)
Health spend: (% GDP) 1.2 (1990–95)

Economy and resources

Currency: Nepalese rupee
GDP: (US$) 4.9 billion (1999)
Real GDP growth: (% change on previous year) 2.7 (1999)
GNP: (US$) 5.1 billion (1999)
GNP per capita (PPP): (US$) 1,219 (1999)
Consumer price inflation: 8.1% (1999)
Unemployment: 4.9% (1990); 40%–50% underemployment (1997)
Foreign debt: (US$) 2.4 billion (1997)
Major trading partners: India, China, Germany, Japan, Singapore, USA, United Arab Emirates, France, Bangladesh, Saudi Arabia
Resources: lignite, talcum, magnesite, limestone, copper, cobalt
Industries: bricks and tiles, carpets, clothing, paper, cotton fabrics, cement, leather, jute goods, electrical cable, soap, edible oils, sugar, tourism
Exports: woollen carpets, clothing, hides and skins, food grains, jute, timber, oil seeds, ghee, potatoes, medicinal herbs, cattle. Principal market: India 32.8% (1998)
Imports: basic manufactures, machinery and transport equipment, chemicals, pharmaceuticals. Principal source: India 30.7% (1998)

Arable land: 20.4% (1996)
Agricultural products: rice, maize, wheat, sugar cane, millet, potatoes, barley, tobacco, cardamoms, fruits, oil seeds; livestock rearing (cattle and pigs)

Population and society

Population: 23,930,000 (2000 est)
Population growth rate: 2.4% (1995–2000)
Population density: (per sq km) 159 (1999 est)
Urban population: (% of total) 12 (2000 est)
Age distribution: (% of total population) 0–14 41%, 15–59 53%, 60+ 6% (2000 est)
Ethnic groups: 80% of Indo-Nepalese origin, including the Gurkhas, Paharis, Newars, and Tharus; 20% of Tibeto-Nepalese descent (concentrated in the north and east)
Language: Nepali (official), Tibetan, numerous local languages
Religion: Hindu 90%; Buddhist 5%, Muslim 3%, Christian
Education: (compulsory years) 5
Literacy rate: 59% (men); 24% (women) (2000 est)
Labour force: 47% of population: 93% agriculture, 1% industry, 6% services (1991)
Life expectancy: 58 (men); 57 (women) (1995–2000)
Child mortality rate: (under 5, per 1,000 live births) 117 (1995–2000)
Physicians: 1 per 13,634 people (1993)
Hospital beds: 1 per 4,305 people (1994)
TV sets: (per 1,000 people) 6 (1997)
Radios: (per 1,000 people) 38 (1997)
Internet users: (per 10,000 people) 15.0 (1999)
Personal computer users: (per 100 people) 0.3 (1999)

Transport

Airports: international airports: Kathmandu (Tribhuvan); 37 domestic airports and airfields; total passenger km: 953 million (1997 est)
Railways: total length: 101 km/63 mi; total passenger km: 18,044 million (1994)
Roads: total road network: 7,700 km/4,785 mi, of which 41.5% paved (1996 est)

Chronology

8th century BC: Kathmandu Valley occupied by Ahirs (shepherd kings), Tibeto-Burman migrants from northern India. *c.* **563 BC:** In Lumbini in far south, Prince Siddhartha Gautama, the historic Buddha, was born. **AD 300:** Licchavis dynasty immigrated from India and introduced caste system. **13th–16th centuries:** Dominated by Malla dynasty, great patrons of the arts. **1768:** Nepal emerged as a unified kingdom after the ruler of the principality of the Gurkhas in the west, King Prithwi Narayan Shah, conquered Kathmandu Valley. **1792:** Nepal's expansion halted by defeat at the hands of Chinese in Tibet; commercial treaty signed with Britain. **1815–16:** Anglo-Nepali 'Gurkha War'; Nepal became British-dependent buffer state with British resident stationed in Kathmandu. **1846:** Fell under sway of Rana family, who became hereditary chief ministers, dominating powerless monarchy and isolating Nepal from outside world. **1923:** Full independence formally recognized by Britain. **1951:** Monarchy restored to power and Ranas overthrown in 'palace revolution' supported by Nepali Congress Party (NCP). **1959:** Constitution created an elected legislature. **1960–61:** Parliament dissolved by King Mahendra; political parties banned after NCP's pro-India socialist leader B P Koirala became prime minister. **1962:** New constitution provided for tiered, traditional system of indirectly elected local councils (*panchayats*) and an appointed prime minister. **1972:** King Mahendra died; succeeded by his son, King Birendra Bikram Shah Dev. **1980:** A constitutional referendum was held, following popular agitation led by B P Koirala, resulted in the introduction of direct, but nonparty, elections to the National Assembly. **1983:** The monarch-supported prime minister was overthrown by directly elected deputies to the National Assembly. **1986:** New assembly elections returned a majority opposed to the *panchayat* system of partyless government. **1988:** Strict curbs were placed on opposition activity; over 100 supporters of the banned NCP were arrested, and censorship was imposed. **1989:** A border blockade was imposed by India during a treaty dispute. **1990:** The *panchayat* system collapsed after mass NCP-led violent prodemocracy demonstrations; a new democratic constitution was introduced, and the ban on political parties lifted. **1991:** The Nepali Congress Party, led by Girija Prasad Koirala, won the general election. **1992:** Communists led antigovernment demonstrations in Kathmandu and Pátan. **1994:** Koirala's government was defeated on a no-confidence motion; parliament was dissolved. A minority communist government was formed under Man Mohan Adhikari. **1995:** Parliament was dissolved by King Birendra at Prime Minister Adhikari's request; fresh elections were called but the Supreme Court ruled the move unconstitutional. **1996:** Maoist guerillas began a violent insurgency aimed at overthrowing the government. **1998:** Krishna Prasad Bhattarai became prime minister and formed a new coalition government. A declared priority was to end the Maoist insurgency. **2000:** A vote of no confidence in Bhattarai led to his replacement by Girija Prasad Koirala. Secret unofficial talks with the Maoist guerillas began, but ended broke off.

Practical information

Visa requirements: UK: visa required. USA: visa required
Time difference: GMT +5.5
Chief tourist attractions: Lumbini, birthplace of Buddha; the Himalayas, including Mount Everest, the world's highest peak; the lake city of Pokhara; wildlife includes tigers, leopards, elephants, buffalo, and gaur
Major holidays: 11 January, 19 February, 8 November, 16, 29 December; variable: New Year (Sinhala/Tamil, April), Maha Shivarata (February/March)

NETHERLANDS, THE OR HOLLAND

Located in Western Europe on the North Sea, bounded east by Germany and south by Belgium.

National name: *Koninkrijk der Nederlanden/Kingdom of the Netherlands*
Area: 41,863 sq km/16,163 sq mi
Capital: Amsterdam
Major towns/cities: Rotterdam, the Hague (seat of government), Utrecht, Eindhoven, Groningen, Tilburg, Maastricht, Apeldoorn, Nijmegen, Breda
Major ports: Rotterdam
Physical features: flat coastal lowland; rivers Rhine, Schelde, Maas; Frisian Islands
Territories: Aruba, Netherlands Antilles (Caribbean)

The number of stripes changed frequently until around 1800. Red, white, and blue became the colours of liberty and an inspiration for other revolutionary flags around the world. Effective date: 19 February 1937.

Government

Head of state: Queen Beatrix Wilhelmina Armgard from 1980
Head of government: Wim Kok from 1994
Political system: liberal democracy
Political executive: parliamentary
Administrative divisions: 12 provinces
Political parties: Christian Democratic Appeal (CDA), Christian, right of centre; Labour Party (PvdA), democratic socialist, left of centre; People's Party for Freedom and Democracy (VVD), liberal, free enterprise; Democrats 66 (D66), ecologist, centrist; Political Reformed Party (SGP), moderate Calvinist; Evangelical Political Federation (RPF), radical Calvinist; Reformed Political Association (GPV), fundamentalist Calvinist; Green Left, ecologist; General League of the Elderly (AOV), pensioner-oriented
Armed forces: 57,200 (1998)
Conscription: military service is voluntary
Death penalty: abolished in 1982
Defence spend: (% GDP) 1.8 (1998)
Education spend: (% GNP) 5.2 (1996)
Health spend: (% GDP) 8.5 (1997)

Economy and resources

Currency: guilder
GDP: (US$) 384.8 billion (1999)
Real GDP growth: (% change on previous year) 3.6 (1999)
GNP: (US$) 384.3 billion (1999)
GNP per capita (PPP): (US$) 23,052 (1999)
Consumer price inflation: 2.2% (1999)
Unemployment: 4.2% (1998)
Major trading partners: EU (principally Germany, Belgium–Luxembourg, UK, France, and Italy), USA
Resources: petroleum, natural gas
Industries: electrical machinery, metal products, food processing, electronic equipment, chemicals, rubber and plastic products, petroleum refining, dairy farming, horticulture, diamond cutting

Exports: machinery and transport equipment, foodstuffs, live animals, petroleum and petroleum products, natural gas, chemicals, plants and cut flowers, plant-derived products. Principal market: Germany 25.9% (1999)
Imports: electrical machinery, cars and other vehicles, mineral fuels, metals and metal products, plastics, paper and cardboard, clothing and accessories. Principal source: Germany 19.5% (1999)
Arable land: 26.1% (1996)
Agricultural products: sugar beet, potatoes, wheat, barley, flax, fruit, vegetables, flowers; dairy farming

Population and society

Population: 15,786,000 (1999 est)
Population growth rate: 0.4% (1995–2000)
Population density: (per sq km) 376 (1999 est)
Urban population: (% of total) 89 (2000 est)
Age distribution: (% of total population) 0–14 18%, 15–59 63%, 60+ 19% (2000 est)
Ethnic groups: primarily Dutch (Germanic, with some Gallo-Celtic mixtures); sizeable Indonesian, Surinamese, and Turkish minorities
Language: Dutch (official)
Religion: atheist 39%, Roman Catholic 31%, Dutch Reformed Church 14%, Calvinist 8%
Education: (compulsory years) 11
Literacy rate: 99% (men); 99% (women) (2000 est)
Labour force: 3.7% agriculture, 22.2% industry, 74.1% services (1997)
Life expectancy: 75 (men); 81 (women) (1995–2000)
Child mortality rate: (under 5, per 1,000 live births) 8 (1995–2000)
Physicians: 1 per 400 people (1996)
Hospital beds: 1 per 88 people (1996)
TV sets: (per 1,000 people) 541 (1997)
Radios: (per 1,000 people) 978 (1997)
Internet users: (per 10,000 people) 1,893.1 (1999)
Personal computer users: (per 100 people) 36.0 (1999)

Transport

Airports: international airports: Amsterdam (Schipol), Rotterdam (Zestienhoven), Eindhoven (Welschap), Maastricht (Beck), Groningen (Eelde), Enschede (Twente); domestic services operate between these; total passenger km: 70,702 million (1997 est)
Railways: total length: 2,739 km/1,702 mi; total passenger km: 14,091 million (1996)
Roads: total road network: 124,530km/77,383 mi, of which 90% paved (1997); passenger cars: 373 per 1,000 people (1997)

Chronology

55 BC: Julius Caesar brought lands south of River Rhine under Roman rule. **4th century AD:** Region overrun by Franks and Saxons. **7th–8th centuries:** Franks subdued Saxons north of Rhine and imposed Christianity. **843–12th centuries:** Division of Holy Roman Empire: the Netherlands repeatedly partitioned, not falling clearly into either French or German kingdoms. **12th–14th centuries:** Local feudal lords, led by count of Holland and bishop of Utrecht, became practically independent; Dutch towns became prosperous trading centres, usually ruled by small groups of merchants. **15th century:** Low Countries (Holland, Belgium, and Flanders) came under rule of dukes of Burgundy. **1477:** Low Countries passed by marriage to Habsburgs. **1555:** The Netherlands passed to Spain upon division of Habsburg domains. **1568:** Dutch rebelled under leadership of William the Silent, Prince of Orange, and fought a long war of independence. **1579:** Union of Utrecht: seven northern rebel provinces formed United Provinces. **17th century:** 'Golden Age': Dutch led world in trade, art, and science, and founded colonies in East and West Indies, primarily through Dutch East India Company, founded in 1602. **1648:** Treaty of Westphalia: United Provinces finally recognized as independent Dutch Republic. **1652–54:** Commercial and colonial rivalries led to naval war with England. **1652–72:** Johann de Witt ruled Dutch Republic as premier after conflict between republicans and House of Orange. **1665–67:** Second Anglo-Dutch war. **1672–74:** Third Anglo-Dutch war. **1672:** William of Orange became stadholder (ruling as chief magistrate) of the Dutch Republic, an office which became hereditary in the Orange family. **1672–78:** The Netherlands fought to prevent domination by King Louis XIV of France. **1688–97 and 1701–13:** War with France resumed. **18th century:** Exhausted by war, the Netherlands ceased to be a Great Power. **1795:** Revolutionary France conquered the Netherlands and established Batavian Republic. **1806:** Napoleon made his brother Louis king of Holland. **1810:** France annexed the Netherlands. **1815:** Northern and southern Netherlands (Holland and Belgium) unified as Kingdom of the Netherlands under King William I of Orange, who also became grand duke of Luxembourg. **1830:** Southern Netherlands rebelled and declared independence as Belgium. **1848:** Liberal constitution adopted. **1890:** Queen Wilhelmina succeeded to throne; dynastic link with Luxembourg broken. **1894–96:** Dutch suppressed colonial revolt in Java. **1914–18:** The Netherlands neutral during World War I. **1940–45:** Occupied by Germany during World War II. **1948:** The Netherlands formed Benelux customs union with Belgium and Luxembourg; Queen Wilhelmina abdicated in favour of her daughter Juliana. **1949:** Became a founding member of the North Atlantic Treaty Organization (NATO); most of Dutch East Indies became independent as Indonesia after four years of war. **1953:** Dykes breached by storm; nearly two thousand people and tens of thousands of cattle died in flood. **1954:** Remaining Dutch colonies achieved internal self-government. **1958:** The Netherlands became a founding member of the European Economic Community (EEC). **1963:** The Dutch colony of Western New Guinea was ceded to Indonesia. **1975:** Dutch Guiana became independent as Suriname. **1980:** Queen Juliana abdicated in favour of her daughter Beatrix. **1994:** Following an inconclusive general election, a three-party coalition was formed under PvdA leader Wim Kok. **1999:** The coalition government resigned in May after the smallest party, Democrats 66 (D-66), withdrew. **2000:** The Netherlands became the first country to legalize euthanasia.

Practical information

Visa requirements: UK/USA: visa not required
Time difference: GMT +1
Chief tourist attractions: the lively, cosmopolitan city of Amsterdam, with its museums and historical buildings; old towns; canals; the bulb fields in spring; art galleries; modern architecture; outlying islands
Major holidays: 1 January, 30 April, 5 May, 25–26 December; variable: Ascension Thursday, Good Friday, Easter Monday, Whit Monday

NEW ZEALAND

Comprises two main islands and other small islands in the southwest Pacific Ocean.

National name: *Aotearoa/New Zealand*
Area: 268,680 sq km/103,737 sq mi
Capital: Wellington
Major towns/cities: Auckland, Hamilton, Christchurch, Manukau, North Shore, Waitakere
Major ports: Auckland, Wellington

The Union Jack marks New Zealand's historical links with Britain. The stars represent the Southern Cross. Effective date: 12 June 1902.

Physical features: comprises North Island, South Island, Stewart Island, Chatham Islands, and minor islands; mainly mountainous; Ruapehu in North Island, 2,797 m/9,180 ft, highest of three active volcanoes; geysers and hot springs of Rotorua district; Lake Taupo (616 sq km/238 sq mi), source of Waikato River; Kaingaroa state forest. In South Island are the Southern Alps and Canterbury Plains
Territories: Tokelau (three atolls transferred in 1926 from former Gilbert and Ellice Islands colony); Niue Island (one of the Cook Islands, separately administered from 1903: chief town Alafi); Cook Islands are internally self-governing but share common citizenship with New Zealand; Ross Dependency in Antarctica

Government

Head of state: Queen Elizabeth II from 1952, represented by Governor General Catherine Tizard from 1990
Head of government: Helen Clark from 1999
Political system: liberal democracy
Political executive: parliamentary
Administrative divisions: 93 counties, nine districts and three towns
Political parties: Labour Party, moderate, left of centre; New Zealand National Party, free enterprise, right of centre; Alliance Party bloc, left of centre, ecologists; New Zealand First Party (NZFP), centrist; United New Zealand Party (UNZ), centrist
Armed forces: 9,600; around 7,000 reserves (1998)
Conscription: military service is voluntary
Death penalty: abolished in 1989
Defence spend: (% GDP) 1.5 (1998)
Education spend: (% GNP) 7.3 (1996)
Health spend: (% GDP) 7.6 (1997)

Economy and resources

Currency: New Zealand dollar
GDP: (US$) 53.9 billion (1999)
Real GDP growth: (% change on previous year) 3.4 (1999)
GNP: (US$) 52.7 billion (1999)
GNP per capita (PPP): (US$) 16,566 (1999)
Consumer price inflation: 1.1% (1999)
Unemployment: 7.5% (1998)
Major trading partners: Australia, USA, Japan, UK
Resources: coal, clay, limestone, dolomite, natural gas, hydroelectric power, pumice, iron ore, gold, forests
Industries: food processing, machinery, textiles and clothing, fisheries, wood and wood products, paper and paper products, metal products; farming, particularly livestock and dairying, cropping, fruit growing, horticulture

Exports: meat, dairy products, wool, fish, timber and wood products, fruit and vegetables, aluminium, machinery. Principal market: Australia 21.7% (1999)
Imports: machinery and mechanical appliances, vehicles and aircraft, petroleum, fertilizer, consumer goods. Principal source: Australia 24.5% (1999)
Arable land: 5.8% (1996)
Agricultural products: barley, wheat, maize, fodder crops, exotic timber, fruit (kiwi fruit and apples); livestock and dairy farming

Population and society

Population: 3,862,000 (2000 est)
Population growth rate: 1% (1995–2000); 0.8% (2000–05)
Population density: (per sq km) 14 (1999 est)
Urban population: (% of total) 86 (2000 est)
Age distribution: (% of total population) 0–14 23%, 15–59 61%, 60+ 16% (2000 est)
Ethnic groups: around 75% of European origin, 15% Maori, 3% Pacific Islander
Language: English (official), Maori
Religion: Christian (Anglican 18%, Roman Catholic 14%, Presbyterian 13%)
Education: (compulsory years) 11
Literacy rate: 99% (men); 99% (women) (2000 est)
Labour force: 8.5% agriculture, 23.8% industry, 67.6% services (1997)
Life expectancy: 74 (men); 80 (women) (1995–2000)
Child mortality rate: (under 5, per 1,000 live births) 8 (1995–2000)
Physicians: 1 per 476 people (1996)
Hospital beds: 1 per 147 people (1994)
TV sets: (per 1,000 people) 508 (1997)
Radios: (per 1,000 people) 990 (1997)
Internet users: (per 10,000 people) 1,828.4 (1999)
Personal computer users: (per 100 people) 32.7 (1999)

Transport

Airports: international airports: Auckland (Mangere), Christchurch, Wellington (Rongotai); 32 domestic airports; total passenger km: 23,020 million (1997 est) **Railways:** total length: 3,973 km/2,469 mi; total

passengers carried: 11.6 million (1997) **Roads:** total road network: 92,200 km/57,293 mi, of which 58.1% paved (1996 est); passenger cars: 470 per 1,000 people (1996 est)

Chronology

1642: Dutch explorer Abel Tasman reached New Zealand but indigenous Maori prevented him from going ashore. **1769:** English explorer James Cook surveyed coastline of islands. **1773 and 1777:** Cook again explored coast. **1815:** First British missionaries arrived in New Zealand. **1826:** New Zealand Company founded in London to establish settlement. **1839:** New Zealand Company relaunched, after initial failure, by Edward Gibbon Wakefield. **1840:** Treaty of Waitangi: Maori accepted British sovereignty; colonization began and large-scale sheep farming developed. **1845–47:** Maori revolt against loss of land. **1851:** Became separate colony (was originally part of the Australian colony of New South Wales). **1852:** Colony procured constitution after dissolution of New Zealand Company; self-government fully implemented in 1856. **1860–72:** Second Maori revolt led to concessions, including representation in parliament. **1891:** New Zealand took part in Australasian Federal Convention in Sydney but rejected the idea of joining the Australian Commonwealth. **1893:** Became the first country to give women the right to vote in parliamentary elections. **1898:** Liberal government under Richard Seddon introduced pioneering old-age pension scheme. **1899–1902:** Volunteers from New Zealand fought alongside imperial forces in Boer War. **1907:** New Zealand achieved dominion status within British Empire. **1912–25:** Government of Reform Party, led by William Massey, reflected interests of North Island farmers and strongly supported imperial unity. **1914–18:** 130,000 New Zealanders fought for the British Empire in World War I. **1916:** Labour Party of New Zealand established. **1931:** Statute of Westminster affirmed equality of status between Britain and dominions, effectively granting independence to New Zealand. **1935–49:** Labour governments of Michael Savage and Peter Fraser introduced social reforms and encouraged state intervention in industry. **1936:** Liberal Party merged with Reform Party to create National Party. **1939–45:** New Zealand troops fought in World War II, notably in Crete, North Africa, and Italy. **1947:** Parliament confirmed independence of New Zealand within British Commonwealth. **1951:** New Zealand joined Australia and USA in ANZUS Pacific security treaty. **1965–72:** New Zealand contingent took part in Vietnam War. **1973:** British entry into European Economic Community (EEC) forced New Zealand to seek closer trading relations with Australia. **1985:** Non-nuclear military policy led to disagreements with France and USA. **1986:** The USA suspended defence obligations to New Zealand after it banned the entry of US warships. **1988:** A free-trade agreement was signed with Australia. **1991:** The Alliance Party was formed to challenge the two-party system. **1998:** The government was ordered to return more than £2 million worth of land confiscated from its Maori owners more than 30 years earlier. **1999:** The conservative government was replaced by a centre-left coalition of the Labour Party and New Zealand Alliance, with Helen Clark, leader of the Labour Party, as the new prime minister. **2000:** Dame Silvia Cartwright was named as next governor-general, to take office in April 2001, at which point all top political offices will be held by women.

Practical information

Visa requirements: UK: visa not required. USA: visa not required **Time difference:** GMT +12 **Chief tourist attractions:** famous for its trout- and

deep-sea fishing and generally idyllic setting, including beaches, hot springs, mountains, lakes, and forests **Major holidays:** 1, 2 January, 6 February, 25 April, 25–26 December; variable: Good Friday, Easter Monday, Queen's Birthday (June), Labour (October)

NICARAGUA

Located in Central America, between the Pacific Ocean and the Caribbean Sea, bounded north by Honduras and south by Costa Rica.

National name: *República de Nicaragua/Republic of Nicaragua*
Area: 127,849 sq km/49,362 sq mi
Capital: Managua
Major towns/cities: León, Chinandega, Masaya, Granada, Estelí
Major ports: Corinto, Puerto Cabezas, El Bluff
Physical features: narrow Pacific coastal plain separated from broad Atlantic coastal plain by volcanic mountains and lakes Managua and Nicaragua; one of the world's most active earthquake regions

The triangle signifies equality. The volcanos recall the five nations of the Central American Federation (CAF). Effective date: 27 August 1971.

Government

Head of state and government: Arnoldo Aleman from 1997
Political system: emergent democracy
Political executive: limited presidency
Administrative divisions: 15 departments and two autonomous regions
Political parties: Sandinista National Liberation Front (FSLN), Marxist–Leninist; Opposition Political Alliance (APO, formerly National Opposition Union: UNO), loose US-backed coalition
Armed forces: 17,000 (1998)
Conscription: military service is voluntary (since 1990)
Death penalty: abolished in 1979
Defence spend: (% GDP) 1.1 (1998)
Education spend: (% GNP) 3.9 (1997)
Health spend: (% GDP) 4.3 (1990–95)

Economy and resources

Currency: cordoba
GDP: (US$) 2.3 billion (1999)
Real GDP growth: (% change on previous year) 7 (1999)
GNP: (US$) 2.1 billion (1999)
GNP per capita (PPP): (US$) 2,154 (1999)
Consumer price inflation: 11.3% (1999)
Unemployment: 13.3% (1997)
Foreign debt: (US$) 6.2 billion (1999)
Major trading partners: USA, El Salvador, Panama, France, Spain, Germany, Costa Rica, Venezuela
Resources: gold, silver, copper, lead, antimony, zinc, iron, limestone, gypsum, marble, bentonite
Industries: food products, beverages, petroleum refining, chemicals, metallic products, processed leather, cement
Exports: coffee, meat, cotton, sugar, seafood, bananas, chemical products. Principal market: USA 37.7% (1999)
Imports: machinery and transport equipment, food and live animals, consumer goods, mineral fuels and lubricants, chemicals, and related products. Principal source: USA 34.5% (1999)

Arable land: 20.2% (1996)
Agricultural products: coffee, cotton, sugar cane, bananas, maize, rice, beans, green tobacco; livestock rearing (cattle and pigs); fishing; forest resources

Population and society

Population: 5,074,000 (2000 est)
Population growth rate: 2.7% (1995–2000)
Population density: (per sq km) 39 (1999 est)
Urban population: (% of total) 56 (2000 est)
Age distribution: (% of total population) 0–14 43%, 15–59 52%, 60+ 5% (2000 est)
Ethnic groups: about 70% of mixed American Indian and Spanish origin; about 15% European origin; about 9% African; 5% American Indian; mixed American Indian and black origin
Language: Spanish (official), English, American Indian languages
Religion: Roman Catholic 95%
Education: (compulsory years) 6
Literacy rate: 67% (men); 70% (women) (2000 est)
Labour force: 41.8% agriculture, 17.4% industry, 40.8% services (1997)
Life expectancy: 66 (men); 71 (women) (1995–2000)
Child mortality rate: (under 5, per 1,000 live births) 58 (1995–2000)
Physicians: 1 per 1,566 people (1993)
Hospital beds: 1 per 856 people (1993)
TV sets: (per 1,000 people) 190 (1997)
Radios: (per 1,000 people) 285 (1997)
Internet users: (per 10,000 people) 40.5 (1999)
Personal computer users: (per 100 people) 0.8 (1999)

Transport

Airports: international airports: Managua (Augusto Cesar Sandino); total passenger km: 80 million (1995)
Railways: total length: 287 km/178 mi; freight services reported withdrawn in 1994
Roads: total road network: 18,000 km/11,185 mi, of which 10.1% paved (1996 est); passenger cars: 18.1 per 1,000 people (1996 est)

Chronology

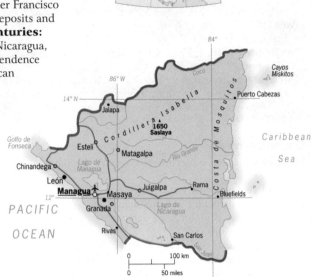

10th century: Indians from Mexico and Mesoamerica migrated to Nicaragua's Pacific lowlands. **1522:** Visited by Spanish explorer Gil Gonzalez de Avila, who named the area Nicaragua after local Indian chief, Nicarao. **1523–24:** Colonized by the Spanish, under Francisco Hernandez de Cordoba, who was attracted by local gold deposits and founded the cities of Granada and León. **17th–18th centuries:** Britain was the dominant force on the Caribbean side of Nicaragua, while Spain controlled the Pacific lowlands. **1821:** Independence achieved from Spain; Nicaragua was initially part of Mexican Empire. **1823:** Became part of United Provinces (Federation) of Central America, also embracing Costa Rica, El Salvador, Guatemala, and Honduras. **1838:** Became fully independent when it seceded from the Federation. **1857–93:** Ruled by succession of Conservative Party governments. **1860:** The British ceded control over the Caribbean ('Mosquito') Coast to Nicaragua. **1893:** Liberal Party leader, José Santos Zelaya, deposed the Conservative president and established a dictatorship which lasted until overthrown by US marines in 1909. **1912–25:** At the Nicaraguan government's request, with the political situation deteriorating, the USA established military bases and stationed marines. **1927–33:** Re-stationed US marines faced opposition from the anti-American guerrilla group led by Augusto César Sandino, who was assassinated in 1934 on the orders of the commander of the US-trained National Guard, Gen Anastasio Somoza Garcia. **1937:** Gen Somoza was elected president; start of near-dictatorial rule by the Somoza family, which amassed a huge personal fortune. **1961:** Left-wing Sandinista National Liberation Front (FSLN) formed to fight the Somoza regime. **1978:** The Nicaraguan Revolution: Pedro Joaquin Chamorro, a popular publisher and leader of the anti-Somoza Democratic Liberation Union (UDEL), was assassinated, sparking a general strike and mass movement in which moderates joined with the FSLN to overthrow the Somoza regime. **1979:** The Somoza government was ousted by the FSLN after a military offensive. **1980:** A FSLN junta took power in Managua, headed by Daniel Ortega Saavedra; lands held by Somozas were nationalized and farming cooperatives established. **1982:** There was subversive activity against the government by right-wing Contra guerrillas, promoted by the USA, attacking from bases in Honduras. A state of emergency was declared. **1984:** US troops mined Nicaraguan harbours. The action was condemned by the World Court in 1986 and $17 billion in reparations ordered. FSLN won the assembly elections. **1985:** The US president Ronald Reagan denounced the Sandinista government, vowing to 'remove it', and imposed a US trade embargo. **1987:** A Central American peace agreement was cosigned by Nicaraguan leaders. **1988:** The peace agreement failed. Nicaragua held talks with the Contra rebel leaders. A hurricane left 180,000 people homeless. **1989:** Demobilization of rebels and release of former Somozan supporters; the ceasefire ended but the economy was in ruins after the Contra war; there was 60% unemployment. **1990:** The FSLN was defeated by right-of-centre National Opposition Union (UNO), a US-backed coalition; Violeta Barrios de Chamorro, widow of the murdered Pedro Joaquin Chamorro, was elected president. There were antigovernment riots. **1992:** Around 16,000 people were made homeless by an earthquake. **1994:** A peace accord was made with the remaining Contra rebels. **1996:** Right-wing candidate Arnoldo Aleman won the presidential elections. **1998:** Daniel Ortega was re-elected FSLN leader.

Practical information

Visa requirements: UK: visa not required for a stay of up to 90 days. USA: visa not required for a stay of up to 90 days
Time difference: GMT –6
Chief tourist attractions: Lake Nicaragua, with its 310 beautiful islands; the Momotombo volcano; the Corn Islands (Islas de Maiz) in the Caribbean, fringed with white coral and palm trees
Major holidays: 1 January, 1 May, 19 July, 14–15 September, 8, 25 December; variable: Good Friday, Holy Thursday

NIGER

Landlocked country in northwest Africa, bounded north by Algeria and Libya, east by Chad, south by Nigeria and Benin, and west by Burkina Faso and Mali.

National name: *République du Niger/Republic of Niger*
Area: 1,186,408 sq km/458,072 sq mi
Capital: Niamey
Major towns/cities: Zinder, Maradi, Tahoua, Agadez, Birui N'Konui, Arlit
Physical features: desert plains between hills in north and savannah in south; River Niger in southwest, Lake Chad in southeast

Orange symbolizes the Sahara Desert. The orange disc represents the sun. Green recalls the country's forests. White stands for the River Niger. Effective date: 23 November 1959.

Government

Head of state: Tandja Mamadou from 1999
Head of government: Hama Amadou from 2000
Political system: military
Political executive: military
Administrative divisions: seven regions and the municipality of Niamey
Political parties: National Movement for a Development Society (MNSD–Nassara), left of centre; Alliance of the Forces for Change (AFC), left-of-centre coalition; Party for Democracy and Socialism–Tarayya (PNDS–Tarayya), left of centre
Armed forces: 5,300; plus paramilitary forces of 5,400 (1998)
Conscription: conscription is selective for two years
Death penalty: retains the death penalty for ordinary crimes but can be considered abolitionist in practice (last known execution in 1976)
Defence spend: (% GDP) 1.5 (1998)
Education spend: (% GNP) 2.3 (1997)
Health spend: (% GDP) 3.5 (1997 est)

Economy and resources

Currency: franc CFA
GDP: (US$) 2.07 billion (1999)
Real GDP growth: (% change on previous year) 2 (1999)
GNP: (US$) 2 billion (1999)
GNP per capita (PPP): (US$) 727 (1999)
Consumer price inflation: –3.6% (1999)
Unemployment: N/A
Foreign debt: (US$) 1.65 billion (1998)
Major trading partners: France, South Korea, Germany, USA, Côte d'Ivoire, Greece, UK, Nigeria
Resources: uranium (one of world's leading producers), phosphates, gypsum, coal, cassiterite, tin, salt, gold; deposits of other minerals (including petroleum, iron ore, copper, lead, diamonds, and tungsten) have been confirmed
Industries: processing of agricultural products, textiles, furniture, chemicals, brewing, cement
Exports: uranium ore, live animals, hides and skins, cow-peas, cotton. Principal market: France 51.9% (1998 est)
Imports: machinery and transport equipment, miscellaneous manufactured articles, cereals, chemicals, refined petroleum products. Principal source: France 17% (1999 est)

Arable land: 3.9% (1996)
Agricultural products: millet, maize, sorghum, groundnuts, cassava, sugar cane, sweet potatoes, cotton; livestock rearing (cattle and sheep) is especially important among the nomadic population; agricultural production is dependent upon adequate rainfall

Population and society

Population: 10,730,000 (2000 est)
Population growth rate: 3.2% (1995–2000)
Population density: (per sq km) 9 (1999 est)
Urban population: (% of total) 21 (2000 est)
Age distribution: (% of total population) 0–14 48%, 15–59 48%, 60+ 4% (2000 est)
Ethnic groups: two ethnic groups make up over 75% of the population: the Hausa (mainly in central areas and the south), and the Djerma-Songhai (southwest); the other principal ethnic groups are the Fulani, Tuareg, and Beriberi-Manga
Language: French (official), Hausa (70%), Djerma, other ethnic languages
Religion: Sunni Muslim 95%; also Christian, and traditional animist beliefs
Education: (compulsory years) 8
Literacy rate: 24% (men); 8% (women) (2000 est)
Labour force: 88.5% agriculture, 4.2% industry, 7.3% services (1997 est)
Life expectancy: 47 (men); 50 (women) (1995–2000)
Child mortality rate: (under 5, per 1,000 live births) 190 (1995–2000)
Physicians: 1 per 53,986 people (1993 est)
TV sets: (per 1,000 people) 26 (1997)
Radios: (per 1,000 people) 69 (1997)
Internet users: (per 10,000 people) 2.9 (1999)

Transport

Airports: international airports: Niamey, Agadez; four major domestic airports; total passenger km: 223 million (1995)
Railways: none
Roads: total road network: 10,100 km/6,276 mi, of which 7.9% paved (1996 est); passenger cars: 3.8 per 1,000 people (1996 est)

Chronology

10th–13th centuries: Kanem-Bornu Empire flourished in southeast, near Lake Chad, spreading Islam from the 11th century. **15th century:** Tuareg sultanate of Agades dominant in the north. **17th century:** Songhai-speaking Djerma established an empire on Niger River. **18th century:** Powerful Gobir kingdom founded by Hausa people, who had migrated from the south in the 14th century. **late 18th–early 19th centuries:** Visited by European explorers, including the Scottish explorer, Mungo Park; Sultanate of Sokoto formed by Islamic revivalist Fulani, who had defeated the Hausa in a jihad (holy war). **1890s:** French conquered the region and ended the local slave trade. **1904:** Became part of French West Africa, although Tuareg resistance continued until 1922. **1946:** Became French overseas territory, with its own territorial assembly and representation in the French parliament. **1958:** Became an autonomous republic within the French community. **1960:** Achieved full independence; Hamani Diori of Niger Progressive Party (NPP) elected president, but maintained close ties with France. **1971:** Uranium production commenced. **1974:** Diori was ousted in an army coup; the military government launched a drive against corruption. **1977:** A cooperation agreement was signed with France. **1984:** There was a partial privatization of state firms due to further drought

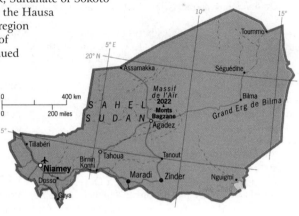

and increased government indebtedness as world uranium prices slumped. **1989:** Ali Saibu was elected president without opposition. **1991:** Saibu was stripped of executive powers, and a transitional government was formed amid student and industrial unrest. **1992:** The transitional government collapsed amid economic problems and ethnic unrest among secessionist Tuareg in the north. A referendum approved of a new multiparty constitution. **1993:** The Alliance of the Forces for Change (AFC), a left-of-centre coalition, won an absolute majority in assembly elections. Mahamane Ousmane, a Muslim Hausa, was elected president in the first free presidential election. **1994:** A peace agreement was signed with northern Tuareg. **1996:** President Ousmane was ousted in a military coup led by Ibrahim Barre Mainassara. Civilian government was restored with Boukary Adji as premier; Mainassara was formally elected president. **1997:** Ibrahim Hassane Mayaki was appointed prime minister. **1999:** President Mainassara was assassinated in a coup; Major Daouda Mallam Wanke, the commander of Niger's presidential guard, assumed power. In the elections which followed, Tandja Mamadou was elected president, and Hama Amadou was appointed prime minister.

Practical information

Visa requirements: UK: visa required. USA: visa required
Time difference: GMT +1
Chief tourist attractions: the Aïr and Ténéré Nature Reserve, covering 77,000 sq km/29,730 sq mi; Agadez, surrounded by green valleys and hot springs, and still a major terminus for trans-Saharan caravans
Major holidays: 1 January, 15 April, 1 May, 3 August, 18, 25 December; variable: Eid-ul-Adha, end of Ramadan, Prophet's Birthday

NIGERIA

Located in west Africa on the Gulf of Guinea, bounded north by Niger, east by Chad and Cameroon, and west by Benin.

National name: *Federal Republic of Nigeria*
Area: 923,773 sq km/356,668 sq mi
Capital: Abuja
Major towns/cities: Ibadan, Lagos, Ogbomosho, Kano, Oshogbo, Ilorin, Abeokuta, Zaria, Port Harcourt
Major ports: Lagos, Port Harcourt, Warri, Calabar
Physical features: arid savannah in north; tropical rainforest in south, with mangrove swamps along coast; River Niger forms wide delta; mountains in southeast

Green stands for Nigeria's forests and agriculture. White represents the River Niger, peace, and unity.
Effective date: 1 October 1960.

Government

Head of state and government: Olusegun Obasanjo from 1999
Political system: emergent democracy
Political executive: limited presidency
Administrative divisions: 30 states and a Federal Capital Territory
Political parties: political parties, suppressed by the military government, were allowed to form in July 1998. Three parties were registered: All People's Party (APP), right of centre; People's Democratic Party (PDP), left of centre; Alliance for Democracy (AD), left of centre
Armed forces: 77,000 (1998)
Conscription: military service is voluntary
Death penalty: retained and used for ordinary crimes
Defence spend: (% GDP) 4.3 (1998)
Education spend: (% GNP) 0.9 (1996)
Health spend: (% GDP) 3.1 (1997 est)

Economy and resources

Currency: naira
GDP: (US$) 43.1 billion (1999)
Real GDP growth: (% change on previous year) 2.5 (1999)
GNP: (US$) 37.9 billion (1999)
GNP per capita (PPP): (US$) 744 (1999)
Consumer price inflation: 6.7% (1999)
Unemployment: 4.5% (1997)
Foreign debt: (US$) 30.1 billion (1999 est)
Major trading partners: USA, UK, Germany, France, Spain, the Netherlands, Italy, India
Resources: petroleum, natural gas, coal, tin, iron ore, uranium, limestone, marble, forest
Industries: food processing, brewing, petroleum refinery, iron and steel, motor vehicles (using imported components), textiles, cigarettes, footwear, pharmaceuticals, pulp and paper, cement
Exports: petroleum, cocoa beans, rubber, palm products, urea and ammonia, fish. Principal market: USA 35.3% (1998 est)
Imports: machinery and transport equipment, basic manufactures, cereals, chemicals, foodstuffs. Principal source: UK 12.7% (1998 est)
Arable land: 33.3% (1996)

Agricultural products: cocoa, groundnuts, oil palm, rubber, rice, maize, taro, yams, cassava, sorghum, millet, plantains; livestock (principally goats, sheep, cattle, and poultry) and fisheries

Population and society

Population: 111,506,000 (2000 est)
Population growth rate: 2.4% (1995–2000)
Population density: (per sq km) 118 (1999 est)
Urban population: (% of total) 44 (2000 est)
Age distribution: (% of total population) 0–14 43%, 15–59 52%, 60+ 5% (2000 est)
Ethnic groups: over 250 tribal groups, ten of which account for over 80% of population: the Hausa-Fulani (in the north), Yoruba (in the south) Ibo (in the east), Tiv, Nupe, Kanuri, Ibibio, Ijaw, and Edo
Language: English, French (both official), Hausa, Ibo, Yoruba
Religion: Sunni Muslim 50% (in north), Christian 35% (in south), local religions 15%
Education: (compulsory years) 6
Literacy rate: 72% (men); 56% (women) (2000 est)
Labour force: 36.1% agriculture, 6.9% industry, 57% services (1997 est)
Life expectancy: 49 (men); 52 (women) (1995–2000)
Child mortality rate: (under 5, per 1,000 live births) 147 (1995–2000)
Physicians: 1 per 5,208 people (1993 est)
Hospital beds: 1 per 157 people (1993 est)
TV sets: (per 1,000 people) 61 (1997)
Radios: (per 1,000 people) 223 (1997)
Internet users: (per 10,000 people) 9.2 (1999)
Personal computer users: (per 100 people) 0.6 (1999)

Transport

Airports: international airports: Lagos (Murtala Mohammed), Kano, Abuja, Port Harcourt, Calabar; 14 domestic airports; total passenger km: 819 million (1995)
Railways: total length: 3,557 km/2,210 mi; total passenger km: 161 million (1995)
Roads: total road network: 193,198 km/120,053 mi, of which 19% paved (1996); passenger cars: 5.9 per 1,000 people (1996 est)

Chronology

4th century BC–2nd century AD: Highly organized Nok culture flourished in northern Nigeria. **9th century:** Northeast Nigeria became part of empire of Kanem-Bornu, based around Lake Chad. **11th century:** Creation of Hausa states, including Kano and Katsina. **13th century:** Arab merchants introduced Islam in the north. **15th century:** Empire of Benin at its height in south; first contact with European traders. **17th century:** Oyo Empire dominant in southwest; development of slave trade in Niger delta. **1804–17:** Islamic Fulani (or Sokoto) Empire established in north. **1861:** British traders procured Lagos; spread of Christian missionary activity in south. **1884–1904:** Britain occupied most of Nigeria by stages. **1914:** North and south protectorates united; growth of railway network and trade. **1946:** Nigerians allowed a limited role in decision-making in three regional councils. **1951:** The introduction of elected representation led to the formation of three regional political parties. **1954:** New constitution increased powers of the regions. **1958:** Oil discovered in the southeast. **1960:** Achieved independence from Britain, within the Commonwealth. **1963:** Became a republic, with Nnamdi Azikiwe as president. **1966:** Gen Aguiyi-Ironsi of Ibo tribe seized power and imposed unitary government; massacre of Ibo by Hausa in north; Gen Gowon seized power and restored federalism. **1967:** Conflict over oil revenues led to secession of eastern region as independent Ibo state of Biafra; ensuing civil war claimed up to a million lives. **1970:** Surrender of Biafra and end of civil war; development of the oil

industry financed more effective central government. **1975:** Gowon ousted in military coup; second coup put Gen Olusegun Obasanjo in power. **1979:** Civilian rule restored under President Shehu Shagari. **1983:** A bloodless coup was staged by Maj-Gen Muhammadu Buhari. **1985:** Buhari was replaced by Maj-Gen Ibrahim Babangida; Islamic northerners were dominant in the regime. **1992:** Multiparty elections were won by Babangida's SDP. **1993:** Moshood Abiola (SDP) won the first free presidential election; the results were suspended. Gen Sani Abacha restored military rule and dissolved political parties. **1995:** Commonwealth membership was suspended in protest at human-rights abuses by the military regime. **1998:** General Abdulsalam Abubakar took over as president. Nigeria's most prominent political prisoner, Moshood Abiola, died suddenly on the eve of his expected release. There were moves towards political liberalization, with the formation of new political parties and the release of some dissidents. **1999:** The People's Democratic Party won a Senate majority. Olusegun Obasanjo was elected president. Nigeria rejoined the Commonwealth. **2000:** Throughout the year, violent clashes between Christians and Muslims accompanied the adoption of Islamic law (sharia) in a number of states throughout Nigeria. Ethnic violence erupted between the militant Yoruba separatists' group Odua People's Congress (OPC) and the Hausas in October. The OPC was outlawed.

Practical information

Visa requirements: UK: visa required. USA: visa required
Time difference: GMT +1
Chief tourist attractions: fine coastal scenery; dense forests; rich diversity of arts
Major holidays: 1 January, 1 May, 1 October, 25–26 December; variable: Eid-ul-Adha (2 days), end of Ramadan (2 days), Good Friday, Easter Monday, Prophet's Birthday

North Korea

Located in East Asia, bounded northeast by Russia, north and northwest by China, east by the Sea of Japan, south by South Korea, and west by the Yellow Sea.

National name: *Chosun Minchu-chui Inmin Konghwa-guk/Democratic People's Republic of Korea*
Area: 120,538 sq km/46,539 sq mi
Capital: Pyongyang
Major towns/cities: Hamhung, Chongjin, Nampo, Wonsan, Sinuiji
Physical features: wide coastal plain in west rising to mountains cut by deep valleys in interior

White stands for purity. Red represents communist revolution. Blue expresses the desire for peace.
Effective date: 8 September 1948.

Government

Head of state: Kim Jong Il from 1994
Head of government: Hong Song Nam from 1997
Political system: communist
Political executive: communist
Administrative divisions: nine provinces and three cities
Political parties: Korean Workers' Party (KWP), Marxist-Leninist (leads Democratic Front for the Reunification of the Fatherland, including Korean Social Democratic Party and Chondoist Chongu Party)
Armed forces: 1,055,000; paramilitary forces of 189,000 (1998)
Conscription: conscription is selective for 3–10 years
Death penalty: retained and used for ordinary crimes
Defence spend: (% GDP) 14.3% (1998)

Economy and resources

Currency: won
GDP: (US$) 22.6 billion (1999 est)
Real GDP growth: (% change on previous year) 1 (1999 est)
GNP: (US$) N/A
GNP per capita (PPP): (US$) 950 (1999 est)
Consumer price inflation: N/A
Foreign debt: (US$) 7.5 billion (1997)
Major trading partners: China, Japan, Russia, South Korea, Germany, Italy, Iran
Resources: coal, iron, lead, copper, zinc, tin, silver, gold, magnesite (has 40–50% of world's deposits of magnesite)
Industries: mining, metallurgy, electricity, machine-building, textiles, cement, chemicals, cotton, silk and rayon weaving, foods
Exports: base metals, textiles, vegetable products, machinery and equipment. Principal market: Japan 27.9% (1995 est)
Imports: petroleum and petroleum products, machinery and equipment, grain, coal, foodstuffs.

Principal source: China 32.6% (1995 est)
Arable land: 14.1% (1996)
Agricultural products: rice, maize, sweet potatoes, soybeans; livestock rearing (cattle and pigs); forestry; fishing

Population and society

Population: 24,039,000 (2000 est)
Population growth rate: 1.6% (1995–2000); 1.3% (2000–05)
Population density: (per sq km) 197 (1999 est)
Urban population: (% of total) 60 (2000 est)
Age distribution: (% of total population) 0–14 28%, 15–59 64%, 60+ 8% (2000 est)
Ethnic groups: entirely Korean, with the exception of a 50,000 Chinese minority
Language: Korean (official)
Religion: Buddhist (predominant religion), Chondoist, Christian, traditional beliefs
Education: (compulsory years) 10
Literacy rate: 99% (men); 99% (women) (1997 est)
Labour force: 50% of population: 32.4% agriculture, 32% industry, 35.6% services (1997 est)
Life expectancy: 69 (men); 75 (women) (1995–2000)
Child mortality rate: (under 5, per 1,000 live births) 26 (1995–2000)
Physicians: 1 per 370 people (1993)
TV sets: (per 1,000 people) 52 (1997)
Radios: (per 1,000 people) 146 (1997)

Transport

Airports: international airport: Pyongyang (Sunan); two domestic airports (which foreigners are not allowed to use); total passenger km: 207 million (1995)
Railways: total length: 8,000 km/4,971 mi; total passenger km: 3,400 million (1993)
Roads: total road network: 31,200 km/19,388 mi, of which 6.4% paved (1996 est)

Chronology

2333 BC: Legendary founding of Korean state by Tangun dynasty.
1122 BC–4th century AD: Period of Chinese Kija dynasty.
668–1000: Peninsula unified by Buddhist Shilla kingdom, with capital at Kyongju. **1392–1910:** Period of Chosun, or Yi, dynasty, during which Korea became a vassal of China and Confucianism became the dominant intellectual force. **1910:** Korea formally annexed by Japan. **1920s and 1930s:** Heavy industries developed in the coal-rich north, with Koreans forcibly conscripted as low-paid labourers; suppression of Korean culture led to the development of a resistance movement. **1945:** Russian and US troops entered Korea at the end of World War II, forced surrender of Japanese, and divided the country in two at the 38th parallel. Soviet troops occupied North Korea. **1946:** Soviet-backed provisional government installed, dominated by Moscow-trained Korean communists, including Kim Il Sung; radical programme of land reform and nationalization launched. **1948:** Democratic People's Republic of Korea declared after pro-USA Republic of Korea founded in the south; Soviet troops withdrew. **1950:** North Korea invaded South Korea to unite the nation, beginning the Korean War. **1953:** Armistice agreed to end the Korean War, which had involved US participation on the side of South Korea, and Chinese on that of North Korea. The war ended in stalemate, at a cost of 2 million lives. **1961:** Friendship and mutual assistance treaty signed with China. **1972:** A new constitution, with an executive president, was adopted. Talks were held with South Korea about possible reunification.

1983: Four South Korean cabinet ministers were assassinated in Rangoon, Burma (Myanmar), by North Korean army officers. **1985:** Relations improved with the Soviet Union. **1990:** Diplomatic contacts with South Korea and Japan suggested a thaw in North Korea's relations with the rest of the world. **1991:** North Korea became a member of the United Nations (UN). A nonaggression agreement with South Korea was signed. **1992:** The Nuclear Safeguards Agreement was signed, allowing international inspection of nuclear facilities. A pact was also signed with South Korea for mutual inspection of nuclear facilities. **1994:** Kim Il Sung died and was succeeded by his son, Kim Jong Il. An agreement was made to halt the nuclear-development programme in return for US aid, resulting in the easing of a 44-year-old US trade embargo. **1996:** US aid was sought in the face of a severe famine caused by floods; rice was imported from South Korea and food aid provided by the UN. **1997:** Kang Song San was replaced as prime minister by Hong Song Nam. Grave food shortages were revealed. **1998:** A UN food-aid operation was instituted in an effort to avert widespread famine. A ballistic missile test was fired over Japan. Deceased former leader Kim Il Sung was declared 'president for perpetuity'. Relations with the USA deteriorated when the USA demanded access to an underground site in Kumchangri suspected of being part of a nuclear-weapons program. **1999:** Japan lifted sanctions, and the USA eased sanctions, against North Korea. **2000:** North Korea forged diplomatic links with Japan, the USA, Italy, and the UK. At a first summit meeting between Kim Jong Il and Kim Dae Jung of South Korea, the two leaders agreed to South Korean economic investment in North Korea, and rail links between the two countries.

Practical information

Visa requirements: UK: visa required. USA: visa required
Time difference: GMT +9

Chief tourist attractions: tourism is permitted only in officially accompanied parties
Major holidays: 1 January, 16 February, 8 March, 15 April, 9 September, 10 October, 27 December

NORWAY

Located in northwest Europe, on the Scandinavian peninsula, bounded east by Sweden, northeast by Finland and Russia, south by the North Sea, west by the Atlantic Ocean, and north by the Arctic Ocean.

National name: *Kongeriket Norge/Kingdom of Norway*
Area: 387,000 sq km/149,420 sq mi (including Svalbard and Jan Mayen)
Capital: Oslo
Major towns/cities: Bergen, Trondheim, Stavanger, Kristiansund, Drammen
Physical features: mountainous with fertile valleys and deeply indented coast; forests cover 25%; extends north of Arctic Circle
Territories: dependencies in the Arctic (Svalbard and Jan Mayen) and in Antarctica (Bouvet and Peter I Island, and Queen Maud Land)

Blue is taken from the Swedish arms. Red and white recall the Danish flag, known as the Dannebrog. Effective date: 15 December 1899.

Government

Head of state: King Harald V from 1991
Head of government: Jens Stoltenberg from 2000
Political system: liberal democracy
Political executive: parliamentary
Administrative divisions: 19 counties
Political parties: Norwegian Labour Party (DNA), moderate left of centre; Conservative Party, progressive, right of centre; Christian People's Party (KrF), Christian, centre left; Centre Party (Sp), left of centre, rural-oriented; Progress Party (FrP), right wing, populist
Armed forces: 28,900; 234,000 reservists (1998)
Conscription: 12 months, with 4–5 refresher training periods
Death penalty: abolished in 1979
Defence spend: (% GDP) 2.2 (1998)
Education spend: (% GNP) 7.5 (1996)
Health spend: (% GDP) 7.5 (1997)

Economy and resources

Currency: Norwegian krone
GDP: (US$) 152.9 billion (1999)
Real GDP growth: (% change on previous year) 0.9 (1999)
GNP: (US$) 146.4 billion (1999)
GNP per capita (PPP): (US$) 26,522 (1999)
Consumer price inflation: 2.3% (1999)
Unemployment: 3.2% (1998)
Major trading partners: UK, Sweden, Germany, the Netherlands, USA, Denmark
Resources: petroleum, natural gas, iron ore, iron pyrites, copper, lead, zinc, forests
Industries: mining, fishery, food processing, non-electrical machinery, metals and metal products, paper products, printing and publishing, shipbuilding, chemicals
Exports: petroleum, natural gas, fish products, non-ferrous metals, wood pulp and paper. Principal market: UK 17.1% (1999 est)
Imports: machinery and transport equipment, chemicals, clothing, fuels and lubricants, iron and steel, office machines and computers, telecommunications and sound apparatus and equipment. Principal source: Sweden

14.7% (1999 est)
Arable land: 3.3% (1996)
Agricultural products: wheat, barley, oats, potatoes, fruit; fishing industry, including fish farming

Population and society

Population: 4,465,000 (2000 est)
Population growth rate: 0.5% (1995–2000)
Population density: (per sq km) 14 (1999 est)
Urban population: (% of total) 76 (2000 est)
Age distribution: (% of total population) 0–14 20%, 15–59 60%, 60+ 20% (2000 est)
Ethnic groups: majority of Nordic descent; Saami minority in far north (approximately 30,000)
Language: Norwegian (official), Saami (Lapp), Finnish
Religion: Evangelical Lutheran (endowed by state) 88%; other Protestant and Roman Catholic 4%
Education: (compulsory years) 9
Literacy rate: 99% (men); 99% (women) (2000 est)
Labour force: 4.8% agriculture, 23.7% industry, 71.6% services (1997)
Life expectancy: 75 (men); 81 (women) (1995–2000)
Child mortality rate: (under 5, per 1,000 live births) 6 (1995–2000)
Physicians: 1 per 338 people (1996)
Hospital beds: 1 per 75 people (1996)
TV sets: (per 1,000 people) 579 (1997)
Radios: (per 1,000 people) 915 (1997)
Internet users: (per 10,000 people) 4,499.1 (1999)
Personal computer users: (per 100 people) 45.0 (1999)

Transport

Airports: international airports: Oslo (Fornebu), Stavanger (Sola), Bergen (Flesland); 54 domestic airports with scheduled services; total passenger km: 14,646 million (1996)
Railways: total length: 4,023 km/2,500 mi; total passenger km: 2,561 million (1997)
Roads: total road network: 91,180 km/56,659 mi, of which 74% paved (1997); passenger cars: 398 per 1,000 people (1997)

Chronology

5th century: First small kingdoms established by Goths. **c. 900:** Harald Fairhair created united Norwegian kingdom; it dissolved after his death. **8th–11th centuries:** Vikings from Norway raided and settled in many parts of Europe. **c. 1016–28:** Olav II (St Olav) reunited the kingdom and introduced Christianity. **1217–63:** Haakon VI established royal authority over nobles and church and made the monarchy hereditary. **1263:** Iceland submitted to the authority of the king of Norway. **1397:** Union of Kalmar: Norway, Denmark, and Sweden united under a single monarch. **15th century:** Norway, the weakest of the three kingdoms, was increasingly treated as an appendage of Denmark. **1523:** Secession of Sweden further undermined Norway's status. **16th century:** Introduction of the sawmill precipitated the development of the timber industry and the growth of export trade. **1661:** Denmark restored formal equality of status to Norway as a twin kingdom. **18th century:** Norwegian merchants profited from foreign wars which increased demand for naval supplies. **1814:** Treaty of Kiel: Denmark ceded Norway (minus Iceland) to Sweden; Norway retained its own parliament but cabinet appointed by the king of Sweden. **19th century:** Economic decline followed slump in timber trade due to Canadian competition; expansion of merchant navy and whaling industry. **1837:** Democratic local government introduced. **1884:** Achieved internal self-government when the king of Sweden made the Norwegian cabinet accountable to the Norwegian parliament. **1895:** Start of constitutional dispute over control of foreign policy: Norway's demand for a separate consular service refused by Sweden. **1905:** Union with Sweden dissolved; Norway achieved independence under King Haakon VII. **1907:** Norway became first the European country to grant women the right to vote in parliamentary elections. **early 20th century:** Development of industry based on hydroelectric power; long period of Liberal government committed to neutrality and moderate social reform. **1940–45:** German occupation with Vidkun Quisling as puppet leader. **1945–65:** Labour governments introduced economic planning and permanent price controls. **1949:** Became a founding member of the North Atlantic Treaty Organization (NATO). **1952:** Joined the Nordic Council. **1957:** Olaf V succeeded his father King Haakon VII. **1960:** Joined European Free Trade Association (EFTA). **1972:** A national referendum rejected membership of European Economic Community (EEC). **1975:** The export of North Sea oil began. **1981:** Gro Harlem Brundtland (Labour) became Norway's first woman prime minister. **1986:** Falling oil prices caused a recession. **1991:** Olaf V was succeeded by his son Harald V. **1994:** A national referendum rejected membership of European Union (EU). **1997:** Kjell Magne Bondevik (KrF) became prime minister. **1998:** There was a decline in the state of the economy. **2000:** Bondevik resigned as prime minister and was succeeded by Jens Stoltenberg.

Practical information

Visa requirements: UK: visa not required. USA: visa not required
Time difference: GMT +1
Chief tourist attractions: rugged landscape with fjords, forests, lakes, and rivers; winter sports
Major holidays: 1 January, 1, 17 May, 25–26 December; variable: Ascension Thursday, Good Friday, Easter Monday, Holy Thursday, Whit Monday

OMAN

Located at the southeastern end of the Arabian peninsula, bounded west by the United Arab Emirates, Saudi Arabia, and Yemen, southeast by the Arabian Sea, and northeast by the Gulf of Oman.

National name: *Saltanat `Uman/Sultanate of Oman*
Area: 272,000 sq km/105,019 sq mi
Capital: Muscat
Major towns/cities: Salalah, Ibra, Suhar, Al-Buraimi, Nazwa, Sur, Matrah
Major ports: Mina Qaboos, Mina Raysut
Physical features: mountains to the north and south of a high arid plateau; fertile coastal strip; Jebel Akhdar highlands; Kuria Muria Islands

The central band was widened in 1995. Red recalls the previous flag of the Kharijite Muslims. Effective date: 18 November

Government

Head of state and government: Qaboos bin Said from 1970
Political system: absolutist
Political executive: absolute
Administrative divisions: eight regional governates and 59 districts
Political parties: none
Armed forces: 43,500 (1998)
Conscription: military service is voluntary
Death penalty: retained and used for ordinary crimes
Defence spend: (% GDP) 13.6 (1998)
Education spend: (% GNP) 4.6 (1995)
Health spend: (% GDP) 3.9 (1997)

Economy and resources

Currency: Omani rial
GDP: (US$) 16.8 billion (1999 est)
Real GDP growth: (% change on previous year) 4 (1999 est)
GNP: (US$) 10.6 billion (1997)
GNP per capita (PPP): (US$) 8,690 (1997)
Consumer price inflation: –0.1% (1999 est)
Unemployment: 11.9% (1993)
Foreign debt: (US$) 4.8 billion (1998 est)
Major trading partners: United Arab Emirates, Japan, South Korea, China
Resources: petroleum, natural gas, copper, chromite, gold, salt, marble, gypsum, limestone
Industries: mining, petroleum refining, cement, construction materials, copper smelting, food processing, chemicals, textiles
Exports: petroleum, metals and metal goods, textiles, animals and products. Principal market: Japan 23.3% (1997)
Imports: machinery and transport equipment, basic manufactures, food and live animals, beverages, tobacco. Principal source: United Arab Emirates 24.2% (1997)
Arable land: 0.1% (1996)
Agricultural products: dates, tomatoes, limes, alfalfa,

mangoes, melons, bananas, coconuts, cucumbers, onions, peppers, frankincense (agricultural production is mainly at subsistence level); livestock; fishing

Population and society

Population: 2,542,000 (2000 est)
Population growth rate: 3.3% (1995–2000)
Population density: (per sq km) 9 (1999 est)
Urban population: (% of total) 84 (2000 est)
Age distribution: (% of total population) 0–14 44%, 15–59 52%, 60+ 4% (2000 est)
Ethnic groups: predominantly Arab, with substantial Iranian, Baluchi, Indo-Pakistani, and East African minorities
Language: Arabic (official), English, Urdu, other Indian languages
Religion: Muslim 75% (predominantly Ibadhi Muslim), about 25% Hindu
Education: not compulsory
Literacy rate: 80% (men); 62% (women) (2000 est)
Labour force: 9.4% agriculture, 27.8% industry, 62.8% services (1993)
Life expectancy: 69 (men); 73 (women) (1995–2000)
Child mortality rate: (under 5, per 1,000 live births) 30 (1995–2000)
Physicians: 1 per 1,265 people (1994)
Hospital beds: 1 per 442 people (1994)
TV sets: (per 1,000 people) 602 (1997)
Radios: (per 1,000 people) 582 (1996)
Internet users: (per 10,000 people) 203.2 (1999)
Personal computer users: (per 100 people) 2.6 (1999)

Transport

Airports: international airports: Muscat (Seeb), Salalah; domestic services operate between these; total passenger km: 3,226 million (1995)
Railways: none
Roads: total road network: 32,800 km/20,382 mi of which 30% paved (1996 est); passenger cars: 152.4 per 1,000 people (1996 est)

Chronology

c. **3000 BC:** Archaeological evidence suggests Oman may have been the semilegendary Magan, a thriving seafaring state at the time of the Sumerian Civilization of Mesopotamia (the Tigris and Euphrates region of Iraq). **9th century BC:** Migration of Arab clans to Oman, notably the Qahtan family from southwest Arabia and the Nizar from northwest Arabia, between whom rivalry has continued. **4th century BC–AD 800:** North Oman under Persian control. **AD 630:** Converted to Islam. **751:** Julanda ibn Masud was elected imam (spiritual leader); Oman remained under imam rule until 1154. **1151:** Dynasty established by Banu Nabhan. **1428:** Dynastic rule came under challenge from the imams. **1507:** Coastal area, including port city of Muscat, fell under Portuguese control. **1650:** Portuguese ousted by Sultan ibn Sayf, a powerful Ya'ariba leader. **early 18th century:** Civil war between the Hinawis (descendents of the Qahtan) and the Ghafiris (descendents of the Nizar). **1749:** Independent Sultanate of Muscat and Oman established by Ahmad ibn Said, founder of the Al Bu Said dynasty that still rules Oman. **first half of 19th century:** Muscat and Oman was the most powerful state in Arabia, ruling Zanzibar until 1861, and coastal parts of Persia, Kenya, and Pakistan; came under British protection. **1951:** The Sultanate of Muscat and Oman achieved full independence from Britain. Treaty of Friendship with Britain signed. **1964:** Discovery of oil led to the transformation of the undeveloped kingdom into a modern state. **1970:** After 38 years' rule, Sultan Said bin Taimur was replaced in a bloodless coup by his son Qaboos bin Said. Name was changed to the Sultanate of Oman and a modernization programme was launched. **1975:** Left-wing rebels in Dhofar in the south, who had been supported by South Yemen, defeated with UK military assistance, ending a ten-year insurrection. **1981:** The Consultative Council was set up; Oman played a key role in the establishment of a six-member Gulf Cooperation Council. **1982:** The Memorandum of Understanding with the UK was signed, providing for regular consultation on international issues. **1991:** Joined the US-led coalition opposing Iraq's occupation of Kuwait.

Practical information

Visa requirements: UK: visa required. USA: visa required

Time difference: GMT +4

Chief tourist attractions: the old towns of Muscat, Nazwa (ancient capital of the interior), and Dhofar; the forts of Nakhl, Ar Rustaq, and Al-Hazm – tourism was introduced in 1985 and is strictly controlled

Major holidays: 18 November (2 days), 31 December; variable: Eid-ul-Adha (5 days), end of Ramadan (4 days), New Year (Muslim), Prophet's Birthday, Lailat al-Miraj (March/April)

Pakistan

Located in southern Asia, stretching from the Himalayas to the Arabian Sea, bounded to the west by Iran, northwest by Afghanistan, and northeast and east by India.

Green represents Islam. The combination of green and white symbolizes peace and prosperity. Effective date: 14 August 1947.

National name: *Islami Jamhuriyya e Pakistan/Islamic Republic of Pakistan*
Area: 803,940 sq km/310,321 sq mi
Capital: Islamabad
Major towns/cities: Lahore, Rawalpindi, Faisalabad, Karachi, Hyderabad, Multan, Peshawar, Gujranwala, Quetta
Major ports: Karachi, Port Qasim
Physical features: fertile Indus plain in east, Baluchistan plateau in west, mountains in north and northwest; the 'five rivers' (Indus, Jhelum, Chenab, Ravi, and Sutlej) feed the world's largest irrigation system; K2 mountain; Khyber Pass

Government

Head of state: Muhammad Rafiq Tarar from 1998
Head of government: Gen Pervez Musharraf from 1999
Political system: military
Political executive: military
Administrative divisions: four provinces, the Federal Capital Territory, and the federally administered tribal areas
Political parties: Islamic Democratic Alliance (IDA), conservative; Pakistan People's Party (PPP), moderate, Islamic, socialist; Pakistan Muslim League (PML), Islamic conservative (contains pro- and anti-government factions); Pakistan Islamic Front (PIF), Islamic fundamentalist, right wing; Awami National Party (ANP), left wing; National Democratic Alliance (NDA) bloc, left of centre; Mohajir National Movement (MQM), Sind-based *mohajir* settlers (Muslims previously living in India); Movement for Justice, reformative, anti-corruption
Armed forces: 587,000; paramilitary forces 247,000 (1998)
Conscription: military service is voluntary
Death penalty: retained and used for ordinary crimes
Defence spend: (% GDP) 6.5 (1998)
Education spend: (% GNP) 2.7 (1997)
Health spend: (% GDP) 4 (1997 est)

Economy and resources

Currency: Pakistan rupee
GDP: (US$) 64 billion (1999)
Real GDP growth: (% change on previous year) 2.7 (1999)
GNP: (US$) 63.9 billion (1999)
GNP per capita (PPP): (US$) 1,757 (1999)
Consumer price inflation: 4.1% (1999)
Unemployment: 5.4% (1996)
Foreign debt: (US$) 35.6 billion (1999 est)
Major trading partners: Japan, USA, Hong Kong, Malaysia, Germany, UK, Saudi Arabia, United Arab Emirates
Resources: iron ore, natural gas, limestone, rock salt, gypsum, silica, coal, petroleum, graphite, copper, manganese, chromite
Industries: textiles (principally cotton), food processing, petroleum refining, leather production, soda ash, sulphuric acid, bicycles
Exports: cotton, textiles, petroleum and petroleum products, clothing and accessories, leather, rice, food and live animals. Principal market: USA 21.4% (1999)
Imports: machinery and transport equipment, mineral

fuels and lubricants, chemicals and related products, edible oil. Principal source: USA 9.8%, Japan 8.0% (1999)
Arable land: 27.3% (1996)
Agricultural products: cotton, rice, wheat, maize, sugar cane

Population and society

Population: 156,483,000 (2000 est)
Population growth rate: 2.8% (1995–2000)
Population density: (per sq km) 189 (1999 est)
Urban population: (% of total) 37 (2000 est)
Age distribution: (% of total population) 0–14 42%, 15–59 53%, 60+ 5% (2000 est)
Ethnic groups: four principal, regionally based, antagonistic communities: Punjabis in the Punjab; Sindhis in Sind; Baluchis in Baluchistan; and the Pathans (Pushtans) in the Northwest Frontier Province
Language: Urdu (official), English, Punjabi, Sindhi, Pashto, Baluchi, other local dialects
Religion: Sunni Muslim 90%, Shiite Muslim 5%; also Hindu, Christian, Parsee, Buddhist
Education: (years) 5–12 (not compulsory, but free)
Literacy rate: 60% (men); 31% (women) (2000 est)
Labour force: 48% agriculture, 18% industry, 34% services (1997 est)
Life expectancy: 63 (men); 65 (women) (1995–2000)
Child mortality rate: (under 5, per 1,000 live births) 106 (1995–2000)
Physicians: 1 per 1,856 people (1996)
Hospital beds: 1 per 1,660 people (1996)
TV sets: (per 1,000 people) 88 (1998)
Radios: (per 1,000 people) 98 (1997)
Internet users: (per 10,000 people) 6.0 (1999)
Personal computer users: (per 100 people) 0.4 (1999)

Transport

Airports: international airports: Karachi (Civil), Lahore, Islamabad, Peshawar, Quetta, Rawalpindi; 30 domestic airports; total passenger km: 10,940 million (1997 est)
Railways: total length: 8,163 km/5,072 mi; total passenger km: 20,476 million (1997 est)
Roads: total road network: 229,934 km/142,881 mi, of which 58% paved (1997); passenger cars: 5.1 per 1,000 people (1997)

Chronology

2500–1600 BC: The area was the site of the Indus Valley civilization, a sophisticated, city-based ancient culture. **327 BC:** Invaded by Alexander the Great of Macedonia. **1st–2nd centuries:** North Pakistan was the heartland of the Kusana Empire, formed by invaders from Central Asia. **8th century:** First Muslim conquests, in Baluchistan and Sind, followed by increasing immigration by Muslims from the west, from the 10th century. **1206:** Establishment of Delhi Sultanate, stretching from northwest Pakistan and across northern India. **16th century:** Sikh religion developed in Punjab. **16th–17th centuries:** Lahore served intermittently as a capital city for the Mogul Empire, which stretched across the northern half of the Indian subcontinent. **1843–49:** Sind and Punjab annexed by British and incorporated within empire of 'British India'. **late 19th century:** Major canal irrigation projects in West Punjab and the northern Indus Valley drew in settlers from the east, as wheat and cotton production expanded. **1933:** The name 'Pakistan' (Urdu for 'Pure Nation') invented by Choudhary Rahmat Ali, as Muslims within British India began to campaign for the establishment of an independent Muslim territory that would embrace the four provinces of Sind, Baluchistan, Punjab, and the Northwest Frontier. **1940:** The All-India Muslim League (established in 1906), led by Karachi-born Muhammad Ali Jinnah, endorsed the concept of a separate nation for Muslims in the Lahore Resolution. **1947:** Independence achieved from Britain, as a dominion within the Commonwealth. Pakistan, which included East Bengal, a Muslim-dominated province more than 1,600 km/1,000 mi from Punjab, was formed following the partition of British India. Large-scale and violent cross-border migrations of Muslims, Hindus, and Sikhs followed, and a brief border war with India over disputed Kashmir. **1956:** Proclaimed a republic. **1958:** Military rule imposed by Gen Ayub Khan. **1965:** Border war with India over disputed territory of Kashmir. **1970:** A general election produced a clear majority in East Pakistan for the pro-autonomy Awami League, and in West Pakistan for Islamic socialist Pakistan People's Party (PPP), led by Zulfiqar Ali Bhutto. **1971:** East Pakistan secured independence, as Bangladesh, following a civil war in which it received decisive military support from India. Power was transferred from the military to the populist Bhutto in Pakistan. **1977:** Bhutto overthrown in military coup by Gen Zia ul-Haq following months of civil unrest; martial law imposed. **1979:** Bhutto executed for alleged murder; tight political restrictions imposed by Zia regime. **1980:** 3 million refugees fled to the Northwest Frontier Province and Baluchistan as a result of the Soviet invasion of Afghanistan. **1981:** The broad-based Opposition Movement for the Restoration of Democracy was formed. **1985:** Martial law and the ban on political parties was lifted. **1986:** Agitation for free elections was launched by Benazir Bhutto, the daughter of Zulfiqar Ali Bhutto. **1988:** An Islamic legal code, the Shari'a, was introduced; Zia was killed in a plane crash. Benazir Bhutto became prime minister after the (now centrist) PPP won the general election. **1989:** Tension with India was increased by outbreaks of civil war in Kashmir. Pakistan rejoined the Commonwealth, which it had left in 1972. **1990:** Bhutto was dismissed as prime minister by President Ghulam Ishaq Khan on charges of incompetence and corruption. The conservative Islamic Democratic Alliance (IDA), led by Nawaz Sharif, won the general election. **1993:** Khan and Sharif resigned. Benazir Bhutto and PPP were re-elected. Farooq Leghari (PPP) was elected president. **1996:** Benazir Bhutto was dismissed amid allegations of corruption. **1997:** The right-of-centre Pakistan Muslim League won the general election, returning Nawaz Sharif to power as prime minister. President Leghari resigned. **1998:** Rafiq Tarar became president. **1999:** Benazir Bhutto and her husband were found guilty of corruption, sentenced to five years in prison, and fined £5.3 million. India agreed to enter peace talks on Kashmir. Pakistan's army overthrew the government after Sharif tried to sack Gen Pervez Musharraf from the top military job. Musharraf, who appointed himself the country's chief executive, declared a state of emergency, and assumed all power, although he maintained Tarar as president. **2000:** Sharif was convicted to two life sentences for hijacking and terrorism, and was also sentenced for corruption. He was later freed and fled to Saudi Arabia. Musharraf announced local elections from December, but decreed that anyone convicted of a criminal offence or of corruption would be disqualified from standing.

Practical information

Visa requirements: UK/USA: visa required
Time difference: GMT +5
Chief tourist attractions: Himalayan scenery; fine climate; mountaineering; trekking; winter sports; archaeological remains and historic buildings
Major holidays: 23 March, 1 May, 1 July, 14 August, 6, 11 September, 9 November, 25, 31 December; variable: Eid-ul-Adha (3 days), Ashora (2 days), end of Ramadan (3 days), Prophet's Birthday, first day of Ramadan

PALAU OR BELAU

more than 350 islands and atolls (mostly uninhabited) in the west Pacific Ocean.

National name: *Belu'u era Belau/Republic of Palau*
Area: 508 sq km/196 sq mi
Capital: Koror (on Koror island)
Physical features: more than 350 (mostly uninhabited) islands, islets, and atolls in the west Pacific; warm, humid climate, susceptible to typhoons

The disc represents the full moon, traditionally the most auspicious time for work and celebration. The disc is set towards the hoist. Effective date: 1 January 1981.

Government

Head of state and government: Kuniwo Nakamura from 1993
Political system: liberal democracy
Political executive: limited presidency
Political parties: Palau Nationalist Party (PVP); Ta Belau Party
Armed forces: under the Compact of Free Association, the USA is responsible for the defence of Palau
Death penalty: laws do not provide for the death penalty for any crime
Health spend: (% GDP) 6 (1997 est)

Economy and resources

Currency: US dollar
GDP: (US$) 129 million (1999)
Unemployment: 7% (1997 est)
Foreign debt: (US$) 100 million (1990)
Major trading partners: USA, UK, Japan
Industries: processing of agricultural products, fish products, handicrafts, tourism
Exports: copra, coconut oil, handicrafts, trochus, tuna

Imports: food and live animals, crude materials, mineral fuels, beverages, tobacco, chemicals, basic manufactures, machinery and transport equipment
Agricultural products: coconuts, cassava, bananas, sweet potatoes

Population and society

Population: 19,000 (2000 est)
Population growth rate: 2.4% (1995–2000)
Population density: (per sq km) 39 (1999 est)
Urban population: (% of total) 72 (2000 est)
Age distribution: (% of total population) 0–14 28%, 15–59 62%, 60+ 10% (2000 est)
Ethnic groups: predominantly Micronesian
Language: Palauan, English (both official in most states)
Religion: Christian, principally Roman Catholic; Modekngei (indigenous religion)
Education: (compulsory years) 8
Literacy rate: 95% (1995 est)
Labour force: 9.3% agriculture, 15.2% industry, 75.5% services (1995)
Life expectancy: 65 (men); 71 (women) (1998 est)

Chronology

c. **1000 BC:** Micronesians first settled the islands. **AD 1543:** First visited by Spanish navigator Ruy Lopez de Villalobos. **16th century:** Colonized by Spain. **1899:** Purchased from Spain by Germany. **1914:** Occupied by Japan at the outbreak of World War I. **1920:** Administered by Japan under League of Nations mandate. **1944:** Occupied by USA after Japanese removed during World War II. **1947:** Became part of United Nations (UN) Pacific Islands Trust Territory, administered by USA. **1981:** Acquired autonomy as the Republic of Belau (Palau) under a constitution which prohibited the entry, storage, or disposal of nuclear or biological weapons. **1982:** The Compact of Free Association signed with the USA, providing for the right to maintain US military facilities in return for economic aid. The compact could not come into force since it contradicted the constitution, which could only be amended by a 75% vote in favour. **1993:** A referendum approved a constitutional amendment allowing the implementation of the Compact of Free Association with the USA. **1994:** Independence was achieved and UN membership granted. **2001:** Tommy Remengesau became president.

Practical information

Visa requirements: UK/USA: visa not required for a stay of up to 30 days

Time difference: GMT +9
Chief tourist attractions: rich marine environment; the myriad Rock Islands, known as the Floating Garden Islands

PANAMA

Located in Central America, on a narrow isthmus between the Caribbean
and the Pacific Ocean, bounded west by Costa Rica and east by Colombia.

National name: *República de Panamá/Republic of Panama*
Area: 77,100 sq km/29,768 sq mi
Capital: Panamá
Major towns/cities: San Miguelito, Colón, David, La Chorrera,
Santiago, Chitré, Changuinola
Major ports: Colón, Cristóbal, Balboa
Physical features: coastal plains and mountainous interior; tropical
rainforest in east and northwest; Archipelago de las Perlas in Gulf of
Panama; Panama Canal

The blue star signifies the purity and honesty of the
life of the country. The red star represents authority
and law. Effective date: 4 June 1904.

Government

Head of state and government: Mireya Moscoso from
1999
Political system: liberal democracy
Political executive: limited presidency
Administrative divisions: nine provinces and three
Autonomous Indian Reservations
Political parties: Democratic Revolutionary Party
(PRD), right wing; Arnulfista Party (PA), left of centre;
Authentic Liberal Party (PLA), left of centre; Nationalist
Liberal Republican Movement (MOLIRENA), right of
centre; Papa Ego Movement (MPE), moderate, left of
centre
Armed forces: army abolished by National Assembly
(1994); paramilitary forces numbered 11,800 (1998)
Conscription: military service is voluntary
Death penalty: laws do not provide for the death
penalty for any crime (last known execution in 1903)
Defence spend: (% GDP) 1.3 (1998)
Education spend: (% GNP) 5.1 (1997)
Health spend: (% GDP) 7.5 (1997)

Economy and resources

Currency: balboa
GDP: (US$) 9.6 billion (1999)
Real GDP growth: (% change on previous year) 3.3
(1999)
GNP: (US$) 8.6 billion (1999)
GNP per capita (PPP): (US$) 5,016 (1999)
Consumer price inflation: 1% (1999)
Unemployment: 13.4% (1997)
Foreign debt: (US$) 7.2 billion (1999)
Major trading partners: USA, Japan, Costa Rica,
Ecuador, Germany, Italy, Venezuela
Resources: limestone, clay, salt; deposits of coal, copper,
and molybdenum have been discovered
Industries: food processing, petroleum refining and
petroleum products, chemicals, paper and paper
products, beverages, textiles and clothing, plastic
products, light assembly, tourism
Exports: bananas, shrimps and lobsters, sugar, clothing,
coffee. Principal market: USA 42.2% (1999)

Imports: machinery and transport equipment,
petroleum and mineral products, chemicals and chemical
products, electrical and electronic equipment, foodstuffs.
Principal source: USA 39.2% (1999)
Arable land: 6.7% (1996)
Agricultural products: rice, maize, dry beans, bananas,
sugar cane, coffee, oranges, mangoes, cocoa; cattle
rearing; tropical timber; fishing (particularly shrimps for
export)

Population and society

Population: 2,856,000 (2000 est)
Population growth rate: 1.6% (1995–2000); 1.4%
(2000–05)
Population density: (per sq km) 36 (1999 est)
Urban population: (% of total) 56 (2000 est)
Age distribution: (% of total population) 0–14 31%,
15–59 61%, 60+ 8% (2000 est)
Ethnic groups: about 70% mestizos (of
Spanish–American and American–Indian descent), 14%
West Indian, 10% white American or European, and 6%
Indian
Language: Spanish (official), English
Religion: Roman Catholic 93%
Education: (compulsory years) 8
Literacy rate: 92% (men); 91% (women) (2000 est)
Labour force: 17.7% agriculture, 18% industry, 64.3%
services (1998)
Life expectancy: 72 (men); 76 (women) (1995–2000)
Child mortality rate: (under 5, per 1,000 live births)
28 (1995–2000)
Physicians: 1 per 725 people (1996 est)
Hospital beds: 1 per 367 people (1996)
TV sets: (per 1,000 people) 187 (1997)
Radios: (per 1,000 people) 299 (1997)
Internet users: (per 10,000 people) 160.0 (1999)
Personal computer users: (per 100 people) 3.2 (1999)

Transport

Airports: international airports: Panamá (Tocumen); two
domestic airports; total passenger km: 1,094 million
(1997 est)

Railways: total length: 583 km/362 mi; total passenger km: 51,250 million (1994)
Roads: total road network: 11,285 km/7,012 mi, of which 33.6% paved (1996); passenger cars: 74.8 per 1,000 people (1996)

Chronology

1502: Visited by Spanish explorer Rodrigo de Bastidas, at which time it was inhabited by Cuna, Choco, Guaymi, and other Indian groups. **1513:** Spanish conquistador Vasco Núñez de Balboa explored Pacific Ocean from Darien isthmus; he was made governor of Panama (meaning 'abundance of fish'). **1519:** Spanish city established at Panama, which became part of the Spanish viceroyalty of New Andalucia (later New Granada). **1572–95 and 1668–71:** Spanish settlements sacked by British buccaneers Francis Drake and Henry Morgan. **1821:** Achieved independence from Spain; joined confederacy of Gran Colombia, which included Colombia, Venezuela, Ecuador, Peru, and Bolivia. **1830:** Gran Colombia split up and Panama became part of Colombia. **1846:** Treaty signed with USA, allowing it to construct a railway across the isthmus. **1880s:** French attempt to build a Panama canal connecting the Atlantic and Pacific Oceans failed as a result of financial difficulties and the death of 22,000 workers from yellow fever and malaria. **1903:** Full independence achieved with US help on separation from Colombia; USA bought rights to build Panama Canal, and were given control of a 10-mile strip, the Canal Zone, in perpetuity. **1914:** Panama Canal opened. **1939:** Panama's status as a US protectorate was terminated by mutual agreement. **1968–81:** Military rule of Gen Omar Torrijos Herrera, leader of the National Guard, who deposed the elected president and launched a costly programme of economic modernization. **1977:** USA–Panama treaties transferred the canal to Panama (effective from 2000), with the USA guaranteeing protection and annual payment. **1987:** Gen Manuel Noriega (head of the National Guard and effective ruler since 1983) resisted calls for his removal, despite suspension of US military and economic aid. **1988:** Noriega, charged with drug smuggling by the USA, declared a state of emergency after a coup against him failed. **1989:** 'State of war' with USA announced, and US invasion (codenamed 'Operation Just Cause') deposed Noriega; 4,000 Panamanians died in the fighting. Guillermo Endara, who had won earlier elections, was installed as president in December. **1991:** Constitutional reforms were approved by the assembly, including the abolition of the standing army; a privatization programme was introduced. **1992:** Noriega was found guilty of drug offences and given a 40-year prison sentence in USA. A referendum rejected the proposed constitutional reforms. **1994:** The constitution was amended by assembly; the army was formally abolished. **1998:** Voters rejected a proposed constitutional change to allow the president to run for a second term. **1999:** Mireya Moscoso, widow of former president Arnulfo Arias, became Panama's first female head of state. Panama formally took control of its canal.

Practical information

Visa requirements: UK: visa not required (business visitors need a business visa). USA: visa required
Time difference: GMT –5
Chief tourist attractions: Panamá; ruins of Portobelo; San Blas Islands, off the Atlantic coast; 800 sandy tropical islands in Gulf of Panama
Major holidays: 1, 9 January, 1 May, 11–12 October, 3–4, 28 November, 8, 25 December; variable: Carnival (2 days), Good Friday

PAPUA NEW GUINEA

Located in the southwest Pacific, comprising the eastern part of the island of New Guinea, the Bismarck Archipelago, and part of the Solomon Islands.

National name: *Gau Hedinarai ai Papua-Matamata Guinea/Independent State of Papua New Guinea*
Area: 462,840 sq km/178,702 sq mi
Capital: Port Moresby (on East New Guinea)
Major towns/cities: Lae, Madang, Arawa, Wewak, Goroka, Rabaul
Major ports: Port Moresby, Rabaul
Physical features: mountainous; swamps and plains; monsoon climate; tropical islands of New Ireland, New Britain, and Bougainville; Admiralty Islands, D'Entrecasteaux Islands, and Louisiade Archipelago; active volcanoes Vulcan and Tavurvur

The bird of paradise represents liberty. The Southern Cross recalls the country's links with Australia.
Effective date: 16 September 1975.

Government

Head of state: Queen Elizabeth II from 1952, represented by Governor General Silas Atopare from 1997
Head of government: Mekere Morauta from 1999
Political system: liberal democracy
Political executive: parliamentary
Administrative divisions: 20 provinces; the National Capital District is administered by an Interim Commission
Political parties: Papua New Guinea Party (Pangu Pati: PP), urban- and coastal-oriented nationalist; People's Democratic Movement (PDM), 1985 breakaway from the PP; National Party (NP), highlands-based, conservative; Melanesian Alliance (MA), Bougainville-based, pro-autonomy, left of centre; People's Progress Party (PPP), conservative; People's Action Party (PAP), right of centre
Armed forces: 4,300 (1998)
Conscription: military service is voluntary
Death penalty: retains the death penalty for ordinary crimes but can be considered abolitionist in practice (last execution 1950)
Defence spend: (% GDP) 1.0 (1998)
Education spend: (% GNP) N/A
Health spend: (% GDP) 3.1 (1997 est)

Economy and resources

Currency: kina
GDP: (US$) 4.3 billion (1999 est)
Real GDP growth: (% change on previous year) 4.1 (1999)
GNP: (US$) 3.7 billion (1999)
GNP per capita (PPP): (US$) 2,263 (1999 est)
Consumer price inflation: 16.2% (1999)
Foreign debt: (US$) 3 billion (1999 est)
Major trading partners: Australia, Japan, USA, Singapore, Germany, South Korea, UK, New Zealand, Malaysia
Resources: copper, gold, silver; deposits of chromite,

cobalt, nickel, quartz; substantial reserves of petroleum and natural gas (petroleum production began in 1992)
Industries: food processing, beverages, tobacco, timber products, metal products, machinery and transport equipment, fish canning
Exports: gold, copper ore and concentrates, crude petroleum, timber, coffee beans, coconut and copra products. Principal market: Australia 30% (1999)
Imports: machinery and transport equipment, manufactured goods, food and live animals, miscellaneous manufactured articles, chemicals, mineral fuels. Principal source: Australia 52.9% (1999)
Arable land: 0.1% (1996)
Agricultural products: coffee, cocoa, coconuts, pineapples, palm oil, rubber, tea, pyrethrum, peanuts, spices, potatoes, maize, taro, bananas, rice, sago, sweet potatoes; livestock; poultry; fishing; timber production

Population and society

Population: 4,807,000 (2000 est)
Population growth rate: 2.2% (1995–2000); 2.1% (2000–05)
Population density: (per sq km) 10 (1999 est)
Urban population: (% of total) 17 (2000 est)
Age distribution: (% of total population) 0–14 39%, 15–59 56%, 60+ 5% (2000 est)
Ethnic groups: mainly Melanesian (95%), particularly in coastal areas; inland (on New Guinea and larger islands), Papuans predominate. On the outer archipelagos and islands, mixed Micronese-Melanesians and Polynesian groups are found. A small Chinese minority also exists
Language: English (official), pidgin English, over 700 local languages
Religion: Christian 97%, of which 3% Roman Catholic; local pantheistic beliefs
Education: not compulsory
Literacy rate: 71% (men); 57% (women) (2000 est)
Labour force: 76% agriculture, 6% industry, 18% services (1997 est)

Life expectancy: 57 (men); 59 (women) (1995–2000)
Child mortality rate: (under 5, per 1,000 live births)
84 (1995–2000)
Physicians: 1 per 12,754 people (1993 est)
Hospital beds: 1 per 290 people (1993 est)
TV sets: (per 1,000 people) 24 (1997)
Radios: (per 1,000 people) 97 (1997)
Internet users: (per 10,000 people) 4.3 (1999)

Transport

Airports: international airports: Port Moresby (Jackson);
177 domestic airports and airstrips with scheduled
services; total passenger km: 830 million (1995)
Railways: none
Roads: total road network: 19,600 km/12,179 mi, of
which 3.5% paved (1996 est); passenger cars: 7.1 per
1,000 people (1996 est)

Chronology

c. **3000 BC:** New settlement of Austronesian (Melanesian) immigrants.
AD 1526: Visited by Portuguese navigator Jorge de Menezes, who named
the main island the Ilhos dos Papua. **1545:** Spanish navigator Ynigo
Ortis de Retez gave the island the name of New Guinea, as a result of a
supposed resemblance of the peoples with those of the Guinea coast of
Africa. **17th century:** Regularly visited by Dutch merchants. **1828:**
Dutch East India Company incorporated the western part of New
Guinea into Netherlands East Indies (becoming Irian Jaya, in
Indonesia). **1884:** Northeast New Guinea annexed by Germany; the
southeast was claimed by Britain. **1870s:** Visits by Western
missionaries and traders increased. **1890s:** Copra plantations
developed in German New Guinea. **1906:** Britain transferred
its rights to Australia, which renamed the lands Papua. **1914:**
German New Guinea occupied by Australia at the outbreak
of World War I; from the merged territories Papua New
Guinea was formed. **1920–42:** Held as League of Nations
mandate by Australia. **1942–45:** Occupied by Japan, who
lost 150,000 troops resisting Allied counterattack. **1947:**
Held as United Nations Trust Territory by Australia. **1951:**
Legislative Council established. **1964:** Elected House of
Assembly formed. **1967:** Pangu Party (Pangu Pati; PP)
formed to campaign for home rule. **1975:** Independence
achieved from Australia, within Commonwealth, with Michael
Somare (PP) as prime minister. **1985:** Somare challenged by
deputy prime minister Paias Wingti, who later left the PP and

formed the People's Democratic Movement (PDM); he became head of a five-party coalition government. **1988:**
Joined Solomon Islands and Vanuatu to form the Spearhead Group, aiming to preserve Melanesian cultural traditions.
1989: State of emergency imposed on copper-rich Bougainville in response to separatist violence. **1990:** The
Bougainville Revolutionary Army (BRA) issued a unilateral declaration of independence. **1991:** There was an
economic boom as gold production doubled. **1994:** There was a short-lived peace agreement with the BRA. **1996:**
The prime minister of Bougainville was murdered, jeopardizing the peace process. Gerard Sinato was elected president
of the transitional Bougainville government. **1997:** The army and police mutinied following the government's use of
mercenaries against secessionist rebels. Silas Atopare was appointed governor general. **1998:** There was a truce with
Bougainville secessionists. At least 1,500 people died and thousands were left homeless when tidal waves destroyed
villages on the north coast. **1999:** A coalition of parties headed by Mekere Morauta won a parliamentary majority to
form a new government. Bougainville Transitional Government (BTG) was replaced by the new interim Bougainville
Reconciliation Government (BRG), headed by former rebel leader Joseph Kabui and BTG leader Gerard Sinato.

Practical information

Visa requirements: UK: visa required. USA: visa
required
Time difference: GMT +10
Chief tourist attractions: spectacular scenery; the
greatest variety of ecosystems in the South Pacific;
abundant wildlife – birds include 38 species of birds of

paradise, also the megapode and cassowary; two-thirds
of the world's orchid species come from Papua New
Guinea
Major holidays: 1 January, 15 August, 16 September,
25–26 December; variable: Good Friday, Easter
Monday, Holy Saturday, Queen's Birthday (June),
Remembrance (July)

PARAGUAY

Landlocked country in South America, bounded northeast by Brazil, south by Argentina, and northwest by Bolivia.

National name: *República del Paraguay/Republic of Paraguay*
Area: 406,752 sq km/157,046 sq mi
Capital: Asunción (and chief port)
Major towns/cities: Ciudad del Este, Pedro Juan Caballero, San Lorenzo, Fernando de la Mora, Lambare, Luque, Capiatá
Major ports: Conceptión
Physical features: low marshy plain and marshlands; divided by Paraguay River; Paraná River forms southeast boundary

The Star of May recalls the declaration of independence on 14 May 1811. The colours were inspired by the French tricolour. Effective date: c. 1990.

Government

Head of state and government: Luis Gonzalez Macchi from 1999
Political system: emergent democracy
Political executive: limited presidency
Administrative divisions: 17 departments and the capital district of Asunción
Political parties: National Republican Association (Colorado Party), right of centre; Authentic Radical Liberal Party (PLRA), centrist; National Encounter, right of centre; Radical Liberal Party (PLR), centrist; Liberal Party (PL), centrist
Armed forces: 20,200 (1998)
Conscription: 12 months (army); 24 months (navy)
Death penalty: abolished in 1992
Defence spend: (% GDP) 1.4 (1998)
Education spend: (% GNP) 4.0 (1997)
Health spend: (% GDP) 5.6 (1997 est)

Economy and resources

Currency: guaraní
GDP: (US$) 7.7 billion (1999)
Real GDP growth: (% change on previous year) 0.5 (1999)
GNP: (US$) 8.5 billion (1999)
GNP per capita (PPP): (US$) 4,193 (1999 est)
Consumer price inflation: 6.8% (1999)
Unemployment: 8.2% (1997)
Foreign debt: (US$) 2.8 billion (1999 est)
Major trading partners: Brazil, Argentina, Uruguay, the Netherlands, Japan, USA, France, UK
Resources: gypsum, kaolin, limestone, salt; deposits (not commercially exploited) of bauxite, iron ore, copper, manganese, uranium; deposits of natural gas discovered in 1994; exploration for petroleum deposits ongoing mid-1990s
Industries: food processing, beverages, tobacco, wood and wood products, textiles (cotton), clothing, leather, chemicals, metal products, machinery
Exports: soybeans (and other oil seeds), cotton, timber and wood manufactures, hides and skins, meat. Principal market: Brazil 31.7% (1999)
Imports: machinery, vehicles and parts, mineral fuels and lubricants, beverages, tobacco, chemicals, foodstuffs.

Principal source: Brazil 28.6% (1999)
Arable land: 5.5% (1996)
Agricultural products: cassava, soybeans, maize, cotton, wheat, rice, tobacco, sugar cane, 'yerba maté' (strongly flavoured tea); livestock rearing; forest resources

Population and society

Population: 5,496,000 (2000 est)
Population growth rate: 2.6% (1995–2000); 2.3% (2000–05)
Population density: (per sq km) 13 (1999 est)
Urban population: (% of total) 56 (2000 est)
Age distribution: (% of total population) 0–14 40%, 15–59 54%, 60+ 6% (2000 est)
Ethnic groups: predominantly mixed-race mestizos (94%); Asian 5%, foreigners and Indian 1%
Language: Spanish (official), Guaraní (an indigenous Indian language)
Religion: Roman Catholic (official religion) 85%; Mennonite, Anglican
Education: (compulsory years) 6
Literacy rate: 94% (men); 92% (women) (2000 est)
Labour force: 45.2% agriculture, 22.5% industry, 32.3% services (1994)
Life expectancy: 68 (men); 72 (women) (1995–2000)
Child mortality rate: (under 5, per 1,000 live births) 48 (1995–2000)
Physicians: 1 per 1,231 people (1993 est)
Hospital beds: 1 per 762 people (1993 est)
TV sets: (per 1,000 people) 101 (1997)
Radios: (per 1,000 people) 182 (1997)
Internet users: (per 10,000 people) 37.3 (1999)
Personal computer users: (per 100 people) 1.1 (1999)

Transport

Airports: international airports: Asunción (Silvio Pettirossi), Ciudad del Este (Guaraní); three domestic airports; total passenger km: 283 million (1995)
Railways: total length: 441 km/274 mi; total passenger km: 1 million (1994)
Roads: total road network: 29,500 km/18,331 mi, of which 9.5% paved (1996 est); passenger cars: 14.3 per 1,000 people (1996 est)

Chronology

1526: Visited by Italian navigator Sebastian Cabot, who travelled up the Paraná River; at this time the east of the country had long been inhabited by Guaraní-speaking Amerindians, who gave the country its name, which means 'land with an important river'. **1537:** Spanish made an alliance with Guaraní Indians against hostile Chaco Indians, enabling them to colonize interior plains; Asunción founded by Spanish. **1609:** Jesuits arrived from Spain to convert local population to Roman Catholicism and administer the country. **1767:** Jesuit missionaries expelled. **1776:** Formerly part of Spanish Viceroyalty of Peru, which covered much of South America, became part of Viceroyalty of La Plata, with capital at Buenos Aires (Argentina). **1808:** With Spanish monarchy overthrown by Napoleon Bonaparte, La Plata Viceroyalty became autonomous, but Paraguayans revolted against rule from Buenos Aires. **1811:** Independence achieved from Spain. **1814:** Under dictator Gen José Gaspar Rodriguez Francia ('El Supremo'), Paraguay became an isolated state. **1840:** Francia was succeeded by his nephew, Carlos Antonio Lopez, who opened country to foreign trade and whose son, Francisco Solano Lopez, as president from 1862, built up a powerful army. **1865–70:** War with Argentina, Brazil, and Uruguay over access to sea; more than half the population died and 150,000 sq km/58,000 sq mi of territory lost. **late 1880s:** Conservative Colorado Party and Liberal Party founded. **1912:** Liberal leader Edvard Schaerer came to power, ending decades of political instability. **1932–35:** Territory in west won from Bolivia during Chaco War (settled by arbitration in 1938). **1940–48:** Presidency of autocratic Gen Higinio Morínigo. **1948–54:** Political instability; six different presidents. **1954:** Gen Alfredo Stroessner seized power in a coup. He ruled as a ruthless autocrat, suppressing civil liberties; the country received initial US backing as the economy expanded. **1989:** Stroessner was ousted in a coup led by Gen Andrés Rodríguez. Rodríguez was elected president; the right-of-centre military-backed Colorado Party won assembly elections. **1992:** A new democratic constitution was adopted. **1993:** The Colorado Party won the most seats in the first free multiparty elections, but no overall majority; its candidate, Juan Carlos Wasmosy, won the first free presidential elections. **1999:** Aenate leader Luis Gonzalez Macchi became president. **2000:** An attempted anti-government coup, led by supporters of the ex-army chief Lino Oviedo, failed after the USA and Brazil put pressure on the army's commanders. The opposition candidate, Julio César Franco, of the Authentic Liberal Radical Party, won the vice-presidential elections, the first national defeat for the Colorado Party. Congress approved the start of a programme of privatization.

Practical information

Visa requirements: UK: visa not required. USA: visa not required
Time difference: GMT –3/4
Chief tourist attractions: the Iguaçu Falls; the sparsely populated Gran Chaco; Asunción (La

Encaración church, the Pantheon of Heroes)
Major holidays: 1 January, 3 February, 1 March, 1, 14–15 May, 12 June, 15, 25 August, 29 September, 12 October, 1 November, 8, 25, 31 December; variable: Corpus Christi, Good Friday, Holy Thursday

PERU

Located in South America, on the Pacific, bounded north by Ecuador and Colombia, east by Brazil and Bolivia, and south by Chile.

National name: *República del Perú/Republic of Peru*
Area: 1,285,200 sq km/496,216 sq mi
Capital: Lima
Major towns/cities: Arequipa, Iquitos, Chiclayo, Trujillo, Huancayo, Piura, Chimbote
Major ports: Callao, Chimbote, Salaverry
Physical features: Andes mountains running northwest–southeast cover 27% of Peru, separating Amazon river-basin jungle in northeast from coastal plain in west; desert along coast north–south (Atacama Desert); Lake Titicaca

Red and white were the colours of the Inca Empire. Red represents the blood shed in the fight for independence. White stands for peace and justice. Effective date: 25 February 1825.

Government

Head of state: Valentin Paniagua from 2000
Head of government: Javier Pérez de Cuéllar from 2000
Political system: liberal democracy
Political executive: limited presidency
Administrative divisions: 24 departments and the constitutional province of Callao
Political parties: American Popular Revolutionary Alliance (APRA), moderate, left wing; United Left (IU), left wing; Change 90 (Cambio 90), centrist; New Majority (Nueva Mayoria), centrist; Popular Christian Party (PPC), right of centre; Liberal Party (PL), right wing
Armed forces: 125,000; plus paramilitary forces numbering 78,000 (1998)
Conscription: conscription is selective for two years
Death penalty: retains the death penalty only for exceptional crimes such as crimes under military law or crimes committed in exceptional circumstances such as wartime (last execution 1979)
Defence spend: (% GDP) 1.6 (1998)
Education spend: (% GNP) 2.9 (1996)
Health spend: (% GDP) 5.6 (1997)

Economy and resources

Currency: nuevo sol
GDP: (US$) 51.7 billion (1999)
Real GDP growth: (% change on previous year) 1.4 (1999)
GNP: (US$) 60.3 billion (1999)
GNP per capita (PPP): (US$) 4,387 (1999)
Consumer price inflation: 3.5% (1999)
Unemployment: 7.7% (1997)
Foreign debt: (US$) 30.7 billion (1999 est)
Major trading partners: USA, Japan, Germany, Italy, China, Chile, Venezuela, Brazil, Colombia
Resources: lead, copper, iron, silver, zinc (world's fourth-largest producer), petroleum
Industries: food processing, textiles and clothing, petroleum refining, metals and metal products, chemicals, machinery and transport equipment, beverages, tourism
Exports: copper, fishmeal, zinc, gold, refined petroleum products. Principal market: USA 28.3% (1999)
Imports: machinery and transport equipment, basic foodstuffs, basic manufactures, chemicals, mineral fuels, consumer goods. Principal source: USA 22.4% (1999)
Arable land: 2.9% (1996)
Agricultural products: potatoes, wheat, seed cotton, coffee, rice, maize, beans, sugar cane; fishing (particularly for South American pilchard and the anchovetta)

Population and society

Population: 25,662,000 (2000 est)
Population growth rate: 1.7% (1995–2000); 1.7% (2000–05)
Population density: (per sq km) 20 (1999 est)
Urban population: (% of total) 73 (2000 est)
Age distribution: (% of total population) 0–14 33%, 15–59 59%, 60+ 8% (2000 est)
Ethnic groups: about 50% South American Indian, 40% mestizo, 7% European, and 3% African origin
Language: Spanish, Quechua (both official), Aymara, many indigenous dialects
Religion: Roman Catholic (state religion) 95%
Education: (compulsory years) 11
Literacy rate: 95% (men); 85% (women) (2000 est)
Labour force: 32% agriculture, 15.9% industry, 52.1% services (1992)
Life expectancy: 66 (men); 71 (women) (1995–2000)
Child mortality rate: (under 5, per 1,000 live births) 65 (1995–2000)
Physicians: 1 per 939 people (1993 est)
Hospital beds: 1 per 664 people (1993 est)
TV sets: (per 1,000 people) 143 (1997)
Radios: (per 1,000 people) 273 (1997)
Internet users: (per 10,000 people) 158.5 (1999)
Personal computer users: (per 100 people) 2.0 (1999)

Transport

Airports: international airports: Lima (Jorge Chávez), Iquitos (Colonel Francisco Secada Vignetta), Cusco (Velasco Astete), Arequipa (Rodríguez Ballón); 27 domestic airports; total passenger km: 2,561 million (1997 est)
Railways: total length: 3,661 km/2,275 mi; total passenger km: 241 million (1994)
Roads: total road network: 72,800 km/45,238 mi, of which 11% paved (1996 est); passenger cars: 58.4 per 1,000 people (1996 est)

Chronology

4000 BC: Evidence of early settled agriculture in Chicama Valley. **AD 700–1100:** Period of Wari Empire, first expansionist militarized empire in Andes. **1200:** Manco Capac became the first emperor of South American Indian Quechua-speaking Incas, who established a growing and sophisticated empire centred on the Andean city of Cusco, and believed their ruler was descended from the Sun. **late 15th century:** At its zenith, the Inca Empire stretched from Quito in Ecuador to beyond Santiago in southern Chile. It superseded the Chimu civilization, which had flourished in Peru 1250–1470. **1532–33:** Incas defeated by Spanish conquistadores, led by Francisco Pizarro. Empire came under Spanish rule, as part of the Viceroyalty of Peru, with capital in Lima, founded in 1535. **1780:** Tupac Amaru, who claimed to be descended from the last Inca chieftain, led a failed native revolt against Spanish. **1810:** Peru became the headquarters for the Spanish government as European settlers rebelled elsewhere in Spanish America. **1820–22:** Fight for liberation from Spanish rule led by Gen José de San Martín and Army of Andes which, after freeing Argentina and Chile, invaded southern Peru. **1824:** Became last colony in Central and South America to achieve independence from Spain after attacks from north by Field Marshal Sucre, acting for freedom fighter Simón Bolívar. **1836–39:** Failed attempts at union with Bolivia. **1849–74:** Around 80,000–100,000 Chinese labourers arrived in Peru to fill menial jobs such as collecting guano. **1866:** Victorious naval war fought with Spain. **1879–83:** Pacific War fought in alliance with Bolivia and Chile over nitrate fields of the Atacama Desert in the south; three provinces along coastal south lost to Chile. **1902:** Boundary dispute with Bolivia settled. **mid-1920s:** After several decades of civilian government, a series of right-wing dictatorships held power. **1927:** Boundary dispute with Colombia settled. **1929:** Tacna province, lost to Chile in 1880, was returned. **1941:** A brief war with Ecuador secured Amazonian territory. **1945:** Civilian government, dominated by left-of-centre American Popular Revolutionary Alliance (APRA, formed 1924), came to power after free elections. **1948:** Army coup installed military government led by Gen Manuel Odría, who remained in power until 1956. **1963:** Return to civilian rule, with centrist Fernando Belaúnde Terry as president. **1968:** Return of military government in bloodless coup by Gen Juan Velasco Alvarado, following industrial unrest. Populist land reform programme introduced. **1980:** Return to civilian rule, with Fernando Belaúnde as president; agrarian and industrial reforms pursued. Sendero Luminoso ('Shining Path') Maoist guerrilla group active. **1981:** Boundary dispute with Ecuador renewed. **1985:** Belaúnde succeeded by Social Democrat Alan García Pérez, who launched campaign to remove military and police 'old guard'. **1988:** García was pressured to seek help from International Monetary Fund (IMF) as the economy deteriorated. Sendero Luminoso increased its campaign of violence. **1990:** Right-of-centre Alberto Fujimori defeated ex-communist writer Vargas Llosa in presidential elections. Inflation rose to 400%; a privatization programme was launched. **1992:** Fujimori allied himself with the army and suspended the constitution, provoking international criticism. The Sendero Luminoso leader was arrested and sentenced to life imprisonment. A new single-chamber legislature was elected. **1993:** A new constitution was adopted. **1994:** 6,000 Sendero Luminoso guerrillas surrendered to the authorities. **1995:** A border dispute with Ecuador was resolved after armed clashes. Fujimori was re-elected. A controversial amnesty was granted to those previously convicted of human-rights abuses. **1996:** Hostages were held in the Japanese embassy by Marxist Tupac Amaru Revolutionary Movement (MRTA) guerrillas. **1997:** The hostage siege ended. **1998:** A border dispute that had lasted for 157 years was settled with Ecuador. **1999:** Alberto Bustamante was made prime minister. **2000:** Fujimori was re-elected as president for a third term in July. In September, the head of the national intelligence service, Vladimiro Montesinos, was proved to have bribed a member of congress. He fled to Panama, which refused to accept him, and returned to Peru, where his arrest was ordered on charges of corruption and abusing human rights. Fujimori sent his resignation from Japan, from where he could not be extradited, and Valentin Paniagua became president. He appointed former United Nations (UN) secretary general Javier Pérez de Cuélla as prime minister.

Practical information

Visa requirements: UK: visa not required for a stay of up to 90 days. USA: visa not required for a stay of up to 90 days
Time difference: GMT –5
Chief tourist attractions: Lima, with its Spanish colonial architecture; Cusco, with its pre-Inca and Inca remains, notably Machu Picchu; Lake Titicaca; Amazon rainforest in northeast
Major holidays: 1 January, 1 May, 29–30 June, 28 July (2 days), 30 August, 8 October, 1 November, 8, 25, 31 December; variable: Good Friday, Holy Thursday

PHILIPPINES

An archipelago of more than 7,000 islands in southeast Asia, west of the Pacific Ocean and south of the Southeast Asian mainland.

National name: *Republika Ñg Pilipinas/Republic of the Philippines*
Area: 300,000 sq km/115,830 sq mi
Capital: Manila (on Luzon island) (and chief port)
Major towns/cities: Quezon City, Davao, Caloocan, Cebu, Bacolod, ·Cagayan, Iloilo
Major ports: Cebu, Davao (on Mindanao), Iloilo, Zamboanga (on Mindanao)

Blue expresses patriotism and noble ideas. Red denotes bravery. White symbolizes peace and purity. The stars stand for the three island groups: Luzon, the Visayan, and Mindanao. Effective date: 25 March 1936.

Physical features: comprises over 7,000 islands; volcanic mountain ranges traverse main chain north–south; 50% still forested. The largest islands are Luzon 108,172 sq km/41,754 sq mi and Mindanao 94,227 sq km/36,372 sq mi; others include Samar, Negros, Palawan, Panay, Mindoro, Leyte, Cebu, and the Sulu group; Pinatubo volcano (1,759 m/5,770 ft); Mindanao has active volcano Apo (2,954 m/9,690 ft) and mountainous rainforest

Government

Head of state and government: Gloria Macapagal Arroyo from 2001
Political system: liberal democracy
Political executive: limited presidency
Administrative divisions: 73 provinces and 60 chartered cities
Political parties: Laban ng Demokratikong Pilipino (Democratic Filipino Struggle Party; LDP–DFSP), centrist, liberal-democrat coalition; Lakas ng Edsa (National Union of Christian Democrats; LNE–NUCD), centrist; Liberal Party, centrist; Nationalist Party (Nacionalista), right wing; New Society Movement (NSM; Kilusan Bagong Lipunan), conservative, pro-Marcos; National Democratic Front, left-wing umbrella grouping, including the Communist Party of the Philippines (CPP); Mindanao Alliance, island-based decentralist body
Armed forces: 117,800; reserve forces 131,000; paramilitary forces around 42,500 (1998)
Conscription: military service is voluntary
Death penalty: retained in law, but considered abolitionist in practice; last execution in 1976
Defence spend: (% GDP) 2.3 (1998)
Education spend: (% GNP) 3.4 (1997)
Health spend: (% GDP) 3.4 (1997)

Economy and resources

Currency: peso
GDP: (US$) 76.6 billion (1999)
Real GDP growth: (% change on previous year) 3.3 (1999)
GNP: (US$) 77.9 billion (1999)
GNP per capita (PPP): (US$) 3,815 (1999)
Consumer price inflation: 6.7% (1999)
Unemployment: 7.9% (1997)
Foreign debt: (US$) 52.2 billion (1999 est)
Major trading partners: USA, Japan, EU, Singapore, Taiwan, South Korea, Hong Kong
Resources: copper ore, gold, silver, chromium, nickel, coal, crude petroleum, natural gas, forests
Industries: food processing, petroleum refining, textiles, chemical products, pharmaceuticals, electrical machinery (mainly telecommunications equipment), metals and metal products, tourism
Exports: electronic products (notably semiconductors and microcircuits), garments, agricultural products (particularly fruit and seafood), woodcraft and furniture, lumber, chemicals, coconut oil. Principal market: USA 29.8% (1999)
Imports: machinery and transport equipment, mineral fuels, basic manufactures, food and live animals, textile yarns, base metals, cereals and cereal preparations. Principal source: USA 20% (1999)
Arable land: 17.5% (1996)
Agricultural products: rice, maize, cassava, coconuts, sugar cane, bananas, pineapples; livestock (chiefly pigs, buffaloes, goats, and poultry) and fisheries

Population and society

Population: 75,967,000 (2000 est)
Population growth rate: 2.1% (1995–2000); 1.8% (2000–05)
Population density: (per sq km) 248 (1999 est)
Urban population: (% of total) 59 (2000 est)
Age distribution: (% of total population) 0–14 37%, 15–59 57%, 60+ 6% (2000 est)
Ethnic groups: comprises more than 50 ethnic communities, although 95% of the population is designated 'Filipino', an Indo-Polynesian ethnic grouping
Language: Filipino, English (both official), Spanish, Cebuano, Ilocano, more than 70 other indigenous languages
Religion: Christian 94%, mainly Roman Catholic (84%), Protestant; Muslim 4%, local religions
Education: (compulsory years) 6
Literacy rate: 95% (men); 95% (women) (2000 est)
Labour force: 40% of population: 46% agriculture, 15% industry, 39% services (1990)
Life expectancy: 67 (men); 70 (women) (1995–2000)
Child mortality rate: (under 5, per 1,000 live births) 44 (1995–2000)
Physicians: 1 per 9,689 people (1996)
Hospital beds: 1 per 934 people (1994)
TV sets: (per 1,000 people) 109 (1997)
Radios: (per 1,000 people) 159 (1997)
Internet users: (per 10,000 people) 67.2 (1999)
Personal computer users: (per 100 people) 1.7 (1999)

Transport

Airports: international airports: Manila (Ninoy Aquino), Cebu (Mactan), Laoag City, Davao, Zamboanga, Puerto Princesa, Subic Bay, Freeport City; comprehensive internal services; total passenger km: 16,872 million (1997 est)

Railways: total length: 897 km/577 mi (of which 492 km/306 mi in operation); total passenger km: 70 million (1996)

Roads: total road network: 161,313 km/100,240 mi, of which 0.18% paved (1997); passenger cars: 10.4 per 1,000 people (1997)

Chronology

14th century: Traders from Malay peninsula introduced Islam and created Muslim principalities of Manila and Jolo. **1521:** Portuguese navigator Ferdinand Magellan reached the islands, but was killed in battle with islanders. **1536:** Philippines named after Charles V's son (later Philip II of Spain) by Spanish navigator Ruy López de Villalobos. **1565:** Philippines conquered by Spanish army led by Miguel López de Lagazpi. **1571:** Manila was made capital of the colony, which was part of the Viceroyalty of Mexico. **17th century:** Spanish missionaries converted much of the lowland population to Roman Catholicism. **1762–63:** British occupied Manila. **1834:** End of Spanish monopoly on trade; British and American merchants bought sugar and tobacco. **1896–97:** Emilio Aguinaldo led a revolt against Spanish rule. **1898:** Spanish-American War: US navy destroyed Spanish fleet in Manila Bay; Aguinaldo declared independence, but Spain ceded Philippines to USA. **1898–1901:** Nationalist uprising suppressed by US troops; 200,000 Filipinos killed. **1907:** Americans set up elected legislative assembly. **1916:** Bicameral legislature introduced based on the US model. **1935:** Philippines gained internal self-government with Manuel Quezon as president. **1942–45:** Occupied by Japan. **1946:** Philippines achieved independence from USA under President Manuel Roxas; USA retained military bases and supplied economic aid. **1957–61:** 'Filipino First' policy introduced by President Carlos García to reduce economic power of Americans and Chinese; official corruption increased. **1972:** President Ferdinand Marcos declared martial law and ended the freedom of the press; economic development financed by foreign loans, of which large sums were diverted by Marcos for personal use. **1981:** Martial law officially ended but Marcos retained sweeping emergency powers, ostensibly needed to combat long-running Muslim and communist insurgencies. **1983:** Opposition leader Benigno Aquino was murdered at Manila airport while surrounded by government troops. **1986:** Corazon Aquino (widow of Benigno Aquino) used 'people's power' to force Marcos to flee the country. **1987:** A 'Freedom constitution' was adopted; Aquino's People's Power won congressional elections. **1989:** A state of emergency was declared after the sixth coup attempt was suppressed with US aid. **1991:** The Philippine senate called for the withdrawal of US forces; US renewal of Subic Bay naval base lease was rejected. **1992:** Fidel Ramos was elected to succeed Aquino; a 'Rainbow Coalition' government was formed. **1995:** Imelda Marcos (the widow of Ferdinand Marcos) was elected to the House of Representatives while on bail from prison on a sentence for corruption. **1996:** The LDP withdrew from the LDP–DFSP coalition. A peace agreement was made between the government and the Moro National Liberation Front (MNLF) after 25 years of civil unrest on Mindanao. **1997:** Preliminary peace talks took place between the government and the Moro Islamic Liberation Front (MILF), fighting for an independent Muslim state on Mindanao. The Supreme Court rejected a proposal to allow a second presidential term. **1998:** Joseph Estrada was inaugurated as president and Gloria Macapagal Arroyo as vice-president. Imelda Marcos was acquitted of corruption charges. A dispute with China over the mineral-rich Spratly Islands was resolved with an agreement on the joint use of the resources. **2000:** The worst fighting since 1996 erupted between government troops and the MILF. In April, another Islamic separatist group, Abu Sayyaf, took 21 foreign hostages from holiday resorts in Malaysia. They were slowly released, the last few in September when Libya paid US$4 million/£2.8 million to the captors. However, further hostages continued to be taken throughout the year. Some were rescued by the Philippine army in September. In November, President Estrada was impeached on corruption charges. **2001:** Estrada's trial was suspended after senators blocked the presentation of vital evidence. This caused mass public demonstrations, and Estrada left office. Former vice-president Gloria Macapagal Arroyo, who had led the call for Estrada's impeachment, became president.

Practical information

Visa requirements: UK/USA: visa not required for a stay of up to 21 days
Time difference: GMT +8
Chief tourist attractions: thousands of islands and islets, some ringed by coral reefs; Mindanao; the Visayas Islands
Major holidays: 1 January, 1 May, 12 June, 4 July, 1, 30 November, 25, 30–31 December; variable: Good Friday, Holy Thursday

POLAND

Located in eastern Europe, bounded north by the Baltic Sea, northeast by Lithuania, east by Belarus and Ukraine, south by the Czech Republic and the Slovak Republic, and west by Germany.

National name: *Rzeczpospolita Polska/Republic of Poland*
Area: 312,683 sq km/120,726 sq mi
Capital: Warsaw
Major towns/cities: Łódź, Kraków, Wroclaw, Poznan, Gdansk, Szczecin, Katowice, Bydgoszcz, Lublin
Major ports: Gdansk (Danzig), Szczecin (Stettin), Gdynia (Gdingen)
Physical features: part of the great plain of Europe; Vistula, Oder, and Neisse rivers; Sudeten, Tatra, and Carpathian mountains on southern frontier

Red and white are the national colours, derived from a 13th-century emblem bearing a white eagle on a red field. Effective date: 23 March 1956.

Government

Head of state: Aleksander Kwasniewski from 1995
Head of government: Jerzy Buzek from 1997
Political system: liberal democracy
Political executive: limited presidency
Administrative divisions: 16 provinces
Political parties: Democratic Left Alliance (SLD), reform socialist (ex-communist); Polish Peasant Party (PSL), moderate, agrarian; Freedom Union (UW), moderate, centrist; Labour Union (UP), left wing; Non-Party Bloc in support of Reforms (BBWR), Christian Democrat, right of centre, pro-Walesa; Confederation for an Independent Poland (KPN), right wing; Solidarity Electoral Action (AWS), Christian, right wing
Armed forces: 240,700 (1998)
Conscription: military service is compulsory
Death penalty: abolished in 1997
Defence spend: (% GDP) 2.2 (1998)
Education spend: (% GNP) 4.6 (1996)
Health spend: (% GDP) 5.2 (1997)

Economy and resources

Currency: zloty
GDP: (US$) 155.5 billion (1999)
Real GDP growth: (% change on previous year) 4.1 (1999)
GNP: (US$) 153.1 billion (1999)
GNP per capita (PPP): (US$) 7,894 (1999)
Consumer price inflation: 7.3% (1999)
Unemployment: 10.3% (1998)
Foreign debt: (US$) 55.8 billion (1999 est)
Major trading partners: Germany, Italy, the Netherlands, Russia, UK, France, USA, Ukraine
Resources: coal (world's fifth-largest producer), copper, sulphur, silver, petroleum and natural gas reserves
Industries: machinery and transport equipment, food products, metals, chemicals, beverages, tobacco, textiles and clothing, petroleum refining, wood and paper products, tourism
Exports: machinery and transport equipment, textiles, chemicals, coal, coke, copper, sulphur, steel, food and agricultural products, clothing and leather products, wood and paper products. Principal market: Germany 36.1% (1999)

Imports: electro-engineering products, fuels and power (notably crude petroleum and natural gas), textiles, food products, iron ore, fertilizers. Principal source: Germany 25.2% (1999)
Arable land: 46.3% (1996)
Agricultural products: wheat, rye, barley, oats, maize, potatoes, sugar beet; livestock rearing; forest resources

Population and society

Population: 38,765,000 (2000 est)
Population growth rate: 0.08% (1995–2000)
Population density: (per sq km) 124 (1999 est)
Urban population: (% of total) 66 (2000 est)
Age distribution: (% of total population) 0–14 19%, 15–59 65%, 60+ 16% (2000 est)
Ethnic groups: 98% ethnic Western-Slav ethnic Poles; small ethnic German, Ukrainian, and Belarussian minorities
Language: Polish (official)
Religion: Roman Catholic 95%
Education: (compulsory years) 8
Literacy rate: 99% (men); 99% (women) (2000 est)
Labour force: 20.6% agriculture, 31.9% industry, 47.5% services (1997)
Life expectancy: 68 (men); 77 (women) (1995–2000)
Child mortality rate: (under 5, per 1,000 live births) 16 (1995–2000)
Physicians: 1 per 436 people (1996)
Hospital beds: 1 per 158 people (1996)
TV sets: (per 1,000 people) 413 (1998)
Radios: (per 1,000 people) 523 (1997)
Internet users: (per 10,000 people) 542.1 (1999)
Personal computer users: (per 100 people) 6.2 (1999)

Transport

Airports: international airports: Warsaw (Okecie), Kraków (Balice), Wroclaw (Strachowice), Gdansk; four domestic airports; total passenger km: 4,204 million (1997 est)
Railways: total length: 24,313 km/15,108 mi; total passenger km: 25,806 million (1997)
Roads: total road network: 377,048 km/234,298 mi, of which 65.7% paved (1997); passenger cars: 221 per 1,000 people (1997)

Chronology

966: Polish Slavic tribes under Mieszko I, leader of Piast dynasty, adopted Christianity and united region around Poznan to form first Polish state. **1241:** Devastated by Mongols. **13th–14th centuries:** German and Jewish refugees settled among Slav population. **1386:** Jagellonian dynasty came to power: golden age for Polish culture. **1569:** Poland united with Lithuania to become the largest state in Europe. **1572:** Jagellonian dynasty became extinct; future kings were elected by nobility and gentry, who formed 10% of the population. **mid-17th century:** Defeat in war against Russia, Sweden, and Brandenburg (in Germany) set in a process of irreversible decline. **1772–95:** Partitioned between Russia, which ruled the northeast; Prussia, the west, including Pomerania; and Austria in the south-centre, including Galicia. **1815:** After Congress of Vienna, Russian eastern portion of Poland re-established as kingdom within Russian Empire. **1830 and 1863:** Uprisings against repressive Russian rule. **1892:** Nationalist Polish Socialist Party (PPS) founded. **1918:** Independent Polish republic established after World War I, with Marshal Józef Pilsudski, founder of the PPS, elected president. **1919–21:** Abortive advance into Lithuania and Ukraine. **1926:** Pilsudski seized full power in coup and established an autocratic regime. **1935:** On Pilsudski's death, a military regime held power under Marshal Smigly-Rydz. **1939:** Invaded by Germany; western Poland incorporated into Nazi Reich (state) and the rest became a German colony; 6 million Poles – half of them Jews – were slaughtered in the next five years. **1944–45:** Liberated from Nazi rule by Soviet Union's Red Army; boundaries redrawn westwards at the Potsdam Conference. One half of 'old Poland', 180,000 sq km/70,000 sq mi in the east, was lost to the USSR; 100,000 sq km/40,000 sq mi of ex-German territory in Silesia, along the Oder and Neisse rivers, was added, shifting the state 240 km/150 mi westwards; millions of Germans were expelled. **1947:** Communist people's republic proclaimed after manipulated election. **1949:** Joined Comecon. **early 1950s:** Harsh Stalinist rule under communist leader Boleslaw Bierut: nationalization; rural collectivization; persecution of Catholic Church members. **1955:** Joined Warsaw Pact defence organization. **1956:** Poznan strikes and riots. The moderate Wladyslaw Gomulka installed as Polish United Workers' Party (PUWP) leader. **1960s:** Private farming reintroduced and Catholicism tolerated. **1970:** Gomulka replaced by Edward Gierek after Gdansk riots against food price rises. **1970s:** Poland heavily indebted to foreign creditors after a failed attempt to boost economic growth. **1980:** Solidarity, led by Lech Walesa, emerged as free trade union following Gdansk disturbances. **1981:** Martial law imposed by General Wojciech Jaruzelski, trade-union activity banned, and Solidarity leaders and supporters arrested. **1983:** Martial law ended. **1984:** Amnesty for 35,000 political prisoners. **1988:** Solidarity-led strikes and demonstrations for pay increases. Reform-communist Mieczyslaw Rakowski became prime minister. **1989:** Agreement to relegalize Solidarity, allow opposition parties, and adopt a more democratic constitution, after round-table talks involving Solidarity, the Communist Party, and the Catholic Church. Widespread success for Solidarity in first open elections for 40 years; noncommunist 'grand coalition' government was formed, headed by Tadeusz Mazowiecki of Solidarity. **1990:** The PUWP was dissolved and re-formed as the Democratic Left Alliance (SLD). Walesa was elected president and Jan Bielecki became prime minister. **1991:** A shock-therapy economic restructuring programme, including large-scale privatization, produced a sharp fall in living standards and a rise in the unemployment rate to 11%. The unpopular Bielecki resigned and, after inconclusive elections, Jan Olszewski formed a fragile centre–right coalition government. **1992:** The political instability continued. **1993:** The economy became the first in Central Europe to grow since the collapse of communism. **1994:** Poland joined the NATO 'partnership for peace' programme; the last Russian troops left the country. **1995:** Aleksander Kwasniewski, leader of the SLD, was elected president. **1997:** Further structural reform and privatization took place and a new constitution was approved. Poland was invited to join NATO and begin negotiations to join the European Union (EU). A general election was won by Solidarity Electoral Action (AWS). A coalition government was formed, led by Jerzy Buzek. **1998:** Full EU membership negotiations commenced. The government was weakened by defections to the opposition. The number of provinces was reduced from 49 to 16. **1999:** Poland became a full member of NATO.

Practical information

Visa requirements: UK: visa not required for a stay of up to six months. USA: visa not required
Time difference: GMT +1
Chief tourist attractions: historic cities of Gdansk, Wroclaw, Kraków, Poznan, and Warsaw; numerous health and climatic resorts; mountain and forest scenery
Major holidays: 1 January, 1, 3, 9 May, 15 August, 1, 11 November, 25–26 December; variable: Corpus Christi, Easter Monday

PORTUGAL

Located in southwestern Europe, on the Atlantic Ocean, bounded north and east by Spain.

National name: *República Portuguesa/Republic of Portugal*
Area: 92,000 sq km/35,521 sq mi (including the Azores and Madeira)
Capital: Lisbon
Major towns/cities: Porto, Coimbra, Amadora, Setúbal, Funchal, Braga, Vila Nova de Gaia
Major ports: Porto, Setúbal
Physical features: mountainous in the north (Serra da Estrêla mountains); plains in the south; rivers Minho, Douro, Tagus (Tejo), Guadiana

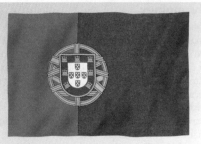

Green and red replaced blue and white as the national colours in 1910. The armillary sphere surrounds the shield of Portugal. Effective date: 30 June 1911.

Government

Head of state: Jorge Sampaio from 1996
Head of government: Antonio Guterres from 1995
Political system: liberal democracy
Political executive: dual executive
Administrative divisions: 18 districts and two autonomous regions
Political parties: Social Democratic Party (PSD), moderate left of centre; Socialist Party (PS), left of centre; People's Party (PP), right wing, anti-European integration
Armed forces: 53,600 (1998)
Conscription: 4–18 months
Death penalty: abolished in 1976
Defence spend: (% GDP) 2.3 (1998)
Education spend: (% GNP) 4.9 (1996)
Health spend: (% GDP) 7.9 (1997)

Economy and resources

Currency: escudo
GDP: (US$) 107.3 billion (1999)
Real GDP growth: (% change on previous year) 3 (1999)
GNP: (US$) 105.9 billion (1999)
GNP per capita (PPP): (US$) 15,147 (1999)
Consumer price inflation: 2.3% (1999)
Unemployment: 4.9% (1998)
Major trading partners: EU (principally Spain, Germany, France, Italy, Belgium–Luxembourg), USA, Japan
Resources: limestone, granite, marble, iron, tungsten, copper, pyrites, gold, uranium, coal, forests
Industries: textiles and clothing, footwear, paper pulp, cork items (world's largest producer of cork), chemicals, petroleum refining, fish processing, viticulture, electrical appliances, ceramics, tourism
Exports: textiles, clothing, footwear, pulp and waste paper, wood and cork manufactures, tinned fish, electrical equipment, wine, refined petroleum. Principal market: Germany 19.9% (1999)
Imports: foodstuffs, machinery and transport equipment, crude petroleum, natural gas, textile yarn, coal, rubber, plastics, tobacco. Principal source: Spain 24.8% (1999)
Arable land: 23.5% (1996)
Agricultural products: wheat, maize, rice, potatoes, tomatoes, grapes, olives, fruit; fishing (1993 sardine catch was the world's largest at 89,914 tonnes)

Population and society

Population: 9,875,000 (2000 est)
Population growth rate: 0.04% (1995–2000)
Population density: (per sq km) 107 (1999 est)
Urban population: (% of total) 64 (2000 est)
Age distribution: (% of total population) 0–14 16%, 15–59 64%, 60+ 20% (2000 est)
Ethnic groups: most of the population is descended from Caucasoid peoples who inhabited the whole of the Iberian peninsula in classical and pre-classical times; there are a number of minorities from Portugal's overseas possessions and former possessions
Language: Portuguese (official)
Religion: Roman Catholic 97%
Education: (compulsory years) 9
Literacy rate: 95% (men); 90% (women) (2000 est)
Labour force: 13.7% agriculture, 31.5% industry, 54.8% services (1997)
Life expectancy: 72 (men); 79 (women) (1995–2000)
Child mortality rate: (under 5, per 1,000 live births) 11 (1995–2000)
Physicians: 1 per 333 people (1996)
Hospital beds: 1 per 244 people (1996)
TV sets: (per 1,000 people) 542 (1998)
Radios: (per 1,000 people) 304 (1997)
Internet users: (per 10,000 people) 701.4 (1999)
Personal computer users: (per 100 people) 9.3 (1999)

Transport

Airports: international airports: Lisbon (Portela de Sacavem), Faro, Oporto (Oporto Sá Carneiro), Madeira (Funchal), Azores (Santa Marta), São Miguel; domestic services operate between these; total passenger km: 9,343 million (1997 est)
Railways: total length: 3,072 km/1,909 mi; total passenger km: 4,503 million (1996)
Roads: total road network: 68,732 km/42,710 mi, of which 87% paved (1995); passenger cars: 257 per 1,000 people (1995)

Chronology

2nd century BC: Romans conquered Iberian peninsula. **5th century AD:** Iberia overrun by Vandals and Visigoths after fall of Roman Empire. **711:** Visigoth kingdom overthrown by Muslims invading from North Africa. **997–1064:** Christians resettled northern area, which came under rule of Léon and Castile. **1139:** Afonso I, son of Henry of Burgundy, defeated Muslims; the area became an independent kingdom. **1340:** Final Muslim invasion defeated. **15th century:** Age of exploration: Portuguese mariners surveyed coast of Africa, opened sea route to India (Vasco da Gama), and reached Brazil (Pedro Cabral). **16th century:** 'Golden Age': Portugal flourished as commercial and colonial power. **1580:** Philip II of Spain took throne of Portugal. **1640:** Spanish rule overthrown in bloodless coup; Duke of Braganza proclaimed as King John IV. **1668:** Spain recognized Portuguese independence. **1755:** Lisbon devastated by earthquake. **1807:** Napoleonic France invaded Portugal; Portuguese court fled to Brazil. **1807–11:** In the Peninsular War British forces played a leading part in liberating Portugal from the French. **1820:** Liberal revolution forced King John VI to return from Brazil and accept constitutional government. **1822:** First Portuguese constitution adopted. **1828:** Dom Miguel blocked the succession of his niece, Queen Maria, and declared himself absolute monarch; civil war ensued between liberals and conservatives. **1834:** Queen Maria regained the throne with British, French, and Brazilian help; constitutional government restored. **1840s:** Severe disputes between supporters of radical 1822 constitution and more conservative 1826 constitution. **late 19th century:** Government faced severe financial difficulties; rise of socialist, anarchist, and republican parties. **1908:** Assassination of King Carlos I. **1910:** Portugal became republic after a three-day insurrection forced King Manuel II to flee. **1911:** New regime adopted liberal constitution, but republic proved unstable, violent, and corrupt. **1916–18:** Portugal fought in World War I on Allied side. **1926–51:** Popular military coup installed Gen António de Fragoso Carmona as president. **1933:** Authoritarian 'Estado Novo' ('New State') constitution adopted. **1949:** Portugal became founding member of North Atlantic Treaty Organization (NATO). **1974:** Army seized power to end stalemate situation in African colonial wars. **1975:** Portuguese colonies achieved independence. **1976:** First free elections in 50 years. **1986:** Soares became the first civilian president in 60 years; Portugal joined the European Community (EC). **1989:** The Social Democrat government started to dismantle the socialist economy and privatize major industries. **1995:** Antonio Gutteres was elected prime minister in the legislative elections. **1996:** Jorge Sampaio (PS) was elected president.

Practical information

Visa requirements: UK: visa not required for a stay of up to three months. USA: visa not required for a stay of up to two months
Time difference: GMT +/–0
Chief tourist attractions: mild climate; historic town of Lisbon; summer resorts in the Algarve; winter resorts on Madeira and the Azores
Major holidays: 1 January, 25 April, 10 June, 15 August, 5 October, 1 November, 1, 8, 24–25 December; variable: Carnival, Corpus Christi, Good Friday

QATAR

Located in the Middle East, occupying Qatar peninsula in the Arabian Gulf, bounded southwest by Saudi Arabia and south by United Arab Emirates.

National name: *Dawlat Qatar/State of Qatar*
Area: 11,400 sq km/4,401 sq mi
Capital: Doha (and chief port)
Major towns/cities: Dukhan, Wakra, ad-Dawhah, ar-Rayyan, Umm Salal, Musay'id, aš-Šahniyah
Physical features: mostly flat desert with salt flats in south

The flags of Qatar and Bahrain are very similar, reflecting the countries' historical links. The proportions of the Qatari flag are unique. Effective date: c. 1949.

Government

Head of state: Sheikh Hamad bin Khalifa al-Thani from 1995
Head of government: Sheikh 'Abd Allah ibn Khalifah Al Thani from 1996
Political system: absolutist
Political executive: absolute
Administrative divisions: nine municipalities
Political parties: none
Armed forces: 11,800 (1998)
Conscription: military service is voluntary
Death penalty: retained and used for ordinary crimes
Defence spend: (% GDP) 12.0 (1998)
Education spend: (% GNP) 3.4 (1996)
Health spend: (% GDP) 3.4 (1995)

Economy and resources

Currency: Qatari riyal
GDP: (US$) 11.8 billion (1999)
Real GDP growth: (% change on previous year) 0.2 (1999)
GNP: (US$) N/A
GNP per capita (PPP): (US$) N/A
Consumer price inflation: 2% (1999)
Unemployment: dependent on immigrant workers – shortage of indigenous labour
Foreign debt: (US$) 11 billion (1999 est)
Major trading partners: Japan, Italy, USA, South Korea, Singapore, UK, Germany, Thailand
Resources: petroleum, natural gas, water resources
Industries: petroleum refining and petroleum products, industrial chemicals, iron and steel, flour, cement, concrete, plastics, paint
Exports: petroleum, liquefied natural gas, petrochemicals. Principal market: Japan 50.3% (1998)
Imports: machinery and transport equipment, basic manufactures, food and live animals, miscellaneous manufactured articles, chemicals. Principal source: Japan 15.4% (1998)

Arable land: 0.7% (1995)
Agricultural products: cereals, vegetables, fruits (especially dates); livestock rearing; fishing

Population and society

Population: 599,000 (2000 est)
Population growth rate: 1.8% (1995–2025)
Population density: (per sq km) 52 (1999 est)
Urban population: (% of total) 93 (2000 est)
Age distribution: (% of total population) 0–14 26%, 15–59 70%, 60+ 4% (2000 est)
Ethnic composition: only about 30% of the population are indigenous Qataris; 40% are Arabs, and the others Pakistanis, Indians, and Iranians
Language: Arabic (official), English
Religion: Sunni Muslim 95%
Education: not compulsory
Literacy rate: 80% (men); 83% (women) (2000 est)
Labour force: 1.6% agriculture, 32.4% industry, 66% services (1997 est)
Life expectancy: 70 (men); 75 (women) (1995–2000)
Child mortality rate: (under 5, per 1,000 live births) 23 (1995–2000)
Physicians: 1 per 664 people (1995)
Hospital beds: 1 per 467 people (1995)
TV sets: (per 1,000 people) 404 (1997)
Radios: (per 1,000 people) 450 (1997)
Internet users: (per 10,000 people) 763.8 (1999)
Personal computer users: (per 100 people) 13.6 (1999)

Transport

Airports: international airports: Doha; total passenger km: 2,766 million (1995)
Railways: none
Roads: total road network: 1,230 km/764 mi, of which 90% paved (1996 est); passenger cars: 218 per 1,000 people (1996 est)

Chronology

7th century AD: Islam introduced. **8th century:**
Developed into important trading centre during time of
Abbasid Empire. **1783:** The al-Khalifa family, who had
migrated to northeast Qatar from west and north of the
Arabian Peninsula, foiled Persian invasion and moved their
headquarters to Bahrain Island, while continuing to rule the
area of Qatar. **1867–68:** After the Bahrain-based al-Khalifa
had suppressed a revolt by their Qatari subjects, destroying the
town of Doha, Britain intervened and installed Muhammad
ibn Thani al-Thani, from the leading family of Qatar, as the
ruling sheikh (or emir). A British Resident was given power to
arbitrate disputes with Qatar's neighbours. **1871–1914:**
Nominally part of Turkish Ottoman Empire, although in 1893
sheik's forces inflicted a defeat on Ottomans. **1916:** Qatar
became British protectorate after treaty signed with Sheikh
Adbullah al-Thani. **1949:** Oil production began at onshore
Dukhan field in west. **1960:** Sheikh Ahmad al-Thani became
new emir. **1968:** Britain's announcement that it would
remove its forces from the Persian Gulf by 1971 led Qatar to
make an abortive attempt to arrange a federation of Gulf
states. **1970:** Constitution adopted, confirming emirate as
absolute monarchy. **1971:** Independence achieved from
Britain. **1991:** Qatar forces joined the United Nations (UN)
coalition in the Gulf War against Iraq. **1995:** Sheikh Khalifa
was ousted by his son, Crown Prince Sheikh Hamad bin
Khalifa al-Thani. **1996:** The announcement of plans to
introduce democracy were followed by an assassination
attempt on Sheikh Hamad. Sheikh 'Abd Allah ibn Khalifah Al
Thani was appointed prime minister.

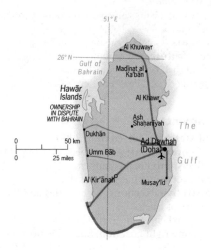

Practical information

Visa requirements: UK: visa not required for a stay
of up to 30 days. USA: visa required

Time difference: GMT +3
Major holidays: 3 September, 31 December; variable:
Eid-ul-Adha (4 days), end of Ramadan (4 days)

ROMANIA

Located in southeast Europe, bounded north and east by Ukraine, east by Moldova, southeast by the Black Sea, south by Bulgaria, southwest by Yugoslavia, and northwest by Hungary.

National name: *România/Romania*
Area: 237,500 sq km/91,698 sq mi
Capital: Bucharest
Major towns/cities: Brasov, Timisoara, Cluj-Napoca, Iasi, Constanta, Galati, Craiova
Major ports: Galati, Constanta, Braila
Physical features: mountains surrounding a plateau, with river plains in south and east. Carpathian Mountains, Transylvanian Alps; River Danube; Black Sea coast; mineral springs

Nowadays the colours are said to stand for Moldavia, Transylvania, and Wallachia.
Effective date: 27 December 1989.

Government

Head of state: Ion Iliescu from 2000
Head of government: Adrian Nastase from 2000
Political system: liberal democracy
Political executive: limited presidency
Administrative divisions: 41 counties and the municipality of Bucharest
Political parties: Democratic Convention of Romania (DCR), centre-right coalition; Social Democratic Union (SDU), reformist; Social Democracy Party of Romania (PSDR), social democrat; Romanian National Unity Party (RNUP), Romanian nationalist, right wing, anti-Hungarian; Greater Romania Party (Romania Mare), far right, ultranationalist, anti-Semitic; Democratic Party–National Salvation Front (DP–NSF), promarket; National Salvation Front (NSF), left of centre; Hungarian Democratic Union of Romania (HDUR), ethnic Hungarian; Christian Democratic–National Peasants' Party (CD–PNC), right of centre, promarket; Socialist Labour Party (SLP), ex-communist
Armed forces: 219,700 (1998)
Conscription: military service is compulsory for 12–18 months
Death penalty: abolished in 1989
Defence spend: (% GDP) 2.3 (1998)
Education spend: (% GNP) 3.6 (1996)
Health spend: (% GDP) 3.8 (1997)

Economy and resources

Currency: leu
GDP: (US$) 33.7 billion (1999)
Real GDP growth: (% change on previous year) –3.2 (1999)
GNP: (US$) 34.2 billion (1999)
GNP per capita (PPP): (US$) 5,647 (1999)
Consumer price inflation: 45.8% (1999)
Unemployment: 8.8% (1997)
Foreign debt: (US$) 9.4 billion (1999)
Major trading partners: Italy, Germany, France, Russia, Iran, China, Turkey
Resources: brown coal, hard coal, iron ore, salt, bauxite, copper, lead, zinc, methane gas, petroleum (reserves

expected to be exhausted by mid- to late 1990s)
Industries: metallurgy, mechanical engineering, chemical products, timber and wood products, textiles and clothing, food processing
Exports: base metals and metallic articles, textiles and clothing, machinery and equipment, mineral products, foodstuffs. Principal market: Italy 23.4% (1999)
Imports: mineral products, machinery and mechanical appliances, textiles, motor cars. Principal source: Italy 19.6% (1999)
Arable land: 40.5% (1996)
Agricultural products: wheat, maize, potatoes, sugar beet, barley, apples, grapes, sunflower seeds; wine production; forestry; fish breeding

Population and society

Population: 22,327,000 (2000 est)
Population growth rate: –0.4% (1995–2000)
Population density: (per sq km) 94 (1999 est)
Urban population: (% of total) 56 (2000 est)
Age distribution: (% of total population) 0–14 18%, 15–59 63%, 60+ 19% (2000 est)
Ethnic groups: 89% non-Slavic ethnic Romanian; substantial Hungarian (7%), Romany (2%), German (0.5%), and Serbian minorities
Language: Romanian (official), Hungarian, German
Religion: Romanian Orthodox 87%; Roman Catholic and Uniate 5%, Reformed/Lutheran 3%, Unitarian 1%
Education: (compulsory years) 8
Literacy rate: 99% (men); 97% (women) (2000 est)
Labour force: 37.5% agriculture, 32% industry, 30.5% services (1997)
Life expectancy: 66 (men); 74 (women) (1995–2000)
Child mortality rate: (under 5, per 1,000 live births) 33 (1995–2000)
Physicians: 1 per 553 people (1996)
Hospital beds: 1 per 131 people (1996)
TV sets: (per 1,000 people) 233 (1997)
Radios: (per 1,000 people) 319 (1997)
Internet users: (per 10,000 people) 267.8 (1999)
Personal computer users: (per 100 people) 2.7 (1999)

Transport

Airports: international airports: Bucharest (Otopeni), Constanta (Mihail Kogainiceanu), Timisoara, Arad; 12 domestic airports; total passenger km: 1,857 million (1997)

Railways: total length: 11,365 km/7,062 mi; total passenger km: 15,794 million (1997)
Roads: total road network: 153,358 km/95,297 mi, of which 51% paved (1996); passenger cars: 106 per 1,000 people (1996 est)

Chronology

106: Formed heartland of ancient region of Dacia, which was conquered by Roman Emperor Trajan and became a province of Roman Empire; Christianity introduced. **275:** Taken from Rome by invading Goths, a Germanic people. **4th–10th centuries:** Invaded by successive waves of Huns, Avars, Bulgars, Magyars, and Mongols. *c.* **1000:** Transylvania, in north, became an autonomous province under Hungarian crown. **mid-14th century:** Two Romanian principalities emerged, Wallachia in south, around Bucharest, and Moldova in northeast. **15th–16th centuries:** The formerly autonomous principalities of Wallachia, Moldova, and Transylvania became tributaries to Ottoman Turks, despite peasant uprisings and resistance from Vlad Tepes ('the Impaler'), ruling prince of Wallachia. **late 17th century:** Transylvania conquered by Austrian Habsburgs. **1829:** Wallachia and Moldova brought under tsarist Russian suzerainty. **1859:** Under Prince Alexandru Ion Cuza, Moldova and Wallachia united to form Romanian state. **1878:** Romania's independence recognized by Great Powers in Congress of Berlin. **1881:** Became kingdom under Carol I. **1916–18:** Fought on Triple Entente side (Britain, France, and Russia) during World War I; acquired Transylvania and Bukovina, in north, from dismembered Austro-Hungarian Empire, and Bessarabia, in east, from Russia. This made it the largest state in Balkans. **1930:** King Carol II abolished democratic institutions and established dictatorship. **1940:** Forced to surrender Bessarabia and northern Bukovina, adjoining Black Sea, to Soviet Union, and northern Transylvania to Hungary; King Carol II abdicated, handing over effective power to Gen Ion Antonescu, who signed Axis Pact with Germany. **1941–44:** Fought on Germany's side against Soviet Union; thousands of Jews massacred. **1944:** Romania joined war against Germany. **1945:** Occupied by Soviet Union; communist-dominated government installed. **1947:** Paris Peace Treaty reclaimed Transylvania for Romania, but lost southern Dobruja to Bulgaria and northern Bukovina and Bessarabia to Soviet Union; King Michael, son of Carol II, abdicated and People's Republic proclaimed. **1955:** Romania joined Warsaw Pact. **1958:** Soviet occupation forces removed. **1965:** Nicolae Ceausescu became Romanian Communist Party leader, and pursued foreign policy autonomous of Moscow. **1975:** Ceausescu made president. **1985–87:** Winter of austerity and power cuts as Ceausescu refused to liberalize the economy. Workers' demonstrations against austerity programme are brutally crushed at Brasov. **1989:** Bloody overthrow of Ceausescu regime in 'Christmas Revolution'; Ceausescu and wife tried and executed; estimated 10,000 dead in civil war. Power assumed by NSF, headed by Ion Iliescu. **1990:** Securitate secret police was replaced by new Romanian Intelligence Service; Eastern Orthodox Church and private farming were re-legalized. **1994:** A military cooperation pact was made with Bulgaria. Far-right parties were brought into the governing coalition. **1996:** There were signs of economic growth; parliamentary elections were won by the DCR, who formed a coalition government with the SDU. **1997:** An economic reform programme and drive against corruption were announced; there was a sharp increase in inflation. Former King Michael returned from exile. **1998:** The Social Democrats withdrew support from ruling coalition, criticizing the slow pace of reform. Full EU membership negotiations commenced. The economy deteriorated sharply. **1999:** Roadblocks were imposed by tanks north of Bucharest to prevent 10,000 striking miners entering Bucharest. Mugur Isarescu became prime minister. **2000:** Former communist president Ion Iliescu was elected president, and his Social Democrats won the largest share of the vote in parliamentary elections.

Practical information

Visa requirements: UK: visa required. USA: visa not required for a stay of up to 30 days
Time difference: GMT +2
Chief tourist attractions: Black Sea resorts (including

Mangalia, Mamaia, and Eforie); Carpathian Mountains; Danube delta
Major holidays: 1–2 January, 15 April, 1 May, 1, 25–26 December

RUSSIA FORMERLY RUSSIAN SOVIET FEDERAL SOCIALIST REPUBLIC (RSFSR; WITHIN THE SOVIET UNION USSR) (UNTIL 1991)

Located in northern Asia and eastern Europe, bounded north by the Arctic Ocean; east by the Bering Sea and the Sea of Okhotsk; west by Norway, Finland, the Baltic States, Belarus, and Ukraine; and south by China, Mongolia, Georgia, Azerbaijan, and Kazakhstan.
National name: *Rossiiskaya Federatsiya/Russian Federation*
Area: 17,075,400 sq km/6,592,811 sq mi
Capital: Moscow
Major towns/cities: St Petersburg, Nizhniy Novgorod, Samara, Yekaterinburg, Novosibirsk, Chelyabinsk, Kazan, Omsk, Perm, Ufa
Physical features: fertile Black Earth district; extensive forests; the Ural Mountains with large mineral resources; Lake Baikal, world's deepest lake

White, blue, and red became known as the pan-Slavic colours, influencing many other Eastern European flags. White, blue, and red are also the colours of the arms of the Duchy of Moscow.
Effective date: 11 December 1993.

Government

Head of state: Vladimir Putin from 2000
Head of government: Mikhail Kasyanov from 2000
Political system: emergent democracy
Political executive: limited presidency
Administrative divisions: 21 republics, 6 provinces, 49 regions, 10 autonomous districts, two cities with federal status (Moscow and St Petersburg), and one autonomous area
Political parties: Russia is Our Home, centrist; Party of Unity and Accord (PRUA), moderate reformist; Communist Party of the Russian Federation (CPRF), left wing, conservative (ex-communist); Agrarian Party, rural-based, centrist; Liberal Democratic Party, far right, ultranationalist; Congress of Russian Communities, populist, nationalist; Russia's Choice, reformist, right of centre; Yabloko, gradualist free market; Russian Social Democratic People's Party (Derzhava), communist-nationalist; Patriotic Popular Union of Russia (PPUR), communist-led; Russian People's Republican Party (RPRP)
Armed forces: 1,159,000; paramilitary forces of 543,000 (1998)
Conscription: two years
Death penalty: retained and used for ordinary crimes
Defence spend: (% GDP) 5.2 (1998)
Education spend: (% GNP) 4.1 (1996)
Health spend: (% GDP) 3.6 (1998)

Economy and resources

Currency: rouble
GDP: (US$) 375.3 billion (1999)
Real GDP growth: (% change on previous year) 3.2 (1999)
GNP: (US$) 332.5 billion (1999)
GNP per capita (PPP): (US$) 6,339 (1999)
Consumer price inflation: 85.8% (1999)
Unemployment: 13.3% (1998)
Foreign debt: (US$) 174.3 billion (1999 est)
Major trading partners: Ukraine, Germany, Belarus, Ukraine, Kazakhstan, USA, Italy, the Netherlands

Resources: petroleum, natural gas, coal, peat, copper (world's fourth-largest producer), iron ore, lead, aluminium, phosphate rock, nickel, manganese, gold, diamonds, platinum, zinc, tin
Industries: cast iron, steel, rolled iron, synthetic fibres, soap, cellulose, paper, cement, machinery and transport equipment, glass, bricks, food processing, confectionery
Exports: mineral fuels, ferrous and non-ferrous metals and derivatives, precious stones, chemical products, machinery and transport equipment, weapons, timber and paper products. Principal market: Ukraine 8.8% (1999)
Imports: machinery and transport equipment, grain and foodstuffs, chemical products, textiles, clothing, footwear, pharmaceuticals, metals. Principal source: Germany 13.8% (1999)
Arable land: 7.8% (1996)
Agricultural products: grain, potatoes, flax, sunflower seed, vegetables, fruit and berries, tea; livestock and dairy farming

Population and society

Population: 146,934,000 (2000 est)
Population growth rate: –0.2% (1995–2000)
Population density: (per sq km) 9 (1999 est)
Urban population: (% of total) 78 (2000 est)
Age distribution: (% of total population) 0–14 18%, 15–59 64%, 60+ 18% (2000 est)
Ethnic groups: predominantly ethnic Russian (eastern Slav); significant Tatar, Ukranian, Chuvash, Belarussian, Bashkir, and Chechen minorities; over 130 nationalities
Language: Russian (official) and many East Slavic, Altaic, Uralic, Caucasian languages
Religion: traditionally Russian Orthodox; significant Muslim and Buddhist communities
Education: (compulsory years) 9
Literacy rate: 99% (men); 99% (women) (2000 est)
Labour force: 52% of population: 14% agriculture, 42% industry, 45% services (1990)
Life expectancy: 61 (men); 73 (women) (1995–2000)
Child mortality rate: (under 5, per 1,000 live births) 22 (1995–2000)

Physicians: 1 per 213 people (1998)
Hospital beds: 1 per 84 people (1998)
TV sets: (per 1,000 people) 420 (1998)
Radios: (per 1,000 people) 418 (1997)
Internet users: (per 10,000 people) 183.4 (1999)
Personal computer users: (per 100 people) 3.7 (1999)

Transport

Airports: international airports: Moscow (Sheremetyevo), St Petersburg (Pulkovo); six principal domestic airports operate services to all major cities; total passenger km: 62,016 million (1997 est)
Railways: total length: 91,116 km/56,619 mi; total passenger km: 170,300 million (1997)
Roads: total road network: 570,719 km/354,645 mi, of which 78.8% paved (1997); passenger cars: 13.7 per 1,000 people (1997)

Chronology

9th–10th centuries: Viking chieftains established own rule in Novgorod, Kiev, and other cities. **10th–12th centuries:** Kiev temporarily united Russian peoples into its empire. Christianity introduced from Constantinople 988. **13th century:** Mongols (Golden Horde) overran the southern steppes in 1223, compelling Russian princes to pay tribute. **14th century:** Byelorussia and Ukraine came under Polish rule. **1462–1505:** Ivan the Great, grand duke of Muscovy, threw off Mongol yoke and united lands in the northwest. **1547–84:** Ivan the Terrible assumed title of tsar and conquered Kazan and Astrakhan; colonization of Siberia began. **1613:** First Romanov tsar, Michael, elected after period of chaos. **1667:** Following Cossack revolt, eastern Ukraine reunited with Russia. **1682–1725:** Peter the Great modernized the bureaucracy and army; he founded a navy and a new capital, St Petersburg, introduced Western education, and wrested the Baltic seaboard from Sweden. By 1700 colonization of Siberia had reached the Pacific. **1762–96:** Catherine the Great annexed the Crimea and part of Poland and recovered western Ukraine and Byelorussia. **1798–1814:** Russia intervened in Revolutionary and Napoleonic Wars (1798–1801, 1805–07), repelled Napoleon, and took part in his overthrow (1812–14). **1827–29:** Russian attempts to dominate the Balkans led to a war with Turkey. **1853–56:** Crimean War. **1856–64:** Caucasian War of conquest completed the annexation of northern Caucasus, causing more than a million people to emigrate. **1858–60:** Treaties of Aigun (1858) and Peking (1860) imposed on China, annexing territories north of the Amur and east of the Ussuri rivers; Vladivostok founded on Pacific coast. **1861:** Serfdom abolished. Rapid growth of industry followed, a working-class movement developed, and revolutionary ideas spread, culminating in the assassination of Alexander II in 1881. **1877–78:** Russo-Turkish War **1898:** Social Democratic Party founded by Russian Marxists; split into Bolshevik and Menshevik factions in 1903. **1904–05:** Russo-Japanese War caused by Russian expansion in Manchuria. **1905:** A revolution, though suppressed, forced tsar to accept parliament (Duma) with limited powers. **1914:** Russo-Austrian rivalry in Balkans was a major cause of outbreak of World War I; Russia fought in alliance with France and Britain. **1917:** Russian Revolution: tsar abdicated, provisional government established; Bolsheviks seized power under Vladimir Lenin. **1918:** Treaty of Brest-Litovsk ended war with Germany; murder of former tsar; Russian Empire collapsed; Finland, Poland, and Baltic States seceded.

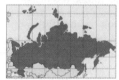

1918–22: Civil War between Red Army, led by Leon Trotsky, and White Russian forces with foreign support; Red Army ultimately victorious; control regained over Ukraine, Caucasus, and Central Asia. **1922:** Former Russian Empire renamed Union of Soviet Socialist Republics. **1924:** Death of Lenin. **1928:** Joseph Stalin emerged as absolute ruler after ousting Trotsky. **1928–33:** First five-year plan collectivized agriculture by force; millions died in famine. **1936–38:** The Great Purge: Stalin executed his critics and imprisoned millions of people on false charges of treason and sabotage. **1939:** Nazi-Soviet nonaggression pact; USSR invaded eastern Poland and attacked Finland. **1940:** USSR annexed Baltic States. **1941–45:** 'Great Patriotic War' against Germany ended with Soviet domination of eastern Europe and led to 'Cold War' with USA and its allies. **1949:** Council for Mutual Economic Assistance (Comecon) created to supervise trade in Soviet bloc. **1953:** Stalin died; 'collective leadership' in power. **1955:** Warsaw Pact created. **1956:** Nikita Khrushchev made 'secret speech' criticizing Stalin; USSR invaded Hungary. **1957–58:** Khrushchev ousted his rivals and became effective leader, introducing limited reforms. **1960:** Rift between USSR and Communist China. **1962:** Cuban missile crisis: Soviet nuclear missiles installed in Cuba but removed after ultimatum from USA. **1964:** Khrushchev ousted by new 'collective leadership' headed by Leonid Brezhnev and Alexei Kosygin. **1968:** USSR and allies invaded Czechoslovakia. **1970s:** 'Détente' with USA and Western Europe. **1979:** USSR invaded Afghanistan; fighting continued until Soviet withdrawal ten years later. **1985:** Mikhail Gorbachev became leader and announced wide-ranging reform programme (*perestroika*). **1986:** Chernobyl nuclear disaster. **1988:** Special All-Union Party Congress approved radical constitutional changes and market reforms; start of open nationalist unrest in Caucasus and Baltic republics. **1989:** Multi-candidate elections held in move towards 'socialist democracy'; collapse of Soviet satellite regimes in eastern Europe; end of Cold War. **1990:** Baltic and Caucasian republics defied central government; Boris Yeltsin became president of Russian Federation and left the Communist Party. **1991:** There was an unsuccessful coup by hardline communists; republics declared independence; communist rule dissolved in the Russian Federation; the USSR was replaced by a loose Commonwealth of Independent States (CIS). Mikhail Gorbachev, president of the USSR, resigned, leaving power to Yeltsin. **1992:** Russia assumed former USSR seat on the United Nations (UN) Security Council; a new constitution was devised; end of price controls. **1993:** There was a power struggle between Yeltsin and the Congress of People's Deputies; congress was dissolved; an attempted coup was foiled; a new parliament was elected. **1994:** Russia joined NATO 'Partnership for Peace'; Russian forces invaded the breakaway republic of Chechnya. **1997:** A peace treaty was signed with Chechnya. Yeltsin signed an agreement on cooperation with NATO. Russia gained effective admission to the G-7 group. **1998:** President Yeltsin sacked the government and appointed Sergei Kiriyenko as prime minister. The rouble was heavily devalued. Yevgeny Primakov replaced Kiriyenko as prime minister and market-centred reform was abandoned. The USA pledged aid of over 3 million tonnes of grain and meat, after a 5% contraction in GDP in 1998. **1999:** Yeltsin dismissed first Primakov's government, and then in August, Stepashin's government, appointing Vladimir Putin as prime minister. Troubles with Chechnya continued and Russian forces claimed to have surrounded the capital, Groznyy, and issued an ultimatum to civilians that they must leave or die. After Western protests the Russian ultimatum was deferred by a week. President Yeltsin resigned on 31 December, and Putin took over as acting president. **2000:** Vladimir Putin was elected president, and sought to reassert central control. Mikhail Kasyanov was appointed prime minister. The Russian army in Chechnya declared it had secured control of the region, despite continuing rebel activity. In August a nuclear-powered submarine, the Kursk, sank after an explosion caused by the misfire of a torpedo. Putin was slow to request Western help in the abortive rescue mission, and all 118 crew died. In December the Duma voted to restore the old Soviet national anthem, though with different words, and re-instate the tsarist flag and double-eagle crest as national emblems. **2001:** Putin announced that control of the war in Chechnya would be transferred to the secret police.

Practical information

Visa requirements: UK: visa required. USA: visa required

Time difference: GMT +2–12

Chief tourist attractions: historic cities of Moscow and St Petersburg, with their cathedrals, fortresses, and art treasures – the Hermitage in St Petersburg has one of the world's largest art collections; the Great Palace at Petrodvorets (formerly Peterhof); Trans-Siberian Railway; the Tver region is famed for its rivers, lakes, reservoirs, and other waterways; the Yaroslavl region has several historic towns including Yaroslavl and Uglich; the Nizhegorodskaya region has the medieval cities of Nizhniy Novgorod and Gordets, Vladimir, and Novgorod, one of Russia's oldest and grandest cities, with its famous early churches; the country's landscape includes forests, lakes, marshes, and pasture, with a rich variety of wildlife and geological formations including the volcanoes of Kamchatka and the cave formations near Archangelsk

Major holidays: 1, 7 January, 8 March, 15 April, 1–2, 9 May, 12 June, 22 August, 7 November

RWANDA FORMERLY RUANDA (UNTIL 1962)

Landlocked country in central Africa, bounded north by Uganda, east by Tanzania, south by Burundi, and west by the Democratic Republic of Congo (formerly Zaire).

National name: *Republika y'u Rwanda/Republic of Rwanda*
Area: 26,338 sq km/10,169 sq mi
Capital: Kigali
Major towns/cities: Butare, Ruhengeri, Gisenyi, Kibungo, Cyangugu
Physical features: high savannah and hills, with volcanic mountains in northwest; part of lake Kivu; highest peak Mount Karisimbi 4,507 m/14,792 ft; Kagera River (whose headwaters are the source of the Nile)

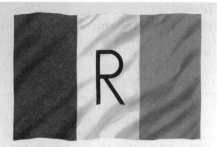

Red stands for the blood shed in the revolution. Yellow represents victory over tyranny. Green expresses hope for the future. 'R' stands for Rwanda, republic, revolution, and referendum. Effective date: c. September 1961.

Government

Head of state: Paul Kagame from 2000
Head of government: Bernard Makuza from 2000
Political system: authoritarian nationalist
Political executive: unlimited presidency
Administrative divisions: 12 prefectures
Political parties: National Revolutionary Development Movement (MRND), nationalist-socialist, Hutu-oriented; Social Democratic Party (PSD), left of centre; Christian Democratic Party (PDC), Christian, centrist; Republican Democratic Movement (MDR), Hutu nationalist; Liberal Party (PL), moderate centrist; Rwanda Patriotic Front (FPR), Tutsi-led but claims to be multi-ethnic
Armed forces: 47,000 (1998)
Conscription: military service is voluntary
Death penalty: retains the death penalty for ordinary crimes but can be considered abolitionist in practice (last execution 1982)
Defence spend: (% GDP) 6.9 (1998)
Education spend: (% GNP) N/A
Health spend: (% GDP) 4.3 (1997 est)

Economy and resources

Currency: Rwandan franc
GDP: (US$) 1.9 billion (1999 est)
Real GDP growth: (% change on previous year) 6.1 (1999 est)
GNP: (US$) 2.1 billion (1999)
GNP per capita (PPP): (US$) 690 (1998)
Consumer price inflation: −2.4% (1999 est)
Foreign debt: (US$) 1.3 billion (1999 est)
Major trading partners: Germany, Kenya, Belgium, France, India, Tanzania, UK, USA
Resources: cassiterite (a tin-bearing ore), wolframite (a tungsten-bearing ore), natural gas, gold, columbo-tantalite, beryl
Industries: food processing, beverages, tobacco, mining, chemicals, rubber and plastic products, metals and metal products, machinery
Exports: coffee, tea, tin ores and concentrates, pyrethrum, quinquina. Principal market: Germany 11.4% (1999 est)
Imports: food, clothing, mineral fuels and lubricants, construction materials, transport equipment, machinery, tools, consumer goods. Principal source: Kenya 24.7% (1999 est)
Arable land: 34.5% (1996)
Agricultural products: sweet potatoes, cassava, dry beans, sorghum, plantains, coffee, tea, pyrethrum; livestock rearing (long-horned Ankole cattle and goats)

Population and society

Population: 7,733,000 (2000 est)
Population growth rate: 7.7% (1995–2000)
Population density: (per sq km) 275 (1999 est)
Urban population: (% of total) 6 (2000 est)
Age distribution: (% of total population) 0–14 45%, 15–59 51%, 60+ 4% (2000 est)
Ethnic groups: 85% belong to the Hutu tribe, most of the remainder being Tutsis (14%); there are also Twa (1%) and Pygmy minorities
Language: Kinyarwanda, French (both official), Kiswahili
Religion: about 50% animist; about 40% Christian, mainly Roman Catholic; 9% Muslim
Education: (compulsory years) 7
Literacy rate: 74% (men); 60% (women) (2000 est)
Labour force: 91% agriculture, 3% industry, 6% services (1996 est)
Life expectancy: 39 (men); 42 (women) (1995–2000)
Child mortality rate: (under 5, per 1,000 live births) 202 (1995–2000)
Physicians: 1 per 24,967 people (1993 est)
Hospital beds: 1 per 1,152 people (1993 est)
Radios: (per 1,000 people) 102 (1997)
Internet users: (per 10,000 people) 1.4 (1999)

Transport

Airports: international airports: Kigali (Kanombe), Kamembe; four domestic airfields; total passenger km: 2 million (1994)
Railways: none
Roads: total road network: 14,900 km/9,259 mi, of which 10% paved (1996 est); passenger cars: 1.4 per 1,000 people (1996 est)

Chronology

10th century onwards: Hutu peoples settled in region formerly inhabited by hunter-gatherer Twa Pygmies, becoming peasant farmers. **14th century onwards:** Majority Hutu community came under dominance of cattle-owning Tutsi peoples, immigrants from the east, who became a semi-aristocracy and established control through land and cattle contracts. **15th century:** Ruganzu Bwimba, a Tutsi leader, founded kingdom near Kigali. **17th century:** Central Rwanda and outlying Hutu communities subdued by Tutsi mwami (king) Ruganzu Ndori. **late 19th century:** Under the great Tutsi king, Kigeri Rwabugiri, a unified state with a centralized military structure was established. **1890:** Known as Ruandi, the Tutsi kingdom, along with neighbouring Burundi, came under nominal German control, as Ruanda-Urundi. **1916:** Occupied by Belgium during World War I. **1923:** Belgium granted League of Nations mandate to administer Ruanda-Urundi; they were to rule 'indirectly' through Tutsi chiefs. **1959:** Inter-ethnic warfare between Hutu and Tutsi, forcing mwami Kigeri V into exile. **1961:** Republic proclaimed after mwami deposed. **1962:** Independence from Belgium achieved as Rwanda, with Hutu Grégoire Kayibanda as president; many Tutsis left the country. **1963:** 20,000 killed in inter-ethnic clashes, after Tutsis exiled in Burundi had launched a raid. **1973:** Kayibanda ousted in military coup led by Hutu Maj-Gen Juvenal Habyarimana; this was caused by resentment of Tutsis, who held some key government posts. **1981:** Elections created civilian legislation, but dominated by Hutu socialist National Revolutionary Development Movement (MRND), in a one-party state. **1988:** Hutu refugees from Burundi massacres streamed into Rwanda. **1990:** The government was attacked by the Rwanda Patriotic Front (FPR), a Tutsi refugee military-political organization based in Uganda, which controlled parts of northern Rwanda. **1993:** A United Nations (UN) mission was sent to monitor the peace agreement made with the FPR in 1992. **1994:** President Habyarimana and Burundian Hutu president Ntaryamira were killed in an air crash; involvement of FPR was suspected. Half a million people were killed in the ensuing civil war, with many Tutsi massacred by Hutu death squads and the exodus of 2 million refugees to neighbouring countries. The government fled as FPR forces closed in. An interim coalition government was installed, with moderate Hutu and FPR leader, Pasteur Bizimungu, as president. **1995:** A war-crimes tribunal opened and government human-rights abuses were reported. Pierre Rwigema was appointed prime minister by President Bizimungu. **1996–97:** Rwanda and Zaire (Democratic Republic of Congo) were on the brink of war after Tutsi killings of Hutu in Zaire. A massive Hutu refugee crisis was narrowly averted as thousands were allowed to return to Rwanda. **1998:** 378 rebels were killed by the Rwandan army. **2000:** President Bizimungu resigned after disagreeing with his party, the Rwanda Patriotic Front. Paul Kagame, the vice-president, was installed as interim president in March, and inaugurated as president the following month. Bernard Makuza became prime minister of a new cabinet.

Practical information

Visa requirements: UK: visa required. USA: visa required
Time difference: GMT +2
Chief tourist attractions: people are advised against all but essential travel to Rwanda; random violence and robbery continue
Major holidays: 1, 28 January, 1 May, 1, 5 July, 1, 15 August, 25 September, 26 October, 1 November, 1, 8, 24–25 December; variable: Carnival, Corpus Christi, Good Friday

St Kitts and Nevis OR St Christopher and Nevis, formerly part of Leeward Islands Federation (until 1956)

Located in the West Indies, in the eastern Caribbean Sea, part of the Leeward Islands.

National name: *Federation of St Christopher and St Nevis*
Area: 262 sq km/101 sq mi (St Kitts 168 sq km/65 sq mi, Nevis 93 sq km/36 sq mi)
Capital: Basseterre (on St Kitts) (and chief port)
Major towns/cities: Charlestown (Nevis), Newcastle, Sandy Point Town, Dieppe Bay Town, Saint Paul
Physical features: both islands are volcanic; fertile plains on coast; black beaches

Green represents fertility. Yellow stands for sunshine. Black recalls the people's African origins. Red symbolizes the struggle for liberty. Effective date: 19 September 1983.

Government

Head of state: Queen Elizabeth II from 1983, represented by Governor General Dr Cuthbert Montraville Sebastian from 1996
Head of government: Denzil Douglas from 1995
Political system: liberal democracy
Political executive: parliamentary
Administrative divisions: 14 parishes
Political parties: People's Action Movement (PAM), right of centre; Nevis Reformation Party (NRP), Nevis-separatist, centrist; Labour Party (SKLP), moderate left of centre
Armed forces: army disbanded in 1981 and absorbed by Volunteer Defence Force; participates in US-sponsored Regional Security System established in 1982
Death penalty: retained and used for ordinary crimes
Education spend: (% GNP) 3.8 (1996)
Health spend: (% GDP) 3.8 (1998)

Economy and resources

Currency: East Caribbean dollar
GDP: (US$) 296 million (1999 est)
Real GDP growth: (% change on previous year) 3 (1999)
GNP: (US$) 262 million (1999)
GNP per capita (PPP): (US$) 9,801 (1999)
Consumer price inflation: 0.4% (1999 est)
Unemployment: 4.5% (1996 est)
Foreign debt: (US$) 115 million (1998)
Major trading partners: USA, UK, Trinidad and Tobago, St Vincent and the Grenadines, Canada, Barbados
Industries: electronic equipment, food and beverage processing (principally sugar and cane spirit), clothing, footwear, tourism
Exports: sugar, manufactures, postage stamps; sugar and sugar products accounted for approximately 40% of export earnings in 1992. Principal market: USA 46.6% (1996)
Imports: foodstuffs, basic manufactures, machinery, mineral fuels. Principal source: USA 45% (1996)
Arable land: 22.2% (1995)
Agricultural products: sugar cane, coconuts, yams, sweet potatoes, groundnuts, sweet peppers, carrots, cabbages, bananas, cotton; fishing

Population and society

Population: 38,000 (2000 est)
Population growth rate: –0.8% (1995–2000)
Population density: (per sq km) 160 (1999 est)
Urban population: (% of total) 34 (2000 est)
Age distribution: (% of total population) 0–14 31%, 15–59 67%, 60+ 12% (2000 est)
Ethnic groups: almost entirely of African descent
Language: English (official)
Religion: Anglican 36%, Methodist 32%, other Protestant 8%, Roman Catholic 10%
Education: (compulsory years) 12
Literacy rate: 99% (men); 97% (women) (1998 est)
Labour force: 14.7% agriculture, 20.9% industry, 64.4% services (1994)
Life expectancy: 65 (men); 71 (women) (1998 est)
Child mortality rate: (under 5, per 1,000 live births) 40 (1995)
Physicians: 1 per 2,200 people (1993)
Hospital beds: 1 per 158 people (1991)
TV sets: (per 1,000 people) 264 (1997)
Radios: (per 1,000 people) 701 (1997)
Internet users: (per 10,000 people) 516.1 (1999)
Personal computer users: (per 100 people) 15.0 (1999)

Transport

Airports: international airports: Basseterre (Golden Rock), Charlestown on Nevis (Newcastle Airfield); domestic services operate between these
Railways: total length: 36 km/22 mi (serving sugar plantations)
Roads: total road network: 320 km/199 mi, of which 42.5% paved (1996 est); passenger cars: 167 per 1,000 people (1993 est)

Chronology

1493: Visited by the explorer Christopher Columbus, after whom the main island is named, but for next two centuries the islands were left in the possession of the indigenous Caribs. **1623 and 1628:** St Kitts and Nevis islands successively settled by British as their first Caribbean colony, with 2,000 Caribs brutally massacred in 1626. **1783:** In the Treaty of Versailles France, which had long disputed British possession, rescinded its claims to the islands, on which sugar cane plantations developed, worked by imported African slaves. **1816:** Anguilla was joined politically to the two islands. **1834:** Abolition of slavery. **1871–1956:** Part of the Leeward Islands Federation. **1937:** Internal self-government granted. **1952:** Universal adult suffrage granted. **1958–62:** Part of the Federation of the West Indies. **1967:** St Kitts, Nevis, and Anguilla achieved internal self-government, within the British Commonwealth, with Robert Bradshaw, Labour Party leader, as prime minister. **1970:** NRP formed, calling for separation for Nevis. **1971:** Anguilla returned to being a British dependency after rebelling against domination by St Kitts. **1980:** People's Action Movement (PAM) and NRP centrist coalition government, led by Kennedy Simmonds, formed after inconclusive general election. **1983:** Full independence was achieved within the Commonwealth. **1994:** A three-week state of emergency was imposed after violent antigovernment riots by Labour Party supporters in Basseterre. **1995:** Labour Party won a general election; Denzil Douglas became prime minister. **1997:** Nevis withdrew from the federation. **1998:** Nevis referendum on secession failed to secure support. **2000:** Denzil Douglas was re-elected as prime minister.

Practical information

Visa requirements: UK: visa not required. USA: visa not required
Time difference: GMT –4
Chief tourist attractions: coral beaches on St Kitts' north and west coasts; coconut forests; spectacular mountain scenery on Nevis; St Kitts' historical

Brimstone Hill Fort and associations with Lord Nelson and Alexander Hamilton
Major holidays: 1 January, 19 September, 25–26, 31 December; variable: Good Friday, Easter Monday, Whit Monday, Labour (May), Queen's Birthday (June), August Monday

St Lucia FORMERLY PART OF WINDWARD ISLANDS FEDERATION (UNTIL 1960)

Located in the West Indies, in the eastern Caribbean Sea, one of the Windward Islands.

Area: 617 sq km/238 sq mi
Capital: Castries
Major towns/cities: Soufrière, Vieux Fort, Choiseul, Gros Islet
Major ports: Vieux Fort
Physical features: mountainous island with fertile valleys; mainly tropical forest; volcanic peaks; Gros and Petit Pitons

Black and white reflect the black and white communities and the harmony between them. Yellow represents the golden beaches. Blue stands for the sea. Effective date: 22 February 1979.

Government

Head of state: Queen Elizabeth II from 1979, represented by Governor General Dr Perlette Louisy from 1997
Head of government: Kenny Anthony from 1997
Political system: liberal democracy
Political executive: parliamentary
Administrative divisions: 11 districts
Political parties: United Workers' Party (UWP), moderate left of centre; St Lucia Labour Party (SLP), moderate left of centre; Progressive Labour Party (PLP), moderate left of centre
Armed forces: none; participates in the US-sponsored Regional Security System established in 1982; police force numbers around 300
Death penalty: retained and used for ordinary crimes
Education spend: (% GNP) 9.8 (1996)
Health spend: (% GDP) 4 (1997 est)

Economy and resources

Currency: East Caribbean dollar
GDP: (US$) 680 million (1999 est)
Real GDP growth: (% change on previous year) 3.1 (1999)
GNP: (US$) 581 million (1999)
GNP per capita (PPP): (US$) 5,022 (1999)
Consumer price inflation: 3.5% (1999)
Unemployment: 15% (1996 est)
Foreign debt: (US$) 131.6 million (1998)
Major trading partners: USA, UK, Trinidad and Tobago (and other CARICOM member states), Japan, Canada, Italy
Resources: geothermal energy
Industries: processing of agricultural products (principally coconut oil, meal, and copra), clothing, rum, beer, and other beverages, plastics, paper and packaging, electronic assembly, tourism
Exports: bananas, coconut oil, cocoa beans, copra, beverages, tobacco, miscellaneous articles. Principal market: UK 50% (1995)
Imports: machinery and transport equipment, foodstuffs, basic manufactures, mineral fuels.

Principal source: USA 36% (1995)
Arable land: 8.2% (1995)
Agricultural products: bananas, cocoa, coconuts, mangoes, citrus fruits, spices, breadfruit

Population and society

Population: 154,000 (2000 est)
Population growth rate: 1.4% (1995–2000)
Population density: (per sq km) 252 (1999 est)
Urban population: (% of total) 38 (2000 est)
Age distribution: (% of total population) 0–14 37%, 15–59 54%, 60+ 9% (2000 est)
Ethnic groups: great majority of African descent; about 6% mixed, 3% Indian
Language: English (official), French patois
Religion: Roman Catholic 85%; Anglican, Protestant
Education: (compulsory years) 10
Literacy rate: 81% (men); 82% (women) (1997 est)
Labour force: 23.5% agriculture, 13.2% industry, 63.3% services (1996 st)
Life expectancy: 68 (men); 75 (women) (1998 est)
Child mortality rate: (under 5, per 1,000 live births) 22 (1995)
Physicians: 1 per 1,777 people (1997)
Hospital beds: 1 per 273 people (1997)
TV sets: (per 1,000 people) 213 (1997)
Radios: (per 1,000 people) 746 (1997)
Internet users: (per 10,000 people) 197.0 (1999)
Personal computer users: (per 100 people) 14.0 (1999)

Transport

Airports: international airports: Castries (Vigie), Vieux Fort (Hewanorra); domestic flights operate between these; aircraft arrivals: 42,436 (1993)
Railways: none
Roads: total road network: 1,210 km/752 mi, of which 5.2% paved (1996 est); passenger cars: 872 per 1,000 people (1996 est)

Chronology

1502: Sighted by the explorer Christopher Columbus on St Lucia's day but not settled for more than a century due to hostility of the island's Carib Indian inhabitants. **1635:** Settled by French, who brought in slaves to work sugar cane plantations as the Carib community was annihilated. **1814:** Ceded to Britain as a crown colony, following Treaty of Paris; African slaves brought in to work sugar cane plantations. **1834:** Slavery abolished. **1860s:** A major coal warehousing centre until the switch to oil and diesel fuels in 1930s. **1871–1960:** Part of Windward Islands Federation. **1951:** Universal adult suffrage granted. **1967:** Acquired internal self-government as a West Indies associated state. **1979:** Independence achieved within Commonwealth with John Compton, leader of United Workers' Party (UWP), as prime minister. **1991:** Integration with other Windward Islands (Dominica, Grenada, and St Vincent) was proposed. **1993:** Unrest and strikes by farmers and agricultural workers arose as a result of depressed prices for the chief cash crop, bananas.

Practical information

Visa requirements: UK: visa not required. USA: visa not required
Time difference: GMT –4
Chief tourist attractions: tropical climate; sandy beaches; mountain scenery; rich birdlife; historical sites; sulphur baths at Soufrière
Major holidays: 1–2 January, 22 February, 1 May, 13, 25–26 December; variable: Carnival, Corpus Christi, Good Friday, Easter Monday, Whit Monday, Emancipation (August), Thanksgiving (October)

St Vincent and the Grenadines

Located in the West Indies, in the eastern Caribbean Sea, part of the Windward Islands.

Area: 388 sq km/150 sq mi (including islets of the Northern Grenadines 43 sq km/17 sq mi)
Capital: Kingstown
Major towns/cities: Georgetown, Châteaubelair, Layon, Dovers
Physical features: volcanic mountains, thickly forested; La Soufrière volcano

Green stands for agriculture, the lush vegetation, and the enduring vitality of the population. Gold symbolizes warmth, the bright spirit of the people, and the golden sands. Blue represents the sky and sea. Effective date: 24 October 1985.

Government

Head of state: Queen Elizabeth II from 1979, represented by Governor General David Jack from 1989
Head of government: Arnhim Eustace from 2000
Political system: liberal democracy
Political executive: parliamentary
Administrative divisions: six parishes
Political parties: New Democratic Party (NDP), right of centre; St Vincent Labour Party (SVLP), moderate left of centre
Armed forces: none – police force only; participates in the US-sponsored Regional Security System established in 1982
Death penalty: retained and used for ordinary crimes
Education spend: (% GNP) N/A
Health spend: (% GDP) 5.9 (1997 est)

Economy and resources

Currency: East Caribbean dollar
GDP: (US$) 326 million (1999 est)
Real GDP growth: (% change on previous year) 4 (1999 est)
GNP: (US$) 307 million (1999)
GNP per capita (PPP): (US$) 4,667 (1999)
Consumer price inflation: –1.1% (1999 est)
Unemployment: 38% (1995 est)
Foreign debt: (US$) 99.3 million (1998)
Major trading partners: USA, UK, Trinidad and Tobago, Antigua and Barbuda, Barbados, Canada, Japan, St Lucia
Industries: clothing, assembly of electronic equipment, processing of agricultural products (including brewing, flour milling, rum distillation, dairy products), industrial gases, plastics, tourism
Exports: bananas, eddoes, dasheen, sweet potatoes, flour, ginger, tannias, plantains. Principal market: UK 38.5% (1996)
Imports: basic manufactures, machinery and transport equipment, food and live animals, mineral fuels, chemicals, miscellaneous manufactured

articles. Principal source: USA 38.6% (1996)
Arable land: 10.3% (1995)
Agricultural products: bananas, cocoa, citrus fruits, mangoes, avocado pears, guavas, sugar cane, vegetables, spices; world's leading producer of arrowroot starch; fishing

Population and society

Population: 114,000 (2000 est)
Population growth rate: 0.7% (1995–2000)
Population density: (per sq km) 355 (1999 est)
Urban population: (% of total) 55 (2000 est)
Age distribution: (% of total population) 0–14 37%, 15–59 54%, 60+ 9% (2000 est)
Ethnic groups: largely of African origin; most of the original indigenous Caribs have disappeared
Language: English (official), French patois
Religion: Anglican, Methodist, Roman Catholic
Education: not compulsory
Literacy rate: 93% (men); 88% (women) (1997 est)
Labour force: 25.1% agriculture, 21.1% industry, 53.8% services (1991)
Life expectancy: 72 (men); 76 (women) (1998 est)
Child mortality rate: (under 5, per 1,000 live births) 23 (1995)
Physicians: 1 per 4,037 people (1992)
Hospital beds: 1 per 213 people (1991)
TV sets: (per 1,000 people) 163 (1997)
Radios: (per 1,000 people) 690 (1997)
Internet users: (per 10,000 people) 176.7 (1999)
Personal computer users: (per 100 people) 10.0 (1999)

Transport

Airports: international airports: Kingstown (E T Joshua); four domestic airports; visitor arrivals: 87,951 (1996)
Railways: none
Roads: total road network: 1,040 km/646 mi, of which 30.7% paved (1996 est)

Chronology

1498: Main island visited by the explorer Christopher Columbus on St Vincent's day.
17th–18th centuries: Possession disputed by France and Britain, with fierce resistance from the indigenous Carib community.
1783: Recognized as British crown colony by Treaty of Versailles. **1795–97:** Carib uprising, with French support, resulted in deportation of 5,000 to Belize and Honduras.
1834: Slavery abolished. **1902:** Over 2,000 killed by the eruption of La Soufrière volcano.
1951: Universal adult suffrage granted.
1958–62: Part of West Indies Federation.
1969: Achieved internal self-government.
1979: Achieved full independence within Commonwealth. **1981:** General strike against new industrial-relations legislation at a time of economic recession. **1984:** James Mitchell, of the centre-right New Democratic Party (NDP) became prime minister. **2000:** Prime Minister James Mitchell gave up his presidency of the ruling New Democratic Party (NDP), and was later replaced as prime minister by Arnhim Eustace.

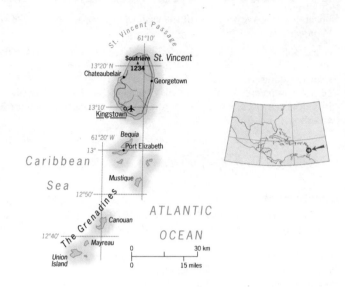

Practical information

Visa requirements: UK: visa not required. USA: visa not required
Time difference: GMT –4
Chief tourist attractions: famous white beaches, clear waters, lush vegetation; excellent yachting facilities
Major holidays: 1, 22 January, 27 October, 25–26 December; variable: Carnival (July), Good Friday, Easter Monday, Whit Monday, Labour (May), Caricom (July), Emancipation (August)

SAMOA FORMERLY WESTERN SAMOA (UNTIL 1997)

Located in the southwest Pacific Ocean, in Polynesia, northeast of Fiji Islands.

National name: *'O la Malo Tu To'atasi o Samoa/Independent State of Samoa*
Area: 2,830 sq km/1,092 sq mi
Capital: Apia (on Upolu island) (and chief port)
Major towns/cities: Lalomanu, Tuasivi, Falealupo, Falelatai, Salotulafai, Taga
Physical features: comprises South Pacific islands of Savai'i and Upolu, with two smaller tropical islands and uninhabited islets; mountain ranges on main islands; coral reefs; over half forested

Red and white are traditional colours, dating back to the flag of 19th-century Samoan king, Malietoa Laupepa. The Southern Cross constellation links Samoa to other countries in the southern hemisphere.
Effective date: 4 July 1997.

Government

Head of state: King Malietoa Tanumafili II from 1962
Head of government: Tuila'epa Sa'ilele Malielegaoi from 1998
Political system: liberal democracy
Political executive: parliamentary
Administrative divisions: 11 districts
Political parties: Human Rights Protection Party (HRPP), led by Tofilau Eti Alesana, centrist; Samoa Democratic Party (SDP), led by Le Tagaloa Pita; Samoa National Development Party (SNDP), led by Tupuola Taisi Efi and Va'ai Kolone, conservative. All 'parties' are personality-based groupings
Armed forces: no standing defence forces; under Treaty of Friendship signed with New Zealand in 1962, the latter acts as sole agent in Samoa's dealings with other countries and international organizations
Death penalty: retains the death penalty for ordinary crimes, but can be considered abolitionist in practice
Education spend: (% GNP) N/A
Health spend: (% GDP) 3.8 (1997 est)

Economy and resources

Currency: tala, or Samoan dollar
GDP: (US$) 209 million (1999 est)
Real GDP growth: (% change on previous year) 4 (1999 est)
GNP: (US$) 181 million (1999)
GNP per capita (PPP): (US$) 3,915 (1999)
Consumer price inflation: 0.3% (1999)
Foreign debt: (US$) 180 million (1998)
Major trading partners: American Samoa, New Zealand, Australia, Fiji Islands, Japan, USA
Industries: coconut-based products, timber, light engineering, construction materials, beer, cigarettes, clothing, leather goods, wire, tourism
Exports: coconut cream, beer, cigarettes, taro, copra, cocoa, bananas, timber. Principal market: American Samoa 32.1% (1998 est)
Imports: food and live animals, machinery and transport equipment, mineral fuel, clothing and other manufactured goods. Principal source: New Zealand 37.9% (1998 est)

Arable land: 19.4% (1995)
Agricultural products: coconuts, taro, copra, bananas, papayas, mangoes, pineapples, cocoa, taamu, breadfruit, maize, yams, passion fruit; livestock rearing (pigs, cattle, poultry, and goats) is important for local consumption; forest resources provide an important export commodity (47% of land was forest and woodland early 1990s)

Population and society

Population: 180,000 (2000 est)
Population growth rate: 1.4% (1995–2000)
Population density: (per sq km) 63 (1999 est)
Urban population: (% of total) 22 (2000 est)
Age distribution: (% of total population) 0–14 38%, 15–59 56%, 60+ 6% (2000 est)
Ethnic groups: 93% of Samoan (Polynesian) origin; 7% Euronesian (mixed European and Polynesian), a small European minority
Language: English, Samoan (both official)
Religion: Congregationalist; also Roman Catholic, Methodist
Education: not compulsory
Literacy rate: 81% (men); 79% (women) (2000 est)
Labour force: 65% agriculture, 5% industry, 30% services (1995 est)
Life expectancy: 69 (men); 74 (women) (1995–2000)
Child mortality rate: (under 5, per 1,000 live births) 27 (1995–2000)
Physicians: 1 per 3,665 people (1992)
Hospital beds: 1 per 311 people (1995)
TV sets: (per 1,000 people) 61 (1997)
Radios: (per 1,000 people) 1,035 (1997)
Internet users: (per 10,000 people) 28.2 (1999)
Personal computer users: (per 100 people) 0.1 (1999)

Transport

Airports: international airports: Apia (Faleolo); two domestic airstrips
Railways: none
Roads: total road network: 790 km/491 mi, of which 42% paved (1996 est); passenger cars: 30 per 1,000 people (1993 est)

Chronology

c. **1000 BC:** Settled by Polynesians from Tonga. **AD 950–1250:** Ruled by Tongan invaders; the Matai (chiefly) system was developed. **15th century:** United under the Samoan Queen Salamasina. **1722:** Visited by Dutch traders. **1768:** Visited by the French navigator Louis Antoine de Bougainville. **1830:** Christian mission established and islanders were soon converted to Christianity. **1887–89:** Samoan rebellion against German attempt to depose paramount ruler and install its own puppet regime. **1889:** Under the terms of the Act of Berlin, Germany took control of the nine islands of Western Samoa, while the USA was granted American Samoa, and Britain Tonga and the Solomon Islands. **1900s:** More than 2,000 Chinese brought in to work coconut plantations. **1914:** Occupied by New Zealand on the outbreak of World War I. **1918:** Nearly a quarter of the population died in an influenza epidemic. **1920s:** Development of nationalist movement, the Mau, which resorted to civil disobedience. **1920–61:** Administered by New Zealand under League of Nations and, later, United Nations mandate. **1959:** Local government established, headed by chief minister Fiame Mata'afa Mulinu'u. **1961:** Referendum favoured independence. **1962:** King Malietoa Tanumafili succeeded to the throne. **1962:** Independence achieved within Commonwealth, with Mata'afa as prime minister, a position he retained (apart from a short break 1970–73) until his death in 1975. **1990:** Universal adult suffrage was introduced and the power of Matai (elected clan leaders) reduced. **1991:** Major damage was caused by 'Cyclone Val'. **1997:** Name was changed officially from Western Samoa to Samoa, despite protests from American Samoa that it would undermine American Samoa's identity. **1998:** Tuila'epa Sa'ilele Malielegaoi, of the HRPP, became the new prime minister.

Practical information

Visa requirements: UK: visa not required. USA: visa not required
Time difference: GMT –11
Chief tourist attractions: pleasant climate; spectacular scenery
Major holidays: 1–2 January, 25 April, 1 June (3 days), 12 October, 25–26 December; variable: Good Friday, Easter Monday, Holy Saturday

SAN MARINO

Small landlocked country within northeast Italy.

National name: *Serenissima Repubblica di San Marino/Most Serene Republic of San Marino*
Area: 61 sq km/24 sq mi
Capital: San Marino
Major towns/cities: Serravalle, Faetano, Fiorentino, Borgo Maggiore, Domagnano
Physical features: the slope of Mount Titano

White represents the snow on Monte Titano and the clouds above. Blue stands for the sky. Effective date: 6 April 1862.

Government

Head of state and government: Gian Franco Terenzi and Enzo Colombini from 2000
Political system: liberal democracy
Political executive: parliamentary
Political parties: San Marino Christian Democrat Party (PDCS), Christian centrist; Progressive Democratic Party (PDP) (formerly the Communist Party: PCS), moderate left wing; Socialist Party (PS), left of centre
Armed forces: voluntary military forces and a paramilitary gendarmerie
Conscription: military service is not compulsory, but all citizens between the ages of 15 and 55 may be enlisted in certain circumstances to defend the state
Death penalty: abolished in 1865
Health spend: (% GDP) 7.5 (1997 est)

Economy and resources

Currency: Italian lira
GDP: (US$) 500 million (1997 est)
Real GDP growth: (% change on previous year) N/A
GNP: (US$) 490 million (1997 est)
GNP per capita (PPP): (US$) 20,000 (1997 est)
Consumer price inflation: 2% (1997)
Unemployment: 4.4% (1997)
Major trading partners: maintains customs union with Italy (for trade data see Italy)
Resources: limestone and other building stone
Industries: cement, synthetic rubber, leather, textiles,

ceramics, tiles, wine, chemicals, olive oil, tourism, postage stamps
Exports: wood machinery, chemicals, wine, olive oil, textiles, tiles, ceramics, varnishes, building stone, lime, chestnuts, hides. Principal market: Italy
Imports: consumer goods, raw materials, energy supply. Principal source: Italy
Arable land: 16.7% (1995)
Agricultural products: wheat, barley, maize, grapes, olives, fruit, vegetables; viticulture; dairy farming

Population and society

Population: 27,000 (2000 est)
Population growth rate: 1.3% (1995–2000)
Population density: (per sq km) 417 (1999 est)
Urban population: (% of total) 89 (2000 est)
Age distribution: (% of total population) 0–14 15%, 15–59 65%, 60+ 20% (2000 est)
Ethnic groups: predominantly Italian, Sanmarinese; about 11% are foreign citizens
Language: Italian (official)
Religion: Roman Catholic 95%
Education: (compulsory years) 8
Literacy rate: 99% (men); 98% (women) (1998 est)
Labour force: 1.4% agriculture, 41% industry, 57.6% services (1998)
Life expectancy: 78 (men); 85 (women) (1998 est)
Child mortality rate: (under 5, per 1,000 live births) 6 (1997 est)

Chronology

c. **AD 301:** Founded as a republic (the world's oldest surviving) by St Marinus and a group of Christians who settled there to escape persecution. **12th century:** Self-governing commune. **1600:** Statutes (constitution) provided for a parliamentary form of government. **1815:** Independent status of the republic recognized by the Congress of Vienna. **1862:** Treaty with Italy signed; independence recognized under Italy's protection.
1945–57: Communist–Socialist administration in power, eventually ousted in a bloodless 'revolution'. **1957–86:** Governed by a series of left-wing and centre-left coalitions. **1992:** San Marino joined the United Nations (UN). **1998:** The ruling PDCS–PSS coalition remained in power after a general election. **2000:** Gian Franco Terenzi and Enzo Colombini were appointed as captains regent.

Practical information

Visa requirements: UK/USA: visa not required
Time difference: GMT +1
Chief tourist attractions: mild climate; varied scenery; well preserved medieval architecture

São Tomé and Príncipe

Located in the Gulf of Guinea, off the coast of West Africa.

National name: *República Democrática de São Tomé e Príncipe/Democratic Republic of São Tomé and Príncipe*
Area: 1,000 sq km/386 sq mi
Capital: São Tomé
Major towns/cities: Santo António, Sant Ana, Porto Alegre, Trinidade, Neves, Santo Amaro
Physical features: comprises two main islands and several smaller ones, all volcanic; thickly forested and fertile

The stars stand for the two islands. Red recalls the struggle for independence. Yellow is said to represent the country's cocoa plantations. Effective date: 12 July 1975.

Government

Head of state: Miguel Trovoada from 1991
Head of government: Guilherme Posser da Costa from 1999
Political system: emergent democracy
Political executive: limited presidency
Administrative divisions: two provinces
Political parties: Movement for the Liberation of São Tomé e Príncipe–Social Democratic Party (MLSTP–PSD), nationalist socialist; Democratic Convergence Party–Reflection Group (PCD–GR), moderate left of centre; Independent Democratic Action (ADI), centrist
Armed forces: no proper army; reorganization of island's armed forces (estimated at 900) and police into two separate police forces (one for public order, the other for criminal investigations) was initiated in 1992
Death penalty: abolished in 1990
Education spend: (% GNP) N/A
Health spend: (% GDP) 4 (1997 est)

Economy and resources

Currency: dobra
GDP: (US$) 47.1 million (1999 est)
Real GDP growth: (% change on previous year) 2.5 (1999)
GNP: (US$) 40 million (1999)
GNP per capita (PPP): (US$) 1,335 (1999)
Consumer price inflation: 16.3% (1999)
Unemployment: 28% (1996 est)
Foreign debt: (US$) 260 million (1999 est)
Major trading partners: Portugal, the Netherlands, Germany, Spain, Belgium, France, Japan, Angola
Industries: agricultural and timber processing, soft drinks, soap, textiles, beer, bricks, ceramics, shirts
Exports: cocoa, copra, coffee, bananas, palm oil. Principal market: the Netherlands 18.2% (1998)
Imports: capital goods, food and live animals (of which 60.7% were donations in 1994), petroleum and petroleum products. Principal source: Portugal

42% (1998)
Arable land: 2.1% (1995)
Agricultural products: cocoa, coconuts, copra, bananas, palm oil, cassava, sweet potatoes, yams, coffee; fishing; forest resources (75% of land area was forest and woodland in the early 1990s)

Population and society

Population: 147,000 (2000 est)
Population growth rate: 2% (1995–2000)
Population density: (per sq km) 161 (1999 est)
Urban population: (% of total) 47 (2000 est)
Age distribution: (% of total population) 0–14 47%, 15–59 46%, 60+ 7% (2000 est)
Ethnic groups: predominantly African; mixed African and Portuguese origin, angolares (descendants of Angolan slaves)
Language: Portuguese (official), Fang (a Bantu language), Lungwa São Tomé (a Portuguese Creole)
Religion: Roman Catholic 80%, animist
Education: (compulsory years) 4
Literacy rate: 85% (men); 62% women (1991 est)
Labour force: 38.1% agriculture, 15.5% industry, 46.4% services (1994)
Life expectancy: 63 (men); 66 (women) (1998 est)
Child mortality rate: (under 5, per 1,000 live births) 81 (1995)
Physicians: 1 per 1,780 people (1992)
Hospital beds: 1 per 214 people (1992)
TV sets: 163 (1997)
Radios: (per 1,000 people) 272 (1997)
Internet users: (per 10,000 people) 34.8 (1999)

Transport

Airports: international airports: São Tomé; one domestic airport (Príncipe); total passenger km: 8 million (1995)
Railways: none
Roads: total road network: 320 km/199 mi, of which 68.1% paved (1996 est); passenger cars: 30 per 1,000 people (1996 est)

Chronology

1471: First visited by the Portuguese, who imported convicts and slaves to work on sugar plantations in the formerly uninhabited islands. **1522:** Became a province of Portugal. **1530:** Slaves successfully revolted, forcing plantation owners to flee to Brazil; thereafter became a key staging post for Congo-Americas slave trade. **19th century:** Forced contract labour used to work coffee and cocoa plantations. **1953:** More than 1,000 striking plantation workers gunned down by Portuguese troops. **1960:** First political party formed, the forerunner of the socialist-nationalist Movement for the Liberation of São Tomé e Príncipe (MLSTP). **1974:** Military coup in Portugal led to strikes, demonstrations, and army mutiny in São Tomé; thousands of Portuguese settlers fled the country. **1975:** Independence achieved, with Manuel Pinto da Costa (MLSTP) as president; close links developed with communist bloc, and plantations nationalized. **1984:** Formally declared a nonaligned state as economy deteriorated. **1988:** Coup attempt against da Costa foiled by Angolan and East European troops. **1990:** Influenced by collapse of communism in Eastern Europe, MLSTP abandoned Marxism; a new pluralist constitution was approved in a referendum. **1991:** First multiparty elections. **1994:** MLSTP returned to power with Carlos da Graca as prime minister. **1998:** MLSTP–PSD won an absolute majority in the assembly. **2000:** Guilherme Posser de Costa became prime minister.

Practical information

Visa requirements: UK: visa required. USA: visa required
Chief tourist attractions: unspoilt beaches; spectacular mountain scenery; unique species of flora and fauna
Major holidays: 1 January, 3 February, 1 May, 12 July, 6, 30 September, 21, 25 December; variable: Corpus Christi, Good Friday, Easter Monday

SAUDI ARABIA

Located on the Arabian peninsula, stretching from the Red Sea in the west to the Arabian Gulf in the east, bounded north by Jordan, Iraq, and Kuwait; east by Qatar and United Arab Emirates; southeast by Oman; and south by Yemen.

The prophet Muhammad is said to have used a green banner. Past variants of the flag show two crossed swords. Effective date: 15 March 1973.

National name: *Al-Mamlaka al-'Arabiyya as-Sa'udiyya/Kingdom of Saudi Arabia*
Area: 2,200,518 sq km/849,620 sq mi
Capital: Riyadh
Major towns/cities: Jiddah, Mecca, Medina, At Taif, Ad Dammam, Al Hufuf, Tabuk, Buraydah
Major ports: Jiddah, Ad Dammam, Al Jubail, Jizan, Yanbu
Physical features: desert, sloping to the Persian Gulf from a height of 2,750 m/9,000 ft in the west

Government

Head of state and government: King Fahd Ibn Abdul Aziz from 1982
Political system: absolutist
Political executive: absolute
Administrative divisions: 13 provinces
Political parties: none
Armed forces: 162,500; paramilitary forces 15,500 (1998)
Conscription: military service is voluntary
Death penalty: retained and used for ordinary crimes
Defence spend: (% GDP) 15.7 (1998)
Education spend: (% GNP) 7.5 (1997)
Health spend: (% GDP) 3.5 (1997 est)

Economy and resources

Currency: riyal
GDP: (US$) 139.4 billion (1999)
Real GDP growth: (% change on previous year) 0.5 (1999)
GNP: (US$) 143.7 billion (1999 est)
GNP per capita (PPP): (US$) 10,472 (1999 est)
Consumer price inflation: –1.4% (1999)
Foreign debt: (US$) 28 billion (1999 est)
Major trading partners: USA, Japan, Germany, South Korea, France, Italy, Singapore, the Netherlands
Resources: petroleum, natural gas, iron ore, limestone, gypsum, marble, clay, salt, gold, uranium, copper, fish
Industries: petroleum and petroleum products, urea and ammonia fertilizers, steel, plastics, cement
Exports: crude and refined petroleum, petrochemicals, wheat. Principal market: USA 18.2% (1999)
Imports: machinery and transport equipment, foodstuffs, beverages, tobacco, chemicals and chemical products, base metals and metal manufactures, textiles and clothing. Principal source: USA 24.6% (1999)
Arable land: 1.7% (1996)
Agricultural products: wheat, barley, sorghum, millet, tomatoes, dates, watermelons, grapes; livestock

(chiefly poultry) and dairy products

Population and society

Population: 21,607,000 (2000 est)
Population growth rate: 3.4% (1995–2000); 3.1% (2000–05)
Population density: (per sq km) 9 (1999 est)
Urban population: (% of total) 86 (2000 est)
Age distribution: (% of total population) 0–14 41%, 15–59 54%, 60+ 5% (2000 est)
Ethnic groups: predominantly Arab; 10% Afro-Asian; over 25% non-nationals
Language: Arabic (official), English
Religion: Sunni Muslim 85%; there is a Shiite minority
Literacy rate: 84% (men); 79% (women) (2000 est)
Labour force: 12.2% agriculture, 23.6% industry, 64.2% services (1997 est)
Life expectancy: 70 (men); 73 (women) (1995–2000)
Child mortality rate: (under 5, per 1,000 live births) 27 (1995–2000)
Physicians: 1 per 650 people (1996)
Hospital beds: 1 per 434 people (1996)
TV sets: (per 1,000 people) 262 (1997)
Radios: (per 1,000 people) 321 (1997)
Internet users: (per 10,000 people) 143.6 (1999)
Personal computer users: (per 100 people) 5.7 (1999)

Transport

Airports: international airports: Riyadh (King Khaled), Dhahran (Al Khobar), Jiddah (King Abdul Aziz); 20 domestic airports; total passenger km: 18,949 million (1997 est)
Railways: total length: 1,390 km/864 mi; total passenger km: 139 million (1993)
Roads: total road network: 162,000 km/100,669 mi, of which 42.7% paved (1996 est); passenger cars: 98.8 per 1,000 people (1996 est)

Chronology

622: Muhammad began to unite Arabs in Muslim faith. **7th–8th centuries:** Muslim Empire expanded, ultimately stretching from India to Spain, with Arabia itself being relegated to a subordinate part. **12th century:** Decline of Muslim Empire; Arabia grew isolated and internal divisions multiplied. **13th century:** Mameluke sultans of Egypt became nominal overlords of Hejaz in western Arabia. **1517:** Hejaz became a nominal part of the Ottoman Empire after the Turks conquered Egypt. **18th century:** Al Saud family united tribes of Nejd in central Arabia in support of the Wahhabi religious movement. **c.1830:** The Al Saud established Riyadh as the Wahhabi capital. **c.1870:** Turks took effective control of Hejaz and also Hasa on Persian Gulf. **late 19th century:** Rival Wahhabi dynasty of Ibn Rashid became leaders of Nejd. **1902:** Ibn Saud organized Bedouin revolt and regained Riyadh. **1913:** Ibn Saud completed the reconquest of Hasa from Turks. **1915:** Britain recognized Ibn Saud as emir of Nejd and Hasa. **1916–18:** British-backed revolt, under aegis of Sharif Hussein of Mecca, expelled Turks from Arabia. **1919–25:** Ibn Saud fought and defeated Sharif Hussein and took control of Hejaz. **1926:** Proclamation of Ibn Saud as king of Hejaz and Nejd. **1932:** Hejaz and Nejd renamed the United Kingdom of Saudi Arabia. **1933:** Saudi Arabia allowed US-owned Standard Oil Company to prospect for oil, which was discovered in Hasa in 1938. **1939–45:** Although officially neutral in World War II, Saudi Arabia received subsidies from USA and Britain. **1940s:** Commercial exploitation of oil began, bringing great prosperity. **1987:** Rioting by Iranian pilgrims caused 400 deaths in Mecca and a breach in diplomatic relations with Iran. **1990:** Iraqi troops invaded Kuwait and massed on the Saudi Arabian border, prompting King Fahd to call for assistance from US and UK forces. **1991:** Saudi Arabia fought on the Allied side against Iraq in the Gulf War. **1992:** Under international pressure to move towards democracy, King Fahd formed a 'consultative council' to assist in the government of the kingdom.

Practical information

Visa requirements: UK: visa required. USA: visa required
Time difference: GMT +3
Chief tourist attractions: the holy cities of Medina, Jiddah, and Mecca – notably Mecca's Al-Harram Mosque and large bazaars, and Medina's numerous mosques and Islamic monuments
Major holidays: 23 September; variable: Eid-ul-Adha (7 days), end of Ramadan (4 days)

SENEGAL

Located in West Africa, on the Atlantic Ocean, bounded north by Mauritania, east by Mali, south by Guinea and Guinea-Bissau, and enclosing the Gambia on three sides.

National name: *République du Sénégal/Republic of Senegal*
Area: 196,200 sq km/75,752 sq mi
Capital: Dakar (and chief port)
Major towns/cities: Thiès, Kaolack, Saint Louis, Ziguinchor, Diourbel, Mbour
Physical features: plains rising to hills in southeast; swamp and tropical forest in southwest; River Senegal; The Gambia forms an enclave within Senegal

The star represents Islam and expresses peace, harmony, hope, and socialism. The tricolour is reminiscent of the flag of France, the former colonial power. The pan-African colours express unity with other African nations. Effective date: 25 August 1960.

Government

Head of state: Abdoulaye Wade from 2000
Head of government: Mustafa Niasse from 2000
Political system: nationalistic socialist
Political executive: unlimited presidency
Administrative divisions: ten regions
Political parties: Senegalese Socialist Party (PS), democratic socialist; Senegalese Democratic Party (PDS), centrist
Armed forces: 11,000 (1998)
Conscription: military services is by selective conscription for two years
Death penalty: retains the death penalty for ordinary crimes but can be considered abolitionist in practice (last execution 1967)
Defence spend: (% GDP) 1.7 (1998)
Education spend: (% GNP) 3.5 (1996)
Health spend: (% GDP) 1.2 (1995)

Economy and resources

Currency: franc CFA
GDP: (US$) 4.7 billion (1999)
Real GDP growth: (% change on previous year) 5 (1999)
GNP: (US$) 4.7 billion (1999)
GNP per capita (PPP): (US$) 1,341 (1999)
Consumer price inflation: 0.8% (1999)
Unemployment: 10.2% (1993)
Foreign debt: (US$) 4.1 billion (1999)
Major trading partners: France, Mali, India, Italy, USA, Côte d'Ivoire, Germany, Nigeria
Resources: calcium phosphates, aluminium phosphates, salt, natural gas; offshore deposits of petroleum to be developed
Industries: food processing (principally fish, groundnuts, palm oil, and sugar), mining, cement, artificial fertilizer, chemicals, textiles, petroleum refining (imported petroleum), tourism
Exports: fresh and processed fish, refined petroleum products, chemicals, groundnuts and related products, calcium phosphates and related products. Principal market: India 26.7% (1998)
Imports: food and live animals, machinery and transport equipment, mineral fuels and lubricants (mainly crude petroleum), basic manufactures, chemicals. Principal

source: France 32.6% (1998)
Arable land: 11.7% (1996)
Agricultural products: groundnuts, cotton, millet, sorghum, rice, maize, cassava, vegetables; fishing

Population and society

Population: 9,481,000 (2000 est)
Population growth rate: 2.6% (1995–2000)
Population density: (per sq km) 47 (1999 est)
Urban population: (% of total) 47 (2000 est)
Age distribution: (% of total population) 0–14 45%, 15–59 50%, 60+ 5% (2000 est)
Ethnic groups: the Wolof group are the most numerous, comprising about 35% of the population; the Fulani comprise about 18%; the Serer 16%; the Diola 3%; and the Mandingo 3%
Language: French (official), Wolof, other ethnic languages
Religion: mainly Sunni Muslim; Christian 4%, animist 1%
Education: (compulsory years) 6
Literacy rate: 43% (men); 27% (women) (2000 est)
Labour force: 74.6% agriculture, 7.2% industry, 18.2% services (1997 est)
Life expectancy: 51 (men); 54 (women) (1995–2000)
Child mortality rate: (under 5, per 1,000 live births) 115 (1995–2000)
Physicians: 1 per 18,192 people (1993 est)
Hospital beds: 1 per 1,923 people (1993 est)
TV sets: (per 1,000 people) 41 (1997)
Radios: (per 1,000 people) 142 (1997)
Internet users: (per 10,000 people) 32.5 (1999)
Personal computer users: (per 100 people) 1.5 (1999)

Transport

Airports: international airports: Dakar (Dakar-Yoff); three domestic airports and 12 smaller airfields; total passenger km: 224 million (1995)
Railways: total length: 904 km/562 mi; total passenger km: 206 million (1993)
Roads: total road network: 14,576 km/9,057 mi, of which 29.3% paved (1996 est); passenger cars: 10 per 1,000 people (1996 est)

Chronology

10th–11th centuries: Links established with North Africa; the Tukolor community was converted to Islam. **1445:** First visited by Portuguese explorers. **1659:** French founded Saint-Louis as a colony. **17th–18th centuries:** Export trades in slaves, gums, ivory, and gold developed by European traders. **1854–65:** Interior occupied by French who checked the expansion of the Islamic Tukolor Empire; Dakar founded. **1902:** Became territory of French West Africa. **1946:** Became French overseas territory, with own territorial assembly and representation in French parliament. **1948:** Leopold Sedar Senghor founded the Senegalese Democratic Bloc to campaign for independence. **1959:** Formed Federation of Mali with French Sudan. **1960:** Achieved independence and withdrew from federation. Senghor, leader of socialist Senegalese Progressive Union (UPS), became president. **1966:** UPS declared only legal party. **1974:** Pluralist system re-established. **1976:** UPS reconstituted as Socialist Party (PS). **1980:** Troops sent to defend The Gambia against suspected Libyan invasion. **1981:** Military help again sent to The Gambia to thwart coup attempt. Abdou Diouf was appointed president. **1982:** Confederation of Senegambia came into effect. **1988:** Mamadou Lamine Loum was appointed prime minister. **1989:** Diplomatic links with Mauritania severed after 450 died in violent clashes; over 50,000 people repatriated from both countries. Senegambia federation abandoned. **1992:** Diplomatic links with Mauritania were re-established. **1993:** Assembly and presidential elections were won by the ruling PS. **1998:** PS won the general election despite claims of fraud. Abdou Diouf became 'president for life'. **1999:** A new 60-member Senate was created as Senegal's second legislative chamber. **2000:** In presidential elections, Abdou Diouf lost to Abdoulaye Wade, who appointed Mustafa Niasse as his prime minister. Diouf later announced his withdrawal from politics.

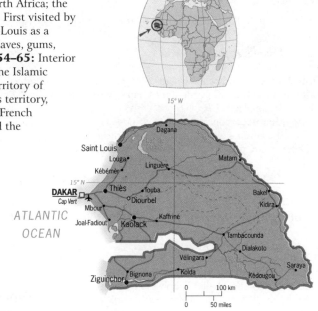

Practical information

Visa requirements: UK: visa not required. USA: visa not required
Time difference: GMT +/–0
Chief tourist attractions: fine beaches; six national parks; the island of Gorée, near Dakar, a former centre for the slave trade
Major holidays: 1 January, 1 February, 4 April, 1 May, 15 August, 1 November, 25 December; variable: Eid-ul-Adha, Easter Monday, end of Ramadan, New Year (Muslim), Prophet's Birthday, Whit Monday

SEYCHELLES

Located in the Indian Ocean, off east Africa, north of Madagascar.

National name: *Republic of Seychelles*
Area: 453 sq km/174 sq mi
Capital: Victoria (on Mahé island) (and chief port)
Major towns/cities: Cascade, Port Glaud, Misere, Anse Boileau, Takamaka
Physical features: comprises two distinct island groups: one, the Granitic group, concentrated, the other, the Outer or Coralline group, widely scattered; totals over 100 islands and islets

Blue represents the sky and the sea. Yellow symbolizes the sun which gives light and life. Red reflects the people and their determination to work in the future in unity and love. White stands for justice and harmony. Green recalls the land and nature. Effective date: 8 January 1996.

Government

Head of state and government: France-Albert René from 1977
Political system: emergent democracy
Political executive: limited presidency
Administrative divisions: 23 districts
Political parties: Seychelles People's Progressive Front (SPPF), nationalist socialist; Democratic Party (DP), left of centre
Armed forces: 200; 300 paramilitary forces; plus 1,000-strong national guard (1998)
Conscription: military service is voluntary
Death penalty: retains death penalty only for exceptional crimes such as crimes under military law or crimes committed in exceptional circumstances such as wartime
Defence spend: (% GDP) 1.2 (1998)
Education spend: (% GNP) 7.6 (1996)
Health spend: (% GDP) 5.9 (1997 est)

Economy and resources

Currency: Seychelles rupee
GDP: (US$) 590.7 million (1999 est)
Real GDP growth: (% change on previous year) 1.8 (1999 est)
GNP: (US$) 520 million (1999)
GNP per capita (PPP): (US$) 10,381 (1999)
Consumer price inflation: 6.3% (1999)
Unemployment: 8.3% (1993)
Foreign debt: (US$) 240 million (1999)
Major trading partners: UK, South Africa, France, Singapore, the Netherlands, Italy
Resources: guano; natural gas and metal deposits were being explored mid-1990s
Industries: food processing (including cinnamon, coconuts, and tuna canning), beer and soft drinks, petroleum refining, cigarettes, paper, metals, chemicals, wood products, paints, tourism
Exports: fresh and frozen fish, canned tuna, shark fins, cinnamon bark, refined petroleum products. Principal market: UK 48.3% (1998 est)
Imports: machinery and transport equipment, food and live animals, petroleum and petroleum products, chemicals, basic manufactures. Principal source: South

Africa 10.7% (1998 est)
Arable land: 2.2% (1995)
Agricultural products: coconuts, copra, cinnamon bark, tea, patchouli, vanilla, limes, sweet potatoes, cassava, yams, sugar cane, bananas; poultry meat and egg production are important for local consumption; fishing

Population and society

Population: 77,000 (2000 est)
Population growth rate: 1.1% (1995–2000)
Population density: (per sq km) 174 (1999 est)
Urban population: (% of total) 64 (2000 est)
Age distribution: (% of total population) 0–14 39%, 15–59 51%, 60+ 10% (2000 est)
Ethnic groups: predominantly Creole (of mixed African, Asian, and European descent); small European minority (mostly French and British)
Language: Creole (an Asian, African, European mixture) (95%), English, French (all official)
Religion: Roman Catholic 90%
Education: (compulsory years) 9
Literacy rate: 86% (men); 82% (women) (1994 est)
Labour force: 6.2% agriculture, 21.5% industry, 72.3% services (1997)
Life expectancy: 66 (men); 76 (women) (1998 est)
Child mortality rate: (under 5, per 1,000 live births) 20 (1995)
Physicians: 1 per 713 people (1996)
Hospital beds: 1 per 175 people (1996)
TV sets: (per 1,000 people) 145 (1997)
Radios: (per 1,000 people) 560 (1997)
Internet users: (per 10,000 people) 652.5 (1999)
Personal computer users: (per 100 people) 13.1 (1999)

Transport

Airports: international airports: Mahé Island (Seychelles); five domestic airports; total passenger km: 847 million (1997 est)
Railways: none
Roads: total road network: 280 km/174 mi, of which 62.9% paved (1996 est); passenger cars: 95.9 per 1,000 people (1996 est)

Chronology

Early 16th century: First sighted by European navigators. **1744:** Became French colony. **1756:** Claimed as French possession and named after an influential French family. **1770s:** French colonists brought African slaves to settle the previously uninhabited islands; plantations established. **1794:** Captured by British during French Revolutionary Wars. **1814:** Ceded by France to Britain; incorporated as dependency of Mauritius. **1835:** Slavery abolished by British, leading to influx of liberated slaves from Mauritius and Chinese and Indian immigrants. **1903:** Became British crown colony, separate from Mauritius. **1963–64:** First political parties formed. **1976:** Independence achieved from Britain as republic within Commonwealth, with a moderate, James Mancham, of the centre-right Seychelles Democratic Party (SDP) as president. **1977:** More radical France-Albert René ousted Mancham in armed bloodless coup and took over presidency; white settlers emigrated. **1979:** Nationalistic socialist Seychelles People's Progressive Front (SPPF) became sole legal party under new constitution; became nonaligned state. **1981:** An attempted coup by South African mercenaries was thwarted. **1993:** A new multiparty constitution was adopted. René defeated Mancham, who had returned from exile, in competitive presidential elections; SPPF won parliamentary elections. **1998:** President René was re-elected. SPUP won assembly elections.

Practical information

Visa requirements: UK: visa not required. USA: visa not required

Time difference: GMT +4

Chief tourist attractions: fine coral beaches; lush tropical vegetation; mountain scenery; rainforest; many unique plant and animal species; national parks and reserves; Aldabra, the largest atoll in the world

Major holidays: 1–2 January, 1 May, 5, 29 June, 15 August, 1 November, 8, 25 December; variable: Corpus Christi, Good Friday, Holy Saturday

SIERRA LEONE

Located in West Africa, on the Atlantic Ocean, bounded north and east by Guinea and southeast by Liberia.

National name: *Republic of Sierra Leone*
Area: 71,740 sq km/27,698 sq mi
Capital: Freetown
Major towns/cities: Koidu, Bo, Kenema, Makeni, Marampa
Major ports: Bonthe
Physical features: mountains in east; hills and forest; coastal mangrove swamps

Green stands for the country's natural resources. White represents peace, justice, and unity. Blue recalls the sea and the capital Freetown's natural harbour. Effective date: 27 April 1961.

Government

Head of state and government: Ahmad Tejan Kabbah from 1996
Political system: transitional
Political executive: transitional
Administrative divisions: four regions comprising 12 districts
Political parties: All People's Congress (APC), moderate socialist; United Front of Political Movements (UNIFORM), left of centre. Party political activity suspended from 1992
Armed forces: 5,000 (1998)
Conscription: military service is voluntary
Death penalty: retained and used for ordinary crimes
Defence spend: (% GDP) 3.3 (1998)
Education spend: (% GNP) N/A
Health spend: (% GDP) 4.9 (1997 est)

Economy and resources

Currency: leone
GDP: (US$) 669 million (1999)
Real GDP growth: (% change on previous year) –8.1 (1999)
GNP: (US$) 653 million (1999)
GNP per capita (PPP): (US$) 414 (1999)
Consumer price inflation: 34.1% (1999)
Unemployment: N/A
Foreign debt: (US$) 1.24 billion (1998)
Major trading partners: Belgium, Luxembourg, UK, USA, Spain, Côte d'Ivoire
Resources: gold, diamonds, bauxite, rutile (titanium dioxide)
Industries: palm oil and other agro-based industries, rice mills, textiles, mining, sawn timber, furniture making
Exports: rutile, diamonds, bauxite, gold, coffee, cocoa beans. Principal market: Belgium 38.4% (1999)
Imports: machinery and transport equipment, food and live animals, basic manufactures, chemicals, miscellaneous manufactured articles. Principal source: UK 33.6% (1999)

Arable land: 6.8% (1996)
Agricultural products: rice, cassava, palm oil, coffee, cocoa, bananas; cattle production

Population and society

Population: 4,854,000 (2000 est)
Population growth rate: 3.0% (1995–2000)
Population density: (per sq km) 66 (1999 est)
Urban population: (% of total) 37 (2000 est)
Age distribution: (% of total population) 0–14 44%, 15–59 51%, 60+ 5% (2000 est)
Ethnic groups: 20 ethnic groups, 3 of which (the Mende, Temne, and Limbe) comprise almost 70% of the population; 10% Creole (descendants of freed Jamaican slaves)
Language: English (official), Krio (a Creole language), Mende, Limba, Temne
Religion: animist 45%, Muslim 44%, Protestant 8%, Roman Catholic 3%
Education: not compulsory
Literacy rate: 51% (men); 23% (women) (2000 est)
Labour force: 63.8% agriculture, 14% industry, 22.2% services (1997 est)
Life expectancy: 36 (men); 39 (women) (1995–2000)
Child mortality rate: (under 5, per 1,000 live births) 263 (1995–2000)
Physicians: 1 per 11,619 people (1990)
Hospital beds: 1 per 873 people (1990)
TV sets: (per 1,000 people) 13 (1998)
Radios: (per 1,000 people) 253 (1997)
Internet users: (per 10,000 people) 4.2 (1999)

Transport

Airports: international airports: Freetown (Lungi); six domestic airports; total passenger km: 24 million (1995)
Railways: total length: 84 km/52 mi (unused since 1985)
Roads: total road network: 11,300 km/7,022 mi, of which 8% paved (1997); passenger cars: 3.9 per 1,000 people (1996 est)

Chronology

15th century: Mende, Temne, and Fulani peoples moved from Senegal into region formerly populated by Bulom, Krim, and Gola peoples. The Portuguese, who named the area Serra Lyoa, established a coastal fort, trading manufactured goods for slaves and ivory. **17th century:** English trading posts established on Bund and York islands. **1787–92:** English abolitionists and philanthropists bought land to establish settlement for liberated and runaway African slaves (including 1,000 rescued from Canada), known as Freetown. **1808:** Became a British colony and Freetown a base for British naval operations against slave trade, after Parliament declared it illegal. **1896:** Hinterland conquered and declared British protectorate. **1951:** First political party, Sierra Leone People's Party (SLPP), formed by Dr Milton Margai, who became 'leader of government business', in 1953. **1961:** Independence achieved within Commonwealth, with Margai as prime minister. **1965:** Free-trade area pact signed with Guinea, Liberia, and the Côte d'Ivoire. **1967:** Election won by All People's Congress (APC), led by Siaka Stevens, but disputed by army, who set up National Reformation Council and forced governor general to leave the country. **1968:** Army revolt brought back Stevens as prime minister. **1971:** New constitution made Sierra Leone a republic, with Stevens as president. **1978:** New constitution made APC the only legal party. **1985:** Stevens retired and was succeeded as president and APC leader by Maj-Gen Joseph Momoh. **1991:** A referendum endorsed multiparty politics and new constitution. A Liberian-based rebel group began guerrilla activities. **1992:** President Momoh was overthrown by the military, and party politics were suspended as the National Provisional Ruling Council was established under Capt Valentine Strasser; 500,000 Liberians fled to Sierra Leone as a result of the civil war. **1995:** The ban on political parties was lifted. **1996:** Ahmad Tejan Kabbah became president after multiparty elections. **1997:** President Kabbah's civilian government was ousted in a bloody coup. Maj Johnny Paul Koroma seized the presidency and the Revolutionary Council was formed. **1998:** A Nigerian-led peacekeeping force drove out Maj Koroma's junta; President Kabbah returned from exile. Former members of military government were executed for treason. **1999:** Fighting between government and rebel forces continued. Diplomatic efforts were spearheaded by the Organization of African Unity; a ceasefire and peace agreement were reached with rebels, and in November the first unit of what would become a 6,000-strong United Nations (UN) peacekeeping force arrived in Sierra Leone. **2000:** As rebel activity continued, the UN force was increased to 11,000, the largest UN force in current operation. Some UN peacekeepers were besieged by rebels but most were released by July and the remainder freed by UN troops. In August, 11 British soldiers were captured. Five were later released and the remaining six rescued in September. In November, the government and rebels signed a 30-day truce at a meeting in Nigeria. **2001:** Fighting between the Guinean army and Sierra Leonean rebels left 170,000 refugees trapped in southern Guinea.

Practical information

Visa requirements: UK: visa required. USA: visa required
Time difference: GMT +/–0
Chief tourist attractions: mountains; game reserves; coastline
Major holidays: 1 January, 19 April, 25–26 December; variable: Eid-ul-Adha, end of Ramadan, Good Friday, Easter Monday, Prophet's Birthday

SINGAPORE FORMERLY PART OF STRAITS SETTLEMENT
(1826–1942), PART OF THE FEDERATION OF MALAYSIA·(1963–65)

Located in southeast Asia, off the tip of the Malay Peninsula.

National name: *Repablik Singapura/Republic of Singapore*
Area: 622 sq km/240 sq mi
Capital: Singapore City
Physical features: comprises Singapore Island, low and flat, and 57 small islands; Singapore Island is joined to the mainland by causeway across Strait of Johore

Red and white are traditional colours in South East Asia. Red represents universal fellowship and equality. White stands for purity and virtue.
Effective date: 3 December 1959.

Government

Head of state: Sellapan Ramanathan Nathan from 1999
Head of government: Goh Chok Tong from 1990
Political system: liberal democracy
Political executive: parliamentary
Administrative divisions: none
Political parties: People's Action Party (PAP), conservative, free market, multi-ethnic; Workers' Party (WP), socialist; Singapore Democratic Party (SDP), liberal pluralist
Armed forces: 72,500; 250,000 reservists (1998)
Conscription: two years
Death penalty: retained and used for ordinary crimes
Defence spend: (% GDP) 5.0 (1998)
Education spend: (% GNP) 3.0 (1996)
Health spend: (% GDP) 3.1 (1997)

Economy and resources

Currency: Singapore dollar
GDP: (US$) 86.5 billion (1999)
Real GDP growth: (% change on previous year) 5.4 (1999)
GNP: (US$) 95.4 billion (1999)
GNP per capita (PPP): (US$) 27,024 (1999)
Consumer price inflation: 0% (1999)
Unemployment: 3.2% (1998)
Foreign debt: (US$) 555 million (1999 est)
Major trading partners: Japan, USA, Malaysia, Hong Kong, Thailand, Germany, China, Saudi Arabia
Resources: granite
Industries: electrical machinery (particularly radios and televisions), petroleum refining and petroleum products, transport equipment (especially shipbuilding), chemicals, metal products, machinery, food processing, clothing, finance and business services
Exports: electrical and nonelectrical machinery, transport equipment, petroleum products, chemicals, rubber, foodstuffs, clothing, metal products, iron and steel, orchids and other plants, aquarium fish. Principal market: USA 19.2% (1999)
Imports: electrical and nonelectrical equipment, crude petroleum, transport equipment, chemicals, food and live animals, textiles, scientific and optical instruments, paper and paper products. Principal source: Japan 16.6% (1999)
Arable land: 1.6% (1996)
Agricultural products: vegetables, plants, orchids; poultry and fish production

Population and society

Population: 3,567,000 (2000 est)
Population growth rate: 1.4% (1995–2000)
Population density: (per sq km) 5,662 (1999 est)
Urban population: (% of total) 100 (2000 est)
Age distribution: (% of total population) 0–14 22%, 15–59 67%, 60+ 11% (2000 est)
Ethnic groups: 77% of Chinese ethnic descent, predominantly Hokkien, Teochew, and Cantonese; 14% Malay; 7% Indian, chiefly Tamil
Language: Malay, Mandarin Chinese, Tamil, English (all official), other Indian languages, Chinese dialects
Religion: Buddhist, Taoist, Muslim, Hindu, Christian
Education: (compulsory years) 6
Literacy rate: 96% (men); 88% (women) (2000 est)
Labour force: 0.2% agriculture, 30.4% industry, 69.4% services (1996)
Life expectancy: 75 (men); 80 (women) (1995–2000)
Child mortality rate: (under 5, per 1,000 live births) 6 (1995–2000)
Physicians: 1 per 641 people (1996)
Hospital beds: 1 per 285 people (1996)
TV sets: (per 1,000 people) 348 (1998)
Radios: (per 1,000 people) 822 (1997)
Internet users: (per 10,000 people) 2,945.9 (1999)
Personal computer users: (per 100 people) 52.7 (1999)

Transport

Airports: international airports: Singapore (Changi); total passenger km: 55,459 million (1997 est)
Railways: total length: 26 km/16 mi
Roads: total road network: 3,017 km/1,875 mi, of which 97.3% paved (1997); passenger cars: 122 per 1,000 people (1997)

Chronology

12th century: First trading settlement established on Singapore Island. **14th century:** Settlement destroyed, probably by Javanese Empire of Mahapahit. **1819:** Stamford Raffles of British East India Company obtained Singapore from sultan of Johore. **1826:** Straits Settlements formed from British possessions of Singapore, Penang, and Malacca ruled by governor of Bengal. **1832:** Singapore became capital of Straits Settlements; the port prospered, attracting Chinese and Indian immigrants. **1851:** Responsibility for Straits Settlements fell to governor general of India. **1858:** British government, through the India Office, took over administration of Straits Settlements. **1867:** Straits Settlements became crown colony of British Empire. **1922:** Singapore chosen as principal British military base in Far East. **1942:** Japan captured Singapore, taking 70,000 British and Australian prisoners. **1945:** British rule restored after defeat of Japan. **1946:** Singapore became separate crown colony. **1959:** Internal self-government achieved as State of Singapore with Lee Kuan Yew (PAP) as prime minister. **1960s:** Rapid development as leading commercial and financial centre. **1963:** Singapore combined with Federation of Malaya, Sabah, and Sarawak to form Federation of Malaysia. **1965:** Became independent republic after withdrawing from Federation of Malaysia in protest at alleged discrimination against ethnic Chinese. **1971:** Last remaining British military bases closed. **1984:** Two opposition members elected to national assembly for first time. **1988:** Ruling PAP won all but one of available assembly seats; increasingly authoritarian rule. **1990:** Lee Kuan Yew retired from the premiership after 31 years and was succeeded by Goh Chok Tong. **1996:** Constitutional change was introduced, allowing better representation of minority races. **1997:** The PAP, led by Prime Minister Goh Chok Tong, won a general election. **1998:** Pay cuts were introduced as Singapore slipped into recession for the first time in 13 years. **1999:** After all other candidates were screened out of the election, Sellapan Ramanathan Nathan, the government's candidate, became the new president.

Practical information

Visa requirements: UK: visa not required. USA: visa not required
Chief tourist attractions: blend of cultures; excellent shopping facilities; Singapore City, with its Tiger Balm Gardens, Sultan Mosque, Buddhist temples, House of Jade, bird park, botanical gardens, and national museum
Major holidays: 1 January, 1 May, 9 August, 25 December; variable: Eid-ul-Adha, Diwali, end of Ramadan, Good Friday, New Year (Chinese, January/February, 2 days), Vesak

SLOVAK REPUBLIC

FORMERLY CZECHOSLOVAKIA (WITH THE CZECH REPUBLIC) (1918–93)

Landlocked country in central Europe, bounded north by Poland, east by the Ukraine, south by Hungary, west by Austria, and northwest by the Czech Republic.

National name: *Slovenská Republika/Slovak Republic*
Area: 49,035 sq km/18,932 sq mi
Capital: Bratislava
Major towns/cities: Košice, Nitra, Prešov, Banská Bystrica, Žilina, Trnava, Martin
Physical features: Western range of Carpathian Mountains, including Tatra and Beskids in north; Danube plain in south; numerous lakes and mineral springs

The arms depict the Carpathian Mountains which traverse the Slovak Republic. The flag uses the pan-Slav colours representing liberation from foreign domination. Effective date: 3 September 1992.

Government

Head of state: Rudolf Schuster from 1999
Head of government: Mikulas Dzurinda from 1998
Political system: emergent democracy
Political executive: parliamentary
Administrative divisions: eight regions and 79 districts
Political parties: Movement for a Democratic Slovakia (MDS), left of centre, nationalist-populist; Democratic Union of Slovakia (DUS), centrist; Christian Democratic Movement (KSDH), right of centre; Slovak National Party (SNP), nationalist; Party of the Democratic Left (PDL), reform socialist, (ex-communist); Association of Workers of Slovakia, left wing; Hungarian Coalition, ethnic Hungarian
Armed forces: 45,500 (1998)
Conscription: military service is compulsory for 18 months
Death penalty: abolished in 1990
Defence spend: (% GDP) 2 (1998)
Education spend: (% GNP) 4.7 (1996)
Health spend: (% GDP) 8.6 (1997)

Economy and resources

Currency: Slovak koruna (based on Czechoslovak koruna)
GDP: (US$) 19.7 billion (1999)
Real GDP growth: (% change on previous year) 1.9 (1999)
GNP: (US$) 19.4 billion (1999)
GNP per capita (PPP): (US$) 9,811 (1999)
Consumer price inflation: 10.5% (1999)
Unemployment: 14% (1998)
Foreign debt: (US$) 10.5 billion (1999 est)
Major trading partners: Germany, Czech Republic, Russia, Austria, Hungary, Italy, Poland
Resources: brown coal, lignite, copper, zinc, lead, iron ore, magnesite
Industries: chemicals, pharmaceuticals, heavy engineering, munitions, mining, textiles, clothing, glass, leather, footwear, construction materials, televisions, transport equipment (cars, lorries, and motorcycles)
Exports: basic manufactures, machinery and transport equipment, miscellaneous manufactured articles. Principal market: Germany 27.7% (1999)
Imports: machinery and transport equipment, mineral fuels and lubricants, basic manufactures, chemicals and related products. Principal source: Germany 26.1% (1999)
Arable land: 30.8% (1996)
Agricultural products: wheat and other grains, sugar beet, potatoes and other vegetables; livestock rearing (cattle, pigs, and poultry)

Population and society

Population: 5,387,000 (2000 est)
Population growth rate: 0.1% (1995–2000)
Population density: (per sq km) 110 (1999 est)
Urban population: (% of total) 57 (2000 est)
Age distribution: (% of total population) 0–14 20%, 15–59 64%, 60+ 16% (2000 est)
Ethnic groups: 86% ethnic Slovak, 11% ethnic Hungarian (Magyar), 2% Romany; small Czech, Moravian, Silesian, and Ukrainian communities
Language: Slovak (official), Hungarian, Czech, other ethnic languages
Religion: Roman Catholic (over 50%), Lutheran, Reformist, Orthodox, atheist 10%
Education: (compulsory years) 9
Literacy rate: 99% (men); 99% (women) (2000 est)
Labour force: 8.9% agriculture, 37.1% industry, 54% services (1997)
Life expectancy: 69 (men); 77 (women) (1995–2000)
Child mortality rate: (under 5, per 1,000 live births) 13 (1995–2000)
Physicians: 1 per 327 people (1996)
Hospital beds: 1 per 131 people (1996)
TV sets: (per 1,000 people) 401 (1997)
Radios: (per 1,000 people) 580 (1997)
Internet users: (per 10,000 people) 1,300.7 (1999)
Personal computer users: (per 100 people) 7.4 (1999)

Transport

Airports: international airports: Bratislava (M R Stefanik), Poprad-Tatry, Košice, Piešt'any, Sliac; domestic services to most major cities; total passenger km: 231 million (1997)

Railways: total length: 3,665 km/2,277 mi; total passenger km: 4,056 million (1996)
Roads: total road network: 17,627 km/11,306 mi, of which 98.9% paved (1997); passenger cars: 211 per 1,000 people (1997)

Chronology

9th century: Part of kingdom of Greater Moravia, in Czech lands to west, founded by Slavic Prince Sviatopluk; Christianity adopted. **906:** Came under Magyar (Hungarian) domination and adopted Roman Catholicism. **1526:** Came under Austrian Habsburg rule. **1867:** With creation of dual Austro-Hungarian monarchy, came under separate Hungarian rule; policy of forced Magyarization stimulated a revival of Slovak national consciousness. **1918:** Austro-Hungarian Empire dismembered; Slovaks joined Czechs to form independent state of Czechoslovakia. Slovak-born Tomas Masaryk remained president until 1935, but political and economic power became concentrated in Czech lands. **1939:** Germany annexed Czechoslovakia, which became Axis puppet state under the Slovak autonomist leader Monsignor Jozef Tiso; Jews persecuted. **1944:** Popular revolt against German rule ('Slovak Uprising'). **1945:** Liberated from German rule by Soviet troops; Czechoslovakia re-established. **1948:** Communists assumed power in Czechoslovakia. **1950s:** Heavy industry introduced into previously rural Slovakia; Slovak nationalism and Catholic Church forcibly suppressed.

1968–69: 'Prague Spring' political reforms introduced by Slovak-born Communist Party leader Alexander Dubcek; Warsaw Pact forces invaded Czechoslovakia to stamp out reforms; Slovak Socialist Republic, with autonomy over local affairs, created under new federal constitution. **1989:** Prodemocracy demonstrations in Bratislava; new political parties, including centre-left People Against Violence (PAV), formed and legalized; Communist Party stripped of powers; new government formed, with ex-dissident playwright Václav Havel as president. **1990:** Slovak nationalists polled strongly in multiparty elections, with Vladimir Meciar (PAV) becoming prime minister. **1991:** There was increasing Slovak separatism as the economy deteriorated. Meciar formed a PAV splinter group, Movement for a Democratic Slovakia (HZDS), pledging greater autonomy for Slovakia. Pro-Meciar rallies in Bratislava followed his dismissal. **1992:** Meciar returned to power following an electoral victory for the HZDS. Slovak parliament's declaration of sovereignty led to Havel's resignation. **1993:** The Slovak Republic joined the United Nations (UN) and Council of Europe as a sovereign state, with Meciar as prime minister and Michal Kovac, formerly of HZDS, as president. **1994:** The Slovak Republic joined NATO's 'Partnership for Peace' programme. **1995:** Slovak was made the sole official language; a Treaty of Friendship and Cooperation was signed with Hungary. **1996:** An anti-Meciar coalition, the Slovak Democratic Coalition, was formed, comprising five opposition parties. **1997:** A referendum on NATO membership and presidential elections was declared invalid after confusion over voting papers. **1998:** Presidential powers were assumed by Meciar after failure to elect new president. The national council chair, Ivan Gasparovic, became acting head of state. Meciar stepped down as prime minister after the opposition Slovak Democratic Coalition (SDC) polled strongly in a general election. A new SDC-led coalition was formed under Mikulas Dzurinda. The koruna was devalued by 6%. **1999:** Rudolf Schuster was elected president. **2000:** Dzurinda formed a new coalition. The death penalty was abolished. Meciar was arrested on charges of corruption.

Practical information

Visa requirements: UK: visa not required. USA: visa not required
Time difference: GMT +1
Chief tourist attractions: ski resorts in the High and Low Tatras and other mountain regions; historic towns,

including Bratislava, Košice, Nitra, and Bardejov; numerous castles and mansions; over 20 spa resorts with thermal and mineral springs
Major holidays: 1, 6 January, 1 May, 5 July, 29 August, 1, 15 September, 1 November, 24–26 December; variable: Good Friday, Easter Monday

SLOVENIA

Located in south-central Europe, bounded north by Austria, east by Hungary, west by Italy, and south by Croatia.

National name: *Republika Slovenija/Republic of Slovenia*
Area: 20,251 sq km/7,818 sq mi
Capital: Ljubljana
Major towns/cities: Maribor, Kranj, Celji, Velenje, Koper, Novo Mesto
Major ports: Koper
Physical features: mountainous; Sava and Drava rivers

The three stars are taken from the arms of the Duchy of Selje. The flag uses the pan-Slav colours. Effective date: 25 June 1991.

Government

Head of state: Milan Kučan from 1990
Head of government: Janez Drnovšek from 2000
Political system: emergent democracy
Political executive: dual executive
Administrative divisions: 136 municipalities and 11 urban municipalities
Political parties: Slovenian Christian Democrats (SKD), right of centre; Slovenian People's Party (SPP), conservative; Liberal Democratic Party of Slovenia (LDS), centrist; Slovenian Nationalist Party (SNS), right-wing nationalist; Democratic Party of Slovenia (LDP), left of centre; United List of Social Democrats (ZLSD) left of centre, ex-communist
Armed forces: 9,600; plus reserve forces of 53,000 and a paramilitary police force of 4,500 (1998)
Conscription: military service is compulsory for seven months
Death penalty: abolished in 1989
Defence spend: (% GDP) 1.7 (1998)
Education spend: (% GNP) 5.8 (1996)
Health spend: (% GDP) 9.4 (1997)

Economy and resources

Currency: tolar
GDP: (US$) 20 billion (1999)
Real GDP growth: (% change on previous year) 4.9 (1999)
GNP: (US$) 19.6 billion (1999)
GNP per capita (PPP): (US$) 15,062 (1999)
Consumer price inflation: 6.1% (1999)
Unemployment: 14% (1997)
Foreign debt: (US$) 5.5 billion (1999)
Major trading partners: Germany, Italy, Croatia, France, Austria, EU
Resources: coal, lead, zinc; small reserves/deposits of natural gas, petroleum, salt, uranium
Industries: metallurgy, furniture making, sports equipment, electrical equipment, food processing, textiles, paper and paper products, chemicals, wood and wood products
Exports: raw materials, semi-finished goods, machinery, electric motors, transport equipment, foodstuffs, clothing, pharmaceuticals, cosmetics. Principal market: Germany 30.7% (1999)
Imports: machinery and transport equipment, raw

materials, semi-finished goods, foodstuffs, chemicals, miscellaneous manufactured articles, mineral fuels and lubricants. Principal source: Germany 20.1% (1999)
Arable land: 11.5% (1996)
Agricultural products: wheat, maize, sugar beet, potatoes, cabbage, fruits (especially grapes); forest resources (approximately 45% of total land area was forest in 1994)

Population and society

Population: 1,986,000 (2000 est)
Population growth rate: –0.05% (1995–2000)
Population density: (per sq km) 98 (1999 est)
Urban population: (% of total) 50 (2000 est)
Age distribution: (% of total population) 0–14 16%, 15–59 65%, 60+ 19% (2000 est)
Ethnic groups: 89% of Slovene origin, 3% ethnic Croat, 2% Serb; small Italian, Hungarian, and Albanian communities
Language: Slovene (related to Serbo-Croat; official), Hungarian, Italian
Religion: Roman Catholic 70%; Eastern Orthodox, Lutheran, Muslim
Education: (compulsory years) 8
Literacy rate: 99% (men); 99% (women) (2000 est)
Labour force: 12% agriculture, 40.6% industry, 47.4% services (1997)
Life expectancy: 71 (men); 78 (women) (1995–2000)
Child mortality rate: (under 5, per 1,000 live births) 8 (1995–2000)
Physicians: 1 per 481 people (1996)
Hospital beds: 1 per 179 people (1996)
TV sets: (per 1,000 people) 356 (1998)
Radios: (per 1,000 people) 406 (1997)
Internet users: (per 10,000 people) 1,257.0 (1999)
Personal computer users: (per 100 people) 25.1 (1999)

Transport

Airports: international airports: Ljubljana (Brnik), Maribor, Portoroz; two domestic airports; total passenger km: 677 million (1997)
Railways: total length: 1,201 km/746 mi; total passenger km: 616 million (1997)
Roads: total road network: 14,830 km/9,215 mi, of which 83% paved (1997); passenger cars: 384 per 1,000 people (1997)

Chronology

1st century BC: Came under Roman rule. **AD 395:** In the division of the
Roman Empire, stayed in the west, along with Croatia and Bosnia. **6th
century:** Settled by the Slovene South Slavs. **7th century:** Adopted
Christianity as Roman Catholics. **8th–9th centuries:** Under successive rule
of Franks and dukes of Bavaria. **907–55:** Came under Hungarian
domination. **1335:** Absorbed in Austro-Hungarian Habsburg Empire,
as part of Austrian crownlands of Carniola, Styria, and
Carinthia. **1848:** Slovene struggle for independence began.
1918: On collapse of Habsburg Empire, Slovenia united
with Serbia, Croatia, and Montenegro to form the
'Kingdom of Serbs, Croats and Slovenes', under
Serbian Karageorgevic dynasty. **1929:** Kingdom
became known as Yugoslavia. **1941–45:** Occupied
by Nazi Germany and Italy during World War II;
anti-Nazi Slovene Liberation Front formed and
became allies of Marshal Tito's communist-led
Partisans. **1945:** Slovenia became a constituent
republic of the Yugoslav Socialist Federal
Republic. **mid-1980s:** The Slovenian
Communist Party liberalized itself and agreed to
free elections. Yugoslav counterintelligence
(KOV) began repression. **1989:** The constitution
was changed to allow secession from the federation.
1990: A Nationalist Democratic Opposition of
Slovenia (DEMOS) coalition secured victory in the first
multiparty parliamentary elections; Milan Kučan, a reform
communist, became president. Sovereignty was declared.

Independence was overwhelmingly approved in a referendum. **1991:** Slovenia seceded from the Yugoslav federation,
along with Croatia; 100 people were killed after the Yugoslav federal army intervened; a ceasefire brokered by the
European Community (EC) brought the withdrawal of the Yugoslav army. **1992:** Janez Drnovšek, a centrist Liberal
Democrat, was appointed prime minister; independence was recognized by the EC and the USA. Slovenia was
admitted into the United Nations (UN). **1997:** A new government was formed by the ruling LDS, led by Prime
Minister Janez Drnovšek. President Kučan was re-elected. The European Union (EU) agreed to open membership
talks with Slovenia. **2000:** A right-wing coalition government, led by Andrej Bajuk, briefly took office, but broke
down over whether to move away from the system of election by proportional representation. New elections were won
by a centre-left coalition led by former prime minister, Janez Drnovšek.

Practical information

Visa requirements: UK: visa not required. USA: visa
not required
Time difference: GMT +1
Chief tourist attractions: Alps in north;
Mediterranean beaches; Karst limestone regions, with

more than 6,000 caves; Ljubljana, with its castle,
cathedral, Tivoli sports park, fairs, and festivals
Major holidays: 1–2 January, 8 February, 27 April, 1–2
May, 25 June, 15 August, 31 October, 1 November,
25–26 December; variable: Good Friday, Easter Monday

SOLOMON ISLANDS

Comprises many hundreds of islands in the southwest Pacific Ocean, east of New Guinea.

Area: 27,600 sq km/10,656 sq mi
Capital: Honiara (on Guadalcanal island) (and chief port)
Major towns/cities: Gizo, Auki, Kirakira, Buala
Major ports: Yandina
Physical features: comprises all but the northernmost islands (which belong to Papua New Guinea) of a Melanesian archipelago stretching nearly 1,500 km/900 mi. The largest is Guadalcanal (area 6,500 sq km/2,510 sq mi); others are Malaita, San Cristobal, New Georgia, Santa Isabel, Choiseul; mainly mountainous and forested

Blue stands for the Pacific Ocean and water on which life depends. Yellow symbolizes the sun. Green represents the lush vegetation.
Effective date: 18 November 1977.

Government

Head of state: Queen Elizabeth II from 1952, represented by Governor General Moses Pitakaka from 1994
Head of government: Mannesseh Sogavare from 2000
Political system: emergent democracy
Political executive: parliamentary
Administrative divisions: nine provinces and the Honiara municipal authority
Political parties: Group for National Unity and Reconciliation (GNUR), centrist coalition; National Coalition Partners (NCP), broad-based coalition; People's Progressive Party (PPP); People's Alliance Party (PAP), socialist
Armed forces: no standing army; 80-strong marine wing of police force (1998)
Death penalty: laws do not provide for the death penalty for any crime
Education spend: (% GNP) N/A
Health spend: (% GDP) 3.2 (1997 est)

Economy and resources

Currency: Solomon Island dollar
GDP: (US$) 301 million (1999 est)
Real GDP growth: (% change on previous year) 1 (1999)
GNP: (US$) 320 million (1999)
GNP per capita (PPP): (US$) 1,793 (1999)
Consumer price inflation: 8% (1999)
Foreign debt: (US$) 152 million (1998)
Major trading partners: Australia, Japan, UK, New Zealand, Singapore, South Korea, USA
Resources: bauxite, phosphates, gold, silver, copper, lead, zinc, cobalt, asbestos, nickel
Industries: food processing (mainly palm oil and rice milling, fish, and coconut-based products), saw milling, logging, tobacco, furniture, handicrafts, boats, clothing, tourism
Exports: timber, fish products, oil palm products, copra, cocoa, coconut oil. Principal market: Japan 59.2% (1997)
Imports: rice, machinery and transport equipment, meat preparations, refined sugar, mineral fuels, basic

manufactures, construction materials. Principal source: Australia 41.5% (1997)
Arable land: 1.5% (1995)
Agricultural products: coconuts, cocoa, rice, cassava, sweet potatoes, yam, taro, banana, palm oil; livestock rearing (pigs and cattle); fishing, sea shells, and seaweed farming; forestry

Population and society

Population: 444,000 (2000 est)
Population growth rate: 3.1% (1995–2000)
Population density: (per sq km) 16 (1999 est)
Urban population: (% of total) 20 (2000 est)
Age distribution: (% of total population) 0–14 43%, 15–59 52%, 60+ 5% (2000 est)
Ethnic groups: 93% Melanesian, 4% Polynesian, 1.5% Micronesian, 0.7% European, 0.2% Chinese
Language: English (official), pidgin English, more than 80 Melanesian dialects (85%), Papuan and Polynesian languages
Religion: more than 80% Christian; Anglican 34%, Roman Catholic 19%, South Sea Evangelical, other Protestant, animist 5%
Education: not compulsory
Literacy rate: 60% (men); 60% (women) (1994 est)
Labour force: 27.4% agriculture, 13.7% industry, 58.9% services (1993)
Life expectancy: 70 (men); 74 (women) (1995–2000)
Child mortality rate: (under 5, per 1,000 live births) 27 (1995–2000)
Physicians: 1 per 6,500 people (1994)
Hospital beds: 1 per 260 people (1994)
TV sets: (per 1,000 people) 6 (1997)
Radios: (per 1,000 people) 141 (1997)
Internet users: (per 10,000 people) 69.7 (1999)

Transport

Airports: international airport: Honiara (Henderson); 26 domestic airports; total passenger km: 89 million (1995)
Railways: none
Roads: total road network: 1,360 km/845 mi, of which 2.5% paved (1996 est)

Chronology

1568: The islands, rumoured in South America to be the legendary gold-rich 'Islands of Solomon', were first sighted by Spanish navigator Alvaro de Mendana, journeying from Peru. **1595 and 1606:** Unsuccessful Spanish efforts to settle the islands, which had long been peopled by Melanesians. **later 18th century:** Visited again by Europeans. **1840s:** Christian missions established. **1870s:** Development of copra export trade and shipment of islanders to work on sugar cane plantations in Australia and Fiji Islands. **1886:** Northern Solomon Islands became German protectorate. **1893:** Southern Solomon Islands placed under British protection. **1899:** Germany ceded Solomon Islands possessions to Britain in return for British recognition of its claims to Western Samoa. **1900:** Unified British Solomon Islands Protectorate

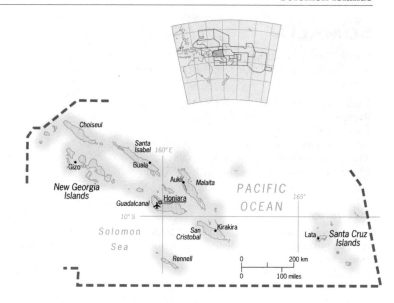

formed and placed under jurisdiction of Western Pacific High Commission (WPHC), with its headquarters in Fiji Islands. **1942–43:** Occupied by Japan. Site of fierce fighting, especially on Guadalcanal, which was recaptured by US forces, with the loss of 21,000 Japanese and 5,000 US troops. **1943–50:** Development of Marching Rule (Ma'asina Ruru) cargo cult populist movement on Malaita island, campaigning for self-rule. **1945:** Headquarters of WPHC moved to Honiara. **1960:** Legislative and executive councils established by constitution. **1974:** Became substantially self-governing, with Solomon Mamaloni of centre-left People's Progressive Party (PPP) as chief minister. **1976:** Became fully self-governing, with Peter Kenilorea of right-of-centre Solomon Islands United Party (SIUPA) as chief minister. **1978:** Independence achieved from Britain within Commonwealth, with Kenilorea as prime minister. **1988:** The Solomon Islands joined Vanuatu and Papua New Guinea to form the Spearhead Group, aiming to preserve Melanesian cultural traditions. **1997:** Bartholomew Ulufa'alu was elected prime minister. **1998:** Ulufa'alu's Alliance for Change government narrowly survived a no-confidence vote. **2000:** A military coup, led by rebel leader, Andrew Nori, in June forced the resignation of Prime Minister Ulufa'alu. The former opposition leader, Mannesseh Sogavare, was elected to become the new prime minister. A peace treaty was signed by rival ethnic militias in October and fighting ceased.

Practical information

Visa requirements: UK: visa not required. USA: visa not required
Time difference: GMT +11
Chief tourist attractions: tourism restricted by relative inaccessibility of country and inadequacy of

tourist facilities
Major holidays: 1 January, 7 July, 25–26 December; variable: Good Friday, Easter Monday, Holy Saturday, Whit Monday, Queen's Birthday (June)

SOMALIA

Located in northeast Africa (the Horn of Africa), on the Indian Ocean, bounded northwest by Djibouti, west by Ethiopia, and southwest by Kenya.

National name: *Jamhuuriyadda Soomaaliya/Republic of Somalia*
Area: 637,700 sq km/246,215 sq mi
Capital: Mogadishu (and chief port)
Major towns/cities: Hargeysa, Berbera, Kismaayo, Marka
Major ports: Berbera, Marka, Kismaayo
Physical features: mainly flat, with hills in north

Blue is said to represent the bright sky. The star stands for freedom. Effective date: 12 October 1954.

Government

Head of state: Abdiqasim Salad Hassan from 2000
Head of government: Ali Khalifa Galaid from 2000
Political system: military
Political executive: military
Administrative divisions: 18 regions
Political parties: parties are mainly clan-based and include the United Somali Congress (USC), Hawiye clan; Somali Patriotic Movement (SPM), Darod clan; Somali Southern Democratic Front (SSDF), Majertein clan; Somali Democratic Alliance (SDA), Gadabursi clan; United Somali Front (USF), Issa clan; Somali National Movement (SNM) based in self-proclaimed Somaliland Republic
Armed forces: 225,000 (1998)
Death penalty: retained and used for ordinary crimes
Defence spend: (% GDP) 4.7 (1998)
Education spend: (% GNP) N/A
Health spend: (% GDP) 1.5 (1997 est)

Economy and resources

Currency: Somali shilling
GDP: (US$) 4.3 billion (1999 est)
Real GDP growth: (% change on previous year) N/A
GNP: (US$) N/A
GNP per capita (PPP): (US$) 600 (1999 est)
Consumer price inflation: N/A
Foreign debt: (US$) 2.6 billion (1999)
Major trading partners: Saudi Arabia, Djibouti, Kenya, Italy, United Arab Emirates, Yemen, Brazil, India, Pakistan
Resources: chromium, coal, salt, tin, zinc, copper, gypsum, manganese, iron ore, uranium, gold, silver; deposits of petroleum and natural gas have been discovered but remain unexploited
Industries: food processing (especially sugar refining), textiles, petroleum refining, processing of hides and skins
Exports: livestock, skins and hides, bananas, fish and fish products, myrrh. Principal market: Saudi Arabia 53% (1999 est)
Imports: petroleum, fertilizers, foodstuffs, machinery and parts, manufacturing raw materials. Principal

source: Djibouti 24% (1999 est)
Arable land: 1.6% (1995)
Agricultural products: bananas, sugar cane, maize, sorghum, grapefruit, seed cotton; agriculture is based on livestock rearing (cattle, sheep, goats, and camels) – 80% of the population depend on this activity

Population and society

Population: 10,097,000 (2000 est)
Population growth rate: 4.2% (1995–2000)
Population density: (per sq km) 15 (1999 est)
Urban population: (% of total) 28 (2000 est)
Age distribution: (% of total population) 0–14 48%, 15–59 48%, 60+ 4% (2000 est)
Ethnic groups: 98% indigenous Somali (about 84% Hamitic and 14% Bantu); population is divided into around 100 clans
Language: Somali, Arabic (both official), Italian, English
Religion: Sunni Muslim; small Christian community, mainly Roman Catholic
Education: (compulsory years) 8
Literacy rate: 36% (men); 14% (women) (1995 est)
Labour force: 72.5% agriculture, 9.5% industry, 18% services (1997 est)
Life expectancy: 45 (men); 49 (women) (1995–2000)
Child mortality rate: (under 5, per 1,000 live births) 204 (1995–2000)
Physicians: 1 per 5,691 people (1990)
Hospital beds: 1 per 897 people (1990)
TV sets: (per 1,000 people) 15 (1997)
Radios: (per 1,000 people) 53 (1997)
Internet users: (per 10,000 people) 0.2 (1999)

Transport

Airports: international airports: Mogadishu, Berbera; seven domestic airports and airfields; passengers carried: 46,000 (1991)
Railways: none
Roads: total road network: 22,100 km/13,733 mi, of which 11.8% paved (1996 est); passenger cars: 0.1 per 1,000 people (1996 est)

Chronology

8th–10th centuries: Arab ancestors of Somali clan families migrated to the region and introduced Sunni Islam; coastal trading cities, including Mogadishu, were formed by Arabian immigrants and developed into sultanates. **11th–14th century:** Southward and westward movement of Somalis and Islamization of Christian Ethiopian interior. **early 16th century:** Portuguese contacts with coastal region. **1820s:** First British contacts with northern Somalia. **1884–87:** British protectorate of Somaliland established in north. **1889:** Italian protectorate of Somalia established in south. **1927:** Italian Somalia became a colony and part of Italian East Africa from 1936. **1941:** Italian Somalia occupied by Britain during World War II. **1943:** Somali Youth League (SYL) formed as nationalist party. **1950:** Italy resumed control over Italian Somalia under UN trusteeship. **1960:** Independence achieved from Italy and Britain as Somalia, with Aden Abdullah Osman as president. **1963:** Border dispute with Kenya; diplomatic relations broken with Britain for five years. **1969:** President Ibrahim Egal assassinated in army coup led by Maj-Gen Muhammad Siad Barre; constitution suspended, political parties banned, Supreme Revolutionary Council set up, and socialist-Islamic state formed. **1972:** 20,000 died in severe drought. **1978:** Defeated in eight-month war with Ethiopia fought on behalf of Somali guerrillas in Ogaden to the southwest. Armed insurrection began in north and hundreds of thousands became refugees. **1979:** New constitution for socialist one-party state dominated by Somali Revolutionary Socialist Party (SRSP). **1982:** The antigovernment Ethiopian-backed Somali National Movement (SNM) was formed in the north, followed by oppressive countermeasures by the government. **late 1980s:** Guerrilla activity increased in the north as the civil war intensified. **1991:** Mogadishu was captured by rebels; Ali Mahdi Muhammad took control of the north of the town, and General Aidid took control of the south; free elections were promised. The secession of northeast Somalia, as the Somaliland Republic, was announced but not recognized internationally. **1992:** There was widespread famine. Western food-aid convoys were hijacked by 'warlords'. United Nations (UN) peacekeeping troops, led by US Marines, were sent in to protect relief operations. **1993:** Leaders of armed factions (except the Somaliland-based faction) agreed to a federal system of government. US-led UN forces destroyed the headquarters of warlord General Aidid after the killing of Pakistani peacekeepers. **1994:** Ali Mahdi Muhammad and Aidid signed a truce. Most Western peacekeeping troops were withdrawn, but clan-based fighting continued. **1996:** Aidid was killed in renewed faction fighting; his son Hussein Aidid succeeded him as interim president. **1998:** A peace plan was agreed. **1999:** In June the Ethiopian army, supporting opponents of Aidid, invaded Somalia. **2000:** The four-month Somali reconciliation conference in Djibouti ended in August after the new transitional parliament elected Abdulkassim Salat Hassan as Somalia's first civilian president since civil war broke out nine years earlier. In October, Ali Khalifa Galad became prime minister.

Practical information

Visa requirements: UK: visa required. USA: visa required
Office hours: 0800–1400 Sat–Thu
Banking hours: 0800–1130 Sat–Thu
Time difference: GMT +3

Chief tourist attractions: beaches protected by a coral reef, among the longest in the world
Major holidays: 1 January, 1 May, 26 June, 1 July, 21 October (2 days); variable: Eid-ul-Adha (2 days), end of Ramadan (2 days), Prophet's Birthday

SOUTH AFRICA

Located on the southern tip of Africa, bounded north by Namibia, Botswana, and Zimbabwe, and northeast by Mozambique and Swaziland.

National name: *Republiek van Suid-Afrika/Republic of South Africa*
Area: 1,222,081 sq km/471,845 sq mi
Capital: Cape Town (legislative), Pretoria (administrative), Bloemfontein (judicial)
Major towns/cities: Johannesburg, Durban, Port Elizabeth, Vereeniging, Pietermaritzburg, Kimberley, Soweto, Tembisa
Major ports: Cape Town, Durban, Port Elizabeth, East London
Physical features: southern end of large plateau, fringed by mountains and lowland coastal margin; Drakensberg Mountains, Table Mountain; Limpopo and Orange rivers
Territories: Marion Island and Prince Edward Island in the Antarctic

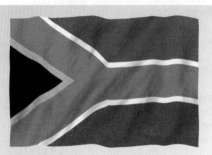

Black, green, and yellow are the African National Congress (ANC) colours. Red, white, and blue are the colours of the former Dutch republics. Effective date: 27 April 1994.

Government

Head of state and government: Thabo Mbeki from 1999
Political system: emergent democracy
Political executive: limited presidency
Administrative divisions: nine provinces
Political parties: African National Congress (ANC), left of centre; National Party (NP), right of centre; Inkatha Freedom Party (IFP), centrist, multiracial (formerly Zulu nationalist); Freedom Front (FF), right wing; Democratic Party (DP), moderate, left of centre, multiracial; Pan-Africanist Congress (PAC), black, left wing; African Christian Democratic Party (ACDP), Christian, right of centre
Armed forces: 82,400 (1998)
Conscription: none
Death penalty: abolished in 1998
Defence spend: (% GDP) 1.6 (1998)
Education spend: (% GNP) 8 (1997 est)
Health spend: (% GDP) 7.1 (1997)

Economy and resources

Currency: rand
GDP: (US$) 131.1 billion (1999)
Real GDP growth: (% change on previous year) 1.2 (1999)
GNP: (US$) 133.2 billion (1999)
GNP per capita (PPP): (US$) 8,318 (1999)
Consumer price inflation: 5.1% (1999)
Unemployment: 5.1% (1996)
Foreign debt: (US$) 24.2 billion (1999)
Major trading partners: USA, Japan, Germany, Italy, UK, Switzerland
Resources: gold (world's largest producer), coal, platinum, iron ore, diamonds, chromium, manganese, limestone, asbestos, fluorspar, uranium, copper, lead, zinc, petroleum, natural gas
Industries: chemicals, petroleum and coal products, gold, diamonds, food processing, transport equipment, iron and steel, metal products, machinery, fertilizers, textiles, paper and paper products, clothing, wood and cork products
Exports: metals and metal products, gold, precious and semiprecious stones, mineral products and chemicals, natural cultured pearls, machinery and mechanical appliances, wool, maize, fruit, sugar. Principal market: USA 8.1% (1998)

Imports: machinery and electrical equipment, transport equipment, chemical products, mechanical appliances, textiles and clothing, vegetable products, wood, pulp, paper and paper products. Principal source: USA 13.1% (1998)
Arable land: 12.3% (1996)
Agricultural products: maize, sugar cane, sorghum, fruits, wheat, groundnuts, grapes, vegetables; livestock rearing, wool production

Population and society

Population: 40,377,000 (2000 est)
Population growth rate: 1.5% (1995–2000)
Population density: (per sq km) 33 (1999 est)
Urban population: (% of total) 50 (2000 est)
Age distribution: (% of total population) 0–14 35%, 15–59 59%, 60+ 6% (2000 est)
Ethnic groups: 77% of the population is black African, 11% white (of European descent), 9% of mixed African–European descent, and 3% Asian
Language: English, Afrikaans, Xhosa, Zulu, Sesotho (all official), other African languages
Religion: Dutch Reformed Church and other Christian denominations 77%, Hindu 2%, Muslim 1%
Education: (compulsory years) 10
Literacy rate: 86% (men); 85% (women) (2000 est)
Labour force: 39% of population: 11.1% agriculture, 25.4% industry, 59.5% services (1995)
Life expectancy: 52 (men); 58 (women) (1995–2000)
Child mortality rate: (under 5, per 1,000 live births) 87 (1995–2000)
Physicians: 1 per 1,049 people (1996)
Hospital beds: 1 per 269 people (1994)
TV sets: (per 1,000 people) 125 (1997)
Radios: (per 1,000 people) 317 (1997)
Internet users: (per 10,000 people) 456.1 (1999)
Personal computer users: (per 100 people) 6.0 (1999)

Transport

Airports: international airports: Cape Town (D F Malan), Durban (Louis Botha), Johannesburg (Jan Smuts); six domestic airports, 212 public aerodromes; total passenger

km: 17,103 million (1997 est)
Railways: total length: 21,595 km/13,419 mi; total passenger km: 71,573 million (1993)

Roads: total road network: 331,265 km/205,848 mi, of which 42% paved (1995); passenger cars: 85.6 per 1,000 people (1996 est)

Chronology

1652: Dutch East India Company established colony at Cape Town as a port of call. **1795:** Britain occupied Cape after France conquered the Netherlands. **1814:** Britain bought Cape Town and hinterland from the Netherlands for £6 million. **1820s:** Zulu people established military kingdom under Shaka. **1836–38:** The Great Trek: 10,000 Dutch settlers (known as Boers, meaning 'farmers') migrated north to escape British rule. **1843:** Britain established colony of Natal on east coast. **1852–54:** Britain recognized Boer republics of Transvaal and Orange Free State. **1872:** The Cape became self-governing colony within British Empire. **1877:** Britain annexed Transvaal. **1879:** Zulu War: Britain destroyed power of Zulus. **1881:** First Boer War: Transvaal Boers defeated British at Majuba Hill and regained independence. **1886:** Discovery of gold on Witwatersrand attracted many migrant miners (uitlanders) to Transvaal, which denied them full citizenship.
1895: Jameson Raid: uitlanders, backed by Cecil Rhodes, tried to overthrow President Paul Kruger of Transvaal. **1899–1902:** Second South African War (also known as Boer War): dispute over rights of uitlanders led to conflict which ended with British annexation of Boer republics. **1907:** Britain granted internal self-government to Transvaal and Orange Free State on a whites-only franchise. **1910:** Cape Colony, Natal, Transvaal, and Orange Free State formed Union of South Africa. **1912:** Gen Barry Hertzog founded (Boer) Nationalist Party; ANC formed to campaign for rights of black majority. **1914:** Boer revolt in Orange Free State suppressed; South African troops fought for British Empire in World War I. **1919:** South West Africa (Namibia) became South African mandate. **1924:** Hertzog became prime minister, aiming to sharpen racial segregation and loosen ties with British Empire. **1939–45:** South Africa entered World War II; South African troops fought with Allies in Middle East, East Africa, and Italy. **1948:** Policy of apartheid ('separateness') adopted when National Party (NP) took power. **1950:** Entire population classified by race; Group Areas Act segregated blacks and whites; ANC responded with campaign of civil disobedience. **1960:** 70 black demonstrators killed at Sharpville; ANC banned. **1961:** South Africa left Commonwealth and became republic. **1964:** ANC leader Nelson Mandela sentenced to life imprisonment. **1967:** Terrorism Act introduced indefinite detention without trial. **1970s:** Over 3 million people forcibly resettled in black 'homelands'. **1976:** Over 600 killed in clashes between black protesters and security forces in Soweto. **1984:** New constitution gave segregated representation to coloureds and Asians, but continued to exclude blacks. **1985:** Growth of violence in black townships led to proclamation of a state of emergency. **1986:** USA and Commonwealth imposed limited economic sanctions against South Africa. **1989:** F W de Klerk succeeded P W Botha as president; public facilities were desegregated; many ANC activists were released. **1990:** The ban on the ANC was lifted; Mandela was released; talks began between the government and the ANC; there was a daily average of 35 murders. **1991:** De Klerk repealed the remaining apartheid laws; sanctions were lifted; however, there was severe fighting between the ANC and the Zulu Inkatha movement. **1993:** An interim majority rule constitution was adopted; de Klerk and Mandela agreed to form a government of national unity after free elections. **1994:** The ANC were victorious in the first nonracial elections; Mandela became president; Commonwealth membership was restored. **1996:** De Klerk withdrew the NP from the coalition after the new constitution failed to provide for power-sharing after 1999. **1997:** A new constitution was signed by President Mandela. De Klerk announced his retirement from politics. **1999:** Mandela retired as state president. ANC won an assembly majority in election.

Practical information

Visa requirements: UK/USA: visa not required
Time difference: GMT +2
Chief tourist attractions: fine climate; varied scenery; wildlife reserves
Major holidays: 1 January, 21 March, 27 April, 1 May, 16 June, 9 August , 24 September, 16, 25–26 December; variable: Good Friday

SOUTH KOREA

Located in East Asia, bounded north by North Korea, east by the Sea of Japan, south by the Korea Strait, and west by the Yellow Sea.

National name: *Daehan Minguk/Republic of Korea*
Area: 98,799 sq km/38,146 sq mi
Capital: Seoul
Major towns/cities: Pusan, Taegu, Inchon, Kwangju, Taejon, Songnam
Major ports: Pusan, Inchon
Physical features: southern end of a mountainous peninsula separating the Sea of Japan from the Yellow Sea

The top left trigram symbolizes summer, south, and heaven. The top right trigram represents autumn, west, and the moon. The bottom right trigram stands for winter, north, and the Earth. The bottom left trigram represents spring, east, and the sun. Effective date: 21 February 1984.

Government

Head of state: Kim Dae Jung from 1998
Head of government: Lee Han Dong from 2000
Political system: liberal democracy
Political executive: limited presidency
Administrative divisions: nine provinces and six cities with provincial status
Political parties: New Korea Party (NKP, formerly Democratic Liberal Party (DLP)), right of centre; National Congress for New Politics (NCNP), left of centre; Democratic Party (DP), left of centre; New Democratic Party (NDP), centrist, pro-private enterprise; United Liberal Democratic Party (ULD), ultra-conservative, pro-private enterprise
Armed forces: 672,000 (1998)
Conscription: 26 months (army); 30 months (navy and air force)
Death penalty: retained and used for ordinary crimes
Defence spend: (% GDP) 3.1 (1998)
Education spend: (% GNP) 3.7 (1996)
Health spend: (% GDP) 6 (1997)

Economy and resources

Currency: won
GDP: (US$) 406.9 billion (1999)
Real GDP growth: (% change on previous year) 10.7 (1999)
GNP: (US$) 397.9 billion (1999)
GNP per capita (PPP): (US$) 14,637 (1999)
Consumer price inflation: 2.4% (1999)
Unemployment: 6.8% (1998)
Foreign debt: (US$) 137.4 billion (1999 est)
Major trading partners: USA, Japan, Germany, Saudi Arabia, Australia, Singapore, China, Taiwan
Resources: coal, iron ore, tungsten, gold, molybdenum, graphite, fluorite, natural gas, hydroelectric power, fish
Industries: electrical machinery, transport equipment (principally motor vehicles and shipbuilding), chemical products, textiles and clothing, iron and steel, electronics equipment, food processing, tourism
Exports: electrical machinery, textiles, clothing, footwear, telecommunications and sound equipment, chemical products, ships ('invisible export' – overseas construction work). Principal market: USA 20.5% (1999)
Imports: machinery and transport equipment (especially electrical machinery), petroleum and petroleum products, grain and foodstuffs, steel, chemical products, basic manufactures. Principal source: USA 20.8% (1999)
Arable land: 17.7% (1996)
Agricultural products: rice, maize, barley, potatoes, sweet potatoes, fruit; livestock (pigs and cattle)

Population and society

Population: 46,844,000 (2000 est)
Population growth rate: 0.8% (1995–2000)
Population density: (per sq km) 473 (1999 est)
Urban population: (% of total) 82 (2000 est)
Age distribution: (% of total population) 0–14 21%, 15–59 68%, 60+ 11% (2000 est)
Ethnic groups: with the exception of a small Nationalist Chinese minority, the population is almost entirely of Korean descent
Language: Korean (official)
Religion: Buddhist 48%, Confucian 3%, Christian 47%, mainly Protestant; Chund Kyo (peculiar to Korea, combining elements of Shaman, Buddhist, and Christian doctrines)
Education: (compulsory years) 9
Literacy rate: 99% (men); 96% (women) (2000 est)
Labour force: 11% agriculture, 31.3% industry, 57.7% services (1997)
Life expectancy: 69 (men); 76 (women) (1995–2000)
Child mortality rate: (under 5, per 1,000 live births) 13 (1995–2000)
Physicians: 1 per 909 people (1995)
Hospital beds: 1 per 227 people (1995)
TV sets: (per 1,000 people) 346 (1998)
Radios: (per 1,000 people) 1,033 (1997)
Internet users: (per 10,000 people) 1,468.0 (1999)
Personal computer users: (per 100 people) 18.3 (1999)

Transport

Airports: international airports: Seoul (Kimpo), Pusan (Kim Hae), Cheju; three principal domestic airports; total passenger km: 61,011 million (1997 est)
Railways: total length: 3,081 km/1,915 mi; total passenger km: 31,912 million (1994)
Roads: total road network: 84,968 km/52,799 mi, of which 76.1% paved (1997); passenger cars: 165 per 1,000 people (1997)

Chronology

2333 BC: Traditional date of founding of Korean state by Tangun (mythical son from union of bear-woman and god). **1122 BC:** Ancient texts record founding of kingdom in Korea by Chinese nobleman Kija. **194 BC:** Northwest Korea united under warlord, Wiman. **108 BC:** Korea conquered by Chinese. **1st–7th centuries AD:** Three Korean kingdoms – Koguryo, Paekche, and Silla – competed for supremacy. **668:** Korean peninsula unified by Buddhist Silla kingdom; culture combining Chinese and Korean elements flourished. **935:** Silla dynasty overthrown by Wang Kon of Koguryo, who founded Koryo dynasty in its place. **1258:** Korea accepted overlordship of Mongol Yüan Empire. **1392:** Yi dynasty founded by Gen Yi Song-gye, vassal of Chinese Ming Empire; Confucianism replaced Buddhism as official creed; extreme conservatism characterized Korean society. **1592 and 1597:** Japanese invasions repulsed by Korea. **1636:** Manchu invasion forced Korea to sever ties with Ming dynasty. **18th–19th centuries:** Korea resisted change in political and economic life and rejected contact with Europeans. **1864:** Attempts to reform government and strengthen army by Taewongun (who ruled in name of his son, King Kojong); converts to Christianity persecuted. **1873:** Taewongun forced to cede power to Queen Min; reforms reversed; government authority collapsed. **1882:** Chinese occupied Seoul and installed governor. **1894–95:** Sino-Japanese War: Japan forced China to recognize independence of Korea; Korea fell to Japanese influence. **1904–05:** Russo-Japanese War: Japan ended Russian influence in Korea. **1910:** Korea formally annexed by Japan; Japanese settlers introduced modern industry and agriculture; Korean language banned. **1919:** 'Samil' nationalist movement suppressed by Japanese. **1945:** After defeat of Japan in World War II, Russia occupied regions of Korea north of 38th parallel (demarcation line agreed at Yalta Conference) and USA occupied regions south of it. **1948:** The USSR refused to permit United Nations (UN) supervision of elections in the northern zone; the southern zone became independent as the Republic of Korea, with Syngman Rhee as president. **1950:** North Korea invaded South Korea; UN forces (mainly from the USA) intervened to defend South Korea; China intervened in support of North Korea. **1953:** The Korean War ended with an armistice which restored the 38th parallel; no peace treaty was agreed and US troops remained in South Korea. **1961:** Military coup placed Gen Park Chung Hee in power; a major programme of industrial development began. **1972:** Martial law was imposed and presidential powers increased. **1979:** The government of President Choi Kyu-Hah introduced liberalizing reforms. **1979:** Gen Chun Doo Hwan assumed power after anti-government riots; Korea emerged as a leading shipbuilding nation and exporter of electronic goods. **1987:** The constitution was made more democratic as a result of Liberal pressure; ruling Democratic Justice Party (DJP) candidate Roh Tae Woo Was elected president amid allegations of fraud. **1988:** The Olympic Games were held in Seoul. **1991:** Large-scale antigovernment protests were forcibly suppressed; South Korea joined the UN. **1992:** South Korea established diplomatic relations with China. **1994:** The US military presence was stepped up in response to the perceived threat from North Korea. **1997:** South Korea was admitted to the OECD. Kim Dae Jung, former dissident and political prisoner, became the first opposition politician to lead South Korea. **1998:** Kim Dae Jung was sworn in as president, with Kim Jong Pil as prime minister. New labour laws ended lifetime employment and the financial system was opened up. More than 2,000 prisoners were released, including 74 political prisoners. There was continuing labour unrest as GDP contracted by 5%. **1999:** Talks on possible reunification with North Korea were suspended. **2000:** Kim Jong Pil resigned as prime minister and was replaced by Park Tae Joon, who in turn resigned after the opposition Grand National Party won a majority in elections. He was replaced by Lee Han Dong. At the first summit meeting between the divided countries, Kim Dae Jung was welcomed by the leader of North Korea, Kim Jong Il, in Pyongyang, North Korea. The two leaders agreed to further economic investment by South Korea investment in North Korea, and to open rail links between the two countries. Kim Dae Jung was awarded the Nobel Peace Prize.

Practical information

Visa requirements: UK: visa not required for a stay of up to 90 days. USA: visa not required
Time difference: GMT +9
Chief tourist attractions: historic sites; mountain scenery; Cheju Island is a popular resort

Major holidays: 1–3 January, 1, 10 March, 5 May, 6 June, 17 July, 15 August, 1, 3, 9 October, 25 December; variable: New Year (Chinese, January/February), Lord Buddha's Birthday (May), Moon Festival (September/October)

SPAIN

Located in southwestern Europe, on the Iberian Peninsula between the Atlantic Ocean and the Mediterranean Sea, bounded north by France and west by Portugal.

National name: *España/Spain*
Area: 504,750 sq km/194,883 sq mi (including the Balearic and Canary islands)
Capital: Madrid
Major towns/cities: Barcelona, Valencia, Zaragoza, Seville, Málaga, Bilbao, Las Palmas (on Gran Canarias island), Murcia, Palma (on Mallorca)
Major ports: Barcelona, Valencia, Cartagena, Málaga, Cádiz, Vigo, Santander, Bilbao
Physical features: central plateau with mountain ranges, lowlands in south; rivers Ebro, Douro, Tagus, Guadiana, Guadalquivir; Iberian Plateau (Meseta); Pyrenees, Cantabrian Mountains, Andalusian Mountains, Sierra Nevada
Territories: Balearic and Canary Islands; in North Africa: Ceuta, Melilla, Alhucemas, Chafarinas Islands, Peñón de Vélez

The Pillars of Hercules represent the promontories of Gibraltar and Ceuta. The shield represents the regions of Castile, Léon, Aragón, Navarre, and Granada. Effective date: 18 December 1981.

Government

Head of state: King Juan Carlos I from 1975
Head of government: José Maria Aznar from 1996
Political system: liberal democracy
Political executive: parliamentary
Administrative divisions: 17 autonomous regions (contain 50 provinces)
Political parties: Socialist Workers' Party (PSOE), democratic socialist; Popular Party (PP), right of centre
Armed forces: 194,000 (1998)
Conscription: nine months
Death penalty: abolished in 1995
Defence spend: (% GDP) 1.3 (1998)
Education spend: (% GNP) 4.9 (1996)
Health spend: (% GDP) 7.4 (1997)

Economy and resources

Currency: peseta
GDP: (US$) 599.9 billion (1999)
Real GDP growth: (% change on previous year) 4 (1999)
GNP: (US$) 551.6 billion (1999)
GNP per capita (PPP): (US$) 16,730 (1999)
Consumer price inflation: 2.3% (1999)
Unemployment: 18.8% (1998)
Major trading partners: EU (principally France, Germany, Italy, and UK), USA, Japan
Resources: coal, lignite, anthracite, copper, iron, zinc, uranium, potassium salts
Industries: machinery, motor vehicles, textiles, footwear, chemicals, electrical appliances, wine, olive oil, fishery products, steel, cement, tourism
Exports: motor vehicles, machinery and electrical equipment, vegetable products, metals and their manufactures, foodstuffs. Principal market: France 19.5% (1999)
Imports: machinery and transport equipment, electrical equipment, petroleum and petroleum products, chemicals, consumer goods. Principal source: France 18.1% (1999)
Arable land: 30.5% (1996)
Agricultural products: barley, wheat, sugar beet, vegetables, citrus fruit, grapes, olives; fishing (one of world's largest fishing fleets)

Population and society

Population: 39,630,000 (2000 est)
Population growth rate: 0.03% (1995–2000)
Population density: (per sq km) 79 (1999 est)
Urban population: (% of total) 78 (2000 est)
Age distribution: (% of total population) 0–14 15%, 15–59 63%, 60+ 22% (2000 est)
Ethnic groups: mostly of Moorish, Roman, and Carthaginian descent
Language: Spanish (Castilian; official), Basque, Catalan, Galician
Religion: Roman Catholic 98%
Education: (compulsory years) 10
Literacy rate: 99% (men); 97% (women) (2000 est)
Labour force: 8.4% agriculture, 30% industry, 61.7% services (1997)
Life expectancy: 75 (men); 82 (women) (1995–2000)
Child mortality rate: (under 5, per 1,000 live births) 8 (1995–2000)
Physicians: 1 per 241 people (1996)
Hospital beds: 1 per 250 people (1996)
TV sets: (per 1,000 people) 506 (1997)
Radios: (per 1,000 people) 333 (1997)
Internet users: (per 10,000 people) 717.9 (1999)
Personal computer users: (per 100 people) 12.2 (1999)

Transport

Airports: international airports: Alicante (Altet), Barcelona (del Prat), Bilbao, Tenerife (2), Madrid (Barajas), Málaga, Santiago de Compostela, Gerona, Gran Canaria, Lanzarote, Palma de Mallorca, Mahon, Valladolid, Seville, Valencia, Zarragoza; domestic services to all major towns; total passenger km: 36,950 million (1997 est)
Railways: total length: 15,372 km/9,552 mi; total passenger km: 16,579 million (1997)
Roads: total road network: 346,858 km/215,966 mi, of which 99% paved (1997); passenger cars: 385 per 1,000 people (1997)

Chronology

2nd century BC: Roman conquest of the Iberian peninsula, which became the province of Hispania. **5th century AD:** After the fall of the Roman Empire, Iberia was overrun by Vandals and Visigoths. **711:** Muslims invaded from North Africa and overthrew Visigoth kingdom.

9th century: Christians in northern Spain formed kingdoms of Asturias, Aragón, Navarre, and Léon, and county of Castile.

10th century: Abd-al-Rahman III established caliphate of Córdoba; Muslim culture at its height in Spain. **1230:** Léon and Castile united under Ferdinand III, who drove the Muslims from most of southern Spain. **14th century:** Spain consisted of Christian kingdoms of Castile, Aragón, and Navarre, and the Muslim emirate of Granada. **1469:** Marriage of Ferdinand of Aragón and Isabella of Castile; kingdoms united on their accession in 1479. **1492:** Conquest of Granada ended Muslim rule in Spain. **1494:** Treaty of Tordesillas; Spain and Portugal divided newly discovered America; Spain became a world power. **1519–56:** Emperor Charles V was both King of Spain and Archduke of Austria. **1555:** Charles V divided his domains between Spain and Austria before retiring; Spain retained the Low Countries and southern Italy as well as South American colonies. **1568:** Dutch rebelled against Spanish rule; Spain recognized independence of Dutch Republic in 1648. **1580:** Philip II of Spain inherited the throne of Portugal, where Spanish rule lasted until 1640. **1588:** Spanish Armada: attempt to invade England defeated. **17th century:** Spanish power declined amid wars, corruption, inflation, and loss of civil and religious freedom. **1701–14:** War of the Spanish Succession: allied powers fought France to prevent Philip of Bourbon inheriting throne of Spain. **1713–14:** Treaties of Utrecht and Rastat: Bourbon dynasty recognized, but Spain lost Gibraltar, southern Italy, and Spanish Netherlands. **1793:** Spain declared war on revolutionary France; reduced to a French client state in 1795. **1808:** Napoleon installed his brother Joseph as King of Spain. **1808–14:** Peninsular War: British forces played a large part in liberating Spain and restoring Bourbon dynasty. **1810–30:** Spain lost control of its South American colonies. **1833–39:** Carlist civil war: Don Carlos (backed by conservatives) unsuccessfully contested the succession of his niece Isabella II (backed by liberals). **1870:** Offer of Spanish throne to Leopold of Hohenzollern-Sigmaringen sparked Franco-Prussian War. **1873–74:** First republic ended by military coup which restored Bourbon dynasty with Alfonso XII. **1898:** Spanish-American War: Spain lost Cuba and Philippines. **1923–30:** Dictatorship of Gen Primo de Rivera with support of Alfonso XIII. **1931:** Proclamation of Second Republic, initially dominated by anticlerical radicals and socialists. **1933:** Moderates and Catholics won elections. **1936:** Left-wing Popular Front narrowly won fresh elections; General Francisco Franco launched military rebellion. **1936–39:** Spanish Civil War: Nationalists (with significant Italian and German support) defeated Republicans (with limited Soviet support); Franco became dictator of nationalist-fascist regime. **1941:** Though officially neutral in World War II, Spain sent 40,000 troops to fight USSR. **1955:** Spain admitted to the United Nations (UN). **1975:** Death of Franco; he was succeeded by King Juan Carlos I. **1978:** A referendum endorsed democratic constitution. **1982:** Socialists took office under Felipe González; Spain joined the North Atlantic Treaty Organization (NATO); Basque separatist organization ETA stepped up its terrorist campaign. **1986:** Spain joined the European Economic Community (EEC). **1997:** 23 Basque nationalist leaders were jailed for terrorist activities. **1998:** ETA announced an indefinite ceasefire. The government announced that it would begin peace talks. **2000:** ETA ended its ceasefire with a bombing in Madrid.

Practical information

Visa requirements: UK/USA: visa not required
Time difference: GMT +1
Chief tourist attractions: climate; beaches; mountain scenery; winter resorts on the Canary Islands; many cities of historical interest, including Madrid, Seville, Córdoba, Barcelona, and Valencia, with their cathedrals, churches, palaces, fortresses, and museums
Major holidays: 1, 6 January, 19 March (most areas), 1 May, 25 July, 15 August, 12 October, 1 November, 8, 25 December; variable: Corpus Christi, Good Friday, Holy Saturday, Holy Thursday

Sri Lanka FORMERLY Ceylon (UNTIL 1972)

Island in the Indian Ocean, off the southeast coast of India.

National name: *Sri Lanka Prajatantrika Samajavadi Janarajaya/Democratic Socialist Republic of Sri Lanka*
Area: 65,610 sq km/25,332 sq mi
Capital: Colombo (and chief port)
Major towns/cities: Kandy, Dehiwala-Mount Lavinia, Moratuwa, Jaffna, Kotte, Galle
Major ports: Jaffna, Galle, Negombo, Trincomalee
Physical features: flat in north and around coast; hills and mountains in south and central interior

Green represents the Islamic minority. Orange stands for the Hindu Tamils. The sword denotes authority. The four pipul leaves symbolize Buddhism. Effective date: 7 September 1978.

Government

Head of state: Chandrika Bandaranaike Kumaratunga from 1994
Head of government: Ratnasiri Wickremanayake from 2000
Political system: liberal democracy
Political executive: dual executive
Administrative divisions: eight provinces
Political parties: United National Party (UNP), right of centre; Sri Lanka Freedom Party (SLFP), left of centre; Democratic United National Front (DUNF), left of centre; Tamil United Liberation Front (TULF), Tamil autonomy (banned from 1983); Eelam People's Revolutionary Liberation Front (EPRLF), Indian-backed Tamil-secessionist 'Tamil Tigers'; People's Liberation Front (JVP), Sinhalese-chauvinist, left wing (banned 1971–77 and 1983–88)
Armed forces: 115,000 plus paramilitary forces numbering around 110,200 (1998)
Conscription: military service is voluntary
Death penalty: retains the death penalty for ordinary crimes but can be considered abolitionist in practice (last execution 1976)
Defence spend: (% GDP) 6.1 (1998)
Education spend: (% GNP) 3.4 (1996)
Health spend: (% GDP) 3 (1997)

Economy and resources

Currency: Sri Lankan rupee
GDP: (US$) 15.8 billion (1999 est)
Real GDP growth: (% change on previous year) 4.3 (1999 est)
GNP: (US$) 15.7 billion (1999)
GNP per capita (PPP): (US$) 3,056 (1999)
Consumer price inflation: 4.7% (1999 est)
Unemployment: 10.7% (1998 est)
Foreign debt: (US$) 9.6 billion (1999)
Major trading partners: Japan, USA, Germany, UK, India, South Korea, Singapore, Hong Kong, Taiwan, China, Iran
Resources: gemstones, graphite, iron ore, monazite, rutile, uranium, iemenite sands, limestone, salt, clay
Industries: food processing, textiles, clothing, petroleum refining, leather goods, chemicals, rubber, plastics, tourism
Exports: clothing and textiles, tea (world's largest exporter and third-largest producer), precious and semi-precious stones, coconuts and coconut products, rubber. Principal market: USA 38.9% (1999)
Imports: machinery and transport equipment, petroleum, food and live animals, beverages, construction materials. Principal source: Japan 9.5% (1999)
Arable land: 13.7% (1996)
Agricultural products: rice, tea, rubber, coconuts; livestock rearing (cattle, buffaloes, pigs, and poultry); fishing

Population and society

Population: 18,827,000 (2000 est)
Population growth rate: 1.0% (1995–2000)
Population density: (per sq km) 284 (1999 est)
Urban population: (% of total) 24 (2000 est)
Age distribution: (% of total population) 0–14 26%, 15–59 64%, 60+ 10% (2000 est)
Ethnic groups: 74% Sinhalese, about 18% Tamil, and 7% Moors or Muslims (concentrated in east); the Tamil community is divided between the long-settled 'Sri Lankan Tamils' (11% of the population), who reside in northern and eastern coastal areas, and the more recent immigrant 'Indian Tamils' (7%), who settled in the Kandyan highlands during the 19th and 20th centuries
Language: Sinhala, Tamil (both official), English
Religion: Buddhist 69%, Hindu 15%, Muslim 8%, Christian 8%
Education: (compulsory years) 10
Literacy rate: 94% (men); 89% (women) (2000 est)
Labour force: 31.2% agriculture, 19.9% industry, 49.9% services (1996)
Life expectancy: 71 (men); 75 (women) (1995–2000)
Child mortality rate: (under 5, per 1,000 live births) 21 (1995–2000)
Physicians: 1 per 4,745 people (1996)
Hospital beds: 1 per 357 people (1996)
TV sets: (per 1,000 people) 92 (1997)
Radios: (per 1,000 people) 209 (1997)
Internet users: (per 10,000 people) 34.9 (1999)
Personal computer users: (per 100 people) 0.6 (1999)

Transport

Airports: international airports: Colombo (Katunayake); five domestic airports; total passenger km: 4,249 million (1997 est)

Railways: total length: 1,484 km/922 mi; total passenger km: 3,265 million (1994)
Roads: total road network: 99, 200 km/61,643 mi, of which 40% paved (1996 est); passenger cars: 5.8 per 1,000 people (1996 est)

Chronology

c. **550 BC:** Arrival of the Sinhalese, from northern India, displacing long-settled Veddas. **5th century BC:** Sinhalese kingdom of Anuradhapura founded. *c.* **250–210 BC:** Buddhism, brought from India, became established in Sri Lanka. **AD 992:** Downfall of Anuradhapura kingdom, defeated by South Indian Colas. **1070:** Overthrow of Colas by Vijayabahu I and establishment of the Sinhalese kingdom of Polonnaruva, which survived for more than two centuries before a number of regional states arose. **late 15th century:** Kingdom of Kandy established in central highlands. **1505:** Arrival of Portuguese navigator Lorenço de Almeida, attracted by spice trade developed by Arab merchants who had called the island Serendip. **1597–1618:** Portuguese controlled most of Sri Lanka, with the exception of Kandy. **1658:** Dutch conquest of Portuguese territories. **1795–98:** British conquest of Dutch territories. **1802:** Treaty of Amiens recognized island as British colony of Ceylon. **1815:** British won control of Kandy, becoming the first European power to rule whole island. **1830s:** Immigration of south Indian Hindu Tamil labourers to work central coffee plantations. **1880s:** Tea and rubber become chief cash crops after blight ended production of coffee. **1919:** Formation of the Ceylon National Congress to campaign for self rule; increasing conflicts between Sinhalese majority community and Tamil minority. **1931:** Universal adult suffrage introduced for elected legislature and executive council in which power was shared with British. **1948:** Ceylon achieved independence from Britain within Commonwealth, with Don Senanayake of conservative United National Party (UNP) as prime minister. **1949:** Indian Tamils disenfranchised. **1956:** Sinhala established as official language. **1960:** Sirimavo Bandaranaike, the widow of assassinated prime minister Solomon Bandaranaike, won general election and formed an SLFP government. **1971:** Sinhalese Marxist uprising, led by students and People's Liberation Army (JVP). **1972:** Socialist Republic of Sri Lanka proclaimed; Buddhism given 'foremost place' in new state, antagonizing Tamils. **1976:** Tamil United Liberation Front formed to fight for independent Tamil state ('Eelam') in north and east Sri Lanka. **1978:** Presidential constitution adopted by new free-market government headed by Junius Jayawardene of UNP. **1983:** Ethnic riots as Tamil guerrilla violence escalated; state of emergency imposed; more than 1,000 Tamils killed by Sinhalese mobs. **1987:** President Jayawardene and Indian prime minister Rajiv Gandhi signed Colombo Accord aimed at creating new provincial councils, disarming Tamil militants ('Tamil Tigers'), and stationing 7,000-strong Indian Peace Keeping Force. **1988:** Left-wing JVP guerrillas campaigned against Indo-Sri Lankan peace pact. Prime Minister Ranasinghe Premadasa elected president. **1989:** Dingiri Banda Wijetunga became prime minister. Leaders of Tamil Tigers and banned Sinhala extremist JVP assassinated. **1990:** The Indian peacekeeping force was withdrawn. Violence continued, with a death toll of over a thousand a month. **1991:** The Sri Lankan army killed 2,552 Tamil Tigers at Elephant Pass in the northern Jaffna region. A new party, the Democratic National United Front (DUNF), was formed by former members of UNP. **1992:** Several hundred Tamil Tiger rebels were killed in an army offensive. **1993:** President Premadasa was assassinated by Tamil Tiger terrorists; he was succeeded by Dingiri Banda Wijetunge. **1994:** The UNP were narrowly defeated in a general election; Chandrika Kumaratunga became prime minister in an SLFP-led left-of-centre coalition (People's Alliance). Peace talks opened with the Tamil Tigers. Kumaratunga was elected the first female president; her mother, Sirimavo Bandaranaike, became prime minister. **1995:** Renewed bombing campaign by Tamil Tigers. A major offensive drove out Tamil Tigers from Jaffna city. **1996:** A state of emergency was extended nationwide after Tamils bombed the capital. **1998:** The Tamil Tigers were outlawed after the bombing of Sri Lanka's holiest Buddhist site. In September over 1,300 Sri Lankan soldiers and Tamil Tiger rebels died in renewed fighting in the north. In October the Tamil Tigers captured the strategic northern town of Kilinochchi. **1999:** The government lost a large amount of territory to Tamil guerrillas. President Kumaratunga was re-elected in December, just days after she survived an attack by a Tamil suicide bomber. **2000:** Terrorist activity continued, and government forces suffered their worst setback in the 17-year civil war when they were forced to surrender Pallai, a key military base, to Tamil guerrillas.

Practical information

Visa requirements: UK: visa only required by business visitors. USA: visa only required by business visitors
Time difference: GMT +5.5
Chief tourist attractions: Buddhist festivals and ancient monuments; scenery

Major holidays: 14 January, 4 February, 1, 22 May, 30 June, 25, 31 December; variable: Eid-ul-Adha, Diwali, end of Ramadan, Good Friday, New Year (Sinhala/Tamil, April), Prophet's Birthday, Maha Sivarathri (February/March), Full Moon (monthly)

SUDAN

Located in northeast Africa, bounded north by Egypt, northeast by the Red Sea, east by Ethiopia and Eritrea, south by Kenya, Uganda, and Congo (formerly Zaire), west by the Central African Republic and Chad, and northwest by Libya. It is the largest country in Africa.

National name: *Al-Jumhuryyat es-Sudan/Republic of Sudan*
Area: 2,505,800 sq km/967,489 sq mi
Capital: Khartoum
Major towns/cities: Omdurman, Port Sudan, Juba, Wadi Medani, El Obeid, Kassala, al-Qadarif, Nyala
Major ports: Port Sudan
Physical features: fertile Nile valley separates Libyan Desert in west from high rocky Nubian Desert in east

White stands for Islam, peace, optimism, light, and love. Red recalls the martyrs of Sudan and the people's struggle. Black stands for Sudan and the Mahdiya revolution of the 1880s. Effective date: 20 May 1970.

Government

Head of state and government: Gen Omar Hassan Ahmed al-Bashir from 1989
Political system: military
Political executive: military
Administrative divisions: 26 states
Political parties: officially banned from 1989, but an influential grouping is the fundamentalist National Islamic Front
Armed forces: 94,700 (1998)
Conscription: military service is compulsory for three years
Death penalty: retained and used for ordinary crimes
Defence spend: (% GDP) 4.8 (1998)
Education spend: (% GNP) 1.4 (1997 est)
Health spend: (% GDP) 3.5 (1997 est)

Economy and resources

Currency: Sudanese dinar
GDP: (US$) 10.7 billion (1999)
Real GDP growth: (% change on previous year) 6 (1999 est)
GNP: (US$) 9.4 billion (1999)
GNP per capita (PPP): (US$) 1,298 (1999)
Consumer price inflation: 16% (1999)
Unemployment: 30% (1993 est)
Foreign debt: (US$) 16.2 billion (1999 est)
Major trading partners: Saudi Arabia, Libya, Thailand, Italy, Germany, UK, China, Japan
Resources: petroleum, marble, mica, chromite, gypsum, gold, graphite, sulphur, iron, manganese, zinc, fluorspar, talc, limestone, dolomite, pumice
Industries: food processing (especially sugar refining), textiles, cement, petroleum refining, hides and skins
Exports: cotton, sesame seed, gum arabic, sorghum, livestock, hides and skins. Principal market: Saudi Arabia 16.3% (1999)
Imports: basic manufacture, crude materials (mainly petroleum and petroleum products), foodstuffs, machinery and equipment. Principal source: Libya 14.7% (1999)

Arable land: 5.4% (1996)
Agricultural products: sorghum, sugar cane, groundnuts, cotton, millet, wheat, sesame, fruits; livestock rearing (cattle, sheep, goats, and poultry)

Population and society

Population: 29,490,000 (2000 est)
Population growth rate: 2.1% (1995–2000)
Population density: (per sq km) 12 (1999 est)
Urban population: (% of total) 36 (2000 est)
Age distribution: (% of total population) 0–14 39%, 15–59 56%, 60+ 5% (2000 est)
Ethnic groups: over 50 ethnic groups and almost 600 subgroups; the population is broadly distributed between Arabs (39%) in the north and black Africans (52%) in the south; Beja (6%), foreigners (2%)
Language: Arabic (51%) (official), 100 local languages
Religion: Sunni Muslim 70%; also animist 25%, and Christian 5%
Education: (compulsory years) 6
Literacy rate: 70% (men); 46% (women) (2000 est)
Labour force: 36% of population: 69% agriculture, 8% industry, 22% services (1990)
Life expectancy: 54 (men); 56 (women) (1995–2000)
Child mortality rate: (under 5, per 1,000 live births) 112 (1995–2000)
Physicians: 1 per 8,979 people (1990)
Hospital beds: 1 per 919 people (1993 est)
TV sets: (per 1,000 people) 86 (1997)
Radios: (per 1,000 people) 272 (1997)
Internet users: (per 10,000 people) 1.7 (1999)
Personal computer users: (per 100 people) 0.2 (1999)

Transport

Airports: international airports: Khartoum (civil); 20 domestic airports; total passenger km: 681 million (1995)
Railways: total length: 4,764 km/2,960 mi; total passenger km: 1,183 million (1993)
Roads: total road network: 11,900 km/7,395 mi, of which 36.3% paved (1996 est); passenger cars: 8.8 per 1,000 people (1996 est)

Chronology

c. **600 BC–AD 350:** Meroë, near Khartoum, was capital of the Nubian Empire, which covered southern Egypt and northern Sudan. **6th century:** Converted to Coptic Christianity. **7th century:** Islam first introduced by Arab invaders, but did not spread widely until the 15th century. **16th–18th centuries:** Arab-African Fur and Fung Empires established in central and northern Sudan. **1820:** Invaded by Muhammad Ali and brought under Egyptian control. **1881–85:** Revolt led to capture of Khartoum by Sheik Muhammad Ahmed, a self-proclaimed Mahdi ('messiah'), and the killing of British general Charles Gordon. **1898:** Anglo-Egyptian offensive led by Lord Kitchener subdued Mahdi revolt at Battle of Omdurman in which 20,000 Sudanese died. **1899:** Sudan administered as Anglo-Egyptian condominium. **1923:** White Flag League formed by Sudanese nationalists in north; British instituted policy of reducing contact between northern and southern Sudan, with the aim that the south would eventually become part of federation of eastern African states. **1955:** Civil war between the dominant Arab Muslim north and black African Christian and animist south broke out. **1956:** Sudan achieved independence from Britain and Egypt as a republic. **1958:** Military coup replaced civilian government with Supreme Council of the Armed Forces. **1964:** Civilian rule reinstated after October Revolution of student demonstrations. **1969:** Coup led by Col Gaafar Mohammed al-Nimeri abolished political institutions and concentrated power in a leftist Revolutionary Command Council. **1971:** Nimeri confirmed as president and the Sudanese Socialist Union (SSU) declared the only legal party by a new constitution. **1972:** Plans to form Federation of Arab Republics, comprising Sudan, Egypt, and Syria, abandoned due to internal opposition. To end 17-year-long civil war, Nimeri agreed to give south greater autonomy. **1974:** National assembly established. **1980:** Country reorganized into six regions, each with own assembly and effective autonomy. **1983:** Shari'a (Islamic law) imposed. Sudan People's Liberation Movement (SPLM) formed in south as civil war broke out again. **1985:** Nimeri deposed in a bloodless coup led by Gen Swar al-Dahab following industrial unrest in north. **1986:** Coalition government formed after general election, with Sadiq al-Mahdi, great-grandson of the Mahdi, as prime minister. **1987:** Civil war with Sudan People's Liberation Army (SPLA); drought and famine in south and refugee influx from Ethiopa and Chad. **1988:** A peace pact was signed with SPLA, but fighting continued. **1989:** Al-Mahdi was overthrown in a coup led by Islamic fundamentalist Gen Omar Hassan Ahmed el-Bashir. All political activity was suspended. **1991:** A federal system was introduced, with division of the country into nine states as the civil war continued. **1998:** Civil war continued between the SPLA and the Islamist government. There was famine in the south, where millions faced starvation. The USA launched a missile attack on a suspected chemical weapons-producing site in retaliation for bombings of US embassies in Nairobi and Dar es Salaam. There was a temporary ceasefire by the SPLA. **1999:** Multiparty politics were reintroduced. Steps to restore diplomatic ties with Uganda were taken when in December an agreement was signed to attempt to end rebel wars across the mutual border by ceasing to support rebel factions in the other's country. In late December, the president declared a state of emergency and dissolved parliament. **2000:** President Bashir dismissed his entire cabinet in January, but then reappointed key ministers. In Khartoum, women were banned from working in public places where they might meet men. Elections re-established a parliament and confirmed Bashir in power, but were boycotted by opposition parties. Famine threatened the lives of 3 million people.

Practical information

Visa requirements: UK: visa required. USA: visa required
Time difference: GMT +2

Major holidays: 1 January, 3 March, 6 April, 25 December; variable: Eid-ul-Adha (5 days), end of Ramadan (5 days), New Year (Muslim), Prophet's Birthday, Sham al-Naseem (April/May)

SURINAME FORMERLY DUTCH GUIANA (1954–75)

Located on the north coast of South America, bounded west by French Guiana, south by Brazil, east by Guyana, and north by the Atlantic Ocean.

National name: *Republiek Suriname/Republic of Suriname*
Area: 163,820 sq km/63,250 sq mi
Capital: Paramaribo
Major towns/cities: Nieuw Nickerie, Moengo, Brokopondo, Nieuw Amsterdam, Albina, Groningen
Physical features: hilly and forested, with flat and narrow coastal plain; Suriname River

Red represents progress and love. White symbolizes justice and freedom. Green stands for hope and fertility. Effective date: 25 November 1975.

Government

Head of state: Ronald Venetiaan from 2000
Head of government: Jules Ajodhia from 2000
Political system: emergent democracy
Political executive: limited presidency
Administrative divisions: ten districts
Political parties: New Front (NF), alliance of four left-of-centre parties: Party for National Unity and Solidarity (KTPI), Suriname National Party (NPS), Progressive Reform Party (VHP), Suriname Labour Party (SPA); National Democratic Party (NDP), left of centre; Democratic Alternative 1991 (DA '91), alliance of three left-of-centre parties
Armed forces: 1,800 (1998)
Conscription: military service is voluntary
Death penalty: retains the death penalty for ordinary crimes but can be considered abolitionist in practice (last execution 1982)
Defence spend: (% GDP) 4.2 (1998)
Education spend: (% GNP) 5.4 (1997 est)
Health spend: (% GDP) 7.6 (1997 est)

Economy and resources

Currency: Suriname guilder
GDP: (US$) 909 million (1999 est)
Real GDP growth: (% change on previous year) –1.2 (1999 est)
GNP: (US$) 685 million (1998)
GNP per capita (PPP): (US$) 3,820 (1998 est)
Consumer price inflation: 98.8% (1999 est)
Unemployment: 20% (1997)
Foreign debt: (US$) 282 million (1999 est)
Major trading partners: USA, Norway, Trinidad and Tobago, the Netherlands, Netherlands Antilles, Brazil, Japan
Resources: petroleum, bauxite (one of the world's leading producers), iron ore, copper, manganese, nickel, platinum, gold, kaolin
Industries: bauxite refining and smelting, food processing, beverages, cigarettes, wood products, chemical products, cement
Exports: alumina, aluminium, shrimps, bananas, plantains, rice, wood and wood products. Principal market: USA 23.2% (1999)
Imports: raw materials and semi-manufactured goods, mineral fuels and lubricants, investment goods, foodstuffs, cars and motorcycles, textiles. Principal source: USA 34.9% (1999)
Arable land: 0.4% (1995)
Agricultural products: rice, citrus fruits, bananas, plantains, vegetables, coconuts, cassava, root crops, sugar cane; forest resources; commercial fishing

Population and society

Population: 417,000 (2000 est)
Population growth rate: 0.4% (1995–2000)
Population density: (per sq km) 3 (1999 est)
Urban population: (% of total) 74 (2000 est)
Age distribution: (% of total population) 0–14 30%, 15–59 62%, 60+ 8% (2000 est)
Ethnic groups: a wide ethnic composition, including Creoles (34%), East Indians (34%), Indonesians (15%), Africans (10%), American Indians (3%), Chinese (3%), European and others (2%)
Language: Dutch (official), Spanish, Sranan (Creole), English, Hindi, Javanese, Chinese, various tribal languages
Religion: Christian 47%, Hindu 28%, Muslim 20%
Education: (compulsory years) 11
Literacy rate: 96% (men); 93% (women) (2000 est)
Labour force: 19% agriculture, 23.6% industry, 57.4% services (1996 est)
Life expectancy: 68 (men); 73 (women) (1995–2000)
Child mortality rate: (under 5, per 1,000 live births) 33 (1995–2000)
Physicians: 1 per 1,605 people (1994)
Hospital beds: 1 per 290 people (1994)
TV sets: (per 1,000 people) 153 (1997)
Radios: (per 1,000 people) 728 (1997)
Internet users: (per 10,000 people) 240.7 (1999)

Transport

Airports: international airport: Paramaribo (Johan A Pengel); one domestic airport and 35 airstrips; total passenger km: 1,068 million (1997 est)
Railways: total length: 301 km/187 mi (private freight railways)
Roads: total road network: 4,470 km/2,778 mi, of which 26% paved (1996); passenger cars: 103 per 1,000 people (1996 est)

Chronology

AD 1593: Visited and claimed by Spanish explorers; the name Suriname derived from the country's earliest inhabitants, the Surinen, who were driven out by other Amerindians in the 16th century. **1602:** Dutch settlements established. **1651:** British colony founded by settlers sent from Barbados. **1667:** Became a Dutch colony, received in exchange for New Amsterdam (New York) by Treaty of Breda. **1682:** Coffee and sugar cane plantations introduced, worked by imported African slaves. **1795–1802 and 1804–16:** Under British rule. **1863:** Slavery abolished and indentured labourers brought in from China, India, and Java. **1915:** Bauxite discovered and gradually became main export. **1954:** Achieved internal self-government as Dutch Guiana. **1958–69:** Politics dominated by Johan Pengel, charismatic leader of the mainly Creole Suriname National Party (NPS). **1975:** Independence achieved, with Dr Johan Ferrier as president and Henck Arron (NPS) as prime minister; 40% of population emigrated to the Netherlands. **1980:** Arron's government overthrown in an army coup. The army replaced Ferrier with Dr Chin A Sen. **1982:** The army, led by Lt Col Desi Bouterse, seized power, setting up a Revolutionary People's Front; economic aid from the Netherlands and US was cut off after opposition leaders, charged with plotting a coup, were executed. **1985:** Ban on political activities lifted. **1989:** Bouterse rejected a peace accord reached by President Shankar with guerrilla insurgents, the Bush Negro (descendants of escaped slaves) maroons, and vowed to continue fighting. **1991:** A New Front opposition alliance won an assembly majority. **1992:** A peace accord was reached with guerrilla groups. **2000:** Ronald Venetiaan was chosen as president in August. He had previously been president 1991–96. Jules Ajodhia was elected prime minister.

Practical information

Visa requirements: UK: visa not required. USA: visa required
Time difference: GMT –3.5
Chief tourist attractions: numerous historical sites; varied cultural activities; unspoiled interior with varied flora and fauna; 13 nature reserves and a nature park
Major holidays: 1 January, 25 February, 1 May, 1 July, 25 November, 25–26 December; variable: Good Friday, Easter Monday, end of Ramadan, Holi (March)

Swaziland

Located in southeast Africa, bounded east by Mozambique and southeast, south, west, and north by South Africa.

National name: *Umbuso wakaNgwane/Kingdom of Swaziland*
Area: 17,400 sq km/6,718 sq mi
Capital: Mbabane
Major towns/cities: Manzini, Big Bend, Mhlume, Havelock Mine, Nhlangano, Lobamba
Physical features: central valley; mountains in west (Highveld); plateau in east (Lowveld and Lubombo plateau)

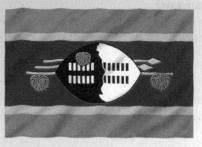

Blue stands for peace. The tassel is a symbol of the monarchy. Yellow represents mineral wealth. Red symbolizes battle. Effective date: 30 October 1967.

Government

Head of state: King Mswati III from 1986
Head of government: Barnabas Sibusiso Dlamini from 1996
Political system: absolutist
Political executive: absolute
Administrative divisions: four regions
Political parties: Imbokodvo National Movement (INM), nationalist monarchist; Swaziland United Front (SUF), left of centre; Swaziland Progressive Party (SPP), left of centre; People's United Democratic Movement, left of centre
Armed forces: 118,500 (1999 est)
Conscription: military service is compulsory for two years
Death penalty: retained and used for ordinary crimes
Defence spend: (% GDP) 1.6 (1997 est)
Education spend: (% GNP) 5.7 (1997 est)
Health spend: (% GDP) 3.4 (1997 est)

Economy and resources

Currency: lilangeni
GDP: (US$) 1.28 billion (1999 est)
Real GDP growth: (% change on previous year) 2.4 (1999)
GNP: (US$) 1.37 billion (1999)
GNP per capita (PPP): (US$) 4,200 (1999)
Consumer price inflation: 6.4% (1999)
Unemployment: 40% (1995 est)
Foreign debt: (US$) 290 million (1999 est)
Major trading partners: South Africa, EU, Mozambique, Japan, USA, Singapore
Resources: coal, asbestos, diamonds, gold, tin, kaolin, iron ore, talc, pyrophyllite, silica
Industries: food processing, paper, textiles, wood products, beverages, metal products
Exports: sugar, wood pulp, cotton yarn, canned fruits, asbestos, coal, diamonds, gold. Principal market: South Africa 65% (1998)
Imports: machinery and transport equipment, minerals, fuels and lubricants, manufactured items, food and live animals. Principal source: South Africa

84% (1998)
Arable land: 10.9% (1995)
Agricultural products: sugar cane, cotton, citrus fruits, pineapples, maize, sorghum, tobacco, tomatoes, rice; livestock rearing (cattle and goats); commercial forestry

Population and society

Population: 1,008,000 (2000 est)
Population growth rate: 2.9% (1995–2000); 2.6% (2000–05)
Population density: (per sq km) 56 (1999 est)
Urban population: (% of total) 26 (2000 est)
Age distribution: (% of total population) 0–14 43%, 15–59 52%, 60+ 5% (2000 est)
Ethnic groups: about 95% indigenous African, comprising the Swazi, Zulu, Tonga, and Shangaan peoples; there are European and Afro-European (Eurafrican) minorities numbering around 22,000
Language: Swazi, English (both official)
Religion: about 60% Christian, animist
Education: (compulsory years) 7
Literacy rate: 81% (men); 79% (women) (2000 est)
Labour force: 35.6% agriculture, 22.3% industry, 42.1% services (1996 est)
Life expectancy: 58 (men); 63 (women) (1995–2000)
Child mortality rate: (under 5, per 1,000 live births) 100 (1995–2000)
Physicians: 1 per 9,091 people (1991)
Hospital beds: 1 per 423 people (1991)
TV sets: (per 1,000 people) 23 (1996)
Radios: (per 1,000 people) 170 (1996)
Internet users: (per 10,000 people) 30.6 (1999)

Transport

Airports: international airport: Manzini (Matsapha); total passenger km: 49 million (1995)
Railways: total length: 301 km/187 mi (freight)
Roads: total road network: 3,810 km/2,368 mi, of which 32% paved (1996 est); passenger cars: 31.6 per 1,000 people (1996 est)

Chronology

Late 16th century: King Ngwane II crossed Lubombo mountains from the east and settled in southeast Swaziland; his successors established a strong centralized Swazi kingdom, dominating the long-settled Nguni and Sothi peoples. **mid-19th century:** Swazi nation was ruled by the warrior King Mswati who, at the height of his power, controlled an area three times the size of the present-day state. **1882:** Gold was discovered in the northwest, attracting European fortune hunters, who coerced Swazi rulers into granting land concessions. **1894:** Came under joint rule of Britain and the Boer republic of Transvaal. **1903:** Following the South African War, Swaziland became a special British protectorate, or High Commission territory, against South Africa's wishes. **1922:** King Sobhuza II succeeded to the Swazi throne. **1968:** Independence achieved within the Commonwealth, as the Kingdom of Swaziland, with King (or Ngwenyama) Sobhuza II as head of state. **1973:** The king suspended the constitution, banned political activity, and assumed absolute powers after the opposition deputies had been elected to parliament. **1977:** The king announced substitution of traditional tribal communities (*tinkhundla*) for the parliamentary system, arguing it was more suited to Swazi values. **1982:** King Sobhuza died; his place was taken by one of his wives, Queen Dzeliwe, until his son, Prince Makhosetive, was old enough to become king. **1983:** Queen Dzeliwe ousted by a younger wife, Queen Ntombi, as real power passed to the prime minister, Prince Bhekimpi Dlamini. **1986:** The crown prince was formally invested as King Mswati III. **1990:** Following demands for greater freedom, King Mswati called for the creation of an *indaba* (popular parliament). **1992:** King Mswati approved further democratic constitutional amendments. **1993:** Direct elections of *tinkhundla* candidates were held for the first time. **1996:** Barnabas Sibusiso Dlamini was appointed prime minister. **2000:** There was further agitation for democratic reform. When the leader of the opposition party Mario Masuku called for an end to the 27-year-old state of emergency, he was arrested for allegedly making seditious comments.

Practical information

Visa requirements: UK: visa not required. USA: visa not required
Time difference: GMT +2
Chief tourist attractions: magnificent mountain scenery; game reserves
Major holidays: 1 January, 25 April, 22 July, 6 September, 24 October, 25–26 December; variable: Ascension Thursday, Good Friday, Easter Monday, Commonwealth (March)

SWEDEN

Located in northern Europe, bounded west by Norway, northeast by Finland and the Gulf of Bothnia, southeast by the Baltic Sea, and southwest by the Kattegat.

National name: *Konungariket Sverige/Kingdom of Sweden*
Area: 450,000 sq km/173,745 sq mi
Capital: Stockholm
Major towns/cities: Göteborg, Malmö, Uppsala, Norrköping, Västerås, Linköping, Orebro, Helsingborg
Major ports: Helsingborg, Malmö, Göteborg, Stockholm
Physical features: mountains in west; plains in south; thickly forested; more than 20,000 islands off the Stockholm coast; lakes, including Vänern, Vättern, Mälaren, and Hjälmaren

The colours are derived from the state coat of arms of 1364. The Scandinavian cross is taken from the Danish flag. Effective date: 22 June 1906.

Government

Head of state: King Carl XVI Gustaf from 1973
Head of government: Göran Persson from 1996
Political system: liberal democracy
Political executive: parliamentary
Administrative divisions: 21 counties
Political parties: Christian Democratic Community Party (KdS), Christian, centrist; Left Party (Vp), European, Marxist; Social Democratic Party (SAP), moderate, left of centre; Moderate Party (M), right of centre; Liberal Party (Fp), left of centre; Centre Party (C), centrist; Ecology Party (MpG), ecological; New Democracy (NG), right wing, populist
Armed forces: 53,100 (1998)
Conscription: 7–15 months (army and navy); 8–12 months (air force)
Death penalty: abolished in 1972
Defence spend: (% GDP) 2.5 (1998)
Education spend: (% GNP) 8.3 (1996)
Health spend: (% GDP) 8.6 (1997)

Economy and resources

Currency: Swedish krona
GDP: (US$) 238.6 billion (1999)
Real GDP growth: (% change on previous year) 3.8 (1999)
GNP: (US$) 221.8 billion (1999)
GNP per capita (PPP): (US$) 20,824 (1999)
Consumer price inflation: 0.5% (1999)
Unemployment: 8.2% (1998)
Major trading partners: Germany, UK, Norway, USA, Denmark, France
Resources: iron ore, uranium, copper, lead, zinc, silver, hydroelectric power, forests
Industries: motor vehicles, foodstuffs, machinery, precision equipment, iron and steel, metal products, wood products, chemicals, shipbuilding, electrical goods
Exports: forestry products (wood, pulp, and paper), machinery, motor vehicles, power-generating non-electrical machinery, chemicals, iron and steel. Principal market: Germany 10.9% (1999)
Imports: machinery and transport equipment, chemicals, mineral fuels and lubricants, textiles, clothing, footwear, food

and live animals. Principal source: Germany 17.7% (1999)
Arable land: 6.8% (1996)
Agricultural products: barley, wheat, oats, potatoes, sugar beet, tame hay, oil seed; livestock and dairy products

Population and society

Population: 8,910,000 (2000 est)
Population growth rate: 0.3% (1995–2000); 0.3% (2000–05)
Population density: (per sq km) 20 (1999 est)
Urban population: (% of total) 83 (2000 est)
Age distribution: (% of total population) 0–14 18%, 15–59 59%, 60+ 23% (2000 est)
Ethnic groups: predominantly of Teutonic descent, with small Saami (Lapp), Finnish, and German minorities
Language: Swedish (official), Finnish, Saami (Lapp)
Religion: Evangelical Lutheran, Church of Sweden (established national church) 90%; Muslim, Jewish
Education: (compulsory years) 9
Literacy rate: 99% (men); 99% (women) (2000 est)
Labour force: 48.4% of population: 2.9% agriculture, 26.1% industry, 71% services (1996)
Life expectancy: 76 (men); 81 (women) (1995–2000)
Child mortality rate: (under 5, per 1,000 live births) 6 (1995–2000)
Physicians: 1 per 323 people (1996)
Hospital beds: 1 per 179 people (1996)
TV sets: (per 1,000 people) 531 (1997)
Radios: (per 1,000 people) 932 (1997)
Internet users: (per 10,000 people) 4,137.0 (1999)
Personal computer users: (per 100 people) 45.1 (1999)

Transport

Airports: international airports: Stockholm (Arlanda), Göteborg (Landvetter), Malmö (Sturup); over 30 domestic airports; total passenger km: 8,615 million (1995)
Railways: total length: 11,269 km/7,003 mi; total passenger km: 6,428 million (1997)
Roads: total road network: 210,760 km/130,966 mi, of which 77.2% paved (1997); passenger cars: 418 per 1,000 people (1997)

Chronology

8th century: Kingdom of the Svear, based near Uppsala, extended its rule across much of southern Sweden. **9th–11th centuries:** Swedish Vikings raided and settled along the rivers of Russia. *c.* **1000:** Olaf Skötkonung, king of the Svear, adopted Christianity and united much of Sweden (except south and west coasts, which remained Danish until 17th century). **11th–13th centuries:** Sweden existed as isolated kingdom under the Stenkil, Sverker, and Folkung dynasties; series of crusades incorporated Finland. **1397:** Union of Kalmar: Sweden, Denmark, and Norway united under a single monarch; Sweden effectively ruled by succession of regents. **1448:** Breach with Denmark: Sweden alone elected Charles VIII as king. **1523:** Gustavus Vasa, leader of insurgents, became king of a fully independent Sweden. **1527:** Swedish Reformation: Gustavus confiscated Church property and encouraged Lutherans. **1544:** Swedish crown became hereditary in House of Vasa. **1592–1604:** Sigismund Vasa, a Catholic, was king of both Sweden and Poland until ousted from Swedish throne by his Lutheran uncle Charles IX. **17th century:** Sweden, a great military power under Gustavus Adolphus 1611–32, Charles X 1654–60, and Charles XI 1660–97, fought lengthy wars with Denmark, Russia, Poland, and Holy Roman Empire. **1720:** Limited monarchy established; political power passed to *Riksdag* (parliament) dominated by nobles. **1721:** Great Northern War ended with Sweden losing nearly all its conquests of the previous century. **1741–43:** Sweden defeated in disastrous war with Russia; further conflict 1788–90. **1771–92:** Gustavus III increased royal power and introduced wide-ranging reforms. **1809:** Russian invaders annexed Finland; Swedish nobles staged coup and restored powers of *Riksdag*. **1810:** Napoleonic marshal, Jean-Baptiste Bernadotte, elected crown prince of Sweden, as Charles XIII had no heir. **1812:** Bernadotte allied Sweden with Russia against France. **1814:** Treaty of Kiel: Sweden obtained Norway from Denmark. **1818–44:** Bernadotte reigned in Sweden as Charles XIV John. **1846:** Free enterprise established by abolition of trade guilds and monopolies. **1866:** Series of liberal reforms culminated in new two-chambered *Riksdag* dominated by bureaucrats and farmers. **late 19th century:** Development of large-scale forestry and iron-ore industry; neutrality adopted in foreign affairs. **1905:** Union with Norway dissolved. **1907:** Adoption of proportional representation and universal suffrage. **1920s:** Economic boom transformed Sweden from an agricultural to an industrial economy. **1932:** Social Democrat government of Per Halbin Hansson introduced radical public-works programme to combat trade slump. **1940–43:** Under duress, neutral Sweden permitted limited transit of German forces through its territory. **1946–69:** Social Democrat government of Tage Erlander developed comprehensive welfare state. **1959:** Sweden joined European Free Trade Association. **1971:** Constitution amended to create single-chamber *Riksdag*. **1975:** Remaining constitutional powers of monarch removed. **1976–82:** Centre–right coalition government under Prime Minister Thorbjörn Fälldin ended 44 years of Social Democrat dominance. **1991:** The leader of the Moderate Party, Carl Bildt, headed up a coalition of the Moderate, Centre, Liberal, and Christian Democratic parties. **1995:** Sweden became a member of the European Union. **1996:** Göran Persson (SAP) became prime minister. **1998:** The SAP were narrowly re-elected in a general election.

Practical information

Visa requirements: UK: visa not required. USA: visa not required

Time difference: GMT +1

Chief tourist attractions: varied landscape – mountains north of Arctic Circle, white sandy beaches in south, lakes, waterfalls, and forests; Stockholm, with its modern architecture and cultural activities

Major holidays: 1, 6 January, 1 May, 1 November, 24–26, 31 December; variable: Ascension Thursday, Good Friday, Easter Monday, Whit Monday, Midsummer Eve and Day (June)

SWITZERLAND

Landlocked country in Western Europe, bounded north by Germany, east by Austria and Liechtenstein, south by Italy, and west by France.

National name: *Schweizerische Eidgenossenschaft* (German)/*Confédération Suisse* (French)/ *Confederazione Svizzera* (Italian)/ *Confederaziun Svizra* (Romansch)/*Swiss Confederation*
Area: 41,300 sq km/15,945 sq mi
Capital: Bern
Major towns/cities: Zürich, Geneva, Basel, Lausanne, Luzern, St Gallen, Winterthur
Major ports: river port Basel (on the Rhine)
Physical features: most mountainous country in Europe (Alps and Jura mountains); highest peak Dufourspitze 4,634 m/15,203 ft in Apennines

The flag may have been based on that of Schwyz, one of the original cantons of the Confederation. While the national flag is square, a rectangular flag is used on Swiss lakes and rivers. Effective date: 12 December 1889.

Government

Head of state and government: Adolf Ogi from 2000
Government: liberal democracy
Political executive: limited presidency
Administrative divisions: 20 cantons and six demi-cantons
Political parties: Radical Democratic Party (FDP/PRD), radical, left of centre; Social Democratic Party (SP/PS), moderate, left of centre; Christian Democratic People's Party (CVP/PDC), Christian, moderate, centrist; Swiss People's Party (SVP/UDC), left of centre; Liberal Party (LPS/PLS), federalist, right of centre; Green Party (GPS/PES), ecological
Armed forces: 26,300 (1998)
Conscription: 17 weeks' recruit training, followed by refresher training of varying length according to age
Death penalty: abolished in 1992
Defence spend: (% GDP) 1.4 (1998)
Education spend: (% GNP) 5.3 (1996)
Health spend: (% GDP) 10.3 (1997)

Economy and resources

Currency: Swiss franc
GDP: (US$) 258.9 billion (1999)
Real GDP growth: (% change on previous year) 1.5 (1999)
GNP: (US$) 273.1 billion (1999)
GNP per capita (PPP): (US$) 27,486 (1999)
Consumer price inflation: 0.8% (1999)
Unemployment: 4.1% (1998 est)
Major trading partners: EU (principally Germany, France, Italy, the Netherlands, and UK), USA, Japan
Resources: salt, hydroelectric power, forest
Industries: heavy engineering, machinery, precision engineering (clocks and watches), jewellery, textiles, chocolate, dairy products, cigarettes, footwear, wine, international finance and insurance services, tourism
Exports: machinery and equipment, pharmaceutical and chemical products, foodstuffs, precision instruments, clocks and watches, metal products. Principal market: Germany 23.3% (1999)
Imports: machinery, motor vehicles, agricultural and forestry products, construction material, fuels and lubricants, chemicals, textiles and clothing. Principal

source: Germany 32.5% (1999)
Arable land: 10.1% (1996)
Agricultural products: sugar beet, potatoes, wheat, apples, pears, tobacco, grapes; livestock and dairy products, notably cheese

Population and society

Population: 7,386,000 (2000 est)
Population growth rate: 0.7% (1995–2000); 0.5% (2000–05)
Population density: (per sq km) 178 (1999 est)
Urban population: (% of total) 68 (2000 est)
Age distribution: (% of total population) 0–14 17%, 15–59 63%, 60+ 20% (2000 est)
Ethnic groups: German 65%, French 18%, Italian 10%, Romansch 1%
Language: German (65%), French (18%), Italian (10%), Romansch (1%) (all official)
Religion: Roman Catholic 46%, Protestant 40%
Education: (compulsory years) 8–9 (depending on canton)
Literacy rate: 99% (men); 99% (women) (2000 est)
Labour force: 4.6% agriculture, 26.8% industry, 68.6% services (1997)
Life expectancy: 75 (men); 82 (women) (1995–2000)
Child mortality rate: (under 5, per 1,000 live births) 8 (1995–2000)
Physicians: 1 per 313 people (1996)
Hospital beds: 1 per 597 people (1993)
TV sets: (per 1,000 people) 536 (1997)
Radios: (per 1,000 people) 1,000 (1997)
Internet users: (per 10,000 people) 2,464.8 (1999)
Personal computer users: (per 100 people) 46.2 (1999)

Transport

Airports: international airports: Zürich (Kloten), Geneva, Bern (Belp), Basel (Basel-Mulhouse); domestic services operate between these; total passenger km: 26,396 million (1997 est)
Railways: total length: 5,208 km/3,236 mi; total passenger km: 11,664 million (1996)
Roads: total road network: 71,048 km/44,149 mi (1997); passenger cars: 469 per 1,000 people (1997)

Chronology

58 BC: Celtic Helvetii tribe submitted to Roman authority after defeat by Julius Caesar. **4th century AD:** Region overrun by Germanic tribes, Burgundians, and Alemannians. **7th century:** Formed part of Frankish kingdom and embraced Christianity. **9th century:** Included in Charlemagne's Holy Roman Empire. **12th century:** Many autonomous feudal holdings developed as power of Holy Roman Empire declined. **13th century:** Habsburgs became dominant as overlords of eastern Switzerland. **1291:** Cantons of Schwyz, Uri, and Lower Unterwalden formed Everlasting League, a loose confederation to resist Habsburg control. **1315:** Battle of Morgarten: Swiss Confederation defeated Habsburgs. **14th century:** Luzern, Zürich, Basel, and other cantons joined Swiss Confederation, which became independent of Habsburgs. **1523–29:** Zürich, Bern, and Basel accepted Reformation but rural cantons remained Roman Catholic. **1648:** Treaty of Westphalia recognized Swiss independence from Holy Roman Empire. **1798:** French invasion established Helvetic Republic, a puppet state with centralized government. **1803:** Napoleon's Act of Mediation restored considerable autonomy to cantons. **1814:** End of French domination; Switzerland reverted to loose confederation of sovereign cantons with a weak federal parliament. **1815:** Great Powers recognized 'Perpetual Neutrality' of Switzerland. **1845:** Seven Catholic cantons founded Sonderbund league to resist any strengthening of central government by Liberals. **1847:** Federal troops defeated Sonderbund in brief civil war. **1848:** New constitution introduced greater centralization; Bern chosen as capital. **1874:** Powers of federal government increased; principle of referendum introduced. **late 19th century:** Development of industry, railways, and tourism led to growing prosperity. **1920:** League of Nations selected Geneva as its headquarters. **1960:** Joined European Free Trade Association (EFTA). **1971:** Women gained right to vote in federal elections. **1986:** A proposal for membership of the United Nations (UN) was rejected in a referendum. **1992:** Closer ties with the European Community (EC) were rejected in a national referendum. **2000:** Adolf Ogi was elected president.

Practical information

Visa requirements: UK: visa not required. USA: visa not required
Time difference: GMT +1
Chief tourist attractions: the Alps; lakes and lake resorts; walking; mountaineering; winter sports

Major holidays: 1 January, 1, 15 August (many cantons), 1 November (many cantons), 24–26 December; variable: Ascension Thursday, Corpus Christi (many cantons), Good Friday, Easter Monday, Whit Monday; many local holidays

SYRIA

Located in western Asia, on the Mediterranean Sea, bounded to the north by Turkey, east by Iraq, south by Jordan, and southwest by Israel and Lebanon.

National name: *al-Jumhuriyya al-Arabiyya as-Suriyya/Syrian Arab Republic*
Area: 185,200 sq km/71,505 sq mi
Capital: Damascus
Major towns/cities: Aleppo, Homs, Latakia, Hamah, Ar Raqqah, Dayr az Zawr
Major ports: Latakia
Physical features: mountains alternate with fertile plains and desert areas; Euphrates River

The stars are said to represent Syria and Iraq. Red, white, black, and green are the pan-Arab colours. Effective date: 29 March 1980.

Government

Head of state: Bashar al-Assad from 2000
Head of government: Muhammad Mustafa Miro from 2000
Political system: nationalistic socialist
Political executive: unlimited presidency
Administrative divisions: 14 provinces
Political parties: National Progressive Front (NPF), pro-Arab, socialist coalition, including the Communist Party of Syria, the Arab Socialist Party, the Arab Socialist Unionist Party, the Syrian Arab Socialist Union Party, the Ba'ath Arab Socialist Party
Armed forces: 320,000; reserve forces 500,000; paramilitary forces 8,000 (1998)
Conscription: 30 months
Death penalty: retained and used for ordinary crimes
Defence spend: (% GDP) 7.3 (1998)
Education spend: (% GNP) 3.1 (1997)
Health spend: (% GDP) 2.5 (1997 est)

Economy and resources

Currency: Syrian pound
GDP: (US$) 16.2 billion (1999 est)
Real GDP growth: (% change on previous year) –1.5 (1999 est)
GNP: (US$) 15.2 billion (1999)
GNP per capita (PPP): (US$) 2,761 (1999)
Consumer price inflation: –2.7% (1999)
Unemployment rate: 12% (1997 est)
Foreign debt: (US$) 22.7 billion (1999 est)
Major trading partners: Germany, Ukraine, Turkey, Italy, France, Lebanon, Japan
Resources: petroleum, natural gas, iron ore, phosphates, salt, gypsum, sodium chloride, bitumen
Industries: petroleum and petroleum products, coal, rubber and plastic products, textiles, clothing, leather products, tobacco, processed food
Exports: crude petroleum, textiles, vegetables, fruit, raw cotton, natural phosphate. Principal market: Germany 13.7% (1998)
Imports: crude petroleum, wheat, base metals, metal products, foodstuffs, machinery, motor vehicles. Principal

source: Ukraine 15.6% (1998)
Arable land: 24.4% (1996)
Agricultural products: cotton, wheat, barley, maize, olives, lentils, sugar beet, fruit, vegetables; livestock (principally sheep and goats)

Population and society

Population: 16,125,000 (2000 est)
Population growth rate: 2.5% (1995–2000)
Population density: (per sq km) 85 (1999 est)
Urban population: (% of total) 55 (2000 est)
Age distribution: (% of total population) 0–14 41%, 15–59 54%, 60+ 5% (2000 est)
Ethnic groups: predominantly Arab, with many differences in language and regional affiliations; Kurds, Armenian
Language: Arabic (89%) (official), Kurdish (6%), Armenian (3%), French, English, Aramaic, Circassian
Religion: Sunni Muslim 74%; other Islamic sects 16%, Christian 10%
Education: (compulsory years) 6
Literacy rate: 88% (men); 60% (women) (2000 est)
Labour force: 40% agriculture, 20% industry, 40% services (1996 est)
Life expectancy: 67 (men); 71 (women) (1995–2000)
Child mortality rate: (under 5, per 1,000 live births) 40 (1995–2000)
Physicians: 1 per 969 people (1994)
Hospital beds: 1 per 918 people (1994)
TV sets: (per 1,000 people) 70 (1997)
Radios: (per 1,000 people) 278 (1997)
Internet users: (per 10,000 people) 12.7 (1999)
Personal computer users: (per 100 people) 1.5 (1999)

Transport

Airports: international airports: Damascus, Aleppo (Nejrab), Latakia (chartered flights); four domestic airports; total passenger km: 1,409 million (1997 est)
Railways: total length: 1,998 km/1,242 mi; total passenger km: 456 million (1996)
Roads: total road network: 40,480 km/25,154 mi, of which 23% paved (1996); passenger cars: 9.1 per 1,000 people (1996 est)

Chronology

*c.*1750 BC: Syria became part of Babylonian Empire; during the next millennium it was successively conquered by Hittites, Assyrians, Chaldeans, and Persians. **333 BC:** Alexander the Great of Macedonia conquered Persia and Syria. **301 BC:** Seleucus I, one of the generals of Alexander the Great, founded the kingdom of Syria, which the Seleucid dynasty ruled for over 200 years. **64 BC:** Syria became part of Roman Empire. **4th century AD:** After division of Roman Empire, Syria came under Byzantine rule. **634:** Arabs conquered most of Syria and introduced Islam. **661–750:** Damascus was the capital of Muslim Empire. **1055:** Seljuk Turks overran Syria.

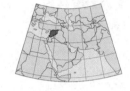

1095–99: First Crusade established Latin states on Syrian coast. **13th century:** Mameluke sultans of Egypt took control. **1516:** Ottoman Turks conquered Syria. **1831:** Egyptians led by Mehemet Ali drove out Turks. **1840:** Turkish rule restored; Syria opened up to European trade. **late 19th century:** French firms built ports, roads, and railways in Syria. **1916:** Sykes-Picot Agreement: secret Anglo-French deal to partition Turkish Empire allotted Syria to France. **1918:** British expelled Turks with help of Arab revolt. **1919:** Syrian national congress called for independence under Emir Faisal and opposed transfer to French rule. **1920:** Syria became League of Nations protectorate, administered by France. **1925:** People's Party founded to campaign for independence and national unity; insurrection by Druze religious sect against French control. **1936:** France promised independence within three years, but martial law imposed in 1939. **1941:** British forces ousted Vichy French regime in Damascus and occupied Syria in conjunction with Free French. **1944:** Syrian independence proclaimed but French military resisted transfer of power. **1946:** Syria achieved effective independence when French forces withdrew. **1948–49:** Arab–Israeli War: Syria joined unsuccessful invasion of newly independent Israel. **1958:** Syria and Egypt merged to form United Arab Republic (UAR). **1959:** USSR agreed to give financial and technical aid to Syria. **1961:** Syria seceded from UAR. **1964:** Ba'ath Socialist Party established military dictatorship. **1967:** Six-Day War: Syria lost Golan Heights to Israel. **1970–71:** Syria invaded Jordan in support of Palestinian guerrillas. **1971:** Hafez al-Assad was elected president. **1973:** Yom Kippur War: Syrian attack on Israel repulsed. **1976:** Start of Syrian military intervention in Lebanese civil war. **1978:** Syria opposed peace deal between Egypt and Israel. **1986:** Britain broke off diplomatic relations, accusing Syria of involvement in international terrorism. **1990:** Diplomatic links with Britain were restored. **1991:** Syria contributed troops to a US-led coalition in the Gulf War against Iraq. A US Middle East peace plan was approved by Assad. **1994:** Israel offered a partial withdrawal from the Golan Heights in return for peace, but Syria remained sceptical. **1995:** A security framework agreement was made with Israel. 1,200 political prisoners, including members of the banned Muslim Brotherhood, were released to commemorate the 25th anniversary of President Assad's seizure of power. **1996:** Syria re-deployed armed forces in southern Lebanon. **1997:** Three border points with Iraq, closed since 1980, were re-opened. **1998:** Relations with Israel deteriorated after Israeli forces seized land cultivated by Arab farmers in the Golan heights. **1999:** Amnesty International charged Syrian authorities with human rights abuses and called for the release of over 300 political prisoners. Peace talks with Israel over Lebanon and the Golan Heights resumed after a break of three years. Relations with Iraq were normalized. **2000:** Further peace talks were held with Israel, and Israel withdrew from the Golan Heights. President Assad appointed Muhammad Mustafa Miro as prime minister. President Assad died in June, and his son Bashar became president. The Iraq–Syria border was reopened. **2001:** Syria signed a free-trade accord with Iraq.

Practical information

Visa requirements: UK: visa required. USA: visa required
Time difference: GMT +2
Chief tourist attractions: antiquities of Damascus and

Palmyra; bazaars; Mediterranean coastline; mountains
Major holidays: 1 January, 8 March, 17 April, 1, 6 May, 23 July, 1 September, 6 October, 25 December; variable: Eid-ul-Adha (3 days), end of Ramadan (4 days), Easter Sunday, New Year (Muslim), Prophet's Birthday

TAIWAN FORMERLY FORMOSA (UNTIL 1949)

Located in east Asia, officially the Republic of China, occupying the island of Taiwan between the East China Sea and the South China Sea, separated from the coast of China by the Taiwan Strait.

National name: *Chung-hua Min-kuo/Republic of China*
Area: 36,179 sq km/13,968 sq mi
Capital: Taipei
Major towns/cities: Kaohsiung, Taichung, Tainan, Panchiao, Chungho, Sanchung
Major ports: Kaohsiung, Keelung
Physical features: island (formerly Formosa) off People's Republic of China; mountainous, with lowlands in west; Penghu (Pescadores), Jinmen (Quemoy), Mazu (Matsu) islands

Known as 'white sun in blue sky', the flag of Chinese revolutionary leader Sun Zhong Shan appears in the canton. The rays of the sun represent 12 traditional Chinese hours (each equalling two hours) symbolizing progress. Red is a traditional Chinese colour.
Effective date: 8 October 1928.

Government

Head of state: Chen Shui-bian from 2000
Head of government: Chang Chun-hsiung from 2000
Political system: emergent democracy
Political executive: limited presidency
Administrative divisions: 16 counties, five municipalities, and two special municipalities (Taipei and Kaohsiung)
Political parties: Nationalist Party of China (Kuomintang: KMT; known as Guomindang outside Taiwan), anticommunist, Chinese nationalist; Democratic Progressive Party (DPP), centrist-pluralist, proself-determination grouping; Workers' Party (Kuntang), left of centre
Armed forces: 376,000; plus paramilitary forces numbering 26,700 and reserves totalling 1,657,500 (1998)
Conscription: military service is compulsory for two years
Death penalty: retained and used for ordinary crimes
Defence spend: (% GDP) 4.6 (1998)
Education spend: (% GDP) 2.5 (1994)

Economy and resources

Currency: New Taiwan dollar
GDP: (US$) 288.6 billion (1999)
Real GDP growth: (% change on previous year) 5.7 (1999)
GNP: (US$) 272 billion (1998 est)
GNP per capita (PPP): (US$) 18,950 (1998 est)
Consumer price inflation: 0.2% (1999)
Unemployment: 2.7% (1998)
Foreign debt: (US$) 31.5 billion (1999)
Major trading partners: USA, Japan, Hong Kong, Germany, Singapore, Malaysia, Indonesia, South Korea, Australia
Resources: coal, copper, marble, dolomite; small reserves of petroleum and natural gas
Industries: electronics, plastic and rubber goods, textiles and clothing, base metals, vehicles, aircraft, ships, footwear, cement, fertilizers, paper
Exports: electronic products, base metals and metal articles, textiles and clothing, machinery, information and communication products, plastic and rubber products, vehicles and transport equipment, footwear, headwear, umbrellas, toys, games, sports equipment. Principal market: USA 25.4% (1999)
Imports: machinery and transport equipment, basic manufactures, chemicals, base metals and metal articles, minerals, textile products, crude petroleum, plastics, precision instruments, clocks and watches, musical instruments. Principal source: Japan 27.6% (1999)
Arable land: 24% (1993)
Agricultural products: rice, tea, bananas, pineapples, sugar cane, maize, sweet potatoes, soybeans, peanuts; fishing; forest resources

Population and society

Population: 22,113,000 (1999 est)
Population growth rate: 1.0% (1995–2000)
Population density: (per sq km) 685 (1999 est)
Urban population: (% of total) 75 (1994)
Age distribution: (% of total population) 0–14 22%, 15–59 70%, 60+ 8% (1999 est)
Ethnic groups: 98% Han Chinese and 2% aboriginal by descent; around 84% are Taiwan-born and 14% are 'mainlanders'
Language: Chinese (dialects include Mandarin (official), Min, and Hakka)
Religion: officially atheist; Buddhist 23%, Taoist 18%, I-Kuan Tao 4%, Christian 3%, Confucian and other 3%
Education: (compulsory years) 9
Literacy rate: 95% (men); 93% (women) (1997 est)
Labour force: 9.6% agriculture, 38.2% industry, 52.2% services (1997)
Life expectancy: 74 (men); 80 (women) (1998 est)
Child mortality rate: (under 5, per 1,000 live births) 7 (1997 est)
Physicians: 1 per 851 people (1995); 3,290 doctors of traditional Chinese medicine (1992)
Hospital beds: 1 per 180 people (1997)
TV sets: (per 1,000 people) 362 (1996)
Radios: (per 1,000 people) 744 (1994)
Internet users: (per 10,000 people) 2,051.2 (1999)
Personal computer users: (per 100 people) 18.1 (1999)

Transport

Airports: international airports: Taipei (Chaing Kai-shek), Kaohsiung; 14 domestic airports; total passenger km: 39,878 million (1997)
Railways: total length: 1,108 km/689 mi; total passenger

km: 9,263 million (1997)
Roads: total road network: 20,189 km/12,545 mi, of which 87.5% paved (1997); passenger cars: 198 per 1,000 people (1997)

Chronology

7th century AD: Island occupied by aboriginal community of Malayan descent; immigration of Chinese from mainland began, but remained limited before 15th century. **1517:** Sighted by Portuguese vessels en route to Japan and named Ilha Formosa ('beautiful island'). **1624:** Occupied and controlled by Dutch. **1662:** Dutch defeated by Chinese Ming general, Cheng Ch'eng-kung (Koxinga), whose family came to rule Formosa for a short period. **1683:** Annexed by China's rulers, the Manchu Qing. **1786:** Major rebellion against Chinese rule. **1860:** Ports opened to Western trade. **1895:** Ceded 'in perpetuity' to Japan under Treaty of Shominoseki at end of Sino-Japanese war. **1945:** Recovered by China's Nationalist Guomindang government at end of World War II. **1947:** Rebellion against Chinese rule brutally suppressed. **1949:** Flight of Nationalist government, led by Generalissimo Jiang Jie Shi (Chiang Kai-shek), to Taiwan after Chinese communist revolution. They retained the designation of Republic of China (ROC), claiming to be the legitimate government for all China, and were recognized by USA and United Nations (UN). Taiwan replaced Formosa as the name of the country. **1950s onwards:** Rapid economic growth as Taiwan became a successful export-orientated Newly Industrializing Country (NIC). **1954:** US–Taiwanese mutual defence treaty. **1971:** Expulsion from UN as USA adopted new policy of détente towards communist China. **1972:** Commencement of legislature elections as a programme of gradual democratization and Taiwanization was launched by the mainlander-dominated Guomindang. **1975:** President Jiang Jie Shi died; replaced as Guomindang leader by his son, Jiang Ching-kuo. **1979:** USA severed diplomatic relations and annulled the 1954 security pact. **1986:** Centrist Democratic Progressive Party (DPP) formed as opposition to nationalist Guomindang. **1987:** Martial law lifted; opposition parties legalized; press restrictions lifted. **1988:** President Jiang Ching-kuo died; replaced by Taiwanese-born Lee Teng-hui. **1990:** Chinese-born Guomindang members became a minority in parliament. **1991:** President Lee Teng-hui declared an end to the civil war with China. The constitution was amended. Guomindang won a landslide victory in elections to the new National Assembly, the 'superparliament'. **1993:** A cooperation pact was signed with China. **1996:** Lee Teng-hui was elected president in the first ever Chinese democratic elections. **1997:** The government narrowly survived a no-confidence motion. Vincent Siew became prime minister. **1998:** President Lee Teng-hui announced that reunion with mainland China was impossible until Beijing adopted democracy. The ruling Guomindang increased its majority in parliamentary and local elections. **2000:** Despite threats of invasion from China if Taiwan made moves towards independence, a pro-independence president, Chen Shui-bian, was elected, who appointed a member of the former government, Tang Fei, as prime minister. Tang Fei was replaced by Chang Chun-hsiung after he resigned in October. **2001:** Taiwan partially lifted its 52-year ban on direct trade and communications with China.

Practical information

Visa requirements: UK: visa not required for a stay of up to 14 days. USA: visa not required for a stay of up to 14 days
Time difference: GMT +8
Chief tourist attractions: island scenery; festivals;

ancient art treasures
Major holidays: 1–3 January, 29 March, 5 April, 1 July, 28 September, 10, 25, 31 October, 12 November, 25 December; variable: New Year (Chinese, January/February, 3 days), Dragon Boat Festival (June), Mid-Autumn Festival (September/October)

TAJIKISTAN

Located in central Asia, bounded north by Kyrgyzstan and Uzbekistan, east by China, and south by Afghanistan and Pakistan.

National name: *Jumhurii Tojikston/Republic of Tajikistan*
Area: 143,100 sq km/55,250 sq mi
Capital: Dushanbe
Major towns/cities: Khujand, Qurghonteppa, Kulob, Uroteppa, Kofarnihon
Physical features: mountainous, more than half of its territory lying above 3,000 m/10,000 ft; huge mountain glaciers, which are the source of many rapid rivers

Red recalls the previous flag. Green represents agricultural produce. Effective date: 24 November 1992.

Government

Head of state: Imamali Rakhmanov from 1994
Head of government: Akil Akilov from 1999
Political system: authoritarian nationalist
Political executive: unlimited presidency
Administrative divisions: two provinces and one autonomous region (Gornyi Badakhstan)
Political parties: Communist Party of Tajikistan (CPT), pro-Rakhmanov; Democratic Party of Tajikistan (DP), anticommunist (banned from 1993); Party of Popular Unity and Justice, anticommunist
Armed forces: 9,000; paramilitary forces around 1,200 (1998)
Death penalty: retained and used for ordinary crimes
Defence spend: (% GDP) 8.3 (1998)
Education spend: (% GNP) 2.2 (1996)
Health spend: (% GDP) 5.8 (1995)

Economy and resources

Currency: Tajik rouble
GDP: (US$) 1.7 billion (1999)
Real GDP growth: (% change on previous year) 3.7 (1999)
GNP: (US$) 1.8 billion (1999)
GNP per capita (PPP): (US$) 981 (1999)
Consumer price inflation: 27.5% (1999)
Unemployment: 2.7% (1997)
Foreign debt: (US$) 1.07 billion (1998)
Major trading partners: Uzbekistan, the Netherlands, Switzerland, Russia, UK, Kazakhstan, Ukraine
Resources: coal, aluminium, lead, zinc, iron, tin, uranium, radium, arsenic, bismuth, gold, mica, asbestos, lapis lazuli; small reserves of petroleum and natural gas
Industries: mining, aluminium production, engineering, food processing, textiles (including silk), carpet making, clothing, footwear, fertilizers
Exports: aluminium, cotton lint. Principal market: Uzbekistan 19.6% (1998)
Imports: industrial products and machinery (principally for aluminium plants), unprocessed agricultural products, food and beverages, petroleum and chemical products, consumer goods. Principal

source: Uzbekistan 28.7% (1998)
Arable land: 5.8% (1996)
Agricultural products: cotton, jute, rice, millet, fruit, vegetables; livestock rearing (cattle, sheep, goats, and pigs)

Population and society

Population: 6,188,000 (2000 est)
Population growth rate: 1.5% (1995–2000)
Population density: (per sq km) 43 (1999 est)
Urban population: (% of total) 28 (2000 est)
Age distribution: (% of total population) 0–14 40%, 15–59 53%, 60+ 7% (2000 est)
Ethnic groups: 62% ethnic Tajik, 24% Uzbek, 8% ethnic Russian, 1% Tatar, 1% Kyrgyz, and 1% Ukrainian
Language: Tajik (related to Farsi; official), Russian
Religion: Sunni Muslim; small Russian Orthodox and Jewish communities
Education: (compulsory years) 9
Literacy rate: 99% (men); 99% (women) (2000 est)
Labour force: 52% agriculture, 17% industry, 31% services (1995)
Life expectancy: 64 (men); 70 (women) (1995–2000)
Child mortality rate: (under 5, per 1,000 live births) 81 (1995–2000)
Physicians: 1 per 442 people (1994)
Hospital beds: 1 per 96 people (1994)
TV sets: (per 1,000 people) 285 (1997)
Radios: (per 1,000 people) 142 (1997)
Internet users: (per 10,000 people) 3.3 (1999)

Transport

Airports: international airport: Dushanbe; three domestic airports; total passenger km: 2,231 million (1995)
Railways: total length: 511 km/318 mi; total passenger km: 144 million (1995)
Roads: total road network: 13,700 km/8,513 mi, of which 82.7% paved (1996 est); passenger cars: 0.1 per 1,000 people (1996 est)

Chronology

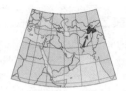

c. **330:** Formed an eastern part of empire of Alexander the Great of Macedonia. **8th century:** Tajiks established as distinct ethnic group, with semi-independent territories under the tutelage of the Uzbeks, to the west; spread of Islam. **13th century:** Conquered by Genghis Khan and became part of Mongol Empire. **1860–1900:** Northern Tajikistan came under tsarist Russian rule, while the south was annexed by Emirate of Bukhara, to the west. **1917–18:** Attempts to establish Soviet control after Bolshevik revolution in Russia resisted initially by armed guerrillas (basmachi). **1921:** Became part of Turkestan Soviet Socialist Autonomous Republic. **1924:** Tajik Autonomous Soviet Socialist Republic formed. **1929:** Became constituent republic of Soviet Union (USSR). **1930s:** Stalinist era of collectivization led to widespread repression of Tajiks. **1978:** 13,000 participated in anti-Russian riots. **late 1980s:** Resurgence in Tajik consciousness, stimulated by the *glasnost* initiative of Soviet leader Mikhail Gorbachev. **1989:** Rastokhez ('Revival') Popular Front established and Tajik declared state language. New mosques constructed. **1990:** Violent interethnic Tajik–Armenian clashes in Dushanbe; a state of emergency was imposed. **1991:** President Kakhar Makhkamov, local communist leader since 1985, was

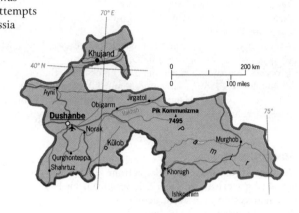

forced to resign after supporting the failed anti-Gorbachev coup in Moscow. Independence was declared. Rakhman Nabiyev, communist leader 1982–85, was elected president. Joined new Commonwealth of Independent States (CIS). **1992:** Joined Muslim Economic Cooperation Organization, the Conference on Security and Cooperation in Europe (CSCE; now the Organization on Security and Cooperation in Europe, OSCE), and the United Nations (UN). Violent demonstrations by Islamic and prodemocracy groups forced Nabiyev to resign. Civil war between pro- and anti-Nabiyev forces claimed 20,000 lives, made 600,000 refugees, and wrecked the economy. Imamali Rakhmanov, a communist sympathetic to Nabiyev, took over as head of state. **1993:** Government forces regained control of most of the country. CIS peacekeeping forces were drafted in to patrol the border with Afghanistan, the base of the pro-Islamic rebels. **1994:** A ceasefire was agreed. Rakhmanov was popularly elected president under a new constitution. **1995:** Parliamentary elections were won by Rakhmanov's supporters. There was renewed fighting on the Afghan border. **1996:** Pro-Islamic rebels captured towns in the southwest. There was a UN-sponsored ceasefire between government and pro-Islamic rebels. **1997:** A four-stage peace plan was signed. There was a peace accord with the Islamic rebel group the United Tajik Opposition (UTO). **1998:** Members of UTO were appointed to the government as part of a peace plan. The UN military observer mission (UNMOT) suspended its operations, following the killing of four UN workers. More than 200 people were killed in clashes in Leninabad between the army and rebel forces loyal to the renegade Tajik army commander Col Makhmud Khudoberdiyev; the deputy leader of the Islamic-led UTO, Ali Akbar Turadzhonzada, was appointed first deputy prime minister. **1999:** Constitutional changes approved the creation of a two-chamber legislature. President Rakhmanov was popularly re-elected and appointed Akil Akilov as his prime minister.

Practical information

Visa requirements: UK: visa required. USA: visa required
Time difference: GMT +5
Chief tourist attractions: spectacular mountain scenery; sites of historical interest in the Fergana Valley, notably the city of Khujand
Major holidays: 1 January, 8, 21 March, 9 May, 9 September, 14 October; variable: end of Ramadan

TANZANIA FORMERLY TANGANYIKA (UNTIL 1964)

Located in east Africa, bounded to the north by Uganda and Kenya; south by Mozambique, Malawi, and Zambia; west by Congo (formerly Zaire), Burundi, and Rwanda; and east by the Indian Ocean.

National name: *Jamhuri ya Muungano wa Tanzania/United Republic of Tanzania*
Area: 945,000 sq km/364,864 sq mi
Capital: Dodoma
Major towns/cities: Zanzibar, Mwanza, Mbeya, Tanga, Morogora
Major ports: (former capital) Dar es Salaam
Physical features: central plateau; lakes in north and west; coastal plains; lakes Victoria, Tanganyika, and Nyasa; half the country is forested; comprises islands of Zanzibar and Pemba; Mount Kilimanjaro, 5,895 m/19,340 ft, the highest peak in Africa; Olduvai Gorge; Ngorongoro Crater, 14.5 km/9 mi across, 762 m/2,500 ft deep

Green stands for the forests and agriculture. Gold symbolizes the country's mineral wealth. Blue represents the sea. Effective date: 30 June 1964.

Government

Head of state: Benjamin Mkapa from 1995
Head of government: Frederick Sumaye from 1995
Political system: emergent democracy
Political executive: limited presidency
Administrative divisions: 25 regions
Political parties: Revolutionary Party of Tanzania (CCM), African, socialist; Civic Party (Chama Cha Wananchi), left of centre; Tanzania People's Party (TPP), left of centre; Democratic Party (DP), left of centre; Justice and Development Party, left of centre; Zanzibar United Front (Kamahuru), Zanzibar-based, centrist
Armed forces: 34,000; citizen's militia of 80,000 (1998)
Conscription: two years
Death penalty: retained and used for ordinary crimes
Defence spend: (% GDP) 3.7 (1998)
Education spend: (% GNP) 5.0 (1993/94)
Health spend: (% GDP) 2.5 (1995)

Economy and resources

Currency: Tanzanian shilling
GDP: (US$) 7.7 billion (1999)
Real GDP growth: (% change on previous year) 4.7 (1999)
GNP: (US$) 8.7 billion (1999)
GNP per capita (PPP): (US$) 478 (1999)
Consumer price inflation: 7.9% (1999)
Foreign debt: (US$) 7 billion (1999)
Major trading partners: India, UK, Germany, South Africa, Japan, the Netherlands, Kenya, Malaysia
Resources: diamonds, other gemstones, gold, salt, phosphates, coal, gypsum, tin, kaolin (exploration for petroleum in progress)
Industries: food processing, textiles, cigarette production, pulp and paper, petroleum refining, diamonds, cement, brewing, fertilizers, clothing, footwear, pharmaceuticals, electrical goods, metalworking, vehicle assembly

Exports: coffee beans, raw cotton, tobacco, tea, cloves, cashew nuts, minerals, petroleum products. Principal market: India 19.5% (1998 est)
Imports: machinery and transport equipment, crude petroleum and petroleum products, construction materials, foodstuffs, consumer goods. Principal source: South Africa 8.3% and Japan 8.3% (1998 est)
Arable land: 3.5% (1996)
Agricultural products: coffee, cotton, tobacco, cloves, tea, cashew nuts, sisal, pyrethrum, sugar cane, coconuts, cardamoms

Population and society

Population: 33,517,000 (2000 est)
Population growth rate: 2.3% (1995–2000)
Population density: (per sq km) 35 (1999 est)
Urban population: (% of total) 33 (2000 est)
Age distribution: (% of total population) 0–14 45%, 15–59 51%, 60+ 4% (2000 est)
Ethnic groups: 99% of the population are Africans, ethnically classified as Bantu, and distributed among over 130 tribes; main tribes are Bantu, Nilotic, Nilo-Hamitic, Khoisan, and Iraqwi
Language: Kiswahili, English (both official), Arabic (in Zanzibar), many local languages
Religion: Muslim, Christian, traditional religions
Education: (compulsory years) 7
Literacy rate: 85% (men); 63% (women) (2000 est)
Labour force: 52% of population: 82% agriculture, 6% industry, 12% services (1996 est)
Life expectancy: 50 (men); 53 (women) (1995–2000)
Child mortality rate: (under 5, per 1,000 live births) 130 (1995–2000)
Physicians: 1 per 20,511 people (1993)
Hospital beds: 1 per 1,044 people (1993)
TV sets: (per 1,000 people) 21 (1997)
Radios: (per 1,000 people) 279 (1997)
Internet users: (per 10,000 people) 7.6 (1999)
Personal computer users: (per 100 people) 0.2 (1999)

Transport

Airports: international airports: Dar es Salaam, Kilimanjaro, Zanzibar; 50 domestic airports and landing strips; total passenger km: 189 million (1995)
Railways: total length: 3,569 km/2,218 mi; total

passenger km: 990 million (1992)
Roads: total road network: 88,200 km/54,807 mi, of which 4.2% paved (1996 est); passenger cars: 0.8 per 1,000 people (1996 est)

Chronology

8th century: Growth of city states along coast after settlement by Arabs from Oman. **1499:** Portuguese navigator Vasco da Gama visited island of Zanzibar. **16th century:** Portuguese occupied Zanzibar, defeated coastal states, and exerted spasmodic control over them. **1699:** Portuguese ousted from Zanzibar by Arabs of Oman. **18th century:** Sultan of Oman reasserted Arab overlordship of East African coast, which became subordinate to Zanzibar. **1744–1837:** Revolt of ruler of Mombasa against Oman spanned 93 years until final victory of Oman. **1822:** Moresby Treaty: Britain recognized regional dominance of Zanzibar, but protested against the slave trade. **1840:** Sultan Seyyid bin Sultan moved his capital from Oman to Zanzibar; trade in slaves and ivory flourished. **1861:** Sultanates of Zanzibar and Oman separated on death of Seyyid. **19th century:** Europeans started to explore inland, closely followed by Christian missionaries. **1884:** German Colonization Society began to acquire territory on mainland in defiance of Zanzibar. **1890:** Britain obtained protectorate over Zanzibar, abolished slave trade, and recognized German claims to mainland. **1897:** German East Africa formally established as colony. **1905–06:** Maji Maji revolt suppressed by German troops. **1916:** Conquest of German East Africa by British and South African forces, led by Gen Jan Smuts. **1919:** Most of German East Africa became British League of Nations mandate of Tanganyika. **1946:** Britain continued to govern Tanganyika as United Nations (UN) trusteeship. **1954:** Julius Nyerere organized the Tanganyikan African National Union (TANU) to campaign for independence. **1961–62:** Tanganyika achieved independence from Britain with Nyerere as prime minister, and became a republic in 1962 with Nyerere as president. **1963:** Zanzibar achieved independence. **1964:** Arab-dominated sultanate of Zanzibar overthrown by Afro-Shirazi Party in violent revolution; Zanzibar merged with Tanganyika to form United Republic of Tanzania. **1967:** East African Community (EAC) formed by Tanzania, Kenya, and Uganda; Nyerere pledged to build socialist state. **1977:** Revolutionary Party of Tanzania (CCM) proclaimed as only legal party; EAC dissolved. **1979:** Tanzanian troops intervened in Uganda to help overthrow President Idi Amin. **1992:** Multiparty politics were permitted. **1995:** Benjamin Mkapa of CCM was elected president. **1998:** A bomb exploded at the US embassy in Dar es Salaam, killing 6 people and injuring 60; an anti-American Islamic group claimed responsibility. **1999:** Tanzania withdrew from Africa's largest trading block, the Common Market for Eastern and Southern Africa. In October, the country's founder, Julius Nyerere, died. **2000:** President Mkapa and the CCM, who had improved the economy over the preceding five years, were re-elected. **2001:** Violence broke out between opposition supporters and troops on Zanzibar after the elections had been partially rerun following claims of corruption.

Practical information

Visa requirements: UK: visa required. USA: visa required
Time difference: GMT +3
Chief tourist attractions: national parks and game

and forest reserves comprise one-third of the country; beaches and coral reefs along the Indian Ocean coast
Major holidays: 1, 12 January, 5 February, 1 May, 7 July, 9, 25 December; variable: Eid-ul-Adha, Good Friday, Easter Monday, end of Ramadan (2 days), Prophet's Birthday

THAILAND FORMERLY SIAM (UNTIL 1939 AND 1945–49)

Located in southeast Asia on the Gulf of Siam, bounded east by Laos and Cambodia, south by Malaysia, and west by Myanmar (Burma).

National name: *Ratcha Anachak Thai/Kingdom of Thailand*
Area: 513,115 sq km/198,113 sq mi
Capital: Bangkok (and chief port)
Major towns/cities: Chiang Mai, Hat Yai, Khon Kaen, Songkhla, Nakhon Ratchasima, Nonthaburi, Udon Thani
Major ports: Nakhon Sawan
Physical features: mountainous, semi-arid plateau in northeast, fertile central region, tropical isthmus in south; rivers Chao Phraya, Mekong, and Salween

The central band was originally red. It was changed to blue to express solidarity with the Allies during World War I. Effective date: 28 September 1917.

Government

Head of state: King Bhumibol Adulyadej from 1946
Head of government: Thaksin Shinawatra from 2001
Political system: emergent democracy
Political executive: parliamentary
Administrative divisions: 76 provinces
Political parties: Democrat Party (DP), left of centre; Thai Nation (Chart Thai), right wing, pro-private enterprise; New Aspiration Party (NAP), centrist; Palang Dharma Party (PDP), anti-corruption, Buddhist; Social Action Party (SAP), moderate, conservative; Chart Pattana (National Development), conservative
Armed forces: 306,000 (1998)
Conscription: two years
Death penalty: retained and used for ordinary crimes
Defence spend: (% GDP) 1.5 (1998)
Education spend: (% GNP) 4.1 (1996)
Health spend: (% GDP) 5.7 (1997)

Economy and resources

Currency: baht
GDP: (US$) 125.9 billion (1999)
Real GDP growth: (% change on previous year) 4.2 (1999)
GNP: (US$) 121.02 billion (1999)
GNP per capita (PPP): (US$) 5,599 (1999)
Consumer price inflation: 0.2% (1999)
Unemployment: 4% (1998 est)
Foreign debt: (US$) 90.1 billion (1999 est)
Major trading partners: Japan, USA, Singapore, Germany, Malaysia, Hong Kong, the Netherlands
Resources: tin ore, lignite, gypsum, antimony, manganese, copper, tungsten, lead, gold, zinc, silver, rubies, sapphires, natural gas, petroleum, fish
Industries: textiles and clothing, electronics, electrical goods, cement, petroleum refining, sugar refining, motor vehicles, agricultural products, beverages, tobacco, metals and metal products, plastics, furniture, tourism
Exports: textiles and clothing, electronic goods, rice, rubber, gemstones, sugar, cassava (tapioca), fish (especially prawns), machinery and manufactures, chemicals. Principal market: USA 21.6% (1999)
Imports: petroleum and petroleum products, machinery, chemicals, iron and steel, consumer goods. Principal source: Japan 24.3% (1999)

Arable land: 33.4% (1996)
Agricultural products: rice, cassava, rubber, sugar cane, maize, kenat (a jute-like fibre), tobacco, coconuts; fishing (especially prawns) and livestock (mainly buffaloes, cattle, pigs, and poultry)

Population and society

Population: 61,399,000 (2000 est)
Population growth rate: 0.9% (1995–2000)
Population density: (per sq km) 119 (1999 est)
Urban population: (% of total) 22 (2000 est)
Age distribution: (% of total population) 0–14 25%, 15–59 66%, 60+ 9% (2000 est)
Ethnic groups: 75% of the population is of Thai descent; 14% ethnic Chinese, one-third of whom live in Bangkok; Thai Malays constitute the next largest minority, followed by hill tribes; a substantial Kampuchean (Khmer) refugee community resides in border camps
Language: Thai, Chinese (both official), English, Lao, Malay, Khmer
Religion: Buddhist 95%; Muslim 5%
Education: (compulsory years) 6
Literacy rate: 97% (men); 94% (women) (2000 est)
Labour force: 57% of population: 64% agriculture, 14% industry, 22% services (1990)
Life expectancy: 66 (men); 72 (women) (1995–2000)
Child mortality rate: (under 5, per 1,000 live births) 35 (1995–2000)
Physicians: 1 per 4,372 people (1996)
Hospital beds: 1 per 765 people (1993 est)
TV sets: (per 1,000 people) 236 (1997)
Radios: (per 1,000 people) 232 (1997)
Internet users: (per 10,000 people) 131.5 (1999)
Personal computer users: (per 100 people) 2.3 (1999)

Transport

Airports: international airports: Bangkok (Don Muang), Chiang Mai, Phuket, Hat Yai, U-tapao; domestic services to all major towns; total passenger km: 30,987 million (1997 est)
Railways: total length: 3,865 km/2,402 mi; total passenger km: 12,300 million (1996)
Roads: total road network: 64,600 km/40,142 mi, of which 97.5 paved (1996 est); passenger cars: 27.4 per 1,000 people (1996 est)

Chronology

13th century: Siamese (Thai) people migrated south and settled in valley of Chao Phraya River in Khmer Empire. **1238:** Siamese ousted Khmer governors and formed new kingdom based at Sukhothai. **14th and 15th centuries:** Siamese expanded at expense of declining Khmer Empire. **1350:** Siamese capital moved to Ayatthaya (which also became name of kingdom). **1511:** Portuguese traders first reached Siam. **1569:** Conquest of Ayatthaya by Burmese ended years of rivalry and conflict. **1589:** Siamese regained independence under King Naresuan. **17th century:** Foreign trade under royal monopoly developed with Chinese, Japanese, and Europeans. **1690s:** Siam expelled European military advisers and missionaries and adopted policy of isolation. **1767:** Burmese invaders destroyed city of Ayatthaya, massacred ruling families, and withdrew, leaving Siam in a state of anarchy. **1782:** Reunification of Siam after civil war under Gen Phraya Chakri, who founded new capital at Bangkok and proclaimed himself King Rama I. **1824–51:** King Rama III reopened Siam to European diplomats and missionaries. **1851–68:** King Mongkut employed European advisers to help modernize the government, legal system, and army. **1856:** Royal monopoly on foreign trade ended. **1868–1910:** King Chulalongkorn continued modernization and developed railway network using Chinese immigrant labour; Siam became major exporter of rice. **1896:** Anglo-French agreement recognized Siam as independent buffer state between British Burma and French Indo-China. **1932:** Bloodless coup forced King Rama VII to grant a constitution with a mixed civilian-military government. **1939:** Siam changed its name to Thailand (briefly reverting to Siam 1945–49). **1941:** Japanese invaded; Thailand became puppet ally of Japan under Field Marshal Phibun Songkhram. **1945:** Japanese withdrawal; Thailand compelled to return territory taken from Laos, Cambodia, and Malaya. **1947:** Phibun regained power in military coup, reducing monarch to figurehead; Thailand adopted strongly pro-American foreign policy. **1955:** Political parties and free speech introduced. **1957:** State of emergency declared; Phibun deposed in bloodless coup; military dictatorship continued under Gen Sarit Thanarat (1957–63) and Gen Thanom Kittikachorn (1963–73). **1967–72:** Thai troops fought in alliance with USA in Vietnam War. **1973:** Military government overthrown by student riots. **1974:** Adoption of democratic constitution, followed by civilian coalition government. **1976:** Military reassumed control in response to mounting strikes and political violence. **1978:** Gen Kriangsak Chomanan introduced constitution with mixed civilian–military government. **1980:** Gen Prem Tinsulanonda assumed power. **1983:** Prem relinquished army office to head civilian government; martial law maintained. **1988:** Chatichai Choonhavan succeeded Prem as prime minister. **1991:** A military coup imposed a new military-oriented constitution despite mass protests. **1992:** A general election produced a five-party coalition; riots forced Prime Minister Suchinda Kraprayoon to flee; Chuan Leekpai formed a new coalition government. **1995–96:** The ruling coalition collapsed. A general election in 1996 resulted in a new six-party coalition led by Chavalit Yongchaiyudh. **1997:** A major financial crisis led to the floating of currency. An austerity rescue plan was agreed with the International Monetary Fund (IMF). Chuan Leekpai was re-elected prime minister. **1998:** Repatriation of foreign workers commenced, as the economy contracted sharply due to the rescue plan. The opposition Chart Patthana party was brought into the coalition government of Chuan Leekpai, increasing its majority to push through economic reforms. **2001:** The Thai Rak Thai party won general elections, but failed to achieve an absolute majority.

Practical information

Visa requirements: UK: visa not required. USA: visa not required
Time difference: GMT +7
Chief tourist attractions: temples, pagodas, palaces; islands
Major holidays: 1 January, 6, 13 April, 1, 5 May, 1 July, 12 August, 23 October, 5, 10, 31 December; variable: end of Ramadan, Makha Bucha (February), Visakha Bucha (May), Buddhist Lent (July)

TOGO FORMERLY TOGOLAND (UNTIL 1956)

Located in West Africa, on the Atlantic Ocean, bounded north by
Burkina Faso, east by Benin, and west by Ghana.

National name: *République Togolaise/Togolese Republic*
Area: 56,800 sq km/21,930 sq mi
Capital: Lomé
Major towns/cities: Sokodé, Kpalimé, Kara, Atakpamé, Bassar, Tsévié
Physical features: two savannah plains, divided by range of hills
northeast–southwest; coastal lagoons and marsh; Mono Tableland, Oti
Plateau, Oti River

The white star is a symbol of hope and national purity.
Green stands for agriculture. Yellow symbolizes mineral
wealth. Effective date: 27 April 1960.

Government

Head of state: Etienne Gnassingbé Eyadéma from 1967
Head of government: Agbeyome Messan Kodjo from 2000
Political system: emergent democracy
Political executive: limited presidency
Administrative divisions: five regions
Political parties: Rally of the Togolese People (RPT),
nationalist, centrist; Action Committee for Renewal
(CAR), left of centre; Togolese Union for Democracy
(UTD), left of centre
Armed forces: 7,000 (1998)
Conscription: military service is by selective conscription
for two years
Death penalty: retains the death penalty for ordinary
crimes, but can be considered abolitionist in practice
Defence spend: (% GDP) 2.4 (1998)
Education spend: (% GNP) 4.7 (1996)
Health spend: (% GDP) 2.8 (1997 est)

Economy and resources

Currency: franc CFA
GDP: (US$) 1.43 billion (1999)
Real GDP growth: (% change on previous year) 3.2 (1999)
GNP: (US$) 1.46 billion (1999)
GNP per capita (PPP): (US$) 1,346 (1999 est)
Consumer price inflation: 0.1% (1999)
Unemployment: N/A
Foreign debt: (US$) 1.45 billion (1998)
Major trading partners: Nigeria, Ghana, Brazil, China,
Canada, France, Philippines, Côte d'Ivoire
Resources: phosphates, limestone, marble, deposits of
iron ore, manganese, chromite, peat; exploration for
petroleum and uranium was under way in the early 1990s
Industries: processing of phosphates, steel rolling,
cement, textiles, processing of agricultural products, beer,
soft drinks
Exports: phosphates (mainly calcium phosphates), ginned
cotton, green coffee, cocoa beans. Principal market:
Nigeria 7.8%, Brazil 6.1% (1999)
Imports: machinery and transport equipment, cotton yarn
and fabrics, cigarettes, antibiotics, food (especially cereals)
and live animals, chemicals, refined petroleum products,
beverages. Principal source: Ghana 23.9% (1999)
Arable land: 38.1% (1996)

Agricultural products: cotton, cocoa, coffee, oil palm,
yams, cassava, maize, millet, sorghum

Population and society

Population: 4,629,000 (2000 est)
Population growth rate: 2.6% (1995–2000)
Population density: (per sq km) 79 (1999 est)
Urban population: (% of total) 33 (2000 est)
Age distribution: (% of total population) 0–14 46%,
15–59 49%, 60+ 5% (2000 est)
Ethnic groups: predominantly of Sudanese Hamitic
origin in the north, and black African in the south; they
are distributed among 37 different ethnic groups. There
are three main ethnic groups: the Ewe, Mina, and Outchi
in the south, the Akposso-Adele in the central region, and
the Kabre in the north. There are also European, Syrian,
and Lebanese minorities
Language: French (official), Ewe, Kabre, Gurma, other
local languages
Religion: animist about 50%, Catholic and Protestant
35%, Muslim 15%
Education: (compulsory years) 6
Literacy rate: 74% (men); 41% (women) (2000 est)
Labour force: 42% of population: 66% agriculture, 10%
industry, 24% services (1990)
Life expectancy: 48 (men); 50 (women) (1995–2000)
Child mortality rate: (under 5, per 1,000 live births)
129 (1995–2000)
Physicians: 1 per 11,385 people (1993 est)
Hospital beds: 1 per 664 people (1993 est)
TV sets: (per 1,000 people) 18 (1998)
Radios: (per 1,000 people) 218 (1997)
Internet users: (per 10,000 people) 22.2 (1999)
Personal computer users: (per 100 people) 0.8 (1999)

Transport

Airports: international airports: Lomé, Niamtougou; four
domestic airports and several smaller airfields; total
passenger km: 223 million (1995)
Railways: total length: 525 km/326 mi; total passenger
km: 16.5 million (1996)
Roads: total road network: 7,520 km/4,673 mi, of which
31.6% paved (1996 est); passenger cars: 19 per 1,000
people (1996 est)

Chronology

15th–17th centuries: Formerly dominated by Kwa peoples in southwest and Gur-speaking Voltaic peoples in north, Ewe clans immigrated from Nigeria and the Ane (Mina) from Ghana and the Côte d'Ivoire. **18th century:** Coastal area held by Danes. **1847:** Arrival of German missionaries. **1884–1914:** Togoland was a German protectorate until captured by Anglo-French forces; cocoa and cotton plantations developed, using forced labour. **1922:** Divided between Britain and France under League of Nations mandate. **1946:** Continued under United Nations trusteeship. **1957:** British Togoland, comprising one-third of the area and situated in the west, integrated with Ghana, following a referendum. **1956:** French Togoland voted to become an autonomous republic within the French union. The new Togolese Republic achieved internal self-government **1960:** French Togoland, situated in the east, achieved full independence from France as the Republic of Togo with Sylvanus Olympio, leader of the United Togolese (UP) party, as head of state. **1967:** Lt-Gen Etienne Gnassingbé Eyadéma became president in a bloodless coup; political parties were banned. **1969:** Assembly of the Togolese People (RPT) formed by Eyadéma as the sole legal political party. **1975:** EEC Lomé convention signed in Lomé, establishing trade links with developing countries. **1977:** An assassination plot against Eyadéma, allegedly involving the Olympio family, was thwarted. **1979:** Eyadéma returned in election. Further EEC Lomé convention signed. **1986:** Attempted coup failed and situation stabilized with help of French troops. **1990:** There were casualties as violent antigovernment demonstrations in Lomé were suppressed; Eyadéma relegalized political parties. **1991:** Eyadéma was forced to call a national conference that limited the president's powers, and elected Joseph Kokou Koffigoh head of an interim government. Three attempts by Eyadéma's troops to unseat the government failed. **1992:** There were strikes in southern Togo. A referendum showed overwhelming support for multiparty politics. A new constitution was adopted. **1993:** Eyadéma won the first multiparty presidential elections amid widespread opposition. **1994:** An antigovernment coup was foiled. The opposition CAR polled strongly in assembly elections. Eyadéma appointed Edem Kodjo of the minority UTD prime minister. **1998:** President Eyadéma was re-elected. **2000:** Agbeyome Messan Kodjo was appointed prime minister.

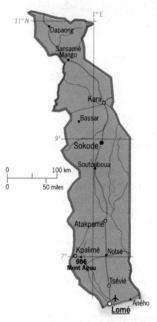

Bight of Benin

ATLANTIC
OCEAN

Practical information

Visa requirements: UK: visa not required. USA: visa not required
Time difference: GMT +/–0
Chief tourist attractions: long sandy beaches shaded by palm trees along the coast between Lomé and Cotonou (Benin)
Major holidays: 1, 13, 24 January, 24, 27 April, 1 May, 15 August, 1 November, 25 December; variable: Ascension Thursday, Eid-ul-Adha, end of Ramadan

TONGA OR FRIENDLY ISLANDS

Located in the southwest Pacific Ocean, in Polynesia.

National name: *Pule'anga Fakatu'i 'o Tonga/Kingdom of Tonga*
Area: 750 sq km/290 sq mi
Capital: Nuku'alofa (on Tongatapu island)
Major towns/cities: Neiafu, Haveloloto, Vaini, Tofoa-Koloua
Physical features: three groups of islands in southwest Pacific, mostly coral formations, but actively volcanic in west; of the 170 islands in the Tonga group, 36 are inhabited

Red represents the blood of Jesus. Effective date: c. 1862.

Government

Head of state: King Taufa'ahau Tupou IV from 1965
Head of government: Prince Ulukalala Lavaka Ata from 2000
Political system: absolutist
Political executive: absolute
Administrative divisions: five divisions comprising 23 districts
Political parties: legally none, but one prodemocracy grouping, the People's Party
Armed forces: 125-strong naval force (1995)
Conscription: military service is voluntary
Death penalty: retains the death penalty for ordinary crimes, but can be considered abolitionist in practice (last execution 1982)
Education spend: (% GNP) N/A
Health spend: (% GDP) 7.8 (1997 est)

Economy and resources

Currency: pa'anga, or Tongan dollar
GDP: (US$) 173 million (1999 est)
Real GDP growth: (% change on previous year) 2.2 (1998 est)
GNP: (US$) 172 million (1999)
GNP per capita (PPP): (US$) 4,281 (1999)
Consumer price inflation: 4.4% (1999 est)
Unemployment: 13.3% (1996)
Foreign debt: (US$) 62 million (1998)
Major trading partners: New Zealand, Japan, Australia, Fiji Islands, USA, UK
Industries: concrete blocks, small excavators, clothing, coconut oil, furniture, handicrafts, sports equipment (including small boats), brewing, sandalwood processing, tourism
Exports: vanilla beans, pumpkins, coconut oil and other coconut products, watermelons, knitted clothes, cassava, yams, sweet potatoes, footwear. Principal market: Japan 52.9% (1997)
Imports: foodstuffs, basic manufactures, machinery and transport equipment, mineral fuels. Principal source: New Zealand 29.7% (1997)
Arable land: 23.6% (1995)
Agricultural products: coconuts, copra, cassava, vanilla, pumpkins, yams, taro, sweet potatoes, watermelons, tomatoes, lemons and limes, oranges, groundnuts, breadfruit; livestock rearing (pigs, goats, poultry, and cattle); fishing

Population and society

Population: 99,000 (2000 est)
Population growth rate: 0.3% (1995–2000)
Population density: (per sq km) 131 (1999 est)
Urban population: (% of total) 38 (2000 est)
Age distribution: (% of total population) 0–14 38%, 15–59 54%, 60+ 8% (2000 est)
Ethnic groups: 98% of Tongan ethnic origin, a Polynesian group with a small mixture of Melanesian; the remainder is European and part-European
Language: Tongan (official), English
Religion: mainly Free Wesleyan Church; Roman Catholic, Anglican
Education: (compulsory years) 8
Literacy rate: 98% (men); 99% (women) (2000 est)
Labour force: 38.1% agriculture, 20.6% industry, 41.3% (1990)
Life expectancy: 68 (men); 72 (women) (1998 est)
Child mortality rate: (under 5, per 1,000 live births) 35 (1998 est)
Physicians: 1 per 2,325 people (1991)
Hospital beds: 1 per 286 people (1991)
TV sets: (per 1,000 people) 20 (1996)
Radios: (per 1,000 people) 612 (1996)
Internet users: (per 10,000 people) 101.8 (1999)

Transport

Airports: international airports: Fua'amotu (15 km/9 mi from Nuku'alofa); five domestic airstrips; total passenger km: 11 million (1995)
Railways: none
Roads: total road network: 680 km/423 mi, of which 27% paved (1996 est); passenger cars: 11 per 1,000 people (1996 est)

Chronology

c. **1000 BC:** Settled by Polynesian immigrants from the Fiji Islands. *c.*
AD 950: The legendary Aho'eitu became the first hereditary Tongan
king (Tu'i Tonga). **13th–14th centuries:** Tu'i Tonga kingdom at
the height of its power. **1643:** Visited by the Dutch navigator,
Abel Tasman. **1773:** Islands visited by British navigator Capt
James Cook, who named them the 'Friendly Islands'. **1826:**
Methodist mission established. **1831:** Tongan dynasty founded
by a Christian convert and chief of Ha'apai, Prince Taufa'ahau
Tupou, who became king 14 years later. **1845–93:** Reign of King
George Tupou I, during which the country was reunited after half a
century of civil war; Christianity was spread and a modern
constitution adopted in 1875. **1900:** Friendship ('Protectorate')
treaty signed between King George Tupou II and Britain, establishing
British control over defence and foreign affairs, but leaving internal
political affairs under Tongan control. **1918:** Queen Salote Tupou
III ascended the throne. **1965:** Queen Salote died; she was
succeeded by her son, King Taufa'ahau Tupou IV, who had been
prime minister since 1949. **1970:** Tonga achieved independence
from Britain, but remained within the Commonwealth. **1991:**
Baron Vaea was appointed prime minister. **1993:** Six
prodemocracy candidates were elected. There were calls for
reform of absolutist power. **1996:** A prodemocracy movement
led by the People's Party won a majority of the 'commoner' seats
in the legislative assembly. Prodemocracy campaigner Akilisis
Pohiva was released after a month's imprisonment. **2000:** Upon
the retirement of Prime Minister Baron Vaea, he was replaced by
Prince Ulakalala Lavaka Ata.

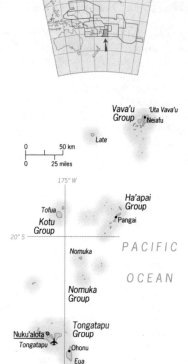

Practical information

Visa requirements: UK: visa required (issued on
arrival). USA: visa required (issued on arrival)
Time difference: GMT +13
Chief tourist attractions: mild climate; scenic
beauty; beautiful beaches; high volcanic and low coral
forms give the islands a unique character
Major holidays: 1 January, 25 April, 5 May, 4 June, 4
July, 4 November, 4, 25–26 December; variable: Good
Friday, Easter Monday

TRINIDAD AND TOBAGO

Located in the West Indies, off the coast of Venezuela.

National name: *Republic of Trinidad and Tobago*
Area: 5,130 sq km/1,980 sq mi (Trinidad 4,828 sq km/1,864 sq mi and Tobago 300 sq km/115 sq mi)
Capital: Port of Spain (and chief port)
Major towns/cities: San Fernando, Arima, Point Fortin
Major ports: Scarborough, Point Lisas
Physical features: comprises two main islands and some smaller ones in Caribbean Sea; coastal swamps and hills east–west

Red stands for the warmth of the sun and of the population. Black represents fortitude and wealth. White symbolizes purity, hope, and the waves. Effective date: 31 August 1962.

Government

Head of state: Arthur Robinson from 1997
Head of government: Basdeo Panday from 1995
Political system: liberal democracy
Political executive: parliamentary
Administrative divisions: eight counties, three municipalities, and one ward (Tobago)
Political parties: National Alliance for Reconstruction (NAR), nationalist, left of centre; People's National Movement (PNM), nationalist, moderate, centrist; United National Congress (UNC), left of centre; Movement for Social Transformation (Motion), left of centre
Armed forces: 2,600; plus a paramlitary force of 4,800 (1998)
Conscription: military service is voluntary
Death penalty: retained and used for ordinary crimes
Defence spend: (% GDP) 0.7 (1998)
Education spend: (% GNP) 3.7 (1996)
Health spend: (% GDP) 4.3 (1997 est)

Economy and resources

Currency: Trinidad and Tobago dollar
GDP: (US$) 6.99 billion (1999)
Real GDP growth: (% change on previous year) 5.2 (1999)
GNP: (US$) 5.66 billion (1999)
GNP per capita (PPP): (US$) 7,262 (1999)
Consumer price inflation: 3.4% (1999)
Unemployment: 16.2% (1997)
Foreign debt: (US$) 2.3 billion (1999)
Major trading partners: USA, Venezuela, UK, Germany, Canada, Barbados, Jamaica, Guyana, Netherlands Antilles
Resources: petroleum, natural gas, asphalt (world's largest deposits of natural asphalt)
Industries: petroleum refining, food processing, iron and steel, beverages, chemicals, cement, beer, cigarettes, motor vehicles, paper, printing and publishing, tourism (third-largest source of foreign exchange)
Exports: mineral fuels and lubricants, chemicals, basic manufactures, food. Principal market: USA 39.3% (1999)
Imports: machinery and transport equipment, manufactured goods, mineral fuel products, food and live animals, chemicals. Principal source: USA 39.8% (1999)
Arable land: 14.6% (1996)
Agricultural products: sugar cane, coffee, cocoa, citrus fruits; fishing

Population and society

Population: 1,295,000 (2000 est)
Population growth rate: 0.8% (1995–2000)
Population density: (per sq km) 251 (1999 est)
Urban population: (% of total) 74 (2000 est)
Age distribution: (% of total population) 0–14 25%, 15–59 65%, 60+ 10% (2000 est)
Ethnic groups: the two main ethnic groups are Africans (40%) and East Indians (40%); 18% are mixed, and there are also European, Afro-European, and Chinese minorities. The original Carib population has largely disappeared
Language: English (official), Hindi, French, Spanish
Religion: Roman Catholic 33%, Hindu 25%, Anglican 15%, Muslim 6%, Presbyterian 4%
Education: (compulsory years) 7
Literacy rate: 99% (men); 98% (women) (2000 est)
Labour force: 9.6% agriculture, 25.5% industry, 64.9% services (1996)
Life expectancy: 72 (men); 76 (women) (1995–2000)
Child mortality rate: (under 5, per 1,000 live births) 16 (1995–2000)
Physicians: 1 per 1,366 people (1997)
Hospital beds: 1 per 272 people (1997)
TV sets: (per 1,000 people) 332 (1997)
Radios: (per 1,000 people) 517 (1996)
Internet users: (per 10,000 people) 232.8 (1999)
Personal computer users: (per 100 people) 5.4 (1999)

Transport

Airports: international airports: Port of Spain, Trinidad (Piarco), Crown Point (near Scarborough, Tobago); domestic services between these; total passenger km: 2,790 million (1997 est)
Railways: railway service discontinued in 1968
Roads: total road network: 8,320 km/5,170 mi, of which 51.1% paved (1996 est); passenger cars: 90.1 per 1,000 people (1996 est)

Chronology

1498: Visited by the explorer Christopher Columbus, who named Trinidad after the three peaks at its southeastern tip and Tobago after the local form of tobacco pipe. Carib and Arawak Indians comprised the indigenous community. **1532:** Trinidad colonized by Spain. **1630s:** Tobago settled by Dutch, who introduced sugarcane growing. **1797:** Trinidad captured by Britain and ceded by Spain five years later under Treaty of Amiens. **1814:** Tobago ceded to Britain by France. **1834:** Abolition of slavery resulted in indentured labourers being brought in from India, rather than Africa, to work sugar plantations. **1889:** Trinidad and Tobago amalgamated as a British colony. **1956:** The People's National Movement (PNM) founded by Eric Williams, a moderate nationalist. **1958–62:** Part of West Indies Federation. **1959:** Achieved internal self-government, with Williams as chief minister. **1962:** Independence achieved within Commonwealth, with Williams as prime minister. **1970:** Army mutiny and violent Black Power riots directed against minority East Indian population; state of emergency imposed for two years. **1976:** Became a republic, with former Governor General Ellis Clarke as president and Williams as prime minister. **1986:** Tobago-based National Alliance for Reconstruction (NAR), headed by A N R Robinson, won the general election. **1990:** An attempted antigovernment coup by Islamic fundamentalists was foiled. **1991:** A general election resulted in victory for PNM, with Patrick Manning as prime minister. **1995:** The UNC and PNM tied in general election; a UNC–NAR coalition was formed, led by Basdeo Panday. **1997:** Former Prime Minister Robinson was elected president. **2000:** The UNC won an absolute majority in parliamentary elections.

Practical information

Visa requirements: UK: visa not required for a stay of up to three months. USA: visa not required for a stay of up to three months
Time difference: GMT –4
Chief tourist attractions: sunny climate; attractive coastline (especially Tobago); the annual pre-Lenten carnival
Major holidays: 1 January, 19 June, 1, 31 August, 24 September, 25–26 December; variable: Corpus Christi, Good Friday, Easter Monday, Whit Monday

TUNISIA

Located in North Africa, on the Mediterranean Sea, bounded southeast by Libya and west by Algeria.

National name: *Al-Jumhuriyya at-Tunisiyya/Tunisian Republic*
Area: 164,150 sq km/63,378 sq mi
Capital: Tunis (and chief port)
Major towns/cities: Sfax, Ariana, Bizerte, Gabès, Sousse, Kairouan, Ettadhamen
Major ports: Sfax, Sousse, Bizerte
Physical features: arable and forested land in north graduates towards desert in south; fertile island of Jerba, linked to mainland by causeway (identified with island of lotus-eaters); Shott el Jerid salt lakes

The flag was introduced by Hassan II, the Bey of Tunisia. Red is an Islamic colour.
Effective date: c. 1835.

Government

Head of state: Zine el-Abidine Ben Ali from 1987
Head of government: Muhammad Ghannouchi from 1999
Political system: nationalistic socialist
Political executive: unlimited presidency
Administrative divisions: 23 governates
Political parties: Constitutional Democratic Rally (RCD), nationalist, moderate, socialist; Popular Unity Movement (MUP), radical, left of centre; Democratic Socialists Movement (MDS), left of centre; Renovation Movement (MR), reformed communists
Armed forces: 35,000; plus paramilitary forces numbering 12,000 (1998)
Conscription: military service is by selective conscription for 12 months
Death penalty: retained and used for ordinary crimes
Defence spend: (% GDP) 1.8 (1998)
Education spend: (% GNP) 7.7 (1997)
Health spend: (% GDP) 3.0 (1990–95)

Economy and resources

Currency: Tunisian dinar
GDP: (US$) 21.18 billion (1999)
Real GDP growth: (% change on previous year) 6.2 (1999)
GNP: (US$) 19.9 billion (1999)
GNP per capita (PPP): (US$) 5,478 (1999)
Consumer price inflation: 2.7% (1999)
Unemployment: 15% (1998 est)
Foreign debt: (US$) 11.3 billion (1999 est)
Major trading partners: France, Italy, Germany, Belgium, USA, Spain, the Netherlands, UK, Libya, Japan
Resources: petroleum, natural gas, phosphates, iron, zinc, lead, aluminium fluoride, fluorspar, sea salt
Industries: processing of agricultural and mineral products (including superphosphate and phosphoric acid), textiles and clothing, machinery, chemicals, paper, wood, motor vehicles, radio and television sets, tourism
Exports: textiles and clothing, crude petroleum, phosphates and fertilizers, olive oil, fruit, leather and shoes, fishery products, machinery and electrical appliances. Principal market: Germany 28% (1999)
Imports: machinery, textiles, food (mainly cereals, dairy produce, meat, and sugar) and live animals, petroleum and

petroleum products. Principal source: France: 23% (1999)
Arable land: 18.3 (1996)
Agricultural products: wheat, barley, olives, citrus fruits, dates, almonds, grapes, melons, apples, apricots and other fruits, chickpeas, sugar beet, tobacco; fishing

Population and society

Population: 9,586,000 (2000 est)
Population growth rate: 1.4% (1995–2000)
Population density: (per sq km) 58 (1999 est est)
Urban population: (% of total) 66 (2000 est)
Age distribution: (% of total population) 0–14 30%, 15–59 62%, 60+ 8% (2000 est)
Ethnic groups: about 10% of the population is Arab; the remainder are of Berber-Arab descent. There are small Jewish and French communities
Language: Arabic (official), French
Religion: Sunni Muslim (state religion); Jewish and Christian minorities
Education: (compulsory years) 9
Literacy rate: 81% (men); 61% (women) (2000 est)
Labour force: 21.6% agriculture, 34.4% industry, 44% services (1994)
Life expectancy: 68 (men); 71 (women) (1995–2000)
Child mortality rate: (under 5, per 1,000 live births) 37 (1995–2000)
Physicians: 1 per 1,549 people (1993 est)
Hospital beds: 1 per 350 people (1993 est)
TV sets: (per 1,000 people) 198 (1998)
Radios: (per 1,000 people) 223 (1997)
Internet users: (per 10,000 people) 31.7 (1999)
Personal computer users: (per 100 people) 1.5 (1999)

Transport

Airports: international airports: Tunis (Carthage), Monastir (Skanes), Djerba (Melita), Sfax, Tozeur (Nefta), Tabarka; domestic services operate between these; total passenger km: 2,539 million (1997 est)
Railways: total length: 2,162 km/1,343 mi; total passenger km: 1,022 million (1997)
Roads: total road network: 23,100 km/14,354 mi, of which 78.9% paved (1996 est); passenger cars: 29.8 per 1,000 people (1996 est)

Chronology

814 BC: Phoenician emigrants from Tyre, in Lebanon, founded Carthage, near modern Tunis, as a trading post. By 6th century BC Carthaginian kingdom dominated western Mediterranean. **146 BC:** Carthage destroyed by Punic Wars with Rome, which began in 264 BC; Carthage became part of Rome's African province. **AD 533:** Came under control of Byzantine Empire. **7th century:** Invaded by Arabs, who introduced Islam. Succession of Islamic dynasties followed, including Aghlabids (9th century), Fatimids (10th century), and Almohads (12th century). **1574:** Became part of Islamic Turkish Ottoman Empire and a base for 'Barbary Pirates' who operated against European shipping until 19th century. **1705:** Husayn Bey founded local dynasty, which held power under rule of Ottomans. **early 19th century:** Ahmad Bey launched programme of economic modernization, which nearly bankrupted the country. **1881:** Became French protectorate, with bey retaining local power. **1920:** Destour (Constitution) Party, named after the original Tunisian constitution of 1861, founded to campaign for equal Tunisian participation in French-dominated government. **1934:** Habib Bourguiba founded a radical splinter party, the Neo-Destour Party, to spearhead the nationalist movement. **1942–43:** Brief German occupation during World War II. **1956:** Independence achieved as monarchy under bey, with Bourguiba as prime minister. **1957:** Bey deposed; Tunisia became one-party republic with Bourguiba as president. **1975:** Bourguiba made president for life. **1979:** Headquarters for Arab League moved to Tunis after Egypt signed Camp David Accords with Israel. **1981:** Multiparty elections held, as a sign of political liberalization, but were won by Bourguiba's Destourian Socialist Party (DSP). **1982:** Allowed Palestine Liberation Organization (PLO) to use Tunis for its headquarters. **1985:** Diplomatic relations with Libya severed; Israel attacked PLO headquarters. **1987:** Zine el-Abidine Ben Ali, the new prime minister, declared Bourguiba (now aged 84) incompetent for government and seized power as president. **1988:** 2,000 political prisoners freed; privatization initiative. Diplomatic relations with Libya restored. DSP renamed RCD. **1990:** The Arab League's headquarters returned to Cairo, Egypt. **1991:** There was opposition to US actions during the Gulf War, and a crackdown on religious fundamentalists. **1992:** Human-rights transgressions provoked Western criticism. **1994:** Ben Ali and the RCD were re-elected. The PLO transferred its headquarters to Gaza City in Palestine. **1999:** In the country's first ever 'competitive' presidential elections, Ben Ali was re-elected president. Muhammad Ghannouchi was elected prime minister.

Practical information

Visa requirements: UK: visa not required. USA: visa not required
Time difference: GMT +1
Chief tourist attractions: Moorish architecture; Roman remains; the ancient Phoenician city of Carthage; sandy beaches
Major holidays: 1, 18 January, 20 March, 9 April, 1 May, 1–2 June, 25 July, 3, 13 August, 3 September, 15 October; variable: Eid-ul-Adha (2 days), end of Ramadan (2 days), New Year (Muslim), Prophet's Birthday

TURKEY

Located between the Black Sea to the north and the Mediterranean Sea to the south, bounded to the east by Armenia, Georgia, and Iran, to the southeast by Iraq and Syria, to the west by Greece and the Aegean Sea, and to the northwest by Bulgaria.

National name: *Türkiye Cumhuriyeti/Republic of Turkey*
Area: 779,500 sq km/300,964 sq mi
Capital: Ankara
Major towns/cities: Istanbul, Izmir, Adana, Bursa, Gaziantep, Konya, Mersin, Antalya, Diyarbakduringr
Major ports: Istanbul and Izmir

The star, which was added to the flag in 1793, initially had more than five points. The star may represent the Morning Star mentioned in the Koran. Effective date: 5 June 1936.

Physical features: central plateau surrounded by mountains, partly in Europe (Thrace) and partly in Asia (Anatolia); Bosporus and Dardanelles; Mount Ararat (highest peak Great Ararat, 5,137 m/16,854 ft); Taurus Mountains in southwest (highest peak Kaldi Dag, 3,734 m/12,255 ft); sources of rivers Euphrates and Tigris in east

Government

Head of state: Ahmet Necdet Sezer from 2000
Head of government: Bülent Ecevit from 1999
Political system: liberal democracy
Political executive: parliamentary
Administrative divisions: 80 provinces
Political parties: Motherland Party (ANAP), Islamic, nationalist, right of centre; Republican People's Party (CHP), left of centre; True Path Party (DYP), right of centre, pro-Western; Virtue Party (FP), Islamic fundamentalist
Armed forces: 639,000 (1998)
Conscription: 18 months
Death penalty: retained for ordinary crimes, but considered abolitionist in practice; last execution in 1984
Defence spend: (% GDP) 4.4 (1998)
Education spend: (% GNP) 2.2 (1996)
Health spend: (% GDP) 4 (1997)

Economy and resources

Currency: Turkish lira
GDP: (US$) 188.4 billion (1999)
Real GDP growth: (% change on previous year) –5.1 (1999)
GNP: (US$) 186.3 billion (1999)
GNP per capita (PPP): (US$) 6,126 (1999)
Consumer price inflation: 65.1% (1999)
Unemployment: 6.3% (1998)
Foreign debt: (US$) 102.1 billion (1999)
Major trading partners: Germany, USA, Italy, France, Saudi Arabia, UK, Russia
Resources: chromium, copper, mercury, antimony, borax, coal, petroleum, natural gas, iron ore, salt
Industries: textiles, food processing, petroleum refining, coal, iron and steel, industrial chemicals, tourism
Exports: textiles and clothing, agricultural products and foodstuffs (including figs, nuts, and dried fruit), tobacco, leather, glass, refined petroleum and petroleum products. Principal market: Germany 20.6% (1999)
Imports: machinery, construction material, motor vehicles, consumer goods, crude petroleum, iron and steel, chemical products, fertilizer, livestock. Principal source: Germany 14.5% (1999)
Arable land: 31.8% (1996)

Agricultural products: barley, wheat, maize, sunflower and other oilseeds, sugar beet, potatoes, tea (world's fifth-largest producer), olives, fruits, tobacco

Population and society

Population: 66,591,000 (2000 est)
Population growth rate: 1.7% (1995–2000)
Population density: (per sq km) 84 (1999 est)
Urban population: (% of total) 75 (2000 est)
Age distribution: (% of total population) 0–14 28%, 15–59 63%, 60+ 9% (2000 est)
Ethnic groups: over 90% of the population are Turks, although only about 5% are of Turkic or Western Mongoloid descent; most are descended from earlier conquerors, such as the Greeks; about 8% Kurds
Language: Turkish (official), Kurdish, Arabic
Religion: Sunni Muslim 99%; Orthodox, Armenian churches
Education: (compulsory years) 5
Literacy rate: 93% (men); 77% (women) (2000 est)
Labour force: 36.3% of population: 44.9% agriculture, 22% industry, 33.1% services (1996)
Life expectancy: 67 (men); 72 (women) (1995–2000)
Child mortality rate: (under 5, per 1,000 live births) 60 (1995–2000)
Physicians: 1 per 830 people (1996)
Hospital beds: 1 per 400 people (1995)
TV sets: (per 1,000 people) 286 (1997)
Radios: (per 1,000 people) 180 (1997)
Internet users: (per 10,000 people) 220.0 (1999)
Personal computer users: (per 100 people) 3.2 (1999)

Transport

Airports: international airports: Ankara (Esenboga), Istanbul (Atatürk), Izmir (Adnan Menderes), Adana, Trabzon, Dalaman, Antalya; 15 domestic airports; total passenger km: 12,379 million (1997 est)
Railways: total length: 10,386 km/6,454 mi; total passenger km: 6,335 million (1994)
Roads: total road network: 382,397 km/237,621 mi, of which 25% paved (1997); passenger cars: 57.7 per 1,000 people (1997)

Chronology

1st century BC: Asia Minor became part of Roman Empire, later passing to Byzantine Empire. **6th century AD:** Turkic peoples spread from Mongolia into Turkestan, where they adopted Islam. **1055:** Seljuk Turks captured Baghdad; their leader Tughrul took the title of sultan. **1071:** Battle of Manzikert: Seljuk Turks defeated Byzantines and conquered Asia Minor. **13th century:** Ottoman Turks, driven west by Mongols, became vassals of Seljuk Turks. *c.* **1299:** Osman I founded small Ottoman kingdom, which quickly displaced Seljuks to include all Asia Minor. **1354:** Ottoman Turks captured Gallipoli and began their conquests in Europe. **1389:** Battle of Kossovo: Turks defeated Serbs to take control of most of Balkan peninsula. **1453:** Constantinople, capital of Byzantine Empire, fell to the Turks; became capital of Ottoman Empire as Istanbul. **16th century:** Ottoman Empire reached its zenith under Suleiman the Magnificent 1520–66; Turks conquered Egypt, Syria, Arabia, Mesopotamia, Tripoli, Cyprus, and most of Hungary. **1683:** Failure of Siege of Vienna marked the start of the decline of the Ottoman Empire. **1699:** Treaty of Karlowitz: Turks forced out of Hungary by Austrians. **1774:** Treaty of Kuchuk Kainarji: Russia drove Turks from Crimea and won the right to intervene on behalf of Christian subjects of the sultan. **19th century:** 'The Eastern Question': Ottoman weakness caused intense rivalry between powers to shape future of Near East. **1821–29:** Greek war of independence: Greeks defeated Turks with help of Russia, Britain, and France. **1854–56:** Crimean War: Britain and France fought to defend Ottoman Empire from further pressure by Russians. **1877–78:** Russo-Turkish War ended with Treaty of Berlin and withdrawal of Turks from Bulgaria. **1908:** Young Turk revolution forced sultan to grant constitution; start of political modernization. **1911–12:** Italo-Turkish War: Turkey lost Tripoli (Libya). **1912–13:** Balkan War: Greece, Serbia, and Bulgaria expelled Turks from Macedonia and Albania. **1914:** Ottoman Empire entered World War I on German side. **1919:** Following Turkish defeat, Mustapha Kemal launched nationalist revolt to resist foreign encroachments. **1920:** Treaty of Sèvres partitioned Ottoman Empire, leaving no part of Turkey fully independent. **1922:** Kemal, having defied Allies, expelled Greeks, French, and Italians from Asia Minor; sultanate abolished. **1923:** Treaty of Lausanne recognized Turkish independence; secular republic established by Kemal, who imposed rapid Westernization. **1935:** Kemal adopted surname Atatürk ('Father of the Turks'). **1938:** Death of Kemal Ataturk; succeeded as president by Ismet Inönü. **1950:** First free elections won by opposition Democratic Party; Adnan Menderes became prime minister. **1952:** Turkey became a member of NATO. **1960:** Military coup led by Gen Cemal Gürsel deposed Menderes, who was executed in 1961. **1961:** Inönü returned as prime minister; politics dominated by the issue of Cyprus. **1965:** Justice Party came to power under Suleyman Demirel. **1971–73:** Prompted by strikes and student unrest, the army imposed military rule. **1974:** Turkey invaded northern Cyprus. **1980–83:** Political violence led to further military rule. **1984:** Kurds began guerrilla war in a quest for greater autonomy. **1989:** Application to join European Community rejected. **1990–91:** Turkey joined the UN coalition against Iraq in the Gulf War. **1995:** Turkish offensives against Kurdish bases in northern Iraq. **1997:** Plans were agreed for the curbing of Muslim fundamentalism. Mesut Yilmaz was appointed prime minister. An agreement was reached with Greece on the peaceful resolution of disputes. **1998:** The Islamic Welfare Party (RP) was banned by Constitutional Court, and regrouped as the Virtue Party (FP). **1999:** Bülent Ecevit became prime minister. Ecevit's ruling centre-left party won the majority of seats in the general election. Turkey suffered two devastating earthquakes, causing extensive loss of life and structural damage. Turkey declared an EU candidate, but to become a full member would first have to settle its territorial dispute with Greece and satisfy EU human rights regulations. **2000:** Ahmet Necdet Sezer was inaugurated as president. He urged reform to push Turkey closer to EU membership and overruled a decree that had allowed the government to dismiss bureaucrats deemed to be too pro-Kurdish or insufficiently secular.

Practical information

Visa requirements: UK/USA: visa not required for a stay of up to three months
Time difference: GMT +3
Chief tourist attractions: sunny climate; fine beaches; ancient monuments; historic Istanbul, with its 15th-century Topkapi Palace, 6th-century Hagia Sophia basilica, Blue Mosque, mosque of Suleiman the Magnificent, covered bazaars, and Roman cisterns
Major holidays: 1 January, 23 April, 19 May, 30 August, 29 October; variable: Eid-ul-Adha (4 days), end of Ramadan (3 days)

TURKMENISTAN

Located in central Asia, bounded north by Kazakhstan and Uzbekistan, west by the Caspian Sea, and south by Iran and Afghanistan.

National name: *Türkmenistan/Turkmenistan*
Area: 488,100 sq km/188,455 sq mi
Capital: Ashgabat
Major towns/cities: Chardzhev, Mary, Nebitdag, Dashkhowuz, Turkmenbashi
Major ports: Turkmenbashi
Physical: about 90% of land is desert including the Karakum 'Black Sands' desert (area 310,800 sq km/120,000 sq mi)

The stars represent the five regions of Turkmenistan. The crescent symbolizes Islam. Effective date: 19 February 1997.

Government

Head of state and government: Saparmurad Niyazov from 1990
Political system: authoritarian nationalist
Political executive: unlimited presidency
Administrative divisions: five regions
Political parties: Democratic Party of Turkmenistan, ex-communist, pro-Niyazov; Turkmen Popular Front (Agzybirlik), nationalist
Armed forces: 19,000 (1998)
Conscription: military service is compulsory for 18 months
Death penalty: retained and used for ordinary crimes
Defence spend: (% GDP) 2.8 (1998)
Education spend: (% GNP) N/A
Health spend: (% GDP) 1.2 (1995)

Economy and resources

Currency: manat
GDP: (US$) 2.7 billion (1999)
Real GDP growth: (% change on previous year) 16 (1999 est)
GNP: (US$) 3.2 billion (1999)
GNP per capita (PPP): (US$) 3,099 (1999)
Consumer price inflation: 24.1% (1999 est)
Unemployment: 5% (1996 est)
Foreign debt: (US$) 2.5 billion (1999 est)
Major trading partners: Ukraine, Turkey, Russia, Iran, Italy, United Arab Emirates, Switzerland, France
Resources: petroleum, natural gas, coal, sulphur, magnesium, iodine-bromine, sodium sulphate and different types of salt
Industries: mining, petroleum refining, energy generation, textiles, chemicals, cement, mineral fertilizer, footwear
Exports: natural gas, cotton yarn, electric energy, petroleum and petroleum products. Principal market: Ukraine 27% (1999)
Imports: machinery and metalwork, light industrial products, processed food, agricultural

products. Principal source: Turkey 17% (1999)
Arable land: 3.1% (1996)

Population and society

Population: 4,459,000 (2000 est)
Population growth rate: 1.8% (1995–2000)
Population density: (per sq km) 9 (1999 est)
Urban population: (% of total) 45 (2000 est)
Age distribution: (% of total population) 0–14 38%, 15–59 56%, 60+ 6% (2000 est)
Ethnic groups: 77% ethnic Turkmen, 7% ethnic Russian, 9% Uzbek, 3% Kazakh, 1% Ukrainian, Armenian, Azeri, and Tartar
Language: Turkmen (a Turkic language; official), Russian, Uzbek, other regional languages
Religion: Sunni Muslim
Education: (compulsory years) 9
Literacy rate: 99% (men); 98% (women) (2000 est)
Labour force: 44% agriculture, 19% industry, 37% services (1996)
Life expectancy: 62 (men); 69 (women) (1995–2000)
Child mortality rate: (under 5, per 1,000 live births) 77 (1995–2000)
Physicians: 1 per 311 people (1994)
Hospital beds: 1 per 95 people (1994)
TV sets: (per 1,000 people) 201 (1998)
Radios: (per 1,000 people) 276 (1997)
Internet users: (per 10,000 people) 4.6 (1999)

Transport

Airports: international airports: Ashgabat; three domestic airports; total passenger km: 1,562 million (1995)
Railways: total length: 2,187 km/1,359 mi; total passenger km: 2,104 million (1996)
Roads: total road network: 24,000 km/14,914 mi, of which 81.2% paved (1996 est)

Chronology

6th century BC: Part of the Persian Empire of Cyrus the Great.
4th century BC: Part of the empire of Alexander the Great of
Macedonia. **7th century:** Spread of Islam into Transcaspian
region, followed by Arab rule from 8th century. **10th–13th
centuries:** Immigration from northeast by nomadic
Oghuz Seljuk and Mongol tribes, whose Turkic-speaking
descendants now dominate the country; conquest
by Genghis Khan. **16th century:** Came under
dominance of Persia, to the south. **1869–81:**
Fell under control of tsarist Russia after
150,000 Turkmen were killed in Battle of
Gok Tepe in 1881; became part of Russia's
Turkestan Governor-Generalship. **1916:**
Turkmen revolted violently against Russian
rule; autonomous Transcaspian government
formed after Russian Revolution of 1917.
1919: Brought back under Russian control
following invasion by the Soviet Red Army.
1921: Part of Turkestan Soviet Socialist
Autonomous Republic. **1925:** Became
constituent republic of USSR. **1920s–30s:** Soviet
programme of agricultural collectivization and
secularization provoked sporadic guerrilla resistance and
popular uprisings. **1960–67:** Lenin Kara-Kum Canal built, leading to dramatic
expansion in cotton production in previously semidesert region. **1985:** Saparmurad
Niyazov replaced Muhammad Gapusov, local communist leader since 1971, whose regime had been viewed
as corrupt. **1989:** Stimulated by the *glasnost* initiative of reformist Soviet leader Mikhail Gorbachev, Agzybirlik
'popular front' formed by Turkmen intellectuals. **1990:** Economic and political sovereignty was declared. Niyazov
was elected state president. **1991:** Niyazov initially supported an attempted anti-Gorbachev coup in Moscow.
Independence was later declared; Turkmenistan joined the new Commonwealth of Independent States (CIS). **1992:**
Joined the Muslim Economic Cooperation Organization and the United Nations; a new constitution was adopted.
1993: A new currency, the manat, was introduced and a programme of cautious economic reform introduced, with
foreign investment in the country's huge oil and gas reserves encouraged. The economy continued to contract. **1994:**
A nationwide referendum overwhelmingly backed Niyazov's presidency. Ex-communists won most seats in
parliamentary elections. **1997:** Private land ownership was legalized.

Practical information

Visa requirements: UK: visa required. USA: visa
required
Time difference: GMT +5
Chief tourist attractions: ruins of the 12th-century
Seljuk capital at Merv (now Mary); hot springs at
Bacharden, on the Iranian border; the bazaar at
Ashgabat
Major holidays: 1, 12 January, 19, 22 February, 8
March, 29 April, 9, 18 May, 27–28 October

TUVALU FORMERLY ELLICE ISLANDS (UNTIL 1978)

Located in the southwest Pacific Ocean.

National name: *Fakavae Aliki-Malo i Tuvalu/Constitutional Monarchy of Tuvalu*
Area: 25 sq km/9.6 sq mi
Capital: Fongafale (on Funafuti atoll)
Physical features: nine low coral atolls forming a
chain of 579 km/650 mi in the Southwest Pacific

The Union Jack signifies the islands'
wish to preserve links with Britain. The
nine stars representing the islands are
placed according to their locations.
Effective date: 11 April 1997.

Government

Head of state: Queen Elizabeth II from 1978, represented
by Governor General Tulaga Manuella from 1994
Head of government: Lagitupu Tuilimu from 2000
Political system: liberal democracy
Political executive: parliamentary
Political parties: none; members are elected to
parliament as independents
Death penalty: laws do not provide for the death penalty
for any crime

Economy and resources

Currency: Australian dollar
GDP: (US$) 7.8 million (1998)
Real GDP growth: (% change on previous year) 2.6 (1998 est)
GNP per capita (PPP): (US$) 970 (1998 est)
Consumer price inflation: 0.8% (1998)
Major trading partners: Australia, Fiji, New Zealand, EU
Industries: processing of agricultural products (principally
coconuts), soap, handicrafts, tourism; a large source of
income is from Tuvaluans working abroad, especially in the
phosphate industry on Nauru
Exports: copra, handicrafts, garments, stamps, fisheries

licences. Principal market: Australia
Imports: food and live animals, beverages, tobacco,
consumer goods, machinery and transport equipment,
mineral fuels. Principal source: Australia
Agricultural products: coconuts, pulaka, taro, papayas, screw-
pine (pandanus), bananas; livestock (pigs, poultry, and goats)

Population and society

Population: 12,000 (2000 est)
Population growth rate: 0.9% (1995–2025)
Population density: (per sq km) 423 (1999 est)
Urban population: (% of total) 51 (1999)
Age distribution: (% of total population) 0–14 35%,
15–59 60%, 60+ 5% (1999 est)
Ethnic groups: almost entirely of Polynesian origin
Language: Tuvaluan, English (both official), a Gilbertese
dialect (on Nui)
Religion: Protestant 96% (Church of Tuvalu)
Education: (compulsory years) 9
Literacy rate: 93% (1997 est)
Life expectancy: 63 (men); 66 (women) (1998 est)
Child mortality rate: (under 5, per 1,000 live births) 29
(1997 est)

Chronology

c. **300 BC:** First settled by Polynesian peoples. **16th
century:** Invaded and occupied by Samoans. **1765:** Islands
first reached by Europeans. **1850–75:** Population decimated
by European slave traders capturing Tuvaluans to work in
South America and by exposure to European diseases. **1856:**
The four southern islands, including Funafuti, claimed by
USA. **1865:** Christian mission established. **1877:** Came
under control of British Western Pacific High Commission
(WPHC), with its headquarters in the Fiji Islands. **1892:**
Known as the Ellice Islands, they were joined with Gilbert
Islands (now Kiribati) to form a British protectorate. **1916:**
Gilbert and Ellice Islands colony formed. **1942–43:** Became
a base for US airforce operations when Japan occupied the
Gilbert Islands during World War II. **1975:** Following a
referendum, the predominantly Melanesian-peopled Ellice
Islands were granted separate status. **1978:** Independence
achieved within Commonwealth, with Toaripi Lauti as prime
minister; reverted to former name Tuvalu ('eight standing
together'). **1979:** The USA signed a friendship treaty,
relinquishing its claim to the four southern atolls in return for
continued access to military bases. **1986:** Islanders rejected
proposal for republican status. **2000:** Tuvalu entered the
United Nations.

Practical information

Visa requirements: UK: visa not required. USA: visa
required
Time difference: GMT +12
Chief tourist attractions: Funafuti lagoon; sandy
beaches; development of tourism has been limited
due to Tuvalu's remote location and lack of amenities

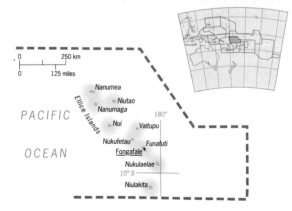

UGANDA

Landlocked country in East Africa, bounded north by Sudan, east by Kenya, south by Tanzania and Rwanda, and west by the Democratic Republic of Congo (formerly Zaire).

National name: *Republic of Uganda*
Area: 236,600 sq km/91,351 sq mi
Capital: Kampala
Major towns/cities: Jinja, Mbale, Entebbe, Masaka, Mbarara, Soroti
Physical features: plateau with mountains in west (Ruwenzori Range, with Mount Margherita, 5,110 m/16,765 ft); forest and grassland; 18% is lakes, rivers, and wetlands (Owen Falls on White Nile where it leaves Lake Victoria; Lake Albert in west); arid in northwest

Red symbolizes the brotherhood of man. Black represents the African people. Yellow stands for sunshine. Effective date: 9 October 1962.

Government

Head of state: Yoweri Museveni from 1986
Head of government: Apolo Nsibambi from 1999
Political system: authoritarian nationalist
Political executive: unlimited presidency
Administrative divisions: 39 districts
Political parties: National Resistance Movement (NRM), left of centre; Democratic Party (DP), left of centre; Conservative Party (CP), right of centre; Uganda People's Congress (UPC), left of centre; Uganda Freedom Movement (UFM), left of centre. From 1986, political parties were forced to suspend activities
Armed forces: 40,000 (1998)
Conscription: military service is voluntary
Death penalty: retained and used for ordinary crimes
Defence spend: (% GDP) 3.1 (1998)
Education spend: (% GNP) 2.6 (1996)
Health spend: (% GDP) 4.1 (1997 est)

Economy and resources

Currency: Ugandan new shilling
GDP: (US$) 6.35 billion (1999)
Real GDP growth: (% change on previous year) 4.6 (1999)
GNP: (US$) 6.8 billion (1999)
GNP per capita (PPP): (US$) 1,136 (1999 est)
Consumer price inflation: 6.4% (1999)
Foreign debt: (US$) 3.6 billion (1999 est)
Major trading partners: Kenya, Spain, UK, Germany, the Netherlands, USA, France
Resources: copper, apatite, limestone; believed to possess the world's second-largest deposit of gold (hitherto unexploited); also reserves of magnetite, tin, tungsten, beryllium, bismuth, asbestos, graphite
Industries: processing of agricultural products, brewing, vehicle assembly, textiles, cement, soap, fertilizers, footwear, metal products, paints, batteries, matches
Exports: coffee, cotton, tea, tobacco, oil seeds and oleaginous fruit; hides and skins, textiles. Principal market: Spain 11.4% (1999 est)
Imports: machinery and transport equipment, basic manufactures, petroleum and petroleum products, chemicals, miscellaneous manufactured articles, iron and steel. Principal source: Kenya 27.5% (1999 est)

Arable land: 25.3% (1996)
Agricultural products: coffee, cotton, tea, maize, tobacco, sugar cane, cocoa, horticulture, plantains, cassava, sweet potatoes, millet, sorghum, beans, groundnuts, rice; livestock rearing (cattle, goats, sheep, and poultry); freshwater fishing

Population and society

Population: 21,778,000 (2000 est)
Population growth rate: 2.8% (1995–2000)
Population density: (per sq km) 89
Urban population: (% of total) 14 (2000 est)
Age distribution: (% of total population) 0–14 50%, 15–59 47%, 60+ 3% (2000 est)
Ethnic groups: about 40 different peoples concentrated into three main groups; the Bantu (the most numerous), the Nilotics, and the Nilo-Hamites; there are also Rwandan, Sudanese, Zairean, and Kenyan minorities
Language: English (official), Kiswahili, other Bantu and Nilotic languages
Religion: Christian 65%, animist 20%, Muslim 15%
Education: not compulsory
Literacy rate: 78% (men); 57% (women) (2000 est)
Labour force: 82% agriculture, 6% industry, 12% services (1997 est)
Life expectancy: 39 (men); 40 (women) (1995–2000)
Child mortality rate: (under 5, per 1,000 live births) 173 (1995–2000)
Physicians: 1 per 22,399 people (1993 est)
Hospital beds: 1 per 760 people (1993 est)
TV sets: (per 1,000 people) 27 (1997)
Radios: (per 1,000 people) 128 (1997)
Internet users: (per 10,000 people) 11.8 (1999)
Personal computer users: (per 100 people) 0.3 (1999)

Transport

Airports: international airports: Entebbe; domestic services operate to all major towns; total passenger km: 103 million (1995)
Railways: total length: 1,241 km/771 mi; total passenger km: 28 million (1996)
Roads: total road network: 26,800 km/16,654 mi, of which 7.7% paved (1995 est); passenger cars: 1.8 per 1,000 people (1996 est)

Chronology

16th century: Bunyoro kingdom founded by immigrants from southeastern Sudan. **17th century:** Rise of kingdom of Buganda people, which became particularly powerful from 17th century. **mid-19th century:** Arabs, trading ivory and slaves, reached Uganda; first visits by European explorers and Christian missionaries. **1885–87:** Uganda Martyrs: Christians persecuted by Buganda ruler, Mwanga. **1890:** Royal Charter granted to British East African Company, a trading company whose agent, Frederick Lugard, concluded treaties with local rulers, including the Buganda and the western states of Ankole and Toro. **1894:** British protectorate established, with Buganda retaining some autonomy under its traditional prince (Kabaka) and other resistance being crushed. **1904:** Cotton growing introduced by Buganda peasants. **1958:** Internal self-government granted. **1962:** Independence achieved from Britain, within Commonwealth, with Milton Obote of Uganda People's Congress (UPC) as prime minister. **1963:** Proclaimed federal republic with King Mutesa II (of Buganda) as president and Obote as prime minister. **1966:** King Mutesa, who opposed creation of a one-party state, ousted in coup led by Obote, who ended federal status and became executive president. **1969:** All opposition parties banned after assassination attempt on Obote; key enterprises nationalized. **1971:** Obote overthrown in army coup led by Maj-Gen Idi Amin Dada; constitution suspended and ruthlessly dictatorial regime established; nearly 49,000 Ugandan Asians expelled; over 300,000 opponents of regime killed. **1976:** Relations with Kenya strained by Amin's claims to parts of Kenya. **1979:** After annexing part of Tanzania, Amin forced to leave the country by opponents backed by Tanzanian troops. Provisional government set up. **1978–79:** Fighting broke out against Tanzanian troops. **1980:** Provisional government overthrown by army. Elections held and Milton Obote returned to power. **1985:** After opposition by pro-Lule National Resistance Army (NRA), and indiscipline in army, Obote ousted by Gen Tito Okello; constitution suspended; power-sharing agreement entered into with NRA leader Yoweri Museveni. **1986:** Museveni became president, heading broad-based coalition government. **1993:** The King of Buganda was reinstated as formal monarch, in the person of Ronald Muwenda Mutebi II. **1996:** A landslide victory was won by Museveni in the first direct presidential elections. **1997:** Allied Democratic Forces (ADF) led uprisings by rebels. **1999:** The leaders of Uganda and Sudan signed an agreement to bring an end to rebel wars across their mutual border by ceasing to support rebel factions in the other's country. President Museveni appointed Apolo Nsibambi as prime minister. **2000:** Rebels attacked towns in northern Uganda and fought the Ugandan army along the border with Congo. A fire at the headquarters of the Restoration of the Ten Commandments of God cult in western Uganda killed up to 500 people. It was later discovered the cult leaders had engaged in mass murder. An outbreak of the ebola virus killed 160 people between September and December.

Practical information

Visa requirements: UK: visa not required. USA: visa not required
Time difference: GMT +3
Chief tourist attractions: good year-round climate; lakes; forests; wildlife; varied scenery includes tropical forest and tea plantations on the slopes of the snow-capped Ruwenzori Mountains and the arid plains of the Karamoja
Major holidays: 1 January, 1 April, 1 May, 9 October, 25–26 December; variable: Good Friday, Easter Monday, Holy Saturday, end of Ramadan

Ukraine

Located in eastern central Europe, bounded to the east by Russia, north by Belarus, south by Moldova, Romania, and the Black Sea, and west by Poland, the Slovak Republic, and Hungary.

National name: *Ukrayina/Ukraine*
Area: 603,700 sq km/233,088 sq mi
Capital: Kiev
Major towns/cities: Kharkiv, Donetsk, Dnipropetrovs'k, Lviv, Kryvyy Rih, Zaporozhye, Odessa
Physical features: Russian plain; Carpathian and Crimean Mountains; rivers: Dnieper (with the Dnieper dam 1932), Donetz, Bug

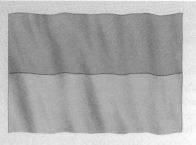

The national colours are taken from the Rusyn arms of 1848. Effective date: 28 January 1992.

Government

Head of state: Leonid Kuchma from 1994
Head of government: Viktor Yushchenko from 1999
Political system: emergent democracy
Political executive: limited presidency
Administrative divisions: 24 provinces, one autonomous republic (Crimea), and two metropolitan areas (Kiev and Sevastopol)
Political parties: Ukrainian Communist Party (UCP), left wing, anti-nationalist (banned 1991–93); Peasants' Party of the Ukraine (PPU), conservative agrarian; Ukrainian Socialist Party (SPU), left wing, anti-nationalist; Ukrainian People's Movement (Rukh); Ukrainian Republican Party (URP), moderate nationalist; Congress of Ukrainian Nationalists (CUN), moderate nationalist; Democratic Party of Ukraine (DPU), moderate nationalist; Social Democratic Party of Ukraine (SDPU), federalist
Armed forces: 346,400 (1997)
Conscription: 18 months (males over 18)
Death penalty: an act was passed to end capital punishment in March 2000, in line with Ukraine's membership of the council of Europe
Defence spend: (% GDP) 2.9 (1998)
Education spend: (% GNP) 7.2 (1996)
Health spend: (% GDP) 4.9 (1995)

Economy and resources

Currency: hryvna
GDP: (US$) 42.4 billion (1999 est)
Real GDP growth: (% change on previous year) –0.4 (1999)
GNP: (US$) 37.5 billion (1999)
GNP per capita (PPP): (US$) 3,142 (1999)
Consumer price inflation: 22.7% (1999)
Unemployment: 4.1% (1998)
Foreign debt: (US$) 12.8 billion (1999 est)
Major trading partners: Russia, Germany, Turkey, Turkmenistan, USA
Resources: coal, iron ore (world's fifth-largest producer), crude oil, natural gas, salt, chemicals, brown coal, alabaster, gypsum
Industries: metallurgy, mechanical engineering, chemicals, machinery products
Exports: grain, coal, oil, various minerals. Principal market: Russia 20.7% (1999)
Imports: mineral fuels, machine-building components, chemicals and chemical products. Principal source: Russia 47.6% (1999)
Arable land: 57.3% (1996)
Agricultural products: wheat, buckwheat, sugar beet, potatoes, fruit and vegetables, sunflowers, cotton, flax, tobacco, hops; animal husbandry accounts for more than 50% of agricultural activity

Population and society

Population: 50,456,000 (2000 est)
Population growth rate: –0.4% (1995–2000)
Population density: (per sq km) 84 (1999 est)
Urban population: (% of total) 68 (2000 est)
Age distribution: (% of total population) 0–14 18%, 15–59 61%, 60+ 21% (2000 est)
Ethnic groups: 73% of the population is of Ukrainian descent; 22% ethnic Russian; 1% Jewish; 4% other races including Belorussian, Moldovan, Hungarian, Bulgarian, Polish and Crimean Tatar
Language: Ukrainian (a Slavonic language; official), Russian (also official in Crimea), other regional languages
Religion: traditionally Ukrainian Orthodox; also Ukrainian Catholic; small Protestant, Jewish, and Muslim communities
Education: (compulsory years) 8 (7–15 age limit)
Literacy rate: 99% (men); 99% (women) (2000 est)
Labour force: 25.1% agriculture, 28% industry, 46.9% services (1997)
Life expectancy: 64 (men); 74 (women) (1995–2000)
Child mortality rate: (under 5, per 1,000 live births) 25 (1995–2000)
Physicians: 1 per 224 people (1996)
Hospital beds: 1 per 85 people (1996)
TV sets: (per 1,000 people) 493 (1997)
Radios: (per 1,000 people) 884 (1997)
Internet users: (per 10,000 people) 39.5 (1999)
Personal computer users: (per 100 people) 1.6 (1999)

Transport

Airports: international airports: Kiev (Borispol); four principal domestic airports; total passenger km: 1,534 million (1995)

Railways: total length: 22,564 km/14,021 mi; total passenger km: 63,759 million (1995)
Roads: total road network: 172,378 km/107,116 mi, of which 95% paved (1997); passenger cars: 97 per 1,000 people (1997)

Chronology

9th century: Rus' people established state centred on Kiev. **988:** Eastern Orthodox Christianity adopted. **1199:** Reunification of southern Rus' lands, after period of fragmentation, under Prince Daniel of Galicia-Volhynia. **13th century:** Mongol-Tatar Golden Horde sacked Kiev and destroyed Rus' state. **14th century:** Poland annexed Galicia; Lithuania absorbed Volhynia and expelled Tatars; Ukraine peasants became serfs of Polish and Lithuanian nobles. **1569:** Poland and Lithuania formed single state; clergy of Ukraine formed Uniate Church, which recognized papal authority but retained Orthodox rites, to avoid Catholic persecution. **16th and 17th centuries:** Runaway serfs known as Cossacks ('outlaws') formed autonomous community in eastern borderlands. **1648:** Cossack revolt led by Gen Bogdan Khmelnitsky drove out Poles from central Ukraine; Khmelnitsky accepted Russian protectorate in 1654. **1660–90:** 'Epoch of Ruins': Ukraine devastated by civil war and invasions by Russians, Poles, and Turks; Poland regained western Ukraine. **1687:** Gen Ivan Mazepa entered into alliance with Sweden in effort to regain Cossack autonomy from Russia. **1709:** Battle of Poltava: Russian victory over Swedes ended hopes of Cossack independence. **1772–95:** Partition of Poland: Austria annexed Galicia, Russian annexations included Volhynia. **1846–47:** Attempt to promote Ukrainian national culture through formation of Cyril and Methodius Society. **1899:** Revolutionary Ukrainian Party founded. **1917:** Revolutionary parliament (Rada) proclaimed Ukrainian autonomy within a federal Russia. **1918:** Ukraine declared full independence; civil war ensued between Rada (backed by Germans) and Reds (backed by Russian Bolsheviks). **1919:** Galicia united with Ukraine; conflict escalated between Ukrainian nationalists, Bolsheviks, anarchists, White Russians, and Poles. **1921:** Treaty of Riga: Russia and Poland partitioned Ukraine. **1921–22:** Several million people perished in famine. **1922:** Ukrainian Soviet Socialist Republic (Ukrainian SSR) became part of Union of Soviet Socialist Republics (USSR). **1932–33:** Enforced collectivization of agriculture caused another catastrophic famine with more than 7.5 million deaths. **1939:** USSR annexed eastern Poland and added Galicia-Volhynia to Ukrainian SSR. **1940:** USSR seized northern Bukhovina from Romania and added it to Ukrainian SSR. **1941–44:** Germany occupied Ukraine; many Ukrainians collaborated; millions of Ukrainians and Ukrainian Jews were enslaved and exterminated by Nazis. **1945:** USSR annexed Ruthenia from Czechoslovakia and added it to Ukrainian SSR, which became a nominal member of the United Nations (UN). **1946:** Uniate Church forcibly merged with Russian Orthodox Church. **1954:** Crimea transferred from Russian Federation to Ukrainian SSR. **1986:** Major environmental and humanitarian disaster caused by explosion of nuclear reactor at Chernobyl, north of Kiev. **1989:** Rukh (nationalist movement) established as political party; ban on Uniate Church lifted. **1990:** Ukraine declared its sovereignty under President Leonid Kravchuk, leader of the CP. **1991:** Ukraine declared its independence from USSR; President Kravchuk left the CP; Ukraine joined the newly formed Commonwealth of Independent States (CIS). **1992:** Crimean sovereignty was declared but then rescinded. **1994:** Election gains were made by radical nationalists in western Ukraine and by Russian unionists in eastern Ukraine; Leonid Kuchma succeeded Kravchuk as president. **1996:** A new constitution replaced the Soviet system, making the presidency stronger; remaining nuclear warheads were returned to Russia for destruction; a new currency was introduced. **1997:** New government appointments were made to speed economic reform. A treaty of friendship was signed with Russia, solving the issue of the Russian Black Sea fleet. A loan of $750 million from the International Monetary Fund (IMF) was approved. **1998:** The communists won the largest number of seats in parliamentary elections, but fell short of an absolute majority. The value of the hryvnya fell by over 50% against the US dollar after the neighbouring Russian currency crisis. The government survived a no-confidence vote tabled by left-wing factions that opposed the government's economic program. **1999:** Kuchma was re-elected as president; Viktor Yushchenko became prime minister. **2000:** The death penalty was abolished. The Chernobyl nuclear power station was closed permanently. **2001:** Protests in Kiev called for Kuchma's resignation on grounds of corruption and mismanagement.

Practical information

Visa requirements: UK: visa required. USA: visa required
Time difference: GMT +2
Chief tourist attractions: popular Black Sea resorts, including Odessa and Yalta; the Crimean peninsula; cities of historical interest, including Kiev and Odessa
Major holidays: 1, 7 January, 8 March, 1–2, 9 May, 24 August

UNITED ARAB EMIRATES

FORMERLY TRUCIAL STATES (UNTIL 1968), FEDERATION OF ARAB EMIRATES (WITH BAHRAIN AND QATAR) (1968–71)

Federation in southwest Asia, on the Arabian Gulf, bounded northwest by Qatar, southwest by Saudi Arabia, and southeast by Oman.

National name: *Dawlat Imarat al-'Arabiyya al Muttahida/State of the Arab Emirates* (UAE)
Area: 83,657 sq km/32,299 sq mi
Capital: Abu Dhabi
Major towns/cities: Dubai, Sharjah, Ras al Khaymah, Ajman, Al 'Ayn
Major ports: Dubai
Physical features: desert and flat coastal plain; mountains in east

Green is a symbol of fertility. White represents neutrality. Black reflects the Emirates' oil wealth. Red recalls the former flags of the Kharijite Muslims. Effective date: 2 December 1971.

Government

Head of state: Sheikh Zayed bin Sultan al-Nahayan of Abu Dhabi from 1971
Head of government: Sheikh Maktum bin Rashid al-Maktum of Dubai from 1990
Supreme council of rulers: *Abu Dhabi* Sheikh Zayed bin Sultan al-Nahayan, president (1966); *Ajman* Sheikh Humaid bin Rashid al-Nuami (1981); *Dubai* Sheikh Maktoum bin Rashid al-Maktoum (1990); *Fujairah* Sheikh Hamad bin Muhammad al-Sharqi (1974); *Ras al Khaymah* Sheikh Saqr bin Muhammad al-Quasimi (1948); *Sharjah* Sheikh Sultan bin Muhammad al-Quasimi (1972); *Umm al Qaiwain* Sheikh Rashid bin Ahmad al-Mu'alla (1981)
Political system: absolutist
Political executive: absolute
Administrative divisions: federation of the emirates of Abu Dhabi, Ajman, Dubai, Fujairah, Ras al Khaymah, Sharjah, Umm al Qaiwain
Political parties: none
Armed forces: 64,500 (1998)
Conscription: military service is voluntary
Death penalty: retained and used for ordinary crimes
Defence spend: (% GDP) 6.5 (1998)
Education spend: (% GNP) 1.8 (1995)
Health spend: (% GDP) 2.0 (1990–95)

Economy and resources

Currency: UAE dirham
GDP: (US$) 52.1 billion (1999)
Real GDP growth: (% change on previous year) 2.5 (1999)
GNP: (US$) 48.7 billion (1999 est)
GNP per capita (PPP): (US$) 18,825 (1999 est)
Consumer price inflation: 4% (1999 est)
Foreign debt: (US$) 12.9 billion (1999 est)
Major trading partners: Japan, USA, UK, India, Singapore, South Korea, Italy
Resources: petroleum and natural gas
Industries: petroleum production and refining, gas handling, petrochemicals and other petroleum products, aluminium products, cable, cement, chemicals, fertilizers, rolled steel, plastics, tools, clothing
Exports: crude petroleum, natural gas, re-exports (mainly machinery and transport equipment). Principal market: Japan 30.3% (1999)
Imports: machinery and transport equipment, food and live animals, fuels and lubricants, chemicals, basic manufactures. Principal source: Japan 8.9% (1999)
Arable land: 0.4% (1996)
Agricultural products: dates, tomatoes, aubergines, other vegetables and fruits; livestock rearing; fishing

Population and society

Population: 2,441,000 (2000 est)
Population growth rate: 2.0% (1995–2000); 1.8% (2000–05)
Population density: (per sq km) 29 (1999 est)
Urban population: (% of total) 86 (2000 est)
Age distribution: (% of total population) 0–14 28%, 15–59 67%, 60+ 5% (2000 est)
Ethnic groups: 75% non-Arab immigrants, mainly Iranians, Indians, and Pakistanis; about 25% Arabs (UAE nationals)
Language: Arabic (official), Farsi, Hindi, Urdu, English
Religion: Muslim 96% (of which 80% Sunni); Christian, Hindu
Education: (compulsory years) 6
Literacy rate: 74% (men); 79% (women) (2000 est)
Labour force: 7.6% agriculture, 34.5% industry, 57.9% services (1997); 93% of workforce were non-UAE nationals (1992 est)
Life expectancy: 74 (men); 77 (women) (1995–2000)
Child mortality rate: (under 5, per 1,000 live births) 19 (1995–2000)
Physicians: 1 per 588 people (1998)
Hospital beds: 1 per 342 people (1998)
TV sets: (per 1,000 people) 294 (1997)
Radios: (per 1,000 people) 354 (1996)
Internet users: (per 10,000 people) 1,668.3 (1999)
Personal computer users: (per 100 people) 12.5 (1999)

Transport

Airports: international airports: Abu Dhabi (Nadia), Dubai, Ras al Khaymah, Sharjah, Fujairah; domestic services operate between Abu Dhabi and Dubai; total passenger km: 9,958 million (1995)

Railways: none
Roads: total road network: 4,835 km/3,004 mi, of which 100% paved (1996 est); passenger cars: 11.4 per 1,000 people (1996 est)

Chronology

7th century AD: Islam introduced. **early 16th century:** Portuguese established trading contacts with Persian Gulf states. **18th century:** Rise of trade and seafaring among Qawasim and Bani Yas, respectively in Ras al Khaymah and Sharjah in north and Abu Dhabi and Dubai in desert of south. Emirates' current ruling families are descended from these peoples. **early 19th century:** Britain signed treaties ('truces') with local rulers, ensuring that British shipping through the Gulf was free from 'pirate' attacks and bringing Emirates under British protection. **1892:** Trucial Sheiks signed Exclusive Agreements with Britain, agreeing not to cede, sell, or mortgage territory to another power. **1952:** Trucial Council established by seven sheikdoms of Abu Dhabi, Ajman, Dubai, Fujairah, Ras al Khaymah, Sharjah, and Umm al Qawain, with a view to later forming a federation. **1958:** Large-scale exploitation of oil reserves led to rapid economic progress. **1968:** Britain's announcement that it would remove its forces from the Persian Gulf by 1971 led to an abortive attempt to arrange federation between seven Trucial States and Bahrain and Qatar. **1971:** Bahrain and Qatar ceded from the Federation of Arab Emirates, which was dissolved. Six Trucial States formed the United Arab Emirates, with the ruler of Abu Dhabi, Sheikh Zayed, as president. A provisional constitution was adopted. The UAE joined the Arab League and the United Nations (UN). **1972:** Seventh state, Ras al Khaymah, joined the federation. **1976:** Sheikh Zayed threatened to relinquish presidency unless progress towards centralization became more rapid. **1985:** Diplomatic and economic links with the Soviet Union and China were established. **1987:** Diplomatic relations with Egypt were restored. **1990:** Sheikh Maktum bin Rashid al-Maktum of Dubai was appointed prime minister. **1990–91:** UAE opposed the Iraqi invasion of Kuwait, and UAE troops fought as part of the UN coalition. **1991:** The Bank of Commerce and Credit International (BCCI), partly owned and controlled by Abu Dhabi's ruler Zayed bin Sultan al-Nahayan, collapsed at a cost to the UAE of $10 billion. **1992:** There was a border dispute with Iran. **1994:** Abu Dhabi agreed to pay BCCI creditors $1.8 billion.

Practical information

Visa requirements: UK: visa not required for a stay of up to 30 days. USA: visa required
Time difference: GMT +4
Chief tourist attractions: Dubai, popularly known as the 'Pearl of the Gulf', with its 16 km/10 mi deep-water creek
Major holidays: 1 January, 6 August, 2 December (2 days); variable: Eid-ul-Adha (3 days), end of Ramadan (4 days), New Year (Muslim), Prophet's Birthday, Lailat al-Miraj (March/April)

UNITED KINGDOM

Located in northwest Europe off the coast of France. Consists of England, Scotland, Wales, and Northern Ireland.

National name: *United Kingdom of Great Britain and Northern Ireland* (UK)
Area: 244,100 sq km/94,247 sq mi
Capital: London
Major towns/cities: Birmingham, Glasgow, Leeds, Sheffield, Liverpool, Manchester, Edinburgh, Bradford, Bristol, Coventry, Belfast, Cardiff
Major ports: London, Grimsby, Southampton, Liverpool
Physical features: became separated from European continent in about 6000 BC; rolling landscape, increasingly mountainous towards the north, with Grampian Mountains in Scotland, Pennines in northern England, Cambrian Mountains in Wales; rivers include Thames, Severn, and Spey
Territories: Anguilla, Bermuda, British Antarctic Territory, British Indian Ocean Territory, British Virgin Islands, Cayman Islands, Falkland Islands, Gibraltar, Montserrat, Pitcairn Islands, St Helena and Dependencies (Ascension, Tristan da Cunha), South Georgia, South Sandwich Islands, Turks and Caicos Islands; the Channel Islands and the Isle of Man are not part of the UK but are direct dependencies of the crown

The white saltire comes from the flag of Scotland. The St Patrick's Cross was, in fact, taken from the arms of the powerful Geraldine family. The red cross of St George is taken from the flag of England. Effective date: 1 January 1801.

Government

Head of state: Queen Elizabeth II from 1952
Head of government: Tony Blair from 1997
Political system: liberal democracy
Political executive: parliamentary
Administrative divisions: England: 34 non-metropolitan counties, 46 unitary authorities, 6 metropolitan counties, (with 36 metropolitan boroughs), 32 London boroughs, and the Corporation of London; Scotland: 9 regions, 29 unitary authorities, and 3 island authorities (from 1996); Wales: 9 counties and 22 unitary authorities/county boroughs (from 1996); Northern Ireland: 26 districts within 6 geographical counties
Political parties: Conservative Party, right of centre; Labour Party, moderate left of centre; Social and Liberal Democrats, left of centre; Scottish National Party (SNP), Scottish nationalist; Plaid Cymru (Welsh Nationalist Party), Welsh nationalist; Official Ulster Unionist Party (OUP), Democratic Unionist Party (DUP), Ulster People's Unionist Party (UPUP), all Northern Ireland right of centre, in favour of remaining part of UK; Social Democratic Labour Party (SDLP), Northern Ireland, moderate left of centre; Green Party, ecological; Sinn Fein, Irish nationalist
Armed forces: 210,900 (1998)
Conscription: military service is voluntary
Death penalty: abolished in 1965, except for treason and piracy; abolished completely in 1998
Defence spend: (% GDP) 2.8 (1998)
Education spend: (% GNP) 5.4 (1996)
Health spend: (% GDP) 6.9 (1997)

Economy and resources

Currency: pound sterling
GDP: (US$) 1,373.6 billion (1999)
Real GDP growth: (% change on previous year) 2.1 (1999)
GNP: (US$) 1,338.1 billion (1999)
GNP per capita (PPP): (US$) 20,883 (1999)
Consumer price inflation: 2.3% (1999)
Unemployment: 6.3% (1998)
Major trading partners: USA, Germany, France, the Netherlands, Japan, Ireland, EU
Resources: coal, limestone, crude petroleum, natural gas, tin, iron, salt, sand and gravel
Industries: machinery and transport equipment, steel, metals and metal products, food processing, shipbuilding, aircraft, petroleum and gas extraction, electronics and communications, chemicals and chemical products, business and financial services, tourism
Exports: industrial and electrical machinery, automatic data-processing equipment, motor vehicles, petroleum, chemicals, finished and semi-finished manufactured products, agricultural products and foodstuffs. Principal market: USA 14.8% (1999)
Imports: industrial and electrical machinery, motor vehicles, food and live animals, petroleum, automatic data processing equipment, consumer goods, textiles, paper, paper board. Principal source: Germany 13.5% (1999)
Arable land: 25.2% (1996)
Agricultural products: wheat, barley, potatoes, sugar beet, fruit, vegetables; livestock rearing (chiefly poultry and cattle), animal products, fishing

Population and society

Population: 58,830,000 (2000 est)
Population growth rate: 0.2% (1995–2000)
Population density: (per sq km) 240 (1999 est)
Urban population: (% of total) 90 (2000 est)
Age distribution: (% of total population) 0–14 19%, 15–59 60%, 60+ 21% (2000 est)
Ethnic groups: 81.5% English; 9.6% Scottish; 2.4% Irish; 1.9% Welsh; about 5% West Indian, Asian, African, and other ethnic minorities

Language: English (official), Welsh (also official in Wales), Gaelic
Religion: about 46% Church of England (established church); other Protestant denominations, Roman Catholic, Muslim, Jewish, Hindu, Sikh
Education: (compulsory years) 11
Literacy rate: 99% (men); 99% (women) (2000 est)
Labour force: 1.9% agriculture, 26.9% industry, 71.3% services (1997)
Life expectancy: 75 (men); 80 (women) (1995–2000)
Child mortality rate: (under 5, per 1,000 live births) 8 (1995–2000)
Physicians: 1 per 625 people (1996)
Hospital beds: 1 per 210 people (1996)
TV sets: (per 1,000 people) 645 (1998)
Radios: (per 1,000 people) 1,436 (1997)

Internet users: (per 10,000 people) 2,127.9 (1999)
Personal computer users: (per 100 people) 30.6 (1999)

Transport

Airports: international airports: London (Heathrow, Gatwick, London City, Stansted, Luton), Birmingham, Manchester, Newcastle, Bristol, Cardiff, Norwich, Derby, Edinburgh, Glasgow, Leeds/Bradford, Liverpool, Southampton; 22 domestic airports; total passenger km: 157,614 million (1997 est)
Railways: total length: 37,849 km/23,519 mi; total passenger km: 34,200 million (1997–98)
Roads: total road network: 369,887 km/229,848 mi, of which 100% paved (1997 est); passenger cars: 371 per 1,000 people (1996)

Chronology

c. **400–200 BC:** British Isles conquered by Celts. **55–54 BC:** Romans led by Julius Caesar raided Britain. **AD 43–60:** Romans conquered England and Wales, which formed the province of Britannia; Picts stopped them penetrating further north. **5th–7th centuries:** After Romans withdrew, Anglo-Saxons overran most of England and formed kingdoms, including Wessex, Northumbria, and Mercia; Wales was the stronghold of Celts. **500:** The Scots, a Gaelic-speaking tribe from Ireland, settled in the kingdom of Dalriada (Argyll). **5th–6th centuries:** British Isles converted to Christianity. **829:** King Egbert of Wessex accepted as overlord of all England. *c.* **843:** Kenneth McAlpin unified Scots and Picts to become the first king of Scotland. **9th–11th centuries:** Vikings raided the British Isles, conquering north and east England and northern Scotland. **1066:** Normans led by William I defeated Anglo-Saxons at Battle of Hastings and conquered England. **12th–13th centuries:** Anglo-Norman adventurers conquered much of Ireland, but effective English rule remained limited to area around Dublin. **1215:** King John of England forced to sign Magna Carta, which placed limits on royal powers. **1265:** Simon de Montfort summoned the first English parliament in which the towns were represented. **1284:** Edward I of England invaded Scotland; Scots defeated English at Battle of Stirling Bridge in 1297. **1314:** Robert the Bruce led Scots to victory over English at Battle of Bannockburn; England recognized Scottish independence in 1328. **1455–85:** Wars of the Roses: House of York and House of Lancaster disputed English throne. **1513:** Battle of Flodden: Scots defeated by English; James IV of Scotland killed. **1529:** Henry VIII founded Church of England after break with Rome; Reformation effective in England and Wales, but not in Ireland. **1536–43:** Acts of Union united Wales with England, with one law, one parliament, and one official language. **1541:** Irish parliament recognized Henry VIII of England as king of Ireland. **1557:** First Covenant established Protestant faith in Scotland. **1603:** Union of crowns: James VI of Scotland became James I of England also. **1607:** First successful English colony in Virginia marked the start of three centuries of overseas expansion. **1610:** James I established plantation of Ulster in Northern Ireland with Protestant settlers from England and Scotland. **1642–52:** English Civil War between king and Parliament, with Scottish intervention and Irish rebellion, resulted in victory for Parliament. **1649:** Execution of Charles I; Oliver Cromwell appointed Lord Protector in 1653; monarchy restored in 1660. **1689:** 'Glorious Revolution' confirmed power of Parliament; replacement of James II by William III resisted by Scottish Highlanders and Catholic Irish. **1707:** Act of Union between England and Scotland created United Kingdom of Great Britain, governed by a single parliament. **1721–42:** Cabinet government developed under Robert Walpole, in effect the first prime minister. **1745:** 'The Forty-Five': rebellion of Scottish Highlanders in support of Jacobite pretender to throne; defeated 1746. *c.* **1760–1850:** Industrial Revolution: Britain became the first industrial nation in the world. **1775–83:** American Revolution: Britain lost 13 American colonies; empire continued to expand in Canada, India, and Australia. **1793–1815:** Britain at war with revolutionary France, except for 1802–03. **1800:** Act of Union created United Kingdom of Great Britain and Ireland, governed by a single parliament; effective 1801. **1832:** Great Reform Act extended franchise; further extensions in 1867, 1884, 1918, and 1928. **1846:** Repeal of Corn Laws reflected shift of power from landowners to industrialists. **1870:** Home Rule Party formed to campaign for restoration of separate Irish parliament. **1880–90s:** Rapid expansion of British Empire in Africa. **1906–14:** Liberal governments introduced social reforms and curbed the power of the House of Lords. **1914–18:** The UK played a leading part in World War I; the British Empire expanded in Middle East. **1919–21:** The Anglo-Irish war ended with the secession of southern Ireland as the Irish Free State; Ulster remained within the United Kingdom of Great Britain and Northern Ireland with some powers devolved to a Northern Irish parliament. **1924:** The first Labour government was led by

Ramsay MacDonald. **1926:** A general strike arose from a coal dispute. Equality of status was recognized between the UK and Dominions of the British Commonwealth. **1931:** A National Government coalition was formed to face a growing economic crisis; unemployment reached 3 million. **1939–45:** The UK played a leading part in World War II. **1945–51:** The Labour government of Clement Attlee created the welfare state and nationalized major industries. **1947–71:** Decolonization brought about the end of the British Empire. **1969:** Start of the Troubles in Northern Ireland; the Northern Irish Parliament was suspended in 1972. **1973:** The UK joined the European Economic Community. **1979–90:** The Conservative government of Margaret Thatcher pursued radical free-market economic policies. **1982:** Unemployment reached over 3 million. The Falklands War with Argentina over the disputed sovereignty of the Falkland Islands cost more than a thousand lives but ended with the UK retaining control of the islands. **1983:** Coal pits were closed by the Conservative government and the miners went on strike. **1991:** British troops took part in a US-led war against Iraq under a United Nations (UN) umbrella. Following the economic successes of the 1980s there was a period of severe economic recession and unemployment. **1993:** A peace proposal for Northern Ireland, the Downing Street Declaration, was issued jointly with the Irish government. **1994:** The IRA and Protestant paramilitary declared a ceasefire in Northern Ireland. **1996:** The IRA renewed its bombing campaign in London. **1997:** The Labour Party won a landslide victory in a general election; Tony Blair became prime minister. Blair launched a new Anglo-Irish peace initiative. Blair met with Sinn Fein leader Gerry Adams; all-party peace talks began in Northern Ireland. Scotland and Wales voted in favour of devolution. Princess Diana was killed in a car crash. **1998:** A historic multiparty agreement (the 'Good Friday Agreement') was reached on the future of Northern Ireland; a peace plan was approved by referenda in Northern Ireland and the Irish Republic. The UUP leader, David Trimble, was elected first minister. **1999:** The Scottish Parliament and the Welsh Assembly opened, with Labour the largest party in both. The IRA agreed to begin decommissioning discussions and a coalition government was established. In December, the British government announced that it would write off its third world debts. **2000:** After it was revealed that there had been no arms handover by the IRA, the Secretary of State for Northern Ireland suspended the Northern Ireland Assembly. After the IRA agreed to allow independent inspectors access to its arms, and to put its weapons out of use, Northern Ireland's power sharing executive resumed work in May. Feuding between loyalist paramilitary groups broke out in Belfast, Northern Ireland. Protesters against the high price of fuel blockaded refineries in September, causing a shortage of petrol throughout the country.

Practical information

Visa requirements: USA: visa not required
Time difference: GMT +/–0
Chief tourist attractions: London, with its many historic monuments, cathedrals, churches, palaces, parks, and museums; historic towns, including York, Bath, Edinburgh, Oxford, and Cambridge; the Lake District in the northwest; the mountains of North Wales; varied coastline includes sandy beaches, cliffs, and the fjord-like inlets of northwest Scotland
Major holidays: 1 January, 25–26 December; variable: Good Friday, Easter Monday (not Scotland), Early May, Late May and Summer (August) Bank Holidays; Northern Ireland also has 17 March, 29 December; Scotland has 2 January

UNITED STATES OF AMERICA

Located in North America, extending from the Atlantic Ocean in the east to the Pacific Ocean in the west, bounded north by Canada and south by Mexico, and including the outlying states of Alaska and Hawaii.

National name: *United States of America* (USA)
Area: 9,372,615 sq km/3,618,766 sq mi
Capital: Washington, DC
Major towns/cities: New York, Los Angeles, Chicago, Philadelphia, Detroit, San Francisco, Dallas, San Diego, San Antonio, Houston, Boston, Phoenix, Indianapolis, Honolulu, San José

Red stands for hardiness and valour. White signifies purity and innocence. Blue represents vigilance, perseverance, and justice. The latest star, representing Hawaii, was added in 1960. Effective date: 4 July 1960.

Physical features: topography and vegetation from tropical (Hawaii) to arctic (Alaska); mountain ranges parallel with east and west coasts; the Rocky Mountains separate rivers emptying into the Pacific from those flowing into the Gulf of Mexico; Great Lakes in north; rivers include Hudson, Mississippi, Missouri, Colorado, Columbia, Snake, Rio Grande, Ohio
Territories: the commonwealths of Puerto Rico and Northern Marianas; Guam, the US Virgin Islands, American Samoa, Wake Island, Midway Islands, Johnston Atoll, Baker Island, Howland Island, Jarvis Island, Kingman Reef, Navassa Island, Palmyra Island

Government

Head of state and government: George W Bush from 2001
Political system: liberal democracy
Political executive: limited presidency
Administrative divisions: 50 states and one district (District of Columbia)
Political parties: Democratic Party, liberal centre; Republican Party, right of centre; Reform Party, prodemocratic
Armed forces: 1,401,600 (1998)
Conscription: military service is voluntary
Death penalty: retained and used for ordinary crimes
Defence spend: (% GDP) 3.2 (1998)
Education spend: (% GNP) 5.4 (1996)
Health spend: (% GDP) 13.9 (1997)

Economy and resources

Currency: US dollar
GDP: (US$) 8,708.9 billion (1999)
Real GDP growth: (% change on previous year) 4.2 (1999)
GNP: (US$) 8,350.9 billion (1999)
GNP per capita (PPP): (US$) 30,600 (1999)
Consumer price inflation: 2.2% (1998)
Unemployment: 4.5% (1998)
Major trading partners: Canada, Japan, Mexico, EU (principally UK and Germany), China
Resources: coal, copper (world's second-largest producer), iron, bauxite, mercury, silver, gold, nickel, zinc (world's fifth-largest producer), tungsten, uranium, phosphate, petroleum, natural gas, timber
Industries: machinery, petroleum refining and products, food processing, motor vehicles, pig iron and steel, chemical products, electrical goods, metal products, printing and publishing, fertilizers, cement

Exports: machinery, motor vehicles, agricultural products and foodstuffs, aircraft, weapons, chemicals, electronics. Principal market: Canada 23.9% (1999)
Imports: machinery and transport equipment, crude and partly refined petroleum, office machinery, textiles and clothing. Principal source: Canada 19.3% (1999)
Arable land: 19.1% (1996)
Agricultural products: hay, potatoes, maize, wheat, barley, oats, sugar beet, soybeans, citrus and other fruit, cotton, tobacco; livestock (principally cattle, pigs, and poultry)

Population and society

Population: 278,357,000 (2000 est)
Population growth rate: 0.8% (1995–2000); 0.8% (2000–05)
Population density: (per sq km) 29 (1999 est)
Urban population: (% of total) 77 (2000 est)
Age distribution: (% of total population) 0–14 21%, 15–59 63%, 60+ 16% (2000 est)
Ethnic groups: approximately three-quarters of the population are of European origin, including 29% who trace their descent from Britain and Ireland, 8% from Germany, 5% from Italy, and 3% each from Scandinavia and Poland; approximately 83% are white (of which over 11% are Hispanic), 13% black, 4% Asian and Pacific Islanders, and about 1% American Indians, Eskimos, and Aleuts (1998); African-Americans form about a third of the population of the states of the 'Deep South', namely Alabama, Georgia, Louisiana, Mississippi, and South Carolina
Language: English, Spanish
Religion: Protestant 58%; Roman Catholic 28%; atheist 10%; Jewish 2%; other 4% (1998)
Education: (compulsory years) 10
Literacy rate: 99% (men); 99% (women) (2000 est)

Labour force: 51% of population: 2.8% agriculture, 23.8% industry, 73.3% services (1996)
Life expectancy: 73 (men); 80 (women) (1995–2000)
Child mortality rate: (under 5, per 1,000 live births) 9 (1995–2000)
Physicians: 1 per 385 people (1996)
Hospital beds: 1 per 238 people (1995)
TV sets: (per 1,000 people) 847 (1997)
Radios: (per 1,000 people) 2,146 (1997)
Internet users: (per 10,000 people) 3,982.4 (1999)
Personal computer users: (per 100 people) 51.1 (1999)

Transport

Airports: international airports: Anchorage, Atlanta (Hartsfield), Baltimore (Baltimore/Washington), Boston (Logan), Chicago (O'Hare), Cincinnati (Northern Kentucky), Cleveland (Hopkins), Dallas/Fort Worth, Denver (Stapleton), Detroit Metropolitan, Honolulu, Houston Intercontinental, Kansas City, Las Vegas (McCarran), Los Angeles, Miami, Minneapolis/St Paul, New Orleans, New York (John F Kennedy, La Guardia, Newark), Orlando, Philadelphia, Phoenix (Sky Harbor), Pittsburgh, Portland, St Louis (Lambert), Salt Lake City, San Diego (Lindbergh Field), San Francisco, Seattle-Tacoma, Tampa, Washington DC (Dulles, National); about 800 domestic airports; total passenger km: 964,430 million (1997 est)
Railways: total length: 225,000 km/139,815 mi; total passenger km: 33,861 million (1997)
Roads: total road network: 6,307,584 km/3,919,533 mi, of which 60.5% paved (1996); passenger cars: 487 per 1,000 people (1996)

Chronology

***c*.15,000 BC:** First evidence of human occupation in North America. **1513:** Ponce de Léon of Spain explored Florida in search of the Fountain of Youth; Francisco Coronado explored southwest region of North America 1540–42. **1565:** Spanish founded St Augustine (Florida), the first permanent European settlement in North America. **1585:** Sir Walter Raleigh tried to establish an English colony on Roanoke Island in what he called Virginia. **1607:** English colonists founded Jamestown, Virginia, and began growing tobacco. **1620:** The Pilgrim Fathers founded Plymouth Colony (near Cape Cod); other English Puritans followed them to New England. **1624:** Dutch formed colony of New Netherlands; Swedes formed New Sweden in 1638; both taken by England in 1664. **17th–18th centuries:** Millions of Africans were sold into slavery on American cotton and tobacco plantations. **1733:** Georgia became thirteenth British colony on east coast. **1763:** British victory over France in Seven Years' War secured territory as far west as Mississippi River. **1765:** British first attempted to levy tax in American colonies with Stamp Act; protest forced repeal in 1767. **1773:** 'Boston Tea Party': colonists boarded ships and threw cargoes of tea into sea in protest at import duty. **1774:** British closed Boston harbour and billeted troops in Massachusetts; colonists formed First Continental Congress. **1775:** American Revolution: colonies raised Continental Army led by George Washington to fight against British rule. **1776:** American colonies declared independence; France and Spain supported them in a war with Britain. **1781:** Americans

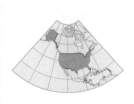

defeated British at Battle of Yorktown; rebel states formed loose confederation, codified in Articles of Confederation. **1783:** Treaty of Paris: Britain accepted loss of colonies. **1787:** 'Founding Fathers' devised new constitution for United States of America. **1789:** Washington elected first president of USA. **1791:** Bill of Rights guaranteed individual freedom. **1803:** Louisiana Purchase: France sold former Spanish lands between Mississippi River and Rocky Mountains to USA. **1812–14:** War with Britain arose from dispute over blockade rights during Napoleonic Wars. **1819:** USA bought Florida from Spain. **19th century:** Mass immigration from Europe; settlers moved westwards, crushing Indian resistance and claiming 'manifest destiny' of USA to control North America. By end of century, the number of states in the Union had increased from 17 to 45. **1846–48:** Mexican War: Mexico ceded vast territory to USA. **1854:** Kansas–Nebraska Act heightened controversy over slavery in southern states; abolitionists formed Republican Party. **1860:** Abraham Lincoln (Republican) elected president. **1861:** Civil war broke out after 11 southern states, wishing to retain slavery, seceded from USA and formed the Confederate States of America under Jefferson Davis. **1865:** USA defeated Confederacy; slavery abolished; President Lincoln assassinated. **1867:** Alaska bought from Russia. **1869:** Railway linked east and west coasts; rapid growth of industry and agriculture 1870–1920 made USA very rich. **1876:** Sioux Indians defeated US troops at Little Big Horn; Indians finally defeated at Wounded Knee in 1890. **1898:** Spanish–American War: USA gained Puerto Rico and Guam; also Philippines (until 1946) and Cuba (until 1901); USA annexed Hawaii. **1917–18:** USA intervened in World War I; President Woodrow Wilson took leading part in peace negotiations in 1919, but USA rejected membership of League of Nations. **1920:** Women received right to vote; sale of alcohol prohibited, until 1933. **1924:** American Indians made citizens of USA by Congress. **1929:** 'Wall Street Crash': stock market collapse led to Great Depression with 13 million unemployed by 1933. **1933:** President Franklin Roosevelt launched the 'New Deal' with public works to rescue the economy. **1941:** Japanese attacked US fleet at Pearl Harbor, Hawaii; USA declared war on Japan; Germany declared war on USA, which henceforth played a leading part in World War II. **1945:** USA ended war in Pacific by dropping two atomic bombs on Hiroshima and Nagasaki, Japan. **1947:** 'Truman Doctrine' pledged US aid for nations threatened by communism; start of Cold War between USA and USSR. **1950–53:** US forces engaged in Korean War. **1954:** Racial segregation in schools deemed unconstitutional; start of campaign to secure civil rights for black Americans. **1962:** Cuban missile crisis: USA forced USSR to withdraw nuclear weapons from Cuba. **1963:** President Kennedy assassinated. **1964–68:** President Lyndon Johnson introduced the 'Great Society' programme of civil-rights and welfare measures. **1961–75:** USA involved in Vietnam War. **1969:** US astronaut Neil Armstrong was first person on the Moon. **1974:** 'Watergate' scandal: evidence of domestic political espionage compelled President Richard Nixon to resign. **1979–80:** Iran held US diplomats hostage, humiliating President Jimmy Carter. **1981–89:** Tax-cutting policies of President Ronald Reagan led to large federal budget deficit. **1986:** 'Irangate' scandal: secret US arms sales to Iran illegally funded Contra guerrillas in Nicaragua. **1990:** President George Bush declared an end to the Cold War. **1991:** USA played leading part in expelling Iraqi forces from Kuwait in the Gulf War. **1992:** Democrat Bill Clinton won presidential elections, beginning his term of office in 1993. **1996:** US launched missile attacks on Iraq in response to Hussein's incursions into Kurdish safe havens. **1998:** The House of Representatives voted to impeach Clinton on the grounds of perjury and obstruction of justice, due to his misleading the public about his relationship with a White House intern. Clinton's national approval rating remained high, and he was acquitted in 1999. In response to bombings of US embassies in Tanzania and Kenya by an Islamic group, the USA bombed suspected sites in Afghanistan and Sudan. The USA also led air strikes against Iraq following the expulsion of UN weapons inspectors by Saddam Hussein. **1999:** US forces led NATO air strikes against Yugoslavia in protest against Serb violence against ethnic Albanians in Kosovo. Three million people fled inland in the largest evacuation in US history as Hurricane Floyd hit the east coast in September. **2000:** In August, the worst wildfires in 30 years consumed 4.4 million acres/6,875 sq mi of land in the west. The presidential elections in November were the closest ever, the result being decided by a few hundred votes in Florida. **2001:** Republican George W Bush, son of former president George Bush, was inaugurated as president. California experienced an electricity crisis, leading to mandatory blackouts in January.

Practical information

Visa requirements: UK: visa not required for a stay of up to 90 days
Time difference: GMT –5–11
Chief tourist attractions: many cities of interest, including New York, with its skyscrapers, Washington DC, with its monuments, Boston, San Francisco, and New Orleans; enormous diversity of geographical features – the Rocky Mountains, the Everglades of Florida, the Grand Canyon; hundreds of national parks, historical

parks, and reserves, including Redwood, Yosemite, and Death Valley (all in California); Disneyland (California) and Walt Disney World (Florida)
Major holidays: 1 January, 4 July, 12 October (not all states), 11 November, 25 December; variable: Martin Luther King's birthday (January, not all states), George Washington's birthday (February), Memorial (May), Labor (first Mon in September), Columbus (October), Thanksgiving (last Thu in November); much local variation

URUGUAY

Located in South America, on the Atlantic coast, bounded north by Brazil and west by Argentina.

National name: *República Oriental del Uruguay/Eastern Republic of Uruguay*
Area: 176,200 sq km/68,030 sq mi
Capital: Montevideo
Major towns/cities: Salto, Paysandú, Las Piedras, Rivera, Tacuarembó
Physical features: grassy plains (pampas) and low hills; rivers Negro, Uruguay, Río de la Plata

The flag is modelled on the Stars and Stripes of the United States' flag. The stripes represent the nine provinces at the time of liberation. Blue and white are the colours of Argentina and also of national hero, José Gervasio Artigas. Effective date: 12 July 1830.

Government

Head of state and government: Jorge Batlle Ibáñez from 2000
Political system: liberal democracy
Political executive: limited presidency
Administrative divisions: 19 departments
Political parties: Colorado Party (PC), progressive, left of centre; National (Blanco) Party (PN), traditionalist, right of centre; New Space (NE), moderate, left wing; Progressive Encounter (EP), left wing
Armed forces: 25,600 (1998)
Conscription: military service is voluntary
Death penalty: abolished in 1907
Defence spend: (% GDP) 2.3 (1998)
Education spend: (% GNP) 3.3 (1996)
Health spend: (% GDP) 10 (1997)

Economy and resources

Currency: Uruguayan peso
GDP: (US$) 20.21 billion (1999)
Real GDP growth: (% change on previous year) –3.2 (1999 est)
GNP: (US$) 19.54 billion (1999)
GNP per capita (PPP): (US$) 8,280 (1999)
Consumer price inflation: 5.6% (1999)
Unemployment: 10.1% (1998 est)
Foreign debt: (US$) 7.7 billion (1999 est)
Major trading partners: Brazil, Argentina, USA, Italy, Germany, Paraguay, France
Resources: small-scale extraction of building materials, industrial minerals, semi-precious stones; gold deposits are being developed
Industries: food processing, textiles and clothing, beverages, cement, chemicals, light engineering and transport equipment, leather products
Exports: textiles, meat (chiefly beef), live animals and by-products (mainly hides and leather products), cereals, footwear. Principal market: Brazil 30% (1999)
Imports: machinery and appliances, transport equipment, chemical products, petroleum and petroleum products, agricultural products. Principal source: Argentina 24.5% (1999)
Arable land: 7.2% (1996)

Agricultural products: rice, sugar cane, sugar beet, wheat, potatoes, barley, maize, sorghum; livestock rearing (sheep and cattle) is traditionally country's major economic activity – exports of animals, meat, skins, and hides accounted for 36.6% of total export revenue (1994)

Population and society

Population: 3,337,000 (2000 est)
Population growth rate: 0.7% (1995–2000)
Population density: (per sq km) 19 (1999 est)
Urban population: (% of total) 91 (2000 est)
Age distribution: (% of total population) 0–14 25%, 15–59 58%, 60+ 17% (2000 est)
Ethnic groups: predominantly of European descent: about 54% Spanish, 22% Italian, with minorities from other European countries; about 8% mestizo, 4% black
Language: Spanish (official), Brazilero (a mixture of Spanish and Portuguese)
Religion: mainly Roman Catholic
Education: (compulsory years) 6
Literacy rate: 97% (men); 98% (women) (2000 est)
Labour force: 13% agriculture, 19% industry, 68% services (1996 est)
Life expectancy: 71 (men); 78 (women) (1995–2000)
Child mortality rate: (under 5, per 1,000 live births) 20 (1995–2000)
Physicians: 1 per 236 people (1993 est)
Hospital beds: 1 per 221 people (1993 est)
TV sets: (per 1,000 people) 241 (1997)
Radios: (per 1,000 people) 607 (1997)
Internet users: (per 10,000 people) 905.5 (1999)
Personal computer users: (per 100 people) 10.0 (1999)

Transport

Airports: international airport: Montevideo (Carrasco); seven domestic airports; total passenger km: 636 million (1995)
Railways: total length: 2,073 km/1,288 mi (passenger service withdrawn in 1988, partially resumed in 1993)
Roads: total road network: 8,983 km/5,582 mi, of which 90% paved (1997); passenger cars: 147 per 1,000 people (1996 est)

Chronology

1516: Río de la Plata visited by Spanish navigator Juan Diaz de Solis, who was killed by native Charrua Amerindians. This discouraged European settlement for more than a century. **1680:** Portuguese from Brazil founded Nova Colonia do Sacramento on Río de la Plata estuary. **1726:** Spanish established fortress at Montevideo and wrested control over Uruguay from Portugal, with much of the Amerindian population being killed. **1776:** Became part of Viceroyalty of La Plata, with capital at Buenos Aires. **1808:** With Spanish monarchy overthrown by Napoleon Bonaparte, La Plata Viceroyalty became autonomous, but Montevideo remained loyal to Spanish Crown and rebelled against Buenos Aires control. **1815:** Dictator José Gervasio Artigas overthrew Spanish and Buenos Aires control. **1820:** Artigas ousted by Brazil, which disputed control of Uruguay with Argentina. **1825:** Independence declared after fight led by Juan Antonio Lavalleja. **1828:** Independence recognized by country's neighbours. **1836:** Civil war between Reds and Whites, after which Colorado and Blanco parties were named. **1840:** Merino sheep introduced by British traders, who later established meat processing factories for export trade. **1865–70:** Fought successfully alongside Argentina and Brazil in war against Paraguay. **1903:** After period of military rule, José Battle y Ordonez, a progressive from centre-left Colorado Party, became president. As president 1903–07 and 1911–15, he gave women the franchise and created an advanced welfare state as a successful ranching economy developed. **1930:** First constitution adopted, but period of military dictatorship followed during Depression period. **1958:** After 93 years out of power, the right-of-centre Blanco Party returned to power. **1967:** The Colorado Party were in power, with Jorge Pacheco Areco as president. A period of labour unrest and urban guerrilla activity by left-wing Tupamaros. **1972:** Juan María Bordaberry Arocena of the Colorado Party became president. **1973:** Parliament dissolved and Bordaberry shared power with military dictatorship, which crushed Tupamaros and banned left-wing groups. **1976:** Bordaberry deposed by army; Dr Aparicio Méndez Manfredini became president. **1981:** Gen Grigorio Alvárez Armellino became new military ruler. **1984:** Violent antigovernment protests after ten years of repressive rule and deteriorating economy. **1985:** Agreement reached between army and political leaders for return to constitutional government and freeing of political prisoners. **1986:** Government of national accord established under President Sanguinetti. **1992:** The public voted against privatization in a national referendum. **2000:** Jorge Batlle Ibáñez, of the Colorado Party, was elected president.

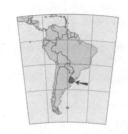

Practical information

Visa requirements: UK: visa not required. USA: visa not required
Time difference: GMT –3
Chief tourist attractions: sandy beaches; forests; tropical swamps on the coast; variety of flora and fauna
Major holidays: 1 January, 19 April, 1, 18 May, 19 June, 18 July, 25 August, 12 October, 2 November, 25 December; variable: Carnival (2 days), Good Friday, Holy Thursday, Mon–Wed of Holy Week

UZBEKISTAN

Located in central Asia, bounded north by Kazakhstan and the Aral Sea, east by Kyrgyzstan and Tajikistan, south by Afghanistan, and west by Turkmenistan.

National name: *Özbekiston Respublikasi/Republic of Uzbekistan*
Area: 447,400 sq km/172,741 sq mi
Capital: Tashkent
Major towns/cities: Samarkand, Bukhara, Namagan, Andijon, Nukus, Qarshi
Physical features: oases in deserts; rivers: Amudar'ya, Syr Darya; Fergana Valley; rich in mineral deposits

Blue stands for the night sky and for water as a source of life. White represents peace. Red indicates the life-force. Green recalls nature and fertility. Effective date: 11 October 1991.

Government

Head of state: Islam Karimov from 1990
Head of government: Otkir Sultonov from 1995
Political system: authoritarian nationalist
Political executive: unlimited presidency
Administrative divisions: 12 regions and one autonomous republic (Karakalpakstan)
Political parties: People's Democratic Party of Uzbekistan (PDP), reform socialist (ex-communist); Fatherland Progress Party (FP; Vatan Taraqioti), pro-private enterprise; Erk (Freedom Democratic Party), mixed economy; Social Democratic Party of Uzbekistan, pro-Islamic; National Revival Democratic Party, centrist, intelligentsia-led
Armed forces: 80,000 (1998)
Conscription: military service is compulsory for 18 months
Death penalty: retained and used for ordinary crimes
Defence spend: (% GDP) 5.4 (1998)
Education spend: (% GNP) 8.1 (1996)
Health spend: (% GDP) 3.5 (1990–95)

Economy and resources

Currency: som
GDP: (US$) 16.8 billion (1999 est)
Real GDP growth: (% change on previous year) 4.1 (1999)
GNP: (US$) 17.6 billion (1999)
GNP per capita (PPP): (US$) 2,092 (1999)
Consumer price inflation: 29% (1999)
Unemployment: 0.4% (officially registered 1997; true level believed to be considerably higher)
Foreign debt: (US$) 3.3 billion (1999 est)
Major trading partners: Russia, Switzerland, South Korea, UK, Germany, Belgium, USA, Kazakhstan, Turkey, Tajikistan
Resources: petroleum, natural gas, coal, gold (world's seventh-largest producer), silver, uranium (world's fourth-largest producer), copper, lead, zinc, tungsten
Industries: processing of agricultural and mineral raw materials, agricultural machinery, chemical products, metallurgy, cement, mineral fertilizer, paper, textiles, footwear, electrical appliances

Exports: cotton fibre, textiles, machinery, food and energy products, gold. Principal market: Russia 13% (1999)
Imports: machinery, light industrial goods, food and raw materials. Principal source: Russia 13.8% (1999)
Arable land: 10.9% (1996)
Agricultural products: cotton (among the world's five largest producers), grain, potatoes, vegetables, fruit and berries; livestock rearing; silkworm breeding

Population and society

Population: 24,318,000 (2000 est)
Population growth rate: 1.6% (1995–2000)
Population density: (per sq km) 54 (1999 est)
Urban population: (% of total) 37 (2000 est)
Age distribution: (% of total population) 0–14 37%, 15–59 56%, 60+ 7% (2000 est)
Ethnic groups: 75% Uzbek, 7% ethnic Russian, 5% Tajik, 4% Kazakh; remaining 9% include Tatar, Karalkalpak, and Korean
Language: Uzbek (a Turkic language; official), Russian, Tajik
Religion: predominantly Sunni Muslim; small Wahhabi, Sufi, and Orthodox Christian communities
Education: (compulsory years) 9
Literacy rate: 93% (men); 85% (women) (2000 est)
Labour force: 44% agriculture, 20% industry, 36% services (1995)
Life expectancy: 64 (men); 71 (women) (1995–2000)
Child mortality rate: (under 5, per 1,000 live births) 63 (1995–2000)
Physicians: 1 per 307 people (1995)
Hospital beds: 1 per 120 people (1995)
TV sets: (per 1,000 people) 275 (1997)
Radios: (per 1,000 people) 465 (1997)
Internet users: (per 10,000 people) 3.1 (1999)

Transport

Airports: international airport: Tashkent; eight domestic airports; total passenger km: 4,855 million (1995)
Railways: total length: 3,380 km/2,100 mi; total passenger km: 2,004 million (1996)
Roads: total road network: 43,463 km/27,008 mi, of which 87.3% paved (1997)

Chronology

6th century BC: Part of the Persian Empire of Cyrus the Great. **4th century BC:** Part of the empire of Alexander the Great of Macedonia. **1st century BC:** Samarkand (Maracanda) developed as transit point on strategic Silk Road trading route between China and Europe. **7th century:** City of Tashkent founded; spread of Islam. **12th century:** Tashkent taken by Turks; Khorezem (Khiva), in northwest, became centre of large Central Asian polity, stretching from Caspian Sea to Samarkand in the east. **13th–14th centuries:** Conquered by Genghis Khan and became part of Mongol Empire, with Samarkand serving as capital for Tamerlane. **18th–19th centuries:** Dominated by independent emirates and khanates (chiefdoms) of Bukhara in southwest, Kokand in east, and Samarkand in centre. **1865–67:** Tashkent was taken by Russia and made capital of Governor-Generalship of Turkestan. **1868–76:** Tsarist Russia annexed emirate of Bukhara (1868); and khanates of Samarkand (1868), Khiva (1873), and Kokand (1876). **1917:** Following Bolshevik revolution in Russia, Tashkent soviet ('people's council') established, which deposed the emir of Bukhara and other khans in 1920. **1918–22:** Mosques closed and Muslim clergy persecuted as part of secularization drive by new communist rulers, despite nationalist guerrilla (basmachi) resistance. **1921:** Part of Turkestan Soviet Socialist Autonomous Republic. **1925:** Became constituent republic of USSR. **1930s:** Skilled ethnic Russians immigrated into urban centres as industries developed. **1944:** About 160,000 Meskhetian Turks forcibly transported from their native Georgia to Uzbekistan by Soviet dictator Joseph Stalin. **1950s–80s:** Major irrigation projects stimulated cotton production, but led to desiccation of Aral Sea. **late 1980s:** Upsurge in Islamic consciousness stimulated by *glasnost* initiative of Soviet Union's reformist leader Mikhail Gorbachev. **1989:** Birlik ('Unity'), nationalist movement, formed. Violent attacks on Meskhetian and other minority communities in Fergana Valley. **1990:** Economic and political sovereignty was declared by the increasingly nationalist UCP, led by Islam Karimov, who became president. **1991:** An attempted anti-Gorbachev coup by conservatives in Moscow was initially supported by President Karimov. Independence was declared. Uzbekistan joined the new Commonwealth of Independent States (CIS); Karimov was re-elected president. **1992:** There were violent food riots in Tashkent. Uzbekistan joined the Economic Cooperation Organization and the United Nations (UN). A new constitution was adopted. **1993:** There was a crackdown on Islamic fundamentalists as the economy deteriorated. **1994:** Economic, military, and social union was forged with Kazakhstan and Kyrgyzstan, and an economic integration treaty was signed with Russia. Links with Turkey were strengthened and foreign inward investment encouraged. **1995:** The ruling PDP (formerly UCP) won a general election, from which the opposition was banned from participating, and Otkir Sultonov was appointed prime minister. Karimov's tenure as president was extended for a further five-year term by national referendum. **1996:** An agreement was made with Kazakhstan and Kyrgyzstan to create a single economic market. **1998:** A treaty of eternal friendship and deepening economic cooperation was signed with Kazakhstan. **1999:** Uzbekistan threatened to end participation in a regional security treaty, accusing Russia of seeking to integrate the former Soviet republics into a superstate. **2000:** President Islam Karimov was re-elected. Islamist rebels crossed into the country from Afghanistan via Tajikistan, reportedly seeking to create an Islamic state in east Uzbekistan.

Practical information

Visa requirements: UK: visa required. USA: visa required
Time difference: GMT +5
Chief tourist attractions: over 4,000 historical monuments, largely associated with the ancient 'Silk Route'; many historical sites and cities, including Samarkand (Tamerlane's capital), Bukhara, and Khiva
Major holidays: 1–2 January, 8, 21 March, 1 September, 8 December

VANUATU FORMERLY NEW HEBRIDES (UNTIL 1980)

Group of islands in the southwest Pacific Ocean, part of Melanesia.

National name: *Ripablik blong Vanuatu/République de Vanuatu/Republic of Vanuatu*
Area: 14,800 sq km/5,714 sq mi
Capital: Port-Vila (on Efate island) (and chief port)
Major towns/cities: Luganville (on Espíritu Santo)
Major ports: Santo
Physical features: comprises around 70 inhabited islands, including Espíritu Santo, Malekula, and Efate; densely forested, mountainous; three active volcanoes; cyclones on average twice a year

Red symbolizes blood. Yellow represents sunshine. Green stands for the islands' riches. Black reflects the Melanesian population. Effective date: 30 July 1980.

Government

Head of state: John Bernard Bani from 1999
Head of government: Barak Sope from 1999
Political system: liberal democracy
Political executive: parliamentary
Administrative divisions: six provinces
Political parties: Union of Moderate Parties (UMP), Francophone centrist; National United Party (NUP), formed by Walter Lini; Vanua'aku Pati (VP), Anglophone centrist; Melanesian Progressive Party (MPP), Melanesian centrist; Fren Melanesian Party,
Armed forces: no standing defence force; paramilitary force of around 300; police naval service of around 50 (1995)
Death penalty: laws do not provide for the death penalty for any crime
Education spend: (% GNP) 4.9 (1996)
Health spend: (% GDP) 3.3 (1997 est)

Economy and resources

Currency: vatu
GDP: (US$) 241 million (1999 est)
Real GDP growth: (% change on previous year) –2 (1999)
GNP: (US$) 221 million (1999)
GNP per capita (PPP): (US$) 2,771 (1999 est)
Consumer price inflation: 2.5% (1999)
Foreign debt: (US$) 48 million (1997)
Major trading partners: Japan, Australia, the Netherlands, New Zealand, New Caledonia, France
Resources: manganese; gold, copper, and large deposits of petroleum have been discovered but have hitherto remained unexploited
Industries: processing of agricultural products (chiefly copra, meat canning, fish freezing, saw milling), soft drinks, building materials, furniture, aluminium, tourism, offshore banking, shipping registry
Exports: copra, beef, timber, cocoa, shells. Principal market: Japan 32.1% (1997)
Imports: machinery and transport equipment, food and live animals, basic manufactures, miscellaneous manufactured articles, mineral fuels, chemicals,

beverages, tobacco. Principal source: Japan 52.6% (1997)
Arable land: 1.6% (1995)
Agricultural products: coconuts and copra, cocoa, coffee, yams, taro, cassava, breadfruit, squash and other vegetables, bananas; livestock rearing (cattle, pigs, goats, and poultry); forest resources

Population and society

Population: 190,000 (2000 est)
Population growth rate: 2.4% (1995–2000)
Population density: (per sq km) 13 (1999 est)
Urban population: (% of total) 20 (2000 est)
Age distribution: (% of total population) 0–14 42%, 15–59 53%, 60+ 5% (2000 est)
Ethnic groups: 94% Melanesian, 4% European or mixed European, 2% Vietnamese, Chinese, or other Pacific islanders
Language: Bislama (82%), English, French (all official)
Religion: Christian 80%, animist about 8%
Education: (compulsory years) 6
Literacy rate: 54% (men); 23% (women) (1995 est)
Labour force: 65% agriculture, 3% industry, 32% services (1995 est)
Life expectancy: 66 (men); 70 (women) (1995–2000)
Child mortality rate: (under 5, per 1,000 live births) 48 (1995–2000)
Physicians: 1 per 14,100 people (1995)
Hospital beds: 1 per 452 people (1995)
TV sets: (per 1,000 people) 13 (1996)
Radios: (per 1,000 people) 345 (1996)
Internet users: (per 10,000 people) 161.4 (1999)

Transport

Airports: international airports: Port-Vila (Banerfield); 28 domestic airports and airstrips; total passenger km: 146 million (1995)
Railways: none
Roads: total road network: 1,070 km/665 mi, of which 23.9% paved (1996 est); passenger cars: 23.3 per 1,000 people (1996 est)

Chronology

1606: First visited by Portuguese navigator Pedro Fernandez de Queiras, who named the islands Espíritu Santo. **1774:** Visited by British navigator Capt James Cook, who named them the New Hebrides, after the Scottish islands. **1830s:** European merchants attracted to islands by sandalwood trade. Christian missionaries arrived, but many were attacked by the indigenous Melanesians who, in turn, were ravaged by exposure to European diseases. **later 19th century:** Britain and France disputed control; islanders were shipped to Australia, the Fiji Islands, Samoa, and New Caledonia to work as plantation labourers. **1906:** The islands were jointly administered by France and Britain as the Condominium of the New Hebrides. **1963:** Indigenous Na-Griamel (NG) political grouping formed on Espíritu Santo to campaign against European acquisition of more than a third of the land area. **1975:** A representative assembly was established following pressure from the VP, formed in 1972 by English-speaking Melanesian Protestants. **1978:** A government of national unity was formed, with Father Gerard Leymang as chief minister. **1980:** A revolt on the island of Espíritu Santo by French settlers and pro-NG plantation workers delayed independence but it was achieved within the Commonwealth, with George Kalkoa (adopted name Sokomanu) as president and left-of-centre Father Walter Lini (VP) as prime minister. **1988:** The dismissal of Lini by Sokomanu led to Sokomanu's arrest for treason. Lini was later reinstated. **1991:** Lini was voted out by party members and replaced by Donald Kalpokas. A general election produced a coalition government of the Francophone Union of Moderate Parties (UMP) and Lini's new National United Party (NUP) under Maxime Carlot Korman. **1993:** A cyclone caused extensive damage. **1995:** The governing UMP–NUP coalition won a general election, but Serge Vohor of the VP-dominated Unity Front became prime minister in place of Carlot Korman. **1996:** The VP, led by Donald Kalpokas, joined the governing coalition. **1997:** Prime Minister Vohor formed a new coalition. The legislature was dissolved and new elections called after a no-confidence motion against Vohor. **1998:** A two-week state of emergency followed rioting in the capital. **1999:** John Bernard Bani was elected president, and Barak Sope was elected prime minister.

Practical information

Visa requirements: UK: visa not required. USA: visa not required
Time difference: GMT +11
Chief tourist attractions: unspoilt landscape

Major holidays: 1 January, 5 March, 1 May, 30 July 15 August, 25–26 December; variable: Ascension, Good Friday, Easter Monday, Constitution (October), Unity (November)

VATICAN CITY STATE

Sovereign area within the city of Rome, Italy.

National name: *Stato della Città del Vaticano/Vatican City State*
Area: 0.4 sq km/0.2 sq mi
Physical features: forms an enclave in the heart of
Rome, Italy

The emblem reflects the Vatican's importance as the headquarters of the Roman Catholic Church. The colours of the flag are based on the gold and silver of the papal keys. Effective date: 8 June 1929.

Government

Head of state: John Paul II from 1978
Head of government: Cardinal Angelo Sodano from 1990
Political system: theocratic
Political executive: theocratic
Death penalty: abolished in 1969

Economy and resources

Currency: Vatican City lira and Italian lira
GDP: see Italy
Real GDP growth: (% change on previous year) see Italy
GNP: (US$) see Italy
GNP per capita (PPP): see Italy
Industries: three main sources of income: the Istituto per le Opere di Religione, 'Peter's pence' (voluntary contributions), and interest on investments managed by the Administration of the Patrimony of the Holy See

Chronology

AD 64: Death of St Peter, a Christian martyr who, by legend, was killed in Rome and became regarded as the first bishop of Rome. The Pope, as head of the Roman Catholic Church, is viewed as the spiritual descendent of St Peter. **756:** The Pope became temporal ruler of the Papal States, which stretched across central Italy, centred around Rome. **11th–13th centuries:** Under Gregory VII and Innocent III the papacy enjoyed its greatest temporal power. **1377:** After seven decades in which the papacy was based in Avignon (France), Rome once again became the headquarters for the Pope, with the Vatican Palace becoming the official residence. **1860:** Umbria, Marche, and much of Emilia Romagna which, along with Lazio formed the Papal States, were annexed by the new unified Italian state. **1870:** First Vatican Council defined as a matter of faith the absolute primacy of the Pope and the infallibility of his pronouncements on 'matters of faith and morals'. **1870–71:** French forces, which had been protecting the Pope, were withdrawn, allowing Italian nationalist forces to capture Rome, which became the capital of Italy; Pope Pius IX retreated into the Vatican Palace, from which no Pope was to emerge until 1929. **1929:** The Lateran Agreement, signed by the Italian fascist leader Benito Mussolini and Pope Pius XI, restored full sovereign jurisdiction over the Vatican City State to the bishopric of Rome (Holy See) and declared the new state to be a neutral and inviolable territory. **1947:** A new Italian constitution confirmed the sovereignty of the Vatican City State. **1962:** The Second Vatican Council was called by Pope John XXIII. **1978:** John Paul II became the first non-Italian pope for more than 400 years. **1985:** A new concordat was signed under which Roman Catholicism ceased to be Italy's state religion. **1992:** Relations with East European states were restored.

Population and society

Population: 1,000 (2000 est)
Population density: (per sq km) 2,500 (2000 est)
Urban population: (% of total) 100 (2000)
Language: Latin (official), Italian
Religion: Roman Catholic
Literacy rate: see Italy
Life expectancy: see Italy

Transport

Airports: international airports: one heliport serves visiting heads of state and Vatican officials; the closest international airport is Rome (see Italy)
Railways: total length: 862 m/2,828 ft (a small railway carrying supplies and goods into the Vatican from Italy)

Practical information

Visa requirements: see Italy. There is free access to certain areas, including St Peter's Church and Square, the Vatican Museum, and Vatican Gardens; special permission is required to visit all other areas
Time difference: GMT +1
Chief tourist attractions: St Peter's Church; St Peter's Square; Vatican Museum; the Pope's Sunday blessing
Major holidays: see Italy

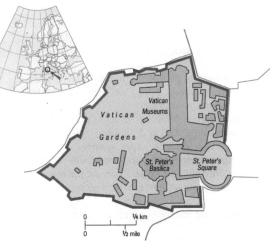

VENEZUELA

Located in northern South America, on the Caribbean Sea, bounded east by Guyana, south by Brazil, and west by Colombia.

National name: *República de Venezuela/Republic of Venezuala*
Area: 912,100 sq km/352,161 sq mi
Capital: Caracas
Major towns/cities: Maracaibo, Maracay, Barquisimeto, Valencia, Ciudad Guayana, Petare
Major ports: Maracaibo
Physical features: Andes Mountains and Lake Maracaibo in northwest; central plains (llanos); delta of River Orinoco in east; Guiana Highlands in southeast

Yellow symbolizes the golden land of South America. Red stands for courage and the blood of the freedom fighters. Blue represents the ocean separating South America from Spain. Effective date: 17 February 1954.

Government

Head of state and government: Hugo Chávez Frías from 1999
Political system: liberal democracy
Political executive: limited presidency
Administrative divisions: 22 states, two territories, one federal district, and one federal dependency; the latter consists of 11 federally controlled island groups with a total of 72 individual islands
Political parties: Democratic Action Party (AD), moderate left of centre; Christian Social Party (COPEI), Christian, right of centre; National Convergence (CN), broad coalition grouping; Movement towards Socialism (MAS), left of centre; Radical Cause (LCR), left wing
Armed forces: 56,000 (1998)
Conscription: selective conscription for 30 months
Death penalty: abolished in 1863
Defence spend: (% GDP) 1.5 (1998)
Education spend: (% GNP) 5.2 (1996)
Health spend: (% GDP) 3.9 (1997)

Economy and resources

Currency: bolívar
GDP: (US$) 103.9 billion (1999)
Real GDP growth: (% change on previous year) –7.2 (1999)
GNP: (US$) 86.96 billion (1999)
GNP per capita (PPP): (US$) 5,268 (1999)
Consumer price inflation: 23.6% (1999)
Unemployment: 12.8% (1998 est)
Foreign debt: (US$) 35.7 billion (1999 est)
Major trading partners: USA, Colombia, Brazil, Japan, Germany, Italy
Resources: petroleum, natural gas, aluminium, iron ore, coal, diamonds, gold, zinc, copper, silver, lead, phosphates, manganese, titanium
Industries: refined petroleum products, metals (mainly aluminium, steel and pig-iron), food products, chemicals, fertilizers, cement, paper, vehicles
Exports: petroleum and petroleum products, metals (mainly aluminium and iron ore), natural gas, chemicals, basic manufactures, motor vehicles and parts. Principal market: USA 47.8% (1998)
Imports: machinery and transport equipment, chemicals,

food and live animals, basic manufactures, crude materials. Principal source: USA 42.1% (1998)
Arable land: 3% (1996)
Agricultural products: coffee, cocoa, sugar cane, bananas, maize, rice, plantains, oranges, sorghum, cassava, wheat, tobacco, cotton, beans, sisal; livestock rearing (cattle)

Population and society

Population: 24,170,000 (2000 est)
Population growth rate: 2.0% (1995–2000); 1.8% (2000–05)
Population density: (per sq km) 26 (1999 est)
Urban population: (% of total) 87 (2000 est)
Age distribution: (% of total population) 0–14 34%, 15–59 59%, 60+ 7% (2000 est)
Ethnic groups: 67% mestizos (of Spanish-American and American-Indian descent), 21% Europeans, 10% Africans, 2% Indians
Language: Spanish (official), Indian languages (2%)
Religion: Roman Catholic 92%
Education: (compulsory years) 10
Literacy rate: 93% (men); 92% (women) (2000 est)
Labour force: 13% agriculture, 23% industry, 64% services (1997 est)
Life expectancy: 70 (men); 76 (women) (1995–2000)
Child mortality rate: (under 5, per 1,000 live births) 25 (1995–2000)
Physicians: 1 per 576 people (1995)
Hospital beds: 1 per 382 people (1995)
TV sets: (per 1,000 people) 185 (1998)
Radios: (per 1,000 people) 468 (1997)
Internet users: (per 10,000 people) 168.7 (1999)
Personal computer users: (per 100 people) 4.2 (1999)

Transport

Airports: international airports: Caracas (Simón Bolívar), Cabello, Maracaibo; domestic services operate to most large towns; total passenger km: 5,800 million (1997 est)
Railways: total length: 336 km/209 mi; total passenger km: 12 million (1995)
Roads: total road network: 84,300 km/52,384 mi, of which 39.4% paved (1996 est); passenger cars: 69.3 per 1,000 people (1996 est)

Chronology

1st millennium BC: Beginnings of settled agriculture. **AD 1498–99:** Visited by explorers Christopher Columbus and Alonso de Ojeda, at which time the principal indigenous Indian communities were the Caribs, Arawaks, and Chibchas; it was named Venezuela ('little Venice') since the coastal Indians lived in stilted thatched houses. **1521:** Spanish settlement established on the northeast coast and was ruled by Spain from Santo Domingo (Dominican Republic). **1567:** Caracas founded by Diego de Losada. **1739:** Became part of newly created Spanish Viceroyalty of New Granada, with capital at Bogotá (Colombia), but, lacking gold mines, retained great autonomy. **1749:** First rebellion against Spanish colonial rule. **1806:** Rebellion against Spain, led by Francisco Miranda. **1811–12:** First Venezuelan Republic declared by patriots, taking advantage of Napoleon Bonaparte's invasion of Spain, but Spanish Royalist forces re-established their authority. **1813–14:** The Venezuelan, Simón Bolívar, 'El Libertador' (the Liberator), created another briefly independent republic, before being forced to withdraw to Colombia. **1821:** After the battle of Carabobo, Venezuelan independence achieved within Republic of Gran Colombia (which also comprised Colombia, Ecuador, and Panama). **1829:** Became separate state of Venezuela after leaving Republic of Gran Colombia. **1830–48:** Gen José Antonio Páez, the first of a series of caudillos (military leaders), established political stability. **1870–88:** Antonio Guzmán Blanco ruled as benevolent liberal–conservative dictator, modernizing infrastructure and developing agriculture (notably coffee) and education. **1899:** International arbitration tribunal found in favour of British Guiana (Guyana) in long-running dispute over border with Venezuela. **1902:** Ports blockaded by British, Italian, and German navies as a result of Venezuela's failure to repay loans. **1908–35:** Harsh rule of dictator Juan Vicente Gómez, during which period Venezuela became world's largest exporter of oil, which had been discovered in 1910. **1947:** First truly democratic elections held, but the new president, Rómulo Gallegos, was removed within eight months by the military in the person of Col Marcos Pérez Jimenez. **1958:** Overthrow of Pérez and establishment of an enduring civilian democracy, headed by left-wing Romulo Betancourt of Democratic Action Party (AD). **1964:** Dr Raúl Leoni (AD) became president in first ever constitutional handover of civilian power. **1974:** Carlos Andrés Pérez (AD) became president, with economy remaining buoyant through oil revenues. Oil and iron industries nationalized. **1984:** Social pact established between government, trade unions, and business; national debt rescheduled as oil revenues plummetted. **1987:** Widespread social unrest triggered by inflation; student demonstrators shot by police. **1989:** An economic austerity programme was instigated. Price increases triggered riots known as 'Caracazo'; 300 people were killed. Martial law was declared and a general strike followed. Elections were boycotted by opposition groups. **1992:** An attempted antigovernment coup failed, at a cost of 120 lives. **1996:** Former President Carlos Andrés Pérez was found guilty on corruption charges and imprisoned. **1999:** Hugo Chávez was inaugurated as president. Flooding and mudslides swamped Venezuela's Caribbean coast in late December, resulting in death tolls as high as 30,000, at least 150,000 homeless civilians from 23,000 destroyed homes, 70,000 evacuees, and 96,000 damaged homes. **2000:** Despite a shrinking economy, Hugo Chávez was re-elected as president, pledging to redistribute oil wealth from the rich to the poor. He later took a leading role in persuading the Organization of Petroleum-Exporting Countries (OPEC) to restrict world oil production to force up prices. In November, Chávez was given powers to legislate on certain issues by decree.

Practical information

Visa requirements: UK/USA: visa not required
Time difference: GMT –4
Major holidays: 1, 6 January, 19 March, 19 April, 1 May, 24, 29 June, 5, 24 July, 15 August, 12 October, 1 November, 8, 25 December; variable: Ascension Thursday, Carnival (2 days), Corpus Christi, Good Friday, Holy Thursday

VIETNAM FORMERLY PART OF FRENCH INDO-CHINA (1884–1945), COMMUNIST DEMOCRATIC REPUBLIC OF VIETNAM (IN NORTH) (1945–75), NON-COMMUNIST REPUBLIC OF VIETNAM (IN SOUTH) (1949–75)

Located in Southeast Asia, on the South China Sea, bounded north by China and west by Cambodia and Laos.

National name: *Công-hòa xã-hôi chu-nghia Viêt Nam/Socialist Republic of Vietnam*
Area: 329,600 sq km/127,258 sq mi
Capital: Hanoi
Major towns/cities: Hô Chi Minh (formerly Saigon), Hai Phong, Da Nang, Can Tho, Nha Trang, Bien Hoa, Hue
Major ports: Hô Chi Minh (formerly Saigon), Da Nang, Hai Phong
Physical features: Red River and Mekong deltas, centre of cultivation and population; tropical rainforest; mountainous in north and northwest

The five points of the star represent the unity of farmers, workers, intellectuals, soldiers, and youth in establishing socialism. Red stands for revolution and bloodshed. Effective date: 2 July 1976.

Government

Head of state: Tran Duc Luong from 1997
Head of government: Phan Van Khai from 1997
Political system: communist
Political executive: communist
Administrative divisions: 58 provinces and three municipalities
Political party: Communist Party
Armed forces: 484,000; plus paramilitary forces numbering 40,000 and around 3 million reserves (1998)
Conscription: military service is compulsory for two years
Death penalty: retained and used for ordinary crimes
Defence spend: (% GDP) 3.4 (1998)
Education spend: (% GNP) 3.0 (1997)
Health spend: (% GDP) 1.1 (1990–95)

Economy and resources

Currency: dong
GDP: (US$) 28.6 billion (1999)
Real GDP growth: (% change on previous year) 4.7 (1999)
GNP: (US$) 28.2 billion (1999)
GNP per capita (PPP): (US$) 1,755 (1999)
Consumer price inflation: 4.3% (1999)
Unemployment: 7% (1994 est)
Foreign debt: (US$) 10.5 billion (1999)
Major trading partners: Singapore, Japan, China, South Korea, Taiwan, Thailand, Australia, Sweden, Germany
Resources: petroleum, coal, tin, zinc, iron, antimony, chromium, phosphate, apatite, bauxite
Industries: food processing, chemicals, machinery, textiles, beer, glass and glassware, cigarettes, crude steel, cement, fertilizers, tourism
Exports: rice (leading exporter), crude petroleum, coal, coffee, marine products, handicrafts, light industrial goods, rubber, nuts, tea, garments, tin. Principal market: Japan 15.5% (1999)
Imports: petroleum products, machinery and spare parts, steel, artificial fertilizers, basic manufactures, consumer goods. Principal source: Singapore 16.2% (1999)
Arable land: 16.9% (1996)

Agricultural products: rice (world's fifth-largest producer), coffee, tea, rubber, cotton, groundnuts, sugar cane, coconuts; livestock rearing; fishing

Population and society

Population: 79,832,000 (2000 est)
Population growth rate: 1.6% (1995–2000)
Population density: (per sq km) 237 (1999 est)
Urban population: (% of total) 20 (2000 est)
Age distribution: (% of total population) 0–14 33%, 15–59 59%, 60+ 8% (2000 est)
Ethnic groups: 84% Viet (also known as Kinh), 2% Chinese, 2% Khmer, 8% consists of more than 50 minority nationalities, including the Hmong, Meo, Muong, Nung, Tay, Thai, and Tho tribal groups
Language: Vietnamese (official), French, English, Khmer, Chinese, local languages
Religion: mainly Buddhist; Christian, Roman Catholic (8–10%); Taoist, Confucian, Hos Hoa, and Cao Dai sects
Education: (compulsory years) 5
Literacy rate: 95% (men); 91% (women) (2000 est)
Labour force: 69.2% agriculture, 12.9% industry, 17% services (1996)
Life expectancy: 65 (men); 70 (women) (1995–2000)
Child mortality rate: (under 5, per 1,000 live births) 56 (1995–2000)
Physicians: 1 per 2,500 people (1994)
Hospital beds: 1 per 261 people (1993 est)
TV sets: (per 1,000 people) 47 (1997)
Radios: (per 1,000 people) 107 (1997)
Internet users: (per 10,000 people) 12.7 (1999)
Personal computer users: (per 100 people) 0.9 (1999)

Transport

Airports: international airports: Hanoi (Noi Bai), Hô Chi Minh (Tan Son Nhat); seven domestic airports; total passenger km: 3,785 million (1997 est)
Railways: total length: 2,605 km/1,619 mi; total passenger km: 2,444 million (1997)
Roads: total road network: 93,300 km/57,977 mi, of which 26% paved (1996 est)

Chronology

300 BC: Rise of Dong Son culture. **111 BC:** Came under Chinese rule. **1st–6th centuries AD:** Southern Mekong delta region controlled by independent Indianized Funan kingdom. **939:** Chinese overthrown by Ngo Quyen at battle of Bach Dang River; first Vietnamese dynasty founded. **11th century:** Theravada Buddhism promoted. **15th century:** North and South Vietnam united, as kingdom of Champa in the south was destroyed in 1471. **16th century:** Contacts with French missionaries and European traders as political power became decentralized. **early 19th century:** Under Emperor Nguyen Anh authority was briefly recentralized. **1858–84:** Conquered by France and divided into protectorates of Tonkin (North Vietnam) and Annam (South Vietnam). **1887:** Became part of French Indo-China Union, which included Cambodia and Laos. **late 19th–early 20th century:** Development of colonial economy based in south on rubber and rice, drawing migrant labourers from north. **1930:** Indo-Chinese Communist Party (ICP) formed by Ho Chi Minh to fight for independence. **1941:** Occupied by Japanese during World War II; ICP formed Vietminh as guerrilla resistance force designed to overthrow Japanese-installed puppet regime headed by Bao Dai, Emperor of Annam. **1945:** Japanese removed from Vietnam at end of World War II; Vietminh, led by Ho Chi Minh, in control of much of the country, declared independence. **1946:** Vietminh war began against French, who tried to reassert colonial control and set up noncommunist state in south in 1949. **1954:** France decisively defeated at Dien Bien Phu. Vietnam divided along 17th parallel between communist-controlled north and US-backed south. **1963:** Ngo Dinh Diem, leader of South Vietnam, overthrown in military coup by Lt-Gen Nguyen Van Thieu. **1964:** US combat troops entered Vietnam War as North Vietnamese army began to attack South and allegedly attacked US destroyers in the Tonkin Gulf. **1969:** Death of Ho Chi Minh, who was succeeded as Communist Party leader by Le Duan. US forces, which numbered 545,000 at their peak, gradually began to be withdrawn from Vietnam as a result of domestic opposition to the rising casualty toll. **1973:** Paris ceasefire agreement provided for the withdrawal of US troops and release of US prisoners of war. **1975:** Saigon captured by North Vietnam, violating Paris Agreements. **1976:** Socialist Republic of Vietnam proclaimed. Hundreds of thousands of southerners became political prisoners; many more fled abroad. Collectivization extended to south. **1978:** Diplomatic relations severed with China. Admission into Comecon. Vietnamese invasion of Cambodia. **1979:** Sino-Vietnamese 17-day border war; 700,000 Chinese and middle-class Vietnamese fled abroad as refugee 'boat people'. **1986:** Death of Le Duan and retirement of 'old guard' leaders; pragmatic Nguyen Van Linh became Communist Party leader. **1987–88:** Over 10,000 political prisoners were released. **1989:** Troops were fully withdrawn from Cambodia. **1991:** A Cambodia peace agreement was signed. Relations with China were normalized. **1992:** A new constitution was adopted, guaranteeing economic freedoms. Relations with South Korea were normalized. **1994:** The US 30-year trade embargo was removed. **1995:** Full diplomatic relations were re-established with the USA. Vietnam became a full member of ASEAN. **1997:** Diplomatic relations with the USA were restored. Tran Duc Luong and Phan Van Khai were elected president and prime minister respectively. The size of the standing army was reduced. **1998:** The Vietnamese currency was devalued. A new emphasis was placed on agricultural development after export and GDP growth slumped to 3%. **1999:** Vietnam signed a trade agreement with the USA, encouraging foreign investment. **2000:** US president Clinton visited Vietnam.

Practical information

Visa requirements: UK: visa required. USA: visa required
Time difference: GMT +7
Chief tourist attractions: Hanoi, with its 11th-century Temple of Literature, Mot Cot Pagoda, 3rd-century Co Loa citadel, museums (many historical sites were destroyed by war); tropical rainforest
Major holidays: 1 January, 30 April, 1 May, 1–2 September; variable: Têt, Lunar New Year (January/February, 3 days)

YEMEN DIVIDED INTO NORTH YEMEN (YEMEN ARAB REPUBLIC) AND SOUTH YEMEN UNTIL 1990

Located in southwest Asia, bounded north by Saudi Arabia, east by Oman, south by the Gulf of Aden, and west by the Red Sea.

National name: *Al-Jumhuriyya al Yamaniyya/Republic of Yemen*
Area: 531,900 sq km/205,366 sq mi
Capital: San'a
Major towns/cities: Aden, Ta'izz, Al Mukalla, Al Hudaydah, Ibb, Dhamar
Major ports: Aden
Physical features: hot, moist coastal plain, rising to plateau and desert

The red, white, and black tricolour formed the basis of the flags of both North and South Yemen.
Effective date: 22 May 1990.

Government

Head of state: Ali Abdullah Saleh from 1990
Head of government: Abdul Ali al-Rahman al-Iryani from 1998
Political system: emergent democracy
Political executive: limited presidency
Administrative divisions: 17 governates
Political parties: General People's Congress (GPC), left of centre; Yemen Socialist Party (YSP), left wing; Yemen Reform Group (al-Islah), Islamic, right of centre; National Opposition Front, left of centre
Armed forces: 66,300; plus paramilitary forces numbering at least 80,000 (1998)
Conscription: military service is compulsory for two years
Death penalty: retained and used for ordinary crimes
Defence spend: (% GDP) 6.6 (1998)
Education spend: (% GNP) 7.0 (1997)
Health spend: (% GDP) 3.4 (1997 est)

Economy and resources

Currency: riyal
GDP: (US$) 6.8 billion (1999)
Real GDP growth: (% change on previous year) 3.8 (1999 est)
GNP: (US$) 5.9 billion (1999)
GNP per capita (PPP): (US$) 688 (1999)
Consumer price inflation: 8% (1999)
Unemployment: 25% (1997 est)
Foreign debt: (US$) 4.2 billion (1999 est)
Major trading partners: Thailand, Saudi Arabia, China, United Arab Emirates, USA, Singapore, France, South Korea
Resources: petroleum, natural gas, gypsum, salt; deposits of copper, gold, lead, zinc, molybdenum
Industries: petroleum refining and petroleum products, building materials, food processing, beverages, tobacco, chemical products, textiles, leather goods, metal goods
Exports: petroleum and petroleum products, cotton, basic manufactures, clothing, live animals, hides and skins, fish, rice, coffee. Principal market: Thailand 35.2% (1999 est)
Imports: textiles and other manufactured consumer goods, petroleum products, sugar, grain, flour, other foodstuffs, cement, machinery, chemicals. Principal source: Saudi Arabia 9.7% (1999 est)
Arable land: 2.7% (1996)
Agricultural products: sorghum, sesame, millet, potatoes, tomatoes, cotton, wheat, grapes, watermelons, coffee, alfalfa, dates, bananas; livestock rearing; fishing

Population and society

Population: 18,112,000 (2000 est)
Population growth rate: 3.7% (1995–2000); 3.1% (2000–05)
Population density: (per sq km) 33 (1999 est)
Urban population: (% of total) 25 (2000 est)
Age distribution: (% of total population) 0–14 48%, 15–59 48%, 60+ 4% (2000 est)
Ethnic groups: predominantly Arab; some of mixed Afro-Arab origin; small Asian and European communities
Language: Arabic (official)
Religion: Sunni Muslim 63%, Shiite Muslim 37%
Education: (compulsory years): 6 (North); 8 (South)
Literacy rate: 67% (men); 25% (women) (2000 est)
Labour force: 54% agriculture, 10% industry, 36% services (1997 est)
Life expectancy: 57 (men); 58 (women) (1995–2000)
Child mortality rate: (under 5, per 1,000 live births) 113 (1995–2000)
Physicians: 1 per 4,498 people (1993 est)
Hospital beds: 1 per 1,196 people (1993 est)
TV sets: (per 1,000 people) 29 (1997)
Radios: (per 1,000 people) 64 (1997)
Internet users: (per 10,000 people) 5.7 (1999)
Personal computer users: (per 100 people) 0.2 (1999)

Transport

Airports: international airports: Sana'a (El-Rahaba), Ta'izz (al-Jahad), Al Hudaydah, Aden (Khormaksar), Al Mukalla (Riyan), Seybun; domestic services operate between these; total passenger km: 486 million (1995)
Railways: none
Roads: total road network: 64,725 km/40,220 mi, of which 8.1% paved (1996); passenger cars: 14.2 per 1,000 people (1996)

Chronology

1st millennium BC: South Yemen (Aden) divided between economically advanced Qataban and Hadramawt kingdoms. ***c.* 5th century BC:** Qataban fell to the Sabaeans (Shebans) of North Yemen (Sana). ***c.* 100 BC–AD 525:** All of Yemen became part of the Himyarite kingdom. **AD 628:** Islam introduced. **1174–1229:** Under control of Egyptian Ayyubids. **1229–1451:** 'Golden age' for arts and sciences under the Rasulids, who had served as governors of Yemen under the Ayyubids.

1538: North Yemen came under control of Turkish Ottoman Empire. **1636:** Ottomans left North Yemen and power fell into hands of Yemeni Imams, based on local Zaydi tribes, who also held South Yemen until 1735. **1839:** Aden became a British territory. Port developed into an important ship refuelling station after opening of Suez Canal in 1869; protectorate was gradually established over 23 Sultanates inland. **1870s:** The Ottomans reestablished control over North Yemen. **1918:** North Yemen became independent, with Imam Yahya from the Hamid al-Din family as king. **1937:** Aden became a British crown colony. **1948:** Imam Yahya assassinated by exiled Free Yemenis nationalist movement, but the uprising was crushed by his son, Imam Ahmad. **1959:** Federation of South Arabia formed by Britain between city of Aden and feudal Sultanates (Aden Protectorate). **1962:** Military coup on death of Imam Ahmad; North Yemen declared Yemen Arab Republic (YAR), with Abdullah al-Sallal as president. Civil war broke out between royalists (supported by Saudi Arabia) and republicans (supported by Egypt). **1963:** Armed rebellion by National Liberation Front (NLF) began against British rule in Aden. **1967:** Civil war ended with republicans victorious. Sallal deposed and replaced by Republican Council. The Independent People's Republic of South Yemen was formed after the British withdrawal from Aden. Many fled to the north as the repressive communist NLF regime took over in south. **1970:** People's Republic of South Yemen renamed People's Democratic Republic of Yemen. **1971–72:** War between South Yemen and YAR; union agreement brokered by Arab League signed but not kept. **1974:** The pro-Saudi Col Ibrahim al-Hamadi seized power in North Yemen; Military Command Council set up. **1977:** Hamadi assassinated; replaced by Col Ahmed ibn Hussein al-Ghashmi. **1978:** Constituent people's assembly appointed in North Yemen and Military Command Council dissolved. Ghashmi killed by envoy from South Yemen; succeeded by Ali Abdullah Saleh. War broke out again between the two Yemens. The South Yemen president was deposed and executed; the Yemen Socialist Party (YSP) was formed in the south by communists. **1979:** A ceasefire was agreed with a commitment to future union. **1986:** There was civil war in South Yemen; the autocratic head of state Ali Nasser was dismissed. A new administration was formed under the more moderate Haydar Abu Bakr al-Attas, who was committed to negotiating union with the north because of the deteriorating economy in the south. **1989:** A draft multiparty constitution for a single Yemen state was published. **1990:** The border between the two Yemens was opened; the countries were formally united on 22 May as the Republic of Yemen. Ali Abdullah Saleh, president of North Yemen since 1978, was appointed president of the new unified Yemen. **1991:** The new constitution was approved; Yemen opposed US-led operations against Iraq in the Gulf War. **1992:** There were antigovernment riots. **1993:** Saleh's General People's Congress (GPC) won most seats in a general election but no overall majority, a five-member presidential council was elected, including Saleh as president, YSP leader Ali Salim al-Baidh as vice-president, and Bakr al-Attas as prime minister. **1994:** Fighting erupted between northern forces, led by President Saleh, and southern forces, led by Vice-president al-Baidh, as southern Yemen announced its secession. Saleh inflicted crushing defeat on al-Baidh and a new GPC coalition was appointed. **1998:** A new government was headed by Abdul Ali al-Rahman al-Iryani. **1999:** In the first ever popular elections for the presidency, Ali Abdullah Saleh, the president for 21 years, was successful. **2000:** A terrorist suicide bomb attack on a US destroyer, *USS Cole*, killed 17 crew members.

Practical information

Visa requirements: UK/USA: visa required
Time difference: GMT +3
Chief tourist attractions: sandy beaches along coastal plains; Hadramaut mountain range; Aden; San'a's old city, with its medieval mosques and other buildings; hillside villages of Kawkaban, Thulla, and Shiban
Major holidays: 1 May, 26 September; variable: Eid-ul-Adha (5 days), end of Ramadan (4 days), New Year (Muslim), Prophet's Birthday

YUGOSLAVIA FORMERLY KINGDOM OF THE SERBS, CROATS, AND SLOVENES (1918–29)

Located in southeast Europe, with a southwest coastline on the Adriatic Sea, bounded west by Bosnia-Herzegovina, northwest by Croatia, north by Hungary, east by Romania and Bulgaria, and south by the Former Yugoslav Republic of Macedonia and Albania.

National name: *Savezna Republika Jugoslavija/Federal Republic of Yugoslavia*
Area: 58,300 sq km/22,509 sq mi
Capital: Belgrade
Major towns/cities: Priština, Novi Sad, Niš, Kragujevac, Podgorica (formerly Titograd), Subotica
Physical features: federation of republics of Serbia and Montenegro and two former autonomous provinces, Kosovo and Vojvodina

Blue, white, and red recall the 19th century Russian tricolour. The 'Partisan Star' was removed in 1991. Effective date: 27 April 1992.

Government

Head of state: Vojislav Koštunica from 2000
Head of government: Zoran Zizic from 2000
Political system: emergent democracy
Political executive: limited presidency
Administrative divisions: two republics (Serbia and Montenegro) and two nominally autonomous provinces (Kosovo and Vojvodina)
Political parties: Socialist Party of Serbia (SPS), Serb nationalist, reform socialist (ex-communist); Montenegrin Social Democratic Party (SDPCG), federalist, reform socialist (ex-communist); Serbian Radical Party (SRS), Serb nationalist, extreme right wing; People's Assembly Party, Christian democrat, centrist; Democratic Party (DS), moderate nationalist; Democratic Party of Serbia (DSS), moderate nationalist; Democratic Community of Vojvodina Hungarians (DZVM), ethnic Hungarian; Democratic Party of Albanians/Party of Democratic Action (DPA/PDA), ethnic Albanian; New Socialist Party of Montenegro (NSPM), left of centre
Armed forces: 114,200 (1998)
Conscription: military service is compulsory for 12–15 months; voluntary military service for women introduced in 1983
Death penalty: retained and used for ordinary crimes
Defence spend: (% GDP) 9.1 (1998)
Education spend: (% GNP) N/A
Health spend: (% GDP) 4.5 (1997 est)

Economy and resources

Currency: new Yugoslav dinar
GDP: (US$) 12.3 billion (1999 est)
Real GDP growth: (% change on previous year) –23.2 (1999 est)
GNP: (US$) 17.9 billion (1997 est)
GNP per capita (PPP): (US$) 5,880 (1997 est)
Consumer price inflation: 42.4% (1999)
Unemployment: 26.1% (1996)
Foreign debt: (US$) 13.1 billion (1999 est)
Major trading partners: Germany, Bosnia-Herzegovina, Russia, Italy, Macedonia
Resources: petroleum, natural gas, coal, copper ore, bauxite, iron ore, lead, zinc
Industries: crude steel, pig-iron, steel castings, cement, machines, passenger cars, electrical appliances, artificial fertilizers, plastics, bicycles, textiles and clothing
Exports: basic manufactures, machinery and transport equipment, clothing, miscellaneous manufactured articles, food and live animals. Principal market: Bosnia-Herzegovina 20.3% (1999)
Imports: machinery and transport equipment, electrical goods, agricultural produce, mineral fuels and lubricants, basic manufactures, foodstuffs, chemicals. Principal source: Germany 12.3% (1999)
Arable land: 36.4% (1996)
Agricultural products: maize, sugar beet, wheat, potatoes, grapes, plums, soybeans, vegetables; livestock production declined 1991–95

Population and society

Population: 10,640,000 (2000 est)
Population growth rate: 0.1% (1995–2000)
Population density: (per sq km) 182 (1999 est)
Urban population: (% of total) 52 (2000 est)
Age distribution: (% of total population) 0–14 20%, 15–59 62%, 60+ 18% (2000 est)
Ethnic groups: 63% Serbs, 14% Albanian, 6% Montenegrin, 4% Hungarian, and 13% other. Serbs predominate in the republic of Serbia, where they form (excluding the autonomous areas of Kosovo and Vojvodina) 85% of the population; in Vojvodina they comprise 55% of the population. Albanians constitute 77% of the population of Kosovo; Montenegrins comprise 69% of the population of the republic of Montenegro; and Muslims predominate in the Sandzak region, which straddles the Serbian and Montenegrin borders. Since 1992 an influx of Serb refugees from Bosnia and Kosovo has increased the proportion of Serbs in Serbia, while many ethnic Hungarians have left Vojvodina, and an estimated 500,000 Albanians have left Kosovo
Language: Serbo-Croat (official), Albanian (in Kosovo)
Religion: Serbian and Montenegrin Orthodox; Muslim in southern Serbia
Education: (compulsory years) 8
Literacy rate: 97% (men); 88% (women) (1995 est)
Labour force: 5% agriculture, 41% industry, 54% services (1994)
Life expectancy: 70 (men); 76 (women) (1995–2000)
Child mortality rate: (under 5, per 1,000 live births) 26 (1995–2000)

Physicians: 1 per 506 people (1993)
Hospital beds: 1 per 183 people (1993)
TV sets: (per 1,000 people) 259 (1997)
Radios: (per 1,000 people) 296 (1997)
Internet users: (per 10,000 people) 75.2 (1999)
Personal computer users: (per 100 people) 2.1 (1999)

Transport

Airports: international airports: Belgrade (Surcin), Podgorica; three domestic airports; total passenger km: 3,443 million (1991)
Railways: total length: 3,987 km/2,478 mi; total passenger km: 2,580 million (1995)
Roads: total road network: 50,414 km/31,327 mi, of which 89.3% paved (1997); passenger cars: 173.1 per 1,000 people (1997)

Chronology

3rd century BC: Serbia (then known as Moesia Superior) conquered by Romans; empire was extended to Belgrade centuries later by Emperor Augustus. **6th century AD:** Slavic tribes, including Serbs, Croats, and Slovenes, settled in Balkan Peninsula. **879:** Serbs converted to Orthodox Church. **mid-10th–11th centuries:** Serbia broke free briefly from Byzantine Empire to establish independent state. **1217:** Independent Serbian kingdom re-established, reaching its height in mid-14th century, when it controlled much of Albania and northern Greece. **1389:** Serbian army defeated by Ottoman Turks at Battle of Kosovo; area became Turkish *pashalik* (province). Montenegro in southwest survived as sovereign principality. Croatia and Slovenia in northwest became part of Habsburg Empire. **18th century:** Vojvodina enjoyed protection from the Austrian Habsburgs. **1815:** Uprisings against Turkish rule secured autonomy for Serbia. **1878:** Independence achieved as Kingdom of Serbia, after Turks defeated by Russians in war over Bulgaria. **1912–13:** During Balkan Wars, Serbia expanded its territory at expense of Turkey and Bulgaria. **1918:** Joined Croatia and Slovenia, formerly under Austrian Habsburg control, to form Kingdom of Serbs, Croats, and Slovenes; Montenegro's citizens voted to depose their ruler, King Nicholas, and join the union. **1929:** New name of Yugoslavia ('Land of the Southern Slavs') adopted; Serbian-dominated military dictatorship established by King Alexander I as opposition mounted from Croatian federalists. **1934:** Nazi Germany and fascist Italy increased their influence. **1941:** Following a coup by pro-Allied air-force officers, Nazi Germany invaded. Armed resistance to German rule began, spearheaded by pro-royalist, Serbian-based Chetniks ('Army of the Fatherland'), led by Gen Draza Mihailovic, and communist Partisans ('National Liberation Army'), led by Marshal Tito. An estimated 900,000 Yugoslavs died in the war, including more than 400,000 Serbs and 200,000 Croats. **1943:** Provisional government formed by Tito at liberated Jajce in Bosnia. **1945:** Yugoslav Federal People's Republic formed under leadership of Tito; communist constitution introduced. **1948:** Split with Soviet Union after Tito objected to Soviet 'hegemonism'; expelled from Cominform. **1953:** Workers' self-management principle enshrined in constitution and private farming supported; Tito became president. **1961:** Nonaligned movement formed under Yugoslavia's leadership. **1971:** In response to mounting separatist demands in Croatia, new system of collective and rotating leadership introduced. **1980:** Tito died; collective leadership assumed power. **1981–82:** Armed forces suppressed demonstrations in Kosovo province, southern Serbia, by Albanians demanding full republic status. **1986:** Slobodan Milošević, a populist-nationalist hardliner who had the ambition of creating a 'Greater Serbia', became leader of communist party in the Serbian republic. **1988:** Economic difficulties: 1,800 strikes, 250% inflation, 20% unemployment. Ethnic unrest in Montenegro and Vojvodina, and separatist demands in rich northwestern republics of Croatia and Slovenia; 'market socialist' reform package, encouraging private sector, inward investment, and liberalizing prices combined with austerity wage freeze. **1989:** Ethnic riots in Kosovo province against Serbian attempt to end autonomous status of Kosovo and Vojvodina; at least 30 were killed and a state of emergency imposed. **1990:** Multiparty systems were established in the republics; Kosovo and Vojvodina were stripped of autonomy. In Croatia, Slovenia, Bosnia, and Macedonia elections brought to power new noncommunist governments seeking a looser confederation. **1991:** Slovenia and Croatia declared their independence, resulting in clashes between federal and republican armies; Slovenia accepted a peace pact sponsored by the European Community (EC), but fighting intensified in Croatia, where Serb militias controlled over a third of the republic. **1992:** There was an EC-brokered ceasefire in Croatia; the EC and the USA recognized Slovenia's and Croatia's independence. Bosnia-Herzegovina and Macedonia then declared their independence, and Bosnia-Herzegovina's independence was recognized by the EC and the USA. A New Federal Republic of Yugoslavia (FRY) was proclaimed by Serbia and Montenegro but not internationally recognized; international sanctions were imposed and UN membership was suspended. Ethnic Albanians proclaimed a new 'Republic of Kosovo', but it was not recognized. **1993:** Macedonia was recognized as independent under the name of the Former Yugoslav Republic of Macedonia. The economy was severely damaged by international sanctions. **1994:** A border blockade was imposed by Yugoslavia against Bosnian Serbs; sanctions were eased as a result. **1995:** Serbia played a key role in the US-brokered Dayton peace accord for Bosnia-Herzegovina and accepted the separate existence of Bosnia and Croatia. **1996:** Diplomatic relations were restored between

Serbia and Croatia, and UN sanctions against Serbia were lifted. Diplomatic relations were established with Bosnia-Herzegovina. There was mounting opposition to Milošević's government following its refusal to accept opposition victories in municipal elections. **1997:** Milošević was elected president and the pro-democracy mayor of Belgrade was ousted. The validity of Serbian presidential elections continued to be questioned. **1998:** A Serb military offensive against ethnic Albanian separatists in Kosovo led to a refugee and humanitarian crisis. The offensive against the Kosovo Liberation Army (KLA) was condemned by the international community and NATO military intervention was threatened. **1999:** Fighting continued between Serbians and Albanian separatists in Kosovo. In March, following the failure of efforts to reach a negotiated settlement, NATO began a bombing campaign against the Serbs; the ethnic cleansing of Kosovars by Serbs intensified and the refugee crisis in neighbouring countries worsened as hundreds of thousands of ethnic Albanians fled Kosovo. In May President Milošević was indicted for crimes against humanity by the International War Crimes Tribunal in The Hague. A peace was agreed on NATO terms in June. Refugees began returning to Kosovo. **2000:** Presidential elections were held in September in which opposition candidate Vojislav Koštunica claimed outright victory against President Slobodan Milošević, but the federal election commission ordered a second round of voting to be held. The opposition claimed ballot-rigging and organized mass demonstrations throughout Yugoslavia, in the face of which Milošević conceded defeat. The UN reinstated Yugoslavia's membership, which had been suspended in 1992, in October.

Practical information

Visa requirements: UK: visa required. USA: visa required
Time difference: GMT +1
Chief tourist attractions: Montenegro's Adriatic coastline and its great lake of Scutari; varied scenery – rich alpine valleys, rolling green hills, bare, rocky gorges, thick forests, limestone mountains
Major holidays: 1–2 January, 1–2 May, 4, 7 (Serbia only), 13 (Montenegro only) July, 29–30 November; Orthodox Christian holidays may also be celebrated throughout much of the region

ZAMBIA FORMERLY NORTHERN RHODESIA (UNTIL 1964)

Landlocked country in southern central Africa, bounded north by the Democratic Republic of Congo (formerly Zaire) and Tanzania, east by Malawi, south by Mozambique, Zimbabwe, Botswana, and Namibia, and west by Angola.

National name: *Republic of Zambia*
Area: 752,600 sq km/290,578 sq mi
Capital: Lusaka
Major towns/cities: Kitwe, Ndola, Kabwe, Mufulira, Chingola, Luanshya, Livingstone
Physical features: forested plateau cut through by rivers; Zambezi River, Victoria Falls, Kariba Dam

Green represents agriculture. Red recalls the struggle for independence. Black stands for the Zambian people. Orange symbolizes Zambia's mineral wealth, particularly the major deposits of copper. Effective date: 24 October 1964.

Government

Head of state and government: Frederick Chiluba from 1991
Political system: emergent democracy
Political executive: limited presidency
Administrative divisions: nine provinces
Political parties: United National Independence Party (UNIP), African socialist; Movement for Multiparty Democracy (MMD), moderate, left of centre; Multiracial Party (MRP), moderate, left of centre, multiracial; National Democratic Alliance (NADA), left of centre; Democratic Party (DP), left of centre
Armed forces: 21,600; plus paramilitary forces of 1,400 (1998)
Conscription: military service is voluntary
Death penalty: retained and used for ordinary crimes
Defence spend: (% GDP) 1.9 (1998)
Education spend: (% GNP) 2.2 (1996)
Health spend: (% GDP) 5.9 (1997 est)

Economy and resources

Currency: Zambian kwacha
GDP: (US$) 3.3 billion (1999)
Real GDP growth: (% change on previous year) 2.2 (1999 est)
GNP: (US$) 3.2 billion (1999)
GNP per capita (PPP): (US$) 686 (1999)
Consumer price inflation: 26.9% (1999)
Foreign debt: (US$) 6.1 billion (1999)
Major trading partners: South Africa, Saudi Arabia, Japan, UK, Thailand, Zimbabwe
Resources: copper (world's fourth-largest producer), cobalt, zinc, lead, coal, gold, emeralds, amethysts and other gemstones, limestone, selenium
Industries: metallurgy (smelting and refining of copper and other metals), food canning, fertilizers, explosives, textiles, bottles, bricks, copper wire, batteries
Exports: copper, zinc, lead, cobalt, tobacco. Principal market: Saudi Arabia 12.5% (1999 est)
Imports: machinery and transport equipment, mineral fuels, lubricants, electricity, basic manufactures, chemicals, food and live animals. Principal source: South Africa 55.5% (1999 est)
Arable land: 7.1% (1996)
Agricultural products: maize, sugar cane, seed cotton, tobacco, groundnuts, wheat, rice, beans, cassava, millet, sorghum, sunflower seeds, horticulture; cattle rearing

Population and society

Population: 9,169,000 (2000 est)
Population growth rate: 2.3% (1995–2000)
Population density: (per sq km) 12 (1999 est)
Urban population: (% of total) 40 (2000 est)
Age distribution: (% of total population) 0–14 47%, 15–59 50%, 60+ 3% (2000 est)
Ethnic groups: over 95% indigenous Africans, belonging to more than 70 different ethnic groups, including the Bantu-Botatwe and the Bemba; about 1% European
Language: English (official), Bantu languages
Religion: about 64% Christian, animist, Hindu, Muslim
Education: (compulsory years) 7
Literacy rate: 85% (men); 71% (women) (2000 est)
Labour force: 42% of population: 75% agriculture, 8% industry, 17% services (1990)
Life expectancy: 40 (men); 41 (women) (1995–2000)
Child mortality rate: (under 5, per 1,000 live births) 147 (1995–2000)
Physicians: 1 per 8,437 people (1995)
Hospital beds: 1 per 349 people (1995)
TV sets: (per 1,000 people) 137 (1998)
Radios: (per 1,000 people) 121 (1997)
Internet users: (per 10,000 people) 16.7 (1999)
Personal computer users: (per 100 people) 0.7 (1999)

Transport

Airports: international airports: Lusaka; over 127 domestic airports, aerodromes, and airstrips; total passenger km: 428 million (1994)
Railways: total length: 1,289 km/801 mi; total passenger km: 268 million (1990)
Roads: total road network: 66,781 km/41,498 mi, of which 18.3% paved (1997); passenger cars: 15 per 1,000 people (1996 est)

Chronology

16th century: Immigration of peoples from Luba and Lunda Empires of Zaire, to the northwest, who set up small kingdoms. **late 18th century:** Visited by Portuguese explorers. **19th century:** Instability with immigration of Ngoni from east, Kololo from west, establishment of Bemba kingdom in north, and slave-trading activities of Portuguese and Arabs from East Africa. **1851:** Visited by British missionary and explorer David Livingstone. **1889:** As Northern Rhodesia, came under administration of British South Africa Company of Cecil Rhodes, and became involved in copper mining, especially from 1920s. **1924:** Became a British protectorate. **1948:** Northern Rhodesia African Congress (NRAC) formed by black Africans to campaign for self-rule. **1953:** Became part of Central African Federation, which included South Rhodesia (Zimbabwe) and Nyasaland (Malawi). **1960:** UNIP was formed by Kenneth Kaunda as a breakaway from NRAC, as African socialist body to campaign for independence and dissolution of federation dominated by South Rhodesia's white minority. **1963:** The federation was dissolved and internal self-government achieved. **1964:** Independence was achieved within the Commonwealth as the Republic of Zambia, with Kaunda of the UNIP as president. **later 1960s:** Key enterprises were brought under state control. **1972:** UNIP was declared the only legal party. **1975:** The opening of the Tan-Zam railway from the Zambian copperbelt, 322 mi/200 km north of Lusaka, to port of Dar es Salaam in Tanzania, reduced Zambia's dependence on the rail route via Rhodesia (Zimbabwe) for its exports. **1976:** Zambia declared its support for Patriotic Front (PF) guerrillas fighting to topple the white-dominated regime in Rhodesia (Zimbabwe). **1980:** There was an unsuccessful South African-promoted coup against President Kaunda; relations with Zimbabwe improved when the PF came to power. **1985:** Kaunda was elected chair of African Front Line States. **1991:** A new multiparty constitution was adopted. The MMD won a landslide election victory, and its leader Frederick Chiluba became president in what was the first democratic change of government in English-speaking black Africa. **1993:** A state of emergency was declared after rumours of a planned antigovernment coup. A privatization programme was launched. **1996:** Kaunda was effectively barred from future elections by an amendment to the constitution. **1997:** There was an abortive antigovernment coup. **1998:** Former president Kaunda was placed under house arrest after alleged involvement in the antigovernment coup. Kaunda was charged but the charges were subsequently dropped.

Practical information

Visa requirements: UK: visa not required. USA: visa required
Time difference: GMT +2
Chief tourist attractions: unspoilt scenery; wildlife; 19 national parks, including the magnificent Luangava and Kafue National Parks, which have some of the most prolific animal populations in Africa; Victoria Falls
Major holidays: 1 January, 1, 25 May, 24 October, 25 December; variable: Good Friday, Holy Saturday, Youth (March), Heroes (July), Unity (July), Farmers (August)

ZIMBABWE FORMERLY SOUTHERN RHODESIA (UNTIL 1980)

Landlocked country in south central Africa, bounded north by Zambia, east by Mozambique, south by South Africa, and west by Botswana.

National name: *Republic of Zimbabwe*
Area: 390,300 sq km/150,694 sq mi
Capital: Harare
Major towns/cities: Bulawayo, Gweru, Kwekwe, Mutare, Kadoma, Chitungwiza
Physical features: high plateau with central high veld and mountains in east; rivers Zambezi, Limpopo; Victoria Falls

Yellow stands for mineral wealth. Green represents the country's vegetation and natural resources. Red recalls the blood spilt during the liberation struggle. Effective date: 18 April 1980.

Government

Head of state and government: Robert Mugabe from 1987
Political system: nationalistic socialist
Political executive: unlimited presidency
Administrative divisions: eight provinces and two cities with provincial status
Political parties: Zimbabwe African National Union–Patriotic Front (ZANU–PF), African socialist; opposition parties exist but none have mounted serious challenge to ruling party
Armed forces: 39,000; 21,800 paramilitary forces (1998)
Conscription: military service is voluntary
Death penalty: retained and used for ordinary crimes
Defence spend: (% GDP) 5.0 (1998)
Education spend: (% GNP) 8.3 (1996)
Health spend: (% GDP) 6.2 (1997)

Economy and resources

Currency: Zimbabwe dollar
GDP: (US$) 5.7 billion (1999)
Real GDP growth: (% change on previous year) –1.4 (1999)
GNP: (US$) 6.1 billion (1999)
GNP per capita (PPP): (US$) 2,470 (1999)
Consumer price inflation: 58.5% (1999)
Unemployment: 45% (1995 est)
Foreign debt: (US$) 4.4 billion (1999)
Major trading partners: South Africa, UK, Germany, Malawi, Botswana, Japan, USA
Resources: gold, nickel, asbestos, coal, chromium, copper, silver, emeralds, lithium, tin, iron ore, cobalt
Industries: metal products, food processing, textiles, furniture and other wood products, chemicals, fertilizers
Exports: tobacco, metals and metal alloys, textiles and clothing, cotton lint. Principal market: South Africa 10.4% (1999 est)
Imports: machinery and transport equipment, basic manufactures, mineral fuels, chemicals, foodstuffs. Principal source: South Africa 46% (1999 est)
Arable land: 8% (1996)
Agricultural products: tobacco, maize, cotton, coffee, sugar cane, wheat, soybeans, groundnuts, horticulture; livestock (cattle)

Population and society

Population: 11,669,000 (2000 est)
Population growth rate: 1.4% (1995–2000)
Population density: (per sq km) 30 (1999 est)
Urban population: (% of total) 35 (2000 est)
Age distribution: (% of total population) 0–14 41%, 15–59 55%, 60+ 4% (2000 est)
Ethnic groups: four distinct ethnic groups: indigenous Africans (mainly Shona 71% and Ndebele 16%),who account for about 95% of the population, Europeans (mainly British), who account for about 3.5%, and Afro-Europeans and Asians, who each comprise about 0.5%
Language: English, Shona, Ndebele (all official)
Religion: 50% follow a syncretic (part Christian, part indigenous beliefs) type of religion, Christian 25%, animist 24%, small Muslim minority
Education: (compulsory years) 8
Literacy rate: 93% (men); 85% (women) (2000 est)
Labour force: 66.2% agriculture, 8.2% industry, 25.6% services (1996 est)
Life expectancy: 44 (men); 45 (women) (1995–2000)
Child mortality rate: (under 5, per 1,000 live births) 117 (1995–2000)
Physicians: 1 per 7,371 people (1995)
TV sets: (per 1,000 people) 30 (1997)
Radios: (per 1,000 people) 93 (1998)
Internet users: (per 10,000 people) 17.4 (1999)
Personal computer users: (per 100 people) 1.3 (1999)

Transport

Airports: international airports: Harare, Bulowayo, Victoria Falls; domestic air services operate between most of the larger towns; total passenger km: 874 million (1997 est)
Railways: total length: 2,759 km/1,714 mi; total passenger km: 545,977 million (1995)
Roads: total road network: 18,462 km/11,472 mi, of which 48% paved (1996); passenger cars: 28 per 1,000 people (1996 est)

Chronology

13th century: Shona people settled Mashonaland (eastern Zimbabwe), erecting stone buildings (hence name Zimbabwe, 'stone house'). **15th century:** Shona Empire reached its greatest extent. **16th–17th centuries:** Portuguese settlers developed trade with Shona states and achieved influence over the kingdom of Mwanamutapa in northern Zimbabwe in 1629. **1837:** Ndebele (or Matabele) people settled in southwest Zimbabwe after being driven north from Transvaal by Boers; Shona defeated by Ndebele led by King Mzilikazi who formed military empire based at Bulawayo. **1870:** King Lobengula succeeded King Mzilikazi. **1889:** Cecil Rhodes's British South Africa Company (SA Co) obtained exclusive rights to exploit mineral resources in Lobengula's domains. **1890:** Creation of white colony in Mashonaland and founding of Salisbury (Harare) by Leander Starr Jameson, associate of Rhodes. **1893:** Matabele War: Jameson defeated Lobengula; white settlers took control of country. **1895:** Matabeleland, Mashonaland, and Zambia named Rhodesia after Cecil Rhodes. **1896:** Matabele revolt suppressed. **1898:** Southern Rhodesia (Zimbabwe) became British protectorate administered by BSA Co; farming, mining, and railways developed. **1922:** Union with South Africa rejected by referendum among white settlers. **1923:** Southern Rhodesia became self-governing colony; Africans progressively disenfranchised. **1933–53:** Prime Minister Godfrey Huggins (later Lord Malvern) pursued 'White Rhodesia' policy of racial segregation. **1950s:** Immigration doubled white population to around 250,000, while indigenous African population stood at around 6 million. **1953:** Southern Rhodesia formed part of Federation of Rhodesia and Nyasaland. **1961:** Zimbabwe African People's Union (ZAPU) formed with Joshua Nkomo as leader; declared illegal a year later. **1962:** Rhodesia Front party of Winston Field took power in Southern Rhodesia, pledging to preserve white rule. **1963:** Federation of Rhodesia and Nyasaland dissolved as Zambia and Malawi moved towards independence; Zimbabwe African National Union (ZANU) formed, with Robert Mugabe as secretary; declared illegal a year later. **1964:** Ian Smith became prime minister; he rejected British terms for independence which required moves towards black majority rule; Nkomo and Mugabe imprisoned. **1965:** Smith made unilateral declaration of independence (UDI); Britain broke off all relations. **1966–68:** United Nations (UN) imposed economic sanctions on Rhodesia, which still received help from South Africa and Portugal. **1969:** Rhodesia declared itself a republic. **1972:** Britain rejected draft independence agreement as unacceptable to African population. **1974:** Nkomo and Mugabe released and jointly formed Patriotic Front to fight Smith regime in mounting civil war. **1975:** Geneva Conference between British, Smith regime, and African nationalists failed to reach agreement. **1978:** At height of civil war, whites were leaving Rhodesia at rate of 1,000 per month. **1979:** Rhodesia became Zimbabwe-Rhodesia with new 'majority' constitution which nevertheless retained special rights for whites; Bishop Abel Muzorewa became premier; Mugabe and Nkomo rejected settlement; Lancaster House Agreement temporarily restored Rhodesia to British rule. **1980:** Zimbabwe achieved independence from Britain with full transition to African majority rule; Mugabe became prime minister with Rev. Canaan Banana as president. **1984:** A ZANU–PF party congress agreed to the principle of a one-party state. **1987:** Mugabe combined the posts of head of state and prime minister as executive president; Nkomo became vice-president. **1989:** ZANU–PF and ZAPU formally merged; the Zimbabwe Unity Movement was founded by Edgar Tekere to oppose the one-party state. **1992:** The United Party was formed to oppose ZANU–PF. Mugabe declared drought and famine a national disaster. **1996:** Mugabe was re-elected president. **1998:** Mugabe issued new rules banning strikes and restricting political and public gatherings. The government's radical land distribution plans were watered down after pressure from aid donors. There were violent antigovernment demonstrations. **1999:** Further violent antigovernment protests took place. In June the human rights group African Rights produced a scathing report on Mugabe's government. **2000:** Veterans of the war of independence, supported by the government, began to invade and claim white-owned farms. To international and internal opposition, the government invoked special powers to seize the farms without compensation. Elections in June, which returned Mugabe's government, were condemned by international observers. In October rioters protested at soaring food and transport costs. In November the high court ruled Mugabe's land acquisition program illegal. **2001:** Mugabe's government agreed to remove all white judges from Zimbabwe's judiciary and decided to replace the five members of the Supreme Court.

Practical information

Visa requirements: UK/USA: visa not required
Time difference: GMT +2
Chief tourist attractions: Victoria Falls; Kariba Dam; mountain scenery, including Mount Inyanganai; some of southern Africa's best wildlife parks, notably Hwange, Matapos, and Nyanga national parks; ruins of old Zimbabwe, near Fort Victoria; World's View in the Matapos Hills
Major holidays: 1 January, 18–19 April, 1, 25 May, 11 August (2 days), 25–26 December; variable: Good Friday, Easter Monday